"This excellent book . . . merits a broad readership, and should endure as a . . . classic." — Frederick Seitz, *Nature*

"It's a rip-snorting techno-thriller that seems at times like a Tom Clancy novel, only it's true." — Peter Carey, *San Jose Mercury News*

"[Buderi's] account of radar may not be as definitive as Rhodes's account of the atomic bomb, but because of the greater complexity of his task Buderi may well have made an equally important contribution." — William Aspray, *Science*

"Masterly . . . Buderi cannot be praised too highly for producing a scholarly and superbly researched book which is also enjoyable to read." — Ian Morison, *New Scientist*

"Buderi's Book can awaken the interest of even the most jaded and non-technical reader. Radar did change the world, and Buderi nicely describes how." — Norman N. Brown, *The Associated Press*

"Much of our view of the technological achievements emerging from World War II is driven by nuclear fetishism, but a more fundamental truth is revealed in the subtitle to this remarkable book. . . . Through extensive interviews and impressive bibliographies, the author has produced a fascinating history." — *Library Journal*

"In a narrative that often reads as compellingly as the best spy fiction . . . Buderi tells the story of how British and American scientists developed microwave radar, a device that helped win World War II and spurred a transformative postwar technical revolution. . . . A fascinating story, well told." — *Kirkus Reviews*

"A diligent, deserving work." — *Booklist*

"A vigorous history." — *Publishers Weekly*

"(A) most marvelous book . . . employing a pleasant blend of anecdote and technical description." — Aleaxander Rose, *The Daily Telegraph*

"As much an adventure of human discovery as it is a paean to engineering, the book illustrates how radar helped to win World War II, navigate through the Cold War, and launch the space program." — *American West Airlines*

"Buderi has hit the "mother load" of largely untold stories of earth shaking proportions. . . . Moreover, Buderi's masterful yarn spinning skills were more than up to the task of bringing to life one of the greatest TRUE stories of our time." — Joseph R. Guerci, *Amazon Books*

The Sloan Technology Series

The
INVENTION

How a Small Group
of Radar Pioneers
Won the Second World War
and Launched a
Technological Revolution

ROBERT BUDERI

That
CHANGED
the
WORLD

A TOUCHSTONE BOOK
Published by Simon & Schuster

⊼⋀⊺

TOUCHSTONE
Rockefeller Center
1230 Avenue of the Americas
New York, NY 10020

First Touchstone Edition 1997

TOUCHSTONE and colophon
are registered trademarks of Simon & Schuster Inc.
Designed by Edith Fowler

Manufactured in the United States of America

10   9   8   7   6   5   4   3   2   1

The Library of Congress has cataloged the
Simon & Schuster edition as follows:

Buderi, Robert.
    The invention that changed the world : how a
small group of radar pioneers won the Second
World War and launched a technological
revolution / Robert Buderi.
        p.   cm. — (Sloan technology series)
    Includes bibliographical references and index.
    1. Radar — History.   2. World War, 1939–1945
— Radar.   I. Title.   II. Series.
TK6574.2.B84   1996
621.3848 — dc20   96-9404   CIP
ISBN 0-684-81021-2
ISBN 0-684-83529-0 (Pbk)

The author and publisher gratefully acknowledge permission to reprint material from the following works:

Henry A. H. Boot Papers, National Archive for Electrical Science and Technology, Institution of Electrical Engineers, London. Reprinted with the permission of the Institution of Electrical Engineers.

E. G. Bowen, Radar Days (Bristol, England: Adam Hilger, 1987). Reprinted with the permission of IOP Publishing Ltd. and the Bowen family.

Lee A. DuBridge, "Memories," Part I; "The Papers of Lee A. DuBridge"; and Lee A. DuBridge, oral history interview with Judith Goodstein, February 19, 1981. Reprinted with the permission of the California Institute of Technology Archives.

Jack H. Scaff, oral history interview with Lillian Hartmann Hoddeson, August 6, 1975. Reprinted with the permission of the American Institute of Physics.

Charles H. Townes, "A Life in Physics," oral history interviews conducted by Suzanne B. Riess, 1991–92. Reprinted by permission of the Regional Oral History Office, the Bancroft Library, University of California at Berkeley.

George E. Valley, Jr., "How the SAGE Development Began." © 1985 IEEE. Reprinted, with permission, from Annals of the History of Computing, vol. 7, no. 3 (July 1985): 196–226.

To the Memory of Betty Lou Krough Buderi,
and for Nancy, Kacey, and Robbie

*The author acknowledges with gratitude
the support of the Alfred P. Sloan Foundation
in the research and writing of this book.*

Contents

Preface to the Sloan Technology Series

TECHNOLOGY is the application of science, engineering, and industrial organization to create a human-built world. It has led, in developed nations, to a standard of living inconceivable a hundred years ago. The process, however, is not free of stress; by its very nature, technology brings change in society and undermines convention. It affects virtually every aspect of human endeavor: private and public institutions, economic systems, communications networks, political structures, international affiliations, the organization of societies, and the condition of human lives. The effects are not one-way; just as technology changes society, so too do societal structures, attitudes, and mores affect technology. But perhaps because technology is so rapidly and completely assimilated, the profound interplay of technology and other social endeavors in modern history has not been sufficiently recognized.

The Sloan Foundation has had a long-standing interest in deepening public understanding about modern technology, its origins, and its impact on our lives. The Sloan Technology Series, of which the present volume is a part, seeks to present to the general reader the story of the development of critical twentieth-century technologies. The aim of the series is to convey both the technical and human dimensions of the subject: the invention and effort entailed in devising the technologies and the comforts and stresses they have introduced into contemporary life. As the century draws to an end, it is hoped that the Series will disclose a past that might provide perspective on the present and inform the future.

The Foundation has been guided in its development of the Sloan Technology Series by a distinguished advisory committee. We express deep gratitude to John Armstrong, Simon Michael Bessie, Samuel Y. Gibbon, Thomas P. Hughes, Victor McElheny, Robert K. Merton, Elting E. Morison (deceased), and Richard Rhodes. The Foundation has been represented on the committee by Ralph E. Gomory, Arthur L. Singer, Jr., Hirsch G. Cohen, A. Frank Mayadas, and Doron Weber.

ALFRED P. SLOAN FOUNDATION

Preface

I FIRST LEARNED about the Radiation Laboratory, the cradle of much World War II radar development and the technological revolution that roared in war's wake, not long after arriving at the Massachusetts Institute of Technology in the fall of 1986 for a year-long fellowship program for science writers. The other fellows and I were on our way to a seminar dealing with the study of ancient Incan life based on an analysis of metals found in various artifacts. We trudged across the campus, past the swank I. M. Pei–designed Media Lab and the biological building, when our eyes came to rest upon a rickety three-story eyesore erected along Vassar Street on the school's northwest border.

I had already become accustomed to some of MIT's physical backwardness. Virtually all the buildings are old and grimy, the linoleum-paved hallways somewhat musty, the laboratories stocked with tottering wooden desks and seemingly built to accommodate a Bunsen burner far more readily than a laser. But nowhere is the contrast between aging physical space and the school's reputation as a leader in science and technology greater than with Building 20, the heart of the Rad Lab. From the wood-and-asbestos shingle exterior to the exposed pipes and wires running down its half-century-old corridors, the low-flung monstrosity fairly screams for demolition.

Building 20 turned out to be the stuff of campus legends. "The edifice is so ugly," espouses Fred Hapgood in *Up the Infinite Corridor*, a book about MIT, "that it is impossible not to admire it, if that makes sense; it has ten times the righteous nerdly swagger of any other building on campus." Stories were rampant. A homeless botanist had lived in a storeroom, apparently for years, turning down a job at a Chicago museum in order to remain the Phantom of Building 20. The anemic structure played host to pioneering communication studies. It is where Harold "Doc" Edgerton, inventor of stroboscopic photography, conducted early experiments in underwater imaging and sonar. Bose Corporation of speaker fame got its start there, as did the acoustic consulting firm of Bolt Beranek and Newman, which evolved into the wide-ranging Internet systems provider BBN Corporation. Since the early 1960s the third floor of E wing has housed the Tech Model Railroad Club, from whose devotees sprang many first-generation computer hackers. "Ah, Building 20?" sighed the neurophysiologist Jerome Lettvin, who himself padded its creaky halls for thirty years. "It is kind of messy, but by God it is procreative."

It all started with radar. To house the rapidly expanding Radiation Lab, Building 20 was designed in less than an afternoon and erected in six furious months at the end of 1943. Fashioned of heavy wood timbers because steel was unavailable during World War II, Cambridge officials had to exempt it from city building codes. Building 20 was a temporary creation, slated to be put out of its misery at war's end. But that hadn't happened forty-odd years later when I first came to campus, and it still hasn't happened—although the historic structure is scheduled for demolition in spring 1998 to make room for a massive engineering complex.

Even before I knew its history, Building 20 fascinated me. What is it doing here? I wanted to know. Why was it ever built? Victor McElheny, the fellowship director and a living data base of history, filled me in on the basics, telling how the Rad Lab had played a critical role in World War II, and a bit of how instrumental the radar work had been in weaving the fabric of contemporary society.

This was just what a writer hopes for—a great, largely untold story. I went straight to the MIT archives and located a copy of *Five Years at the Radiation Laboratory*, a kind of yearbook for Rad Lab veterans put out in 1946, not long after the war ended. The fascinating book lists staff members and many of the wartime projects, though sections had been intriguingly blanked out for security reasons. Along with it, I found a copy of the *Radiation Laboratory Handbook*, which spells out work hours, security procedures, and the like.

The discoveries brought the enterprise more alive in my mind. In the summer of 1987, after the fellowship ended, I grabbed a flight to New York and interviewed I. I. Rabi, a Nobel Prize winner who had served as the Rad Lab's research director. He was to the lab what Building 20 became to MIT — the physical embodiment of its spirit. His biography, containing two chapters on the radar work, had recently been published. But I wanted to know far more. I met the legendary physicist in his Riverside Drive apartment. I was largely winging it, unsure which questions to ask. He was old and somewhat feeble, as it turned out only a few months from his death; it was not a very good interview. By the time I began writing this book, the notes from our conversation had been hopelessly misplaced. My tape recording turned out so garbled that Rabi could not be understood. But I remember well Rabi's keen sense of pride in the laboratory's accomplishments and in being a part of it. In the course of writing this book, in talking with many of his colleagues, I have encountered this feeling again and again.

I returned from New York with a vague hope that someday I would tell the lab's story and chronicle radar's critical importance to the war. At first, the book I envisioned started on the eve of World War II and ended with the Japanese surrender. I didn't imagine as I got on the plane to come home, later sat down at my computer to type some notes, made a phone call — even warmed up a cup of coffee — that I was benefiting from the technological children of radar. It is this greater legacy — going far beyond the war to include radar's tremendous importance to postwar science and technology — that this book seeks to capture.

What arose from the World War II microwave radar work was a dramatic, concentrated example of the endless intertwinement of science and technology. Scientists, many of them nuclear physicists, were forced to put research on hold during the war years for the sake of a technology — microwave radar — that almost no one knew anything about. They became experts in the field, drawing on physics to help develop the technology, which in turn taught them techniques critical to science. With these came breakthroughs that gave rise to new technologies.

The Radiation Lab, coupled with related microwave radar endeavors in the United States and Britain, emerged as a science and technology incubator on a scale probably unprecedented in history. At least two Nobel Prizes — for nuclear magnetic resonance and the maser — can be traced directly to wartime radar work. Every day several thousand commercial air carriers take to the skies. Virtually all the aircraft are tracked continuously by radar, sometimes to and from their gates. It takes more than three hundred radars, backed up by a ponderous array of display screens, communications nets, and personnel, just to keep the U.S. skies organized. Thousands more radar sets provide extra eyes for ships, boats, and pleasure craft. Many vessels also draw on the global navigational network Loran. Virtually all of these sprang directly from the Rad Lab, as do much of the world's storm-watching systems and the TV weather report.

The transistor — via the legacy of the solid-state semiconductor crystals that formed the heart of radar receivers — is largely a product of work contracted by the Rad Lab to places like Purdue University, the University of Pennsylvania, and Bell Labs, the second largest of the wartime radar houses. Digital computers, including their cathode ray displays and memories, owe a great debt to radar: they are the offspring of World War II systems. Microwave telephones and early television networks got critical boosts from wartime radar. The technology made a huge impact on astronomy by opening a region of the electromagnetic spectrum — radio as opposed to optical — that ultimately brought on the discovery of pulsars, quasars, and a plethora of hidden galaxies. And the list hardly stops there. Early particle accelerators owe a great debt to the Building 20 eyesore. So does microwave spectroscopy. So, too, do the microwave ovens common in today's homes, for a secret radar transmitter carried from Britain to America in fall 1940 forms the very core of these time-saving appliances.

Early in the book's planning stages, it became clear that writing a definitive history of radar and its almost panoramic impact on science and technology was impossible if one sought to maintain a strong and effective narrative. I wanted to tell the story of people, not technological processes, bringing out their escapades, from the silly to the deadly serious, and flushing out their thoughts and motivations, fears, and rivalries. The downside of focusing on relatively few individuals is that the contributions of others on the narrative's periphery occasionally must be omitted. At a place like the Rad Lab, everybody stood on the shoulders of many others: it was truly an interdisciplinary, cooperative effort. Still, I have attempted to identify the central personages, most of whom I have interviewed.

In a similar vein, while the book describes critical events and areas of pursuit, I have sometimes had to brush over less compelling, but still noteworthy areas. For the World War II period, since the Rad Lab concerned itself mainly with the European conflict, it is the Pacific front that gets relatively short shrift. For the postwar years, when radar's technological and scientific payoffs exploded on a vast number of fronts, I have only touched on such important subjects as air traffic control, weather forecasting, and maritime radar. Happily, just before the publication of the paperback edition of this book, Louis Brown unveiled his own radar manuscript, *Technical and Military Imperatives: A Radar History of World War II*, which fills in many of the holes in the World War II period — and brings to light many of the finer points of German radar. Hopefully, one day others will do the same for radar's post-war legacy.

Exacting pains were taken to get things right. Besides consulting books and archives on three continents, I have sought out the radar pioneers themselves to mine their recollections and fill holes in the written record. When thoughts or emotions are presented, it is what people said (or wrote) they were thinking and feeling. Memories are tricky things, especially after fifty years. Accounts of events were cross-checked with the paper trail, as well as the recollections of others. In cases where conflicts proved irreconcilable, words such as "probably" or "apparently" are used to alert the reader that this is merely a best estimate, and my reasoning is often explained in the notes section.

I learned much about history from this project, enough to give serious pause. Sometimes I uncovered mistakes that had been passed on, even by the principals themselves, until they became accepted as fact. This caused me to wonder how many similar, undetected errors are contained in my own work. Everything I have written is attributable to a source, but beyond my own slipups in understanding and synthesis, what if the sources were wrong? I have been left with a profound sense that so much of history is not truth, only what is accepted as truth. I will never regard history in quite the same light.

This is not a statement of disillusionment, but rather enlightenment. It is entirely possible to miss certain details but paint the overall picture with the right hues, just as a straightforward account of the paper trail often misses the heart of the story: the people. So, while what is contained in these pages may be something less than a total account, by seeking out the human dimension I hope it illuminates something much greater — the unmistakable spirit of innovation and discovery that sprang from a critical technology and a cause people believed in, and changed the face of the modern world.

Vannevar Bush at the Carnegie Institution of Washington, D.C., home of the National Defense Research Committee during World War II.

During a 1958 get-together in Washington, D.C., Sir Robert Watson-Watt (center), father of the British radar effort, converses with MIT president James Killian Jr. (left) and former Radiation Laboratory director Lee DuBridge, then president of the California Institute of Technology (right).

Sir Henry Tizard (center), the main force behind the British decision to share technological secrets with the United States, visiting MIT after World War II and inspecting a cavity magnetron with Alfred Loomis (left) and Lee DuBridge.

Outside the "glass house" of financier and amateur scientist Alfred Loomis in Tuxedo Park, New York, on October 12, 1940. British and American scientists gathered there to lay plans for creating what became the MIT Radiation Laboratory. From left to right: Carroll Wilson, Frank Lewis, Edward Bowles, Edward G. "Taffy" Bowen, E. O. Lawrence, Loomis. The photograph was taken by Bowen's fellow Tizard Mission member John Cockcroft.

The brain trust of the Rad Lab's pioneering SCR-584 gun-laying radar, Lee Davenport (left) and Ivan Getting, meet with an Army colonel atop a truck carrying an experimental model of the set during 1942 anti-aircraft trials. Just two years later, the radar proved instrumental in shooting down V-1 buzz bombs aimed at London.

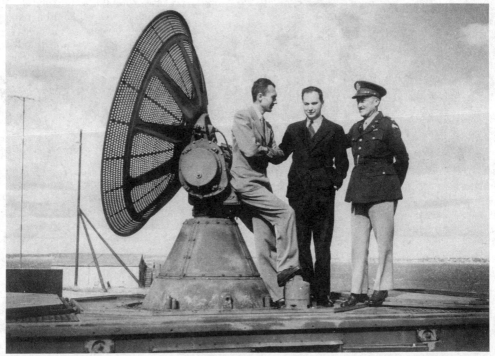

Two photos: The MIT Museum

Famous for his enthusiastic tackling of almost any scientific puzzle, Edward Purcell turned the mysterious atmospheric absorption of radar waves during World War II into a Nobel Prize–winning discovery of nuclear magnetic resonance.

British engineer Denis Robinson, who pushed the Americans to build radars for hunting U-boats, also set the Allies on the path to using semiconductor crystals as microwave radar detectors, a crucial step in the later invention of the transistor.

Luis Alvarez, a powerfully creative thinker and future Nobel laureate who would also shed light on John F. Kennedy's assassination and the extinction of the dinosaurs, spends his last day at the Rad Lab in September 1943, bound for Los Alamos and atomic bomb work.

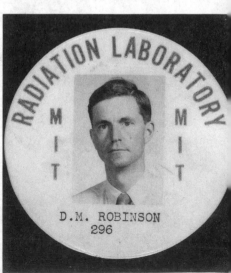

The central figure in the Wizard War, R. V. Jones almost single-handedly unraveled the secrets of German radar.

Screen of one of Luis Alvarez's Microwave Early Warning radars captured the D-day invasion on June 6, 1944.

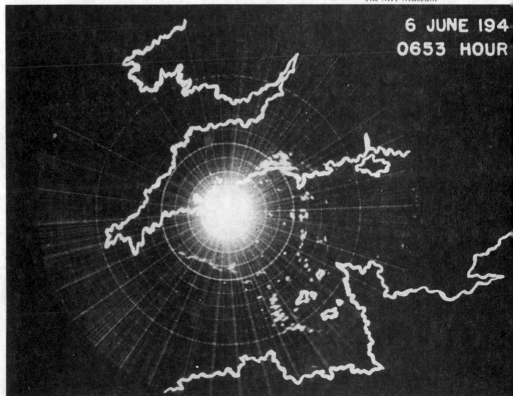

6 JUNE 194
0653 HOUR

Above left: Women's Auxiliary Air Force members track friendly and enemy planes in the RAF operations room at Uxbridge, where Churchill watched his forces seal victory in the Battle of Britain on September 15, 1940.

Above right: Taffy Bowen used his wartime contacts in America to raise funds for the gigantic Parkes radio telescope in Australia.

A Royal Air Force Coastal Command plane destroys a U-boat on July 8, 1944. Airborne radar, combined with better weapons and tactics, had already broken the back of the German submarine force a year earlier.

Courtesy of Harold Ewen

Leftover wartime radar equipment helped Bowen's Australian colleagues study Centaurus A (NGC 5128), one of the first three radio sources identified with an optical object.

Doc Ewen inspects the horn antenna he built on a parapet outside Harvard's physics building in consultation with advisor Ed Purcell. On Easter Sunday 1951, the antenna detected the faint signal of interstellar hydrogen.

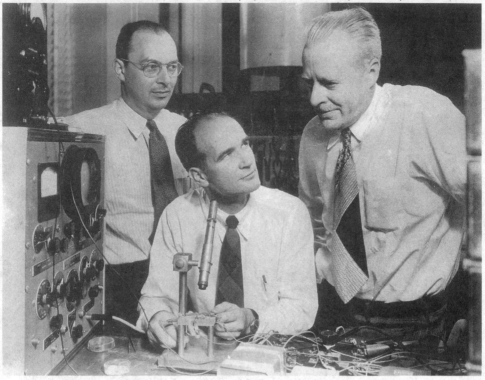

Bell Labs scientists John Bardeen, William Shockley, and Walter Brattain in 1948. Eight years later, the three would share the Nobel Prize in physics for their invention of the transistor, a major payoff of the wartime semiconductor work.

Charles Townes and James Gordon, in late 1954 or early 1955, beside the second maser ever built. Townes's insights into the same radar wave absorption problem that led Purcell to nuclear magnetic resonance helped bring him the 1964 Nobel Prize in physics.

Two photos: The MIT Museum

Jay Forrester and Bob Everett standing inside the Whirlwind control room in 1951. Whirlwind was developed largely as a prototype digital computer to control U.S. air defenses during the early Cold War years; its needs spurred Forrester to invent the core memories essential to computers until the rise of semiconductor chips in the 1970s.

Courtesy of Gordon Pettengill

George Valley (second from left) led the Rad Lab's main radar bombing project during World War II, then rescued Whirlwind from near oblivion for the postwar air defense project. In 1956, he showed MIT and Air Force officials the computer-controlled Semi-Automatic Ground Environment, or SAGE, network, designed to give warning of Soviet bombers.

Gordon Pettengill, in his office at Millstone Hill in the late 1950s, was one of a small group of Lincoln Laboratory employees who usurped the prototype missile-tracking radar for radar astronomy. Nearly three decades later, he headed up the Magellan radar team.

Magellan was launched to map Venus on May 4, 1989.

Magellan radar image showing an impact crater in the Thetis region of Venus.

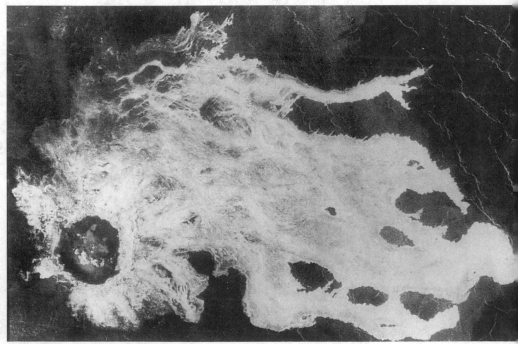

Two photos: Jet Propulsion Laboratory

1 • The Most Valuable Cargo

"When the members of the Tizard Mission brought one to America in 1940, they carried the most valuable cargo ever brought to our shores."
JAMES PHINNEY BAXTER III,
Official Historian of the
Office of Scientific Research and Development

THE BLACK JAPANNED METAL deed box could just be seen above the wartime throngs on the shoulder of a railway porter. The small container bobbed along frustratingly out of reach, as Eddie Bowen zigzagged through the crowd in hot pursuit. Only moments before, sometime around 8:15 the morning of August 29, 1940, the Welshman had arrived at London's Euston Station with the box safely in his possession. Innocently, Bowen had handed it to the porter while gathering up his remaining luggage, then watched helplessly as the man headed off to find the 8:30 train to Liverpool without waiting for his customer.

As he struggled to keep the porter in sight, Bowen would not have drawn much attention from busy Londoners. In stature and build he blended into a crowd and would have seemed like any other young man in a hurry. Only his face set him slightly apart. Wavy hair cut short crowned a wide forehead and jaw and gave his head a squared-off look. Old photographs often show an infectious grin spanning the broad tableau. But one could also imagine the weathered visage locked in determination, and that August morning Bowen had reason to be concerned. Just five days short of the war's first anniversary, Britain faced one of its most desperate hours. Bombs were falling nightly on Liverpool, Nazi armies ringed the country from the Norwegian coast down to France, and an invasion was expected within weeks. As Bowen knew, the seemingly ordinary solicitor's deed box — now visible, now not in Euston's morning rush — held the power to change the course of the conflict.

Inside lay nothing less than the military secrets of Britain, virtually every single technological item the country could bring to bear on the war. Had some freak accident burst the lock off the chest, the platform would have been awash in blueprints and circuit diagrams for rockets, explosives, superchargers, gyroscopic gunsights, submarine detection devices, self-sealing fuel tanks, and even the initial germs of the jet engine and the atomic bomb.

Among these treasures, nothing carried the all-pervasive importance of the resonant cavity magnetron, Britain's most closely guarded secret. The black box contained one of the first twelve production copies of the mysterious device, probably the only piece of hardware it sheltered. Small enough to fit in the palm of a hand, the magnetron looked like a clay pigeon used in skeet shooting, with a few couplings thrown in. Yet, it could spit out pulses of microwave radio energy on a wavelength of about ten centimeters, so powerful conventional scientific wisdom still put anything like it years off.

The magnetron was a radar transmitter, one with the potential to bolster British military capabilities almost across the board and give the country the upper hand in what already seemed like a technological war: no one in the country knew it, but the Germans were generally ahead in the radar race until the device arrived on the scene. More immediately to the point, as Bowen chased the porter across the Euston platform, the largely copper disk offered a way to invigorate the strapped British defenses that had been coping with Luftwaffe bombing onslaughts the past six weeks — a softening up before Hitler's planned invasion. Radar, or radio distance finding as Bowen's countrymen called the technology, formed the backbone of these defenses. Imposing towers up to 350 feet tall — the Chain Home station network — lined the country's south and east coasts to provide the only effective early warning of German attacks. These electronic sentries operated around the clock, rain or shine, sending out pulses of radio energy and picking up the faint echo from enemy aircraft more than a hundred miles away. Radar was basically all the outgunned country had that enabled Fighter Command to husband its too-thin air resources. Without it, planners would have had to consider keeping standing patrols aloft, wasting fuel, needlessly fatiguing pilots, and risking being in the wrong place at the wrong time.

Magnetrons represented the next crucial step — a leap, really — in the evolution. The Chain Home stations worked well in daylight, when a pilot's sharp eyes could correct for the several-mile error range inherent in their long operating wavelengths of between ten and thirteen meters. But to cut losses, the Germans were widely expected to move soon to concentrated night attacks, when visibility was slashed dramatically. The British had tried to supplement the chain by installing short-range systems inside fighter aircraft. The idea behind this was that once the main network got the interceptors close, airborne radars could carry them the rest of the way, but these remained clumsy and inaccurate. Only the magnetron seemed certain to keep the British well ahead of the game. Its ten-centimeter transmissions ran a mere fifteenth those of standard airborne radars. Fitted into night-fighters, such a device would generate sharper pulses in a tightly concentrated parcel of energy that would fan out far less during the brief journey to an enemy aircraft and back, making it immensely easier for pilots to home in on their quarry even on the darkest nights.

That, though, was only the beginning. Although the magnetron had been invented just eight months earlier by two physicists at the University of Birmingham, its portability and versatility soon summoned visions of putting the beleaguered nation on the offensive. Aircraft equipped with centimeter radar might pick out U-boat periscopes rising under cover of darkness. Lancasters and other bombers could use the extremely short waves the magnetron produced to illuminate the way through the thick cloud cover obscuring Hitler's forces and factories on the European continent, keeping planes flying on days the Royal Air Force would normally be grounded.

Yet for all the device's promise, a series of technical glitches continued to plague its development, the most serious stumbling block being uneven power performance. British industry, with its limited production capacity, and already under the threat of bombardment and invasion, simply could not trust that it alone possessed the capability for correcting the problems and churning out magnetrons in the numbers needed for war.

This overriding concern, not just in regard to the cavity magnetron but extended to all the devices in the black box, brought Bowen to the Euston platform that August morning. Though still four months shy of his thirtieth birthday, the Welsh physicist ranked as one of Britain's defense pioneers. For the past five years he had labored in some of the island's most isolated spots, sometimes night and day, to develop the Chain Home network and the country's first crude airborne radar systems. As a leading defense scientist he had been tapped to join a top-secret government mission aimed largely at convincing the still-uncommitted American government and key industrial officials to pick up where British resources left off. The mission was to sail from Liverpool that night.

To pave the way for the venture, a special team had spent the first two weeks of August rounding up the black box's contents. Bowen himself had visited the General Electric Company research laboratory in the London suburb of Wembley, where he picked out the best working model of the first dozen magnetrons made. He had then carried his selection unescorted on the underground to the Ministry of Supply headquarters between London's Victoria Embankment and the Strand. At the ministry, the precious cargo had been placed safely in the black box, remaining under lock and key until the evening of the twenty-eighth, when Bowen returned to escort the entire booty to Liverpool. A guard delivered it via the arched doorway on the ministry's back steps. From there, Bowen hailed a taxi to whisk him to the Cumberland Hotel, not far from Euston at historic Marble Arch.

Because the box would not fit in the hotel safe, Bowen had spent the night with England's greatest military secrets wedged under his bed. In the morning, to add to his discomfort, the cabby taking him to the train station would not allow the small chest inside the taxi, insisting it be placed on the roof. The Welshman had thought all was well when the cab finally reached Euston, but then the fast-footed porter had prolonged his unease.

Bowen didn't catch up with the man until they reached the train. At this point he knew only that a first-class seat had been reserved. But when he found his place, it appeared an entire compartment had been set aside: the blinds were drawn and reserved notices placed on the windows. Intrigued, Bowen sat down to wait for the train to leave, figuring all would become clear on the other end.

A few minutes before departure, a well-dressed and exceptionally trim man with a school tie entered the compartment. With scarcely a glance around, the man took up the seat diagonally across from Bowen and began reading a newspaper. The mysterious companion didn't speak until a few minutes after the train began edging out of the station, when some late-comers opened the door, happy to have found an empty cabin.

"Out," he ordered. "Don't you see this is specially reserved?"

Bowen was struck not so much by the man's words as the commanding tone of the delivery. "The would-be intruders wilted," he later recalled, "and we had no further interruptions." At that moment, for the first time in a harrowing sixteen hours or so, Bowen realized, too, that his precious cargo carried some form of protection.

The journey passed in silence. When the train finally pulled into Liverpool's dockside station, Bowen, following instructions to stay put until an Army escort arrived to pick up the box, didn't budge from his seat. His compartment mate also remained in place, ostensibly absorbed in the paper.

At last, a dozen fully armed soldiers marched down the platform and came to a glorious, rifle-slapping halt alongside the car. A sergeant barked some orders, put the group at ease, and dispatched three men to collect the cargo. Bowen watched as Britain's technological pride and joy was carried outside, hoisted onto some shoulders, and marched back down the platform. The display of military exactitude eased the young physicist's mind, but not totally. Telling the story later, he joked, "I was beginning to feel that things were well looked after. Alternatively, if this was the enemy making off with Britain's secrets, they were making a spectacular job of it."

Through all the commands and gesturings, Bowen's mysterious cabin mate still had not uttered a companionable word. Now the man rolled up his paper, and with a slight nod at his fellow traveler, took his leave.

Bowen also roused himself and shuffled off along Gladstone Dock to find his ship, the *Duchess of Richmond.* On board the Canadian liner, he joined the main body of what was formally called the British Technical and Scientific Mission to the United States. Informally, and far more commonly, the venture was known as the Tizard Mission, after its organizer, Sir Henry Tizard, rector of the Imperial College of Science and Technology and chairman of the government's key scientific committee on air defense.

Tizard, an Oxford-trained chemist, had already made his name as one of Britain's shrewdest scientific visionaries. Beginning in 1935, his Committee for the Scientific Survey of Air Defence had pushed radio direction finding over all other competitors — sound mirrors, infrared detection, balloon barrages. In late 1939, recognizing the need for American assistance in developing radar and other military technologies, he had conceived the idea of an exchange mission with the United States. His proposal had received strong support from Archibald Vivian Hill, the influential Nobel laureate and joint secretary of the Royal Society, who had gone to America early in 1940 to grease the wheels on the other side of the Atlantic.

The plan hinged on making a full disclosure of the kingdom's technical secrets in the hopes that America, even if it stayed neutral, would gear up its immense industrial machine to help develop and produce them. Initially, many British authorities wanted to trade secret for secret, seeing the exchange as a way to pry loose details of the coveted American Norden bombsight. But after months of infighting and wrangling, new Prime Minister Winston Churchill, who had taken over the governmental reins in May 1940 on the heels of the German blitz into western Europe, decided to make the offer with no strings attached. American cooperation would be more complete, the prevailing view held, if there was no attempt to barter secret for secret.

So complete was the offering that by the time Eddie Bowen walked along the Liverpool docks that August afternoon, only two items of any note had been held back: some particulars of the jet engine and details of the latest German magnetic mines used to block British harbors. Besides the crucial cavity magnetron, nearly everything about radar could be found in the black box; and several containers of working sets and components apparently had been sent through separate channels to supplement its contents.

Tizard deliberately restricted the mission to just seven members, counting himself. Bowen was his handpicked radar expert. Cambridge University physicist John Cockcroft, architect of one of the world's first proton accelerators, would brief the Americans on the remainder of the technological booty, as well as a few isolated aspects of radar. In addition to the two scientists, each of Britain's three services — the Royal Air Force, the Admiralty, and the Army — contributed an officer with recent combat experience who could talk about military needs. The last member was Arthur Edgar Woodward-Nutt, an Air Ministry official who served as the mission's secretary.

Tizard and one of the mission's military representatives, Group Captain F. L. Pearce of the Royal Air Force, had flown across the Atlantic a few days ahead of the main body to pave the way for the exchange. But the other members would make the crossing with Bowen on the *Duchess*.

With the black box safely escorted off the train, the Welshman's responsibility had ended. Aboard ship, Woodward-Nutt, the sole member of the entourage allowed access to the chest during the voyage, saw the secret cargo locked in the strong room. He arranged to meet the third officer, who held the keys, in the event of a German attack, so that they could dump the rich bounty overboard.

The ship left its mooring that evening, inching down the Mersey River toward the Irish Sea. An air raid hit Liverpool, with a few bomb splashes rocking the boat right after dinner, so the crew anchored down for the night near the river mouth. The *Duchess* finally set sail the next morning, Friday, August 30. Minesweepers escorted the liner the rest of the way through the Mersey, which was littered with wrecked boats. Later, two destroyers took over, shepherding the vessel through heavy seas for a few hours until she built speed and opened a zigzag course to elude any lurking U-boats.

Tizard Mission members passed time aboard ship in the usual way: reading, listening to BBC broadcasts, playing deck games and bingo, watching films in the ship's cinema, and taking brisk walks in the cold North Atlantic air. About a thousand sailors also took berths on the *Duchess*, bound to pick up the first aged U.S. destroyers consigned to Britain in exchange for the rights to various naval and air bases. The well-known Cockcroft lectured the bored servicemen on a scientific subject he felt safe to discuss, since it couldn't possibly have a bearing on the war: nuclear energy. He impressed his audience by pronouncing that a cupful of water held enough atomic power to blow a battleship a foot out of the sea. In a more private exercise, Cockcroft also calculated the black box's chances of sinking with the ship should they be struck by an enemy torpedo, and concluded the buoyant cargo would stay afloat. Holes were drilled in each end.

On the evening of September 5, the ship pulled off Newfoundland's

Cape Race. The following morning dawned calm and misty as she slipped into Halifax harbor. Bowen remembered spying an American armored vehicle, "submachine guns bristling from every orifice. . . ." Woodward-Nutt, though, recorded spending several hours on the phone with the British embassy in Washington, arranging for a Canadian military guard to take the secret equipment to the U.S. border, where it would be turned over to American authorities and transported to the embassy. He personally saw the equipment off early the next day.

At Halifax, Bowen split off from the rest of the group, heading to Ottawa to arrange for officials from Canada's National Research Council to join the exchange, and to locate some of the equipment presumably shipped over earlier. He would catch up with the others in Washington a few days later. The rest of the mission left Nova Scotia by rail at 8:45 the morning of the seventh, changing trains in Boston and arriving in Washington at 5:30 the next evening.

The group met Tizard at the Shoreham Hotel, overlooking Rock Creek Park near the British embassy in northwest Washington. "I was a bit shaken," writes Woodward-Nutt, "to find that the samples and documents that I had seen off so carefully at Halifax had not yet arrived." It took a series of telephone calls to locate the cargo. The precious container, bearing the cavity magnetron and the technological hopes of an entire nation, finally arrived at the embassy on Monday, September 9. There, it was locked in the wine cellar and given to the care of the ambassador's butler, who as far as could be determined possessed the only key.

•

The Americans anxiously awaited the Tizard Mission. It hadn't seemed that way at first. Sir Henry had arrived in Washington on August 22, expecting a welcome mat arranged by A. V. Hill. Instead, he complained to his diary: "No administrative arrangements made for my Mission. No office, no typists, etc. Felt rather annoyed."

The bad taste had not lingered, however. The next day Tizard huddled with Navy Secretary Franklin Knox to establish the ground rules for the exchange. On the twenty-sixth he received an audience with Franklin D. Roosevelt, who welcomed him but explained that political considerations prevented the United States from sharing details of the Norden bombsight. Most important of all, two days later over dinner at the Cosmos Club, an exclusive Lafayette Square haven for the inner circles of science, art, and literature, Tizard met Vannevar Bush.

With his raw-boned face, wire-rim glasses, and piercing gaze, the charismatic Bush in many ways formed Tizard's mirror image on the western side of the Atlantic. Scion of seven or eight generations of Cape Cod Yankees, and equipped with a telltale northeastern twang, he could be confidently stamped Made in America, just as Tizard, with his accent, pince-nez, and somewhat aloof manner, left no doubt of his origins. Like Tizard, Bush hid a hard edge behind a calm demeanor. Like Tizard, too, he was a scientist — an MIT electrical engineer who had pioneered early computing — responsible for marshaling civilian science and technology for war. Few men would match his power during the war years, as his dominion grew to include medical research, the atomic bomb, and virtually all forms of chemical and conventional warfare. "Of the men whose death in the summer of 1940 would have been the greatest calamity for America, the President is first, and Dr. Bush would be second or third," noted the multimillionaire investment banker Alfred Loomis, a Bush friend destined to play a pivotal role in the radar story.

Bush had been in the nation's capital since late 1938, when he became president of the prestigious Carnegie Institution of Washington, a private research organization endowed by steel baron Andrew Carnegie. However, he dined with Tizard as chairman of the National Defense Research Committee, established by presidential order two months earlier to mobilize civilian scientists for war. Bush had created the NDRC almost through sheer personal will. During World War I, working on submarine detection, he had seen firsthand the distinct lack of cooperation between civilian scientists and the military. So he conceived the idea of establishing a new national committee to bridge the gap. Maneuvering deftly through the Washington maze, he drew on the influential lawyer Oscar Cox and Commerce Secretary Harry Hopkins to negotiate an interview with the President. Bush entered the Oval Office on June 12, 1940, carrying a single sheet of paper with a four-paragraph sketch of the proposed agency. Less than ten minutes later, Roosevelt had signed on: "That's okay," he told the feisty engineer. "Put 'OK, FDR' on it."

Some Washingtonians complained that the NDRC represented a power grab by a small band of scientists and engineers working outside established channels. Bush made no bones about it. "That, in fact, is exactly what it was," he once admitted. But his personal mandate from Roosevelt extended to helping the country "excel in the arts of war if that be necessary." And while he respected the military's turf, Bush made certain people never forgot who had issued his orders.

The Carnegie president moved quickly to solidify his power base, bringing in as key lieutenants some old friends and confreres: MIT president Karl Compton, Harvard University president James B. Conant, and Frank B. Jewett, president of the National Academy of Sciences and Bell Telephone

Laboratories. The scientific cabal, Bush co-conspirators in conceiving the NDRC, immediately launched a survey of Army and Navy research activities and began compiling a list of technical jobs to take over, either because the work had not yet gotten under way or because once the United States abandoned its neutrality the military would have to drop them to meet more pressing demands. At the same time, the men contacted some 775 universities, industrial labs, and nonprofit institutions, compiling a roster of personnel and facilities in scientific arenas likely to affect the war. This was "the bible."

By the time Bush dined with Tizard on August 28, he had mustered his forces into several main divisions, covering everything from armor and ordnance to communications, explosives, and patents. Radar matters fell to Karl Compton's Division D — instruments and controls. Since the military survey showed that the Army and Navy both had already made great strides in meter wave radar, the NDRC adopted as its domain the vague promise of microwaves, naming Alfred Loomis chairman of a special Microwave Committee, Section D-1. It was a natural, insider's choice. Loomis sat on the MIT board and had contributed funds to the institution's general microwave research. Moreover, he was a noted amateur physicist who conducted his own fledgling microwave radar studies on a private estate outside New York City, and therefore appreciated the challenges in store.

While the various NDRC divisions could probably all delve into the British black box and find interesting treasures, it was on the microwave radar front, a top priority for both groups, that Bush and Tizard found their perfect match. The American possessed the presidential authority to develop the technology. The Englishman had the cavity magnetron.

When the two men met at the Cosmos Club, Bush remained unaware of the magnetron's existence. But he made it his business to know what was going on, and had been tipped off, probably by A. V. Hill, to certain generalities of the British radar bonanza long before the mission arrived. Face-to-face at last, however, he felt compelled to advise Tizard that although the NDRC welcomed a meeting with the British mission, the two groups should keep their distance until the U.S. military opened the talks; that way, Washington insiders could not accuse them of plotting some sort of conspiracy. Once the exchange was formally under way, Bush would take steps to correct the situation.

Tizard took the cue. While waiting for the NDRC to be let in on the talks, he and Bush met several times "behind the barn," as the wily engineer called it. It is not clear what transpired between the two men, so alike and so seemingly destined to forge a new bond. Most likely they covered general logistics, hinting at the shape of things to come in the clubby ways at which both were so adept. In any case, as his entourage began sharing extensive details on longwave radar and other subjects with U.S. military representatives in early September, Tizard managed to give the impression of an extraordinary advance without revealing the secret of the cavity magnetron, even when the Navy showed its visitors an experimental, extremely low-powered, ten-centimeter radar system. It wasn't until September 16 that Vannevar Bush won formal approval from both the Army and Navy for the NDRC to join the exchange. Only then did Sir Henry play his trump card.

•

The British disclosed the existence of the cavity magnetron at the first extensive contact between the Tizard Mission and NDRC members, a party hosted by Alfred Loomis the night of September 19 at the Wardman Park Hotel. The rambling 1800-room megacomplex dominated the southeast corner of Connecticut Avenue and Woodley Road, just a stone's throw from the Shoreham, where Tizard had set up shop in an office suite swept daily for bugs.

Eddie Bowen and John Cockcroft showed up at Loomis's rooms around nine o'clock. Bowen had returned from Canada the night of the eleventh, and the two men had spent the past week detailing British meter wave radar accomplishments to American military officials at the War Department and nearby Naval Research Laboratory. Among the disclosures were technical details of the Chain Home early warning stations already doing yeoman's service in the Battle of Britain; radio homing beacons; submarine-hunting radars; and Identification, Friend or Foe (IFF), a radio beacon carried in planes designed to help radar operators distinguish "friendlies" from the enemy.

The exchange had proven interesting, but only marginally useful to the British. Going into the meetings, both sides were convinced the other could not possibly possess radar. But as they quickly discovered, each had invented the technology independently in the mid-1930s, within a few months of each other: in fact, the British Chain Home Low, which guarded against low-flying planes, turned out to be virtually identical to the U.S. Navy's CXAM radar, operating on the same frequency and sharing several other technical features. As far as anything the British could use in the war effort, however, pickings were slim. The Americans did enjoy an edge in receiver technology. But at the same time, the United States had not developed airborne radars or anything like IFF, and the few other systems in existence had seen little operational use.

If Bowen and Cockcroft were hoping for more on the microwave front from contact with Loomis's group, they were not disappointed. Vannevar

Bush himself was not on hand; he preferred to delegate authority and leave his lieutenants alone. However, besides the host the small gathering included Carroll Wilson, Bush's personal assistant and alter ego, Karl Compton, and Admiral Harold Bowen, director of the Naval Research Lab. The admiral, who had earlier authored an internal memo discounting the idea of British radar, apparently harbored ongoing doubts about the exchange. He appeared to drink heavily at the party, but Compton suspected his colleague of feigning to be farther gone than he really was in order to avoid sharing information.

The British sensed such misgivings. "I still remember the rather doubtful opening with the U.S. officers suspicious as to whether we were putting all our cards on the table," Cockcroft related. The Americans showed their hand first, though, detailing an exhaustive survey of the nation's general microwave research that Loomis and Compton had conducted over the summer. It soon became clear to Bowen and Cockcroft that for the ten-centimeter waves emitted by the cavity magnetron, Bell Telephone Laboratories and General Electric both could contribute a lot to receiver technology. Bell Labs, Stanford University, and the Massachusetts Institute of Technology, they were told, also conducted advanced research in microwave waveguides and horn-shaped antennas. The British physicists found the information exceedingly helpful in pinpointing areas to visit.

Their hosts, however, confessed to being at loose ends trying to find a transmitter tube able to generate enough power to make for a feasible centimeter radar system. By the time of Loomis's party, a stymied Microwave Committee had steeled itself to write a report—a sure sign, as one member explained, "that we didn't know what to do next."

Bowen and Cockcroft quietly pulled out the cavity magnetron—by one account, they typically carried the device in a small wooden box whose lid was fastened by thumbscrews—and told their dumbfounded listeners that it could generate ten kilowatts of power at ten centimeters, roughly a thousand times the output of the best U.S. tube on the same wavelength. In one fell swoop, the disclosure dispelled any tension left in the room, and from that point on things went smoothly.

"It was a gift from the gods we disclosed to Alfred Loomis and Karl Compton," Bowen boasted late in life. The financier swiftly embraced the offering, inviting his newfound friends to Tuxedo Park, the posh retreat about thirty-five miles northwest of New York City where he had built his private laboratory. It was time for mere mortals to get to work.

2 • Radiation Laboratory

"[It was] the greatest cooperative research establishment in the history of the world."

KARL COMPTON

ALFRED LEE LOOMIS'S scientific playground, a stone mansion with medieval-style tower and battlements, overlooked a small lake in the foothills of New York's Ramapo Mountains. The investment banker–turned-physicist lived a few miles away in a spacious stucco colonial. He had purchased the second estate in 1926, when his extracurricular research outgrew his garage. He overhauled the tower and basement into a first-class scientific facility, complete with machine shop, forty-foot spectrograph, and a dozen or more labs. The main house, with its panoply of bedrooms and expansive living quarters, was reserved for the small staff of Loomis Laboratories and the many guests invited for an idyllic working vacation. In the late 1930s, hoping to move closer to the action, Loomis built, just behind the labs, a unique third house that featured double-glass walls for insulation and ventilation. His wife, Ellen, refused to leave their longtime home, so this quarters, designed by the noted Swiss architect William Lascaze and shielded by oak, maple, and hemlock, was often reserved for guests or private liaisons.

One of the giant behind-the-scenes figures of the day, Loomis had graduated cum laude from Harvard Law School in 1912 and gone straight to Wall Street, where he made a fortune financing public utilities sprouting up around the country. Along with his brother-in-law Landon K. Thorne, he owned outright the modern-day resort of Hilton Head Island, which the two partners set aside as a private hunting and riding reserve. On the eve of World War II, Loomis enjoyed an almost unrivaled reach deep into the country's power structure—in government, science, and industry. His favorite cousin was the veteran diplomat Henry L. Stimson, who would serve as war secretary throughout the coming conflict. An intimate of the heads of major corporations and universities, the financier could pick up the phone and cut through red tape: when his friend Ernest O. Lawrence, the University of California at Berkeley physicist, complained about the price of iron for his new cyclotron, Loomis simply dialed the chairman of U.S. Steel

Corporation, Edward Stettinius. "Hello, Ed, this is Alfred. I'm with a young chap that I know you would like to meet—when can we come over?"

A gadgeteer since childhood, Loomis gradually took on more serious scientific studies in his spare time. Beginning in the late 1920s and continuing for more than a decade, the private hilltop laboratory became the fountainhead of a series of investigations ranging from ultrasonics to the study of brain waves and the construction of the world's most accurate clocks. Tuxedo Park also was the scene of noted international colloquia. The list of guests brought in, all expenses paid with a special train waiting at New York's Pennsylvania Station, swelled to include Albert Einstein, Werner Heisenberg, and James Franck.

In the late 1930s, already retired from business life and thinking about the applicability of science to the looming war, Loomis turned his attention to microwaves and radio detection. Over the summer of 1940 a small research team assembled at Tuxedo and installed a crude device in a Loomis Labs truck, making what may have been the first radar gun. Positioned to track cars on a nearby highway, the contraption proved so successful at catching speeders one researcher remarked: "For the Lord's sake, don't let the cops know about this."

Eddie Bowen and John Cockcroft made the hejira to Loomis's research haven on Saturday, September 28, 1940. There was a sense of crisis in their world back home. In the just-ended week, 1500 of their countrymen had been killed in German air raids, bringing the month's total to nearly 7000. In return, the British were bombing Berlin every night. And a day earlier, on the twenty-seventh, Japan had formally joined Italy and Germany in the Tripartite Pact, which stipulated that if one of the countries got into a war with the United States the other two would help.

One of Loomis's cars met the pair at La Guardia Airport and whisked them to Tuxedo Park, a pleasant drive through countryside resplendent with the red and yellow hues of the changing autumn leaves. To take part in the seminal meeting with the two Tizard Mission representatives, Loomis had invited a handful of other dignitaries. Notable among them were Carroll Wilson and Microwave Committee secretary Edward Bowles, an MIT professor who had received his master's degree under Vannevar Bush, then gone on to become one of the country's foremost authorities in high-frequency and communications research.

The men were gathered primarily to explore the cavity magnetron's full ramifications. As if savoring the anticipation, however, Loomis spent a good part of the day showing his visitors the American state of the art. At a nearby airport, he demonstrated his own primitive continuous wave microwave radar, based on a klystron transmitter, which detected strong reflections from the Goodyear blimp and tracked two-seater planes up to a mile away. After dinner the suspense was further built by passing hours going over American developments and the benefits of pushing the technology into the centimetric wavelengths.

It wasn't until Sunday night that Bowen and Cockcroft finally reached into a small brown bag and extracted the cavity magnetron, spreading the accompanying blueprints and other technical drawings on Loomis's sitting room floor. Earlier that day the group had been joined by another Microwave Committee member, Hugh Willis, director of research for Sperry Gyroscope Company. For three hours, the high-powered group huddled around the device in animated conversation, discussing it and a few related British developments. As Bowen later wrote: "The atmosphere was electric — they found it hard to believe that such a small device could produce so much power and that what lay on the table in front of us might prove to be the salvation of the Allied cause." Ed Bowles, for one, concurred. "All we could do was sit in admiration and gasp," he recalled.

The disclosure, far more complete than the cursory overview presented at the Washington hotel party, gave the Microwave Committee a new life. Loomis returned to New York City and called a special luncheon meeting to bring those industrial members of the committee not able to attend the Tuxedo Park confab up to speed.

Within a few days a demonstration of the magnetron was in the works. Bowen and Cockcroft headed back to Washington to clear the arrangements with Tizard. Sir Henry returned to Britain via the Pan Am Clipper early in October, but not before settling a £5 bet with Cockcroft that their country would have been invaded before his homecoming. Bowen accompanied the mission head on the short flight from Washington to La Guardia, going over the final details. Tizard told the Welshman to push the Americans to manufacture the magnetron "for all you are worth."

On the third, Bowen brought the device to the Bell Labs headquarters at 463 West Street, at the corner of Bethune on the western edge of Greenwich Village. Waiting were the lab's research director, Mervin Kelly; radio and television research head Ralph Bown; and a host of valve experts whose names the Welshman recognized from research papers: J. G. Wilson, A. L. Samuel, J. O. McNally. The men discussed the magnetron at length, as Bowen produced blueprints and went over the rough specifications for running it — a 10,000-volt anode potential to extract electrons from the cathode, and a 1500-gauss magnetic field to bend and shape the electron flow. The Americans grew excited: if it worked as indicated, Bell would mass-produce the device. Kelly and his associates wanted to conduct a test right away, but Bowen needed to go to MIT on other business. He left the magnetron at

West Street for safekeeping, and the men arranged to meet three days later at Bell's laboratory in nearby Whippany, New Jersey.

Bowen returned to New York on Saturday, October 5, the day before the big event, spending the night as Ralph Bown's guest in Summit, New Jersey. Across the Atlantic back on September 7, Hitler had begun daily attacks on London, opening the period of war later called the Blitz. October 5 saw the beginning of German air raids that came only at night: hundreds of thousands of Londoners began sleeping in the underground. In New Jersey, Bowen and Bown passed the evening drinking martinis and watching a Mickey Rooney movie at a local cinema. But on the sixth the mood turned more serious. In nervous anticipation the pair drove the few miles to the rambling Whippany research and development facility, passing up a long road into an experimental station marked by low-flung buildings and scattered huts.

Bown ushered his passenger into one of the buildings and a room filled with electronic equipment: among other engineering paraphernalia, Bowen took note of a large electromagnet common in university physics departments. Wilson, Samuel, and other key members of Bell's staff were already assembled for the magnetron's first demonstration on American soil.

Aware of the stakes, Bowen felt increasingly nervous as someone checked the magnetic field and ran the filament up to the required potential. Who could say what would happen? The magnetron had journeyed three thousand miles and hadn't been tested in two months. "Very gingerly, we switched on the anode potential and were immediately rewarded with a glow discharge about an inch long coming from the output terminal," the Welshman relates. "I had not seen anything like this before and we looked at it in amazement."

On-the-spot estimates put the power output at between ten and fifteen kilowatts. This was some seven times the power generated by Bell's own microwave radar, a forty-centimeter naval set then under development. And it was likely better than the promised factor of a thousand greater than anything available on such a short wavelength in the United States.

Or the rest of the world for that matter. Bowen left for Washington walking on air.

•

The magnetron's successful debut engendered a firestorm. In the wake of the Tuxedo Park tête-à-tête, Loomis had already excitedly told his cousin the war secretary that the British invention had pushed the American radar development program ahead two years. With the Whippany demonstration erasing any remaining doubts about the power of the British invention,

Loomis, Karl Compton, and Vannevar Bush were unanimous that a central laboratory should be created to develop radar systems around the device.

The men felt an acute sense of urgency. Close to home, the NDRC needed to bulldoze past any opposition it might encounter from the services, which still turned a skeptical eye toward a civilian agency taking up wartime radar work. Against this lay the backdrop of the war. As Tizard and others had predicted, the Luftwaffe's switch to nightly bombardments had overwhelmed the Chain Home system, which did not possess the resolution needed for nighttime interceptions. Supplementing the system with centimetric radars, installed in fighters themselves, could prove decisive.

Yet as was abundantly clear from meeting with the Tizard Mission representatives, the magnetron, still an experimental device, had to be geared up for mass production. It needed to be combined with receivers, antennas, and a myriad of other components, and tailored to meet the various needs of the services — with different systems for airborne radars, submarine hunters, anti-aircraft guns, and bombers. By almost any measure the job would take at least two years, and nearly unlimited resources. With no time to waste, Loomis called a second Tuxedo Park meeting for the weekend of October 12 and 13 to nail down details of the revitalized American centimeter radar program. But just as Bell Labs got busy copying the transmitter for mass production, an unexpected finding almost brought the endeavor to a screeching halt.

On October 7, the day after the Whippany demonstration, Eddie Bowen received a call from Mervin Kelly, who forcefully summoned the Welshman to New York. The Bell research chief would not be specific over the telephone, but something was drastically amiss. "Oh my God," Bowen thought, "the magnetron must have blown up." He couldn't make the trip that day, but awoke at 5:30 the next morning to grab an early flight. He was met in the West Street foyer and ushered to a top-floor conference room. Kelly and several staff members greeted him coolly, laid the magnetron on the table, rolled out the blueprints from the black box, then produced an X-ray of the device in their possession. Looking it all over, Bowen grew astonished. The blueprints showed a magnetron with six resonant cavities, just as he believed the design called for. However, the X-ray revealed Bell's copy of the transmitter had eight cylindrical holes.

"There was little I could say," Bowen recalled. He asked to send a cable to the General Electric Company labs in Wembley, where initial production versions of the magnetron had been fabricated. But he had forgotten that he sat at the heart of telephony. It was not long before Kelly's crew had GEC engineering wizard Eric C.S. Megaw on the line. At first Megaw was also bewildered. Then the penny dropped. Ten of the first dozen devices had been made with the standard six cavities. But the remaining pair had involved novel designs — one with seven holes, the other with eight. In the haste to prepare for the Tizard Mission no one had remembered this fact, and when Bowen had picked out the best one, he had unknowingly selected the eight-cavity version. The realization broke the tension with Kelly. But then the men had to consider whether Bell should build copies from the blueprints or from the device they held. "We agreed that there was only one way to go — to copy the one which they knew worked," relates Bowen. After that, all the early British magnetrons were made with six holes, the American models with eight.

In the wake of that heart-stopping episode, everything rushed ahead madly. That weekend, as planned, a number of NDRC specialists showed up at Tuxedo Park to learn more specifics of the British radar organization, as well as glean clues to their guests' thoughts for the American lab. In addition to Bowles and Wilson, two new important faces joined the group. One belonged to Frank Lewis, a Loomis employee recently cleared to hear microwave radar details. The second newcomer was none other than Ernest O. Lawrence, still riding the wave of winning the 1939 Nobel Prize for his invention of the cyclotron, which had opened the door on the mysteries and delights of quantum physics by allowing scientists to bombard atomic nuclei with particle beams of unheard-of energies and intensities. Lawrence had climbed on board the Microwave Committee train early in October after an urgent summons from Bush. On a resonant note, too, the first day coincided with Franklin Roosevelt's stirring speech in Dayton, Ohio: "Our course is clear," the President advised, seeking an unprecedented third term in the coming November elections. "Our decision is made. We will continue to pile up our defense and our armaments. We will continue to help those who resist aggression, and who now hold the aggressors far from our shores."

Loomis's gathering started out easy. Bowen remembered losing badly to Lawrence at tennis and having a casual drink with Averell Harriman, a future U.S. ambassador to the Soviet Union, who had a house nearby and stopped in to say hello. But the next day things took a more serious turn. At their host's urging, Bowen and Cockcroft laid out the British needs, giving highest priority to a ten-centimeter airborne interception radar to be fitted in nightfighters: the British considered such a set absolutely vital to stemming the nightly Luftwaffe raids on London. Second on the list was some sort of long-range navigation system for guiding bombers to and from targets without depending on a signal from the plane itself. Third in priority came a ten-centimeter gunlaying radar to direct anti-aircraft fire.

In response to American prodding, the two Tizard Mission representatives began laying out the technical requirements for their wish list. Cockcroft detailed the full specifications necessary for anti-aircraft radar. Bowen wrote out his own specs for airborne radar and outlined a proposed British navigation system in which ground stations transmitted a grid of precise radio beams that enabled aircraft to fix their locations. The two men also described how the British had set up civilian-run labs and drafted university people for the research.

The disclosures struck a chord with the Americans. Lawrence, for one, had joined the Microwave Committee extremely pessimistic about the British ability to survive the German onslaught. But upon hearing Bowen and Cockcroft describe British radar in detail, he saw things in a new light. "This changes the odds," he said. "Let's get going as fast as possible and do anything we can to help the British."

Loomis hardly needed encouragement. Back at his New York apartment at 21 East Seventy-ninth Street the following Monday, he set about twisting the arms of key industrial contacts—Mervin Kelly at Bell Labs, Radio Corporation of America research boss Ralph Beal, and Sperry's Hugh Willis—to deliver five working sets of magnetrons and other needed equipment within a month. Under the agreement hammered out, Bell promised to produce magnetrons, Sperry the parabolic reflectors, and RCA contracted to deliver cathode ray tubes, modulated power supplies, and intermediate frequency amplifiers. Representatives of General Electric and Westinghouse could not make the hastily arranged meeting. But GE, an American company not related to its British namesake, soon took on the job of supplying magnets for the magnetrons, while Westinghouse joined RCA in producing antennas.

If anyone harbored doubts as to the urgency of the task at hand, it was only necessary to look to Britain. The night of October 15 was the worst of the war so far. Bombers pounded London from eight in the evening until five the next morning, giving rise to nine hundred fires. A bomb broke through the platform at Balham underground station, creating an avalanche of debris that buried alive sixty-four of the six hundred sleeping below. All told, four hundred Londoners died that night. By that Friday, October 18, with Eddie Bowen imploring them on, the Microwave Committee had approved the contracts, along with the idea of establishing a central lab, adopting as the first order of business the three projects set out at Tuxedo Park.

The enterprise marked a novel departure from past American wartime ventures: the lab would be administered by civilian scientists, and it would not work directly for the military. Not only had the British sung the praises of such a setup, NDRC members had grown frustrated in their early dealings with the services. "We were quick to learn that everytime you mentioned anything, they'd say, 'Well, we've done that; or we're trying it . . .'" Bowles recounted. "And they'd open a file cabinet and show you a piece of paper. That led us to the next step, to the conclusion that we must have a laboratory of our own. In other words, not a committee going around and helping the services by telling them what we think they ought to do. Because we knew damn well that we had no control over them."

Money would come from a civilian agency, the NDRC, which in turn had special access to the presidential tap. The idea sailed along. A week later, on October 25, Bush and his committee gave the plan their final approval, allocating a budget of $455,000 for the lab's first year of operation.

A site for the lab was not chosen until the last minute. Apparently seeking proximity to Washington, Loomis held out for the Carnegie Institution. But Bush feared disciplinary and administrative problems in his own shop. "I protested, and we had a hell of an argument that took half the night and a bottle of scotch," the NDRC chairman recalled. After considering several other possibilities, MIT finally seemed the best bet. Not only did the school have a rich history in microwave research, the group also reasoned that opening a new lab at a top research university would attract a minimum of attention.

The only potential snag came from Frank Jewett. Both Loomis and Ed Bowles suspected the Bell Labs chief might try to snare the proposed radar facility for his own organization. They maneuvered to feel him out during a meeting in Bush's office on October 16. Sure enough, Jewett balked at the idea of an MIT-based lab. Bell had run a similar effort on submarine detection in the first war, he pointed out, near its West Street headquarters. A laboratory of this character needed Bell-type management; it was too complex for an institution like MIT.

There was a pause. Apparently realizing he had raised the hackles of Bush and Bowles, both MIT men, the Bell head beat a tactical retreat. He did not mean, Jewett stressed, to denigrate the school's overall ability.

Before he could continue, Loomis broke in: "I am so glad, Dr. Jewett, that you approve the idea of having the laboratory at MIT." The decision had been made.

The next day the three men, now of a single mind, sandbagged Karl Compton when he showed up in Washington for a Microwave Committee meeting and virtually wrung an agreement from the MIT president. Compton made an emergency phone call back to Cambridge to ensure the needed 10,000 square feet of laboratory space could be freed up. Later, he met with Massachusetts Governor Leverett Saltonstall to arrange space for flight tests of the planned airborne radar at a National Guard hangar at East Boston Airport, predecessor to Logan Airport.

Initially, the upstart facility simply bore the moniker Microwave Laboratory. Within the first few weeks, however, to further mislead unwanted observers, it was christened the Radiation Laboratory. The Microwave Committee members hoped that enemy spies would merely consider it an East Coast version of E. O. Lawrence's Berkeley Radiation Laboratory, which specialized in the esoteric field of nuclear research — still not considered a viable military pursuit.

Recruitment for the lab had begun immediately, even before all these plans crystallized. Following the British example, it was decided that the Radiation Lab would draw on university scientists, primarily nuclear physicists who were familiar with high-frequency radio waves from their accelerator work. Lawrence turned down the director's position to pursue construction of his latest brainchild, a gigantic 184-inch cyclotron. But no better headhunter could be found than the legendary Berkeley physicist, whose face had already graced the cover of *Time* and whose lab served as a mecca for nuclear physicists wanting to learn how to build accelerators. A man of sudden bursts of excitement, Lawrence zealously got to work. On October 8 he stopped by Harvard to sound out Kenneth Bainbridge, a native of Cooperstown, New York, who had designed and built the school's first cyclotron. Six days later, with the noted mass spectroscopist firmly on board, the two had met with Loomis in New York to discuss the character of the proposed laboratory; it was the same day Loomis spent hashing out industry commitments.

Another early feeler went to Lee A. DuBridge, a Lawrence protégé who had led the University of Rochester physics department to national prominence partly by building one of the first East Coast cyclotrons and partly by daring to hire the school's first Jewish faculty, most notably the Austrian Victor Weisskopf. Besides his keen administrative abilities — DuBridge then served as chairman of Rochester's physics department and dean of the Faculty of Arts and Sciences — the popular Hoosier exhibited an unbridled desire to take part in the war effort.

Lawrence called from Loomis's apartment on October 15. "I can't tell you about it, but I assure you it's very important," he informed DuBridge. The recruit took the train to New York and the next day accepted a job as technical director, soon to be changed to overall director. If Lawrence backed the program, the Rochester dean once noted, "that was what I wanted to be in."

DuBridge and Lawrence immediately began beating the bushes for additional conscripts. They visted Louis A. Turner at Princeton and made the theoretician a consultant. Next, DuBridge attended a convention at

Indiana University to sound out other researchers, including the highly regarded Columbia University physicist I. I. Rabi and University of Illinois physics chairman Wheeler Loomis, no relation to Alfred. Both men ultimately signed on.

A perfect drawing card appeared at the end of October, when some six hundred scientists gathered in Boston for a conference on applied nuclear physics. At Loomis's suggestion, Bowles arranged a series of lab visits and special seminars—MIT waveguide pioneer Wilmer L. Barrow lectured on his own ultrahigh frequency work—to bring a select few up to speed on general microwave problems. Loomis and Compton also hosted a historic luncheon meeting at the exclusive Algonquin Club on Commonwealth Avenue, bringing together Bowen, who had used a secret entrance, with about two dozen American counterparts. In an upstairs dining room, recruits signed an agreement of secrecy before receiving a primer on the new lab and its mission. "It was a grand pep-talk and got the whole group keen to be in on it," DuBridge wrote E. O. Lawrence a few days later.

Not everyone could come immediately, but about twenty physicists agreed to start by December 1. "So the lab will start off with a bang," DuBridge continued in his update to Lawrence. "All of these people can go to work on some project even before they are cleared by FBI. MIT will take the responsibility and they have all signed (for Compton's MIT files) a copy of the espionage act. Only the cleared ones can know the *whole* story of course." He noted, "I am still dizzy at the pace things are going but at least feel that things are under control." And then, "Everyone is working together beautifully and it's going to be great fun."

DuBridge had good reason to be optimistic, for before him lay the nucleus of an extraordinary group. Almost none of the men had any experience with microwaves or basic engineering, but sharp minds abounded. Some, acutely aware that the uranium nucleus had been split by German scientists the year before, fretted over the possibility of atomic power in Hitler's hands. Many in this fairly tight-knit physics community were Jews or had studied abroad and maintained deep ties to the imperiled Jewish-European scientific community. Still others felt a strong sense of kinship with their beleaguered British cousins—the sense of outrage brought home by the nightly bombardments suffered by Londoners. Whatever their reasons, almost to a man the Radiation Lab's new staffers did not take well to events in Germany.

No one characterized the fervor more than Isidor Isaac Rabi, one of those invited to the Algonquin Club. Born in the old Austro-Hungarian town of Rymanow but raised since infancy in Brooklyn, he hated the Nazis. At forty-two, one of the elder statesmen of physics, he would permeate the lab with his driving force. "The war in Europe was going so badly, I was dying to get into something," he later explained. "When I heard of this, I said, 'I want to be in on it.'"

Rabi joined the lab on November 6, the day after Roosevelt's reelection victory over Wendell Willkie. Within a few days, he enticed two of his top students, Jerrold Zacharias and Norman Ramsey. It took Zacharias three weeks to get the necessary clearance. Ramsey, who had started work five weeks earlier at the University of Illinois, where he had expected to spend his professional life, checked in around midmonth. Elinor, his wife of a mere two months, won the distinction of becoming the first out-of-town spouse to arrive.

Turning to nearby Harvard, the fledgling enterprise conscripted J. Curry Street, co-discoverer of the muon. He was joined by two up-and-comers — cherub-faced former Rhodes scholar Ivan Getting and a young instructor named Edward Purcell, who had shown up at his office one night to get in some extra work and stumbled across Bainbridge and Rabi conducting an early Rad Lab planning session in the Faculty Room. Getting started work in mid-November, but teaching obligations prevented Purcell from joining the lab full-time until the next January.

E. O. Lawrence contributed two of his brightest protégés, Edwin McMillan and Luis "Luie" Alvarez. A bon voyage party saw the young physicists off as they boarded a cross-country train in the second week of November. Robert Oppenheimer, another Berkeley physicist, gave each a bottle of whiskey; he had abandoned his usual practice of handing out books to those about to take long trips after the physicist Paul Dirac had politely refused the offer, saying, "I never read books — they interfere with thought." The two arrived in Cambridge four days later. Alvarez received Notebook No. 11, his colleague No. 12.

These men represented the cream of the crop of American physics. Five would bear the distinction in later years of winning the Nobel Prize. Ultimately five other Rad Lab staffers or consultants would also garner science's top honor. But for the moment, these scientific stars were expected to abjure individualism, absorb a brief indoctrination, and get to work as an engineering team. They dropped their research, sometimes almost literally: Nuclear physicist Ernest Pollard, a Yale recruit, received a telegram from DuBridge asking him to come to Cambridge. He stuffed the slip of paper in his lab coat, and when he moved, hung the jacket up and left it at Yale.

As these and other recruits streamed in, DuBridge and a few close advisors toiled furiously on general logistics. Ken Bainbridge set about making salary arrangements and running down rooming and housing facilities. Melville Eastham, president of the General Radio Corporation and the

Radiation Lab's early business manager, laid preliminary plans for library resources, power supplies, and the like.

By Armistice Day, November 11, the bare bones were in place. A tiny nucleus of physicists gathered in their new headquarters in 4-133—the obscure MIT code to designate building and room number—for the first official Rad Lab meeting. Karl Compton provided attendees with a situational overview before giving way to Alfred Loomis, who filled everyone in on the basics of microwave radar. By the next day, when Bell Labs delivered five magnetrons right on schedule, the main organizational outlines had been agreed upon and problems parceled out among seven technical sections arranged along component lines: transmitter, receiver, antenna, and so on. "We chose up just like a baseball team," remembered Rabi. "We chose up sides. What would we take? Well, I took the magnetron."

Even those not working directly with the remarkable device were stunned by its potential. "A sudden improvement by a factor of three thousand may not surprise physicists, but it is almost unheard of in engineering," Luis Alvarez recalled in the mid-1980s. "If automobiles had been similarly improved, modern cars would cost about a dollar and go a thousand miles on a gallon of gas. We were correspondingly awed by the cavity magnetron. Suddenly it was clear that microwave radar was there for the asking."

For all its power and promise, however, no one on either side of the Atlantic really fundamentally understood how the strange transmitter worked: somehow in the magnetron's cylindrical cavities, electrons were accelerated and bunched into tight coherent packets. In the Rad Lab's early days, a group of theorists gathered around the tube's disassembled parts.

"It's simple," Rabi told the group. "It's just a kind of whistle."

"Okay, Rabi," said Edward U. Condon, another Berkeley-trained physicist brought in from Westinghouse Research Laboratory, "how does a whistle work?"

At night, when no one remained to engage in such debates, the magnetrons were prudently locked in a safe in DuBridge's office. To keep away prying eyes, the Rad Lab's windows were painted black.

The war took on fresh fury as the Rad Lab hummed to life. The night of November 3, no German air raids had materialized over London for the first time since September 7. But eleven days later, five hundred bombers sparked a firestorm in Coventry that killed 568 and reduced thousands of buildings to rubble: production at a host of vital war factories was shut down for months. The British retaliation two days later killed 233 civilians in Hamburg.

On November 17, in the midst of the maelstrom, Eddie Bowen drove

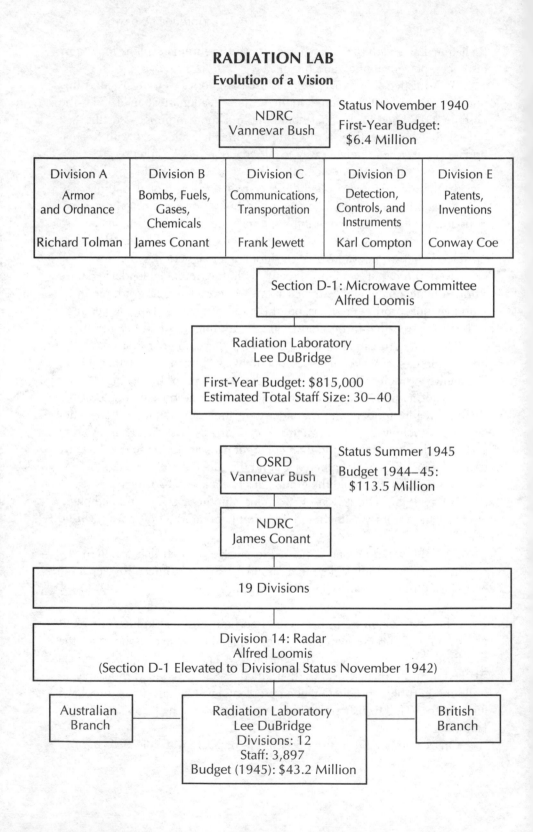

RADIATION LAB
Evolution of a Vision

NDRC
Vannevar Bush

Status November 1940

First-Year Budget:
$6.4 Million

Division A	Division B	Division C	Division D	Division E
Armor and Ordnance	Bombs, Fuels, Gases, Chemicals	Communications, Transportation	Detection, Controls, and Instruments	Patents, Inventions
Richard Tolman	James Conant	Frank Jewett	Karl Compton	Conway Coe

Section D-1: Microwave Committee
Alfred Loomis

Radiation Laboratory
Lee DuBridge

First-Year Budget: $815,000
Estimated Total Staff Size: 30–40

OSRD
Vannevar Bush

Status Summer 1945

Budget 1944–45:
$113.5 Million

NDRC
James Conant

19 Divisions

Division 14: Radar
Alfred Loomis
(Section D-1 Elevated to Divisional Status November 1942)

Australian Branch

Radiation Laboratory
Lee DuBridge
Divisions: 12
Staff: 3,897
Budget (1945): $43.2 Million

British Branch

up from Tuxedo Park with Lawrence and Loomis, who had already transferred his own lab's work to MIT and was giving demonstrations at East Boston Airport. That afternoon, to those cleared, the Welshman gave a talk on military tactics and airborne interception. Lawrence soon left to press on with the recruiting, taking a leisurely train trip across the country and stopping at the University of Chicago, Purdue, and other major universities along the way.

By mid-December, the Rad Lab had swollen to include thirty physicists, three guards, two stockboys, and one secretary—an eleven-year veteran of MIT's electrical engineering department named Edythe Baker, who would follow DuBridge for the next fifty years. A wooden penthouse, comprised of gray-green tarpaper, sprang up on the roof of Building 6. By January 6, barely three weeks away, this was to hold a working microwave radar system. The rest of the basic goals had also been set: Alfred Loomis scrawled them on a chalkboard. A breadboard version of the first airborne radar should be installed in an Army B-18 aircraft by February 1. And within a month after that, the lab was to have met its top priority—a working centimeter set in a nightfighter.

No one knew for certain if any of these goals could be met; no one except Bowen had much experience even building a radar system. But that was all right. It was a talented bunch, and the energetic Welshman didn't mind sharing what he knew. In fact, he thrived on bringing the Yanks up to speed, providing the details of British radar, and mesmerizing them with wild tales of how it all began.

3 • Beginnings

"The flying peril is not a peril from which one can fly."
WINSTON CHURCHILL

IN THE FOURTEEN YEARS following the Great War, during what Winston Churchill termed "that period of exhaustion which has been described as Peace," military minds increasingly looked toward the sky and wondered what kinds of horrors might fall out. Britain had already experienced a taste of the awful future. During World War I, between Zeppelins and bombers, the country suffered a total of 103 air raids that claimed 1413 lives and left another 3400 injured, more than half the casualties coming in the greater London area. The German raids were too puny to affect the war's course. But the cumbersome Gothas, especially, thump-a-thumping to the attack, ushered in a new era of strategic bombing that for future wars promised a rain of destruction on largely defenseless civilian areas.

The kingdom considered almost any fanciful scheme to warn of an impending attack and repel it. In that first global conflict, blind people with acute senses of hearing listened through stethoscopes to large binaural gramophone horns. At least one sound-focusing cavity was carved into a sea cliff. Ground crews wheeled gigantic white arrows suspended on pivots to show radioless fighter pilots the way to the attackers. Barrage balloons ringed London, carrying aloft long strands of steel cable designed to ensnare propellers. All this was supplemented by gun and searchlight stations, night aircraft patrols, and a complex observer organization.

The system was barely effective against snail-like airships and early-vintage bombers. As airspeeds rose meteorically after the armistice, however, any warning system based on a medium as slow as sound did not provide time to scramble fighters for a successful intercept. By the mid-1930s, the British had effectively abandoned the latest idea for a state-of-the-art network of gigantic acoustic mirrors to warn of invaders approaching the Thames estuary. For those worrying about air defense it was not a pleasant time. Hitler had become the German chancellor in January 1933, and already the implications of the swelling Nazi military mobilization could be imagined. Almost the only option left, besides saying a prayer for the anti-aircraft guns and barrage balloons, was the expensive and impractical idea of keeping aloft standing fighter patrols.

Many had already abandoned hope of a successful defense network. As early as November 10, 1932, in a chilling speech to the House of Commons, Stanley Baldwin articulated the nation's fears. "I think it is well for the man in the street to realize that there is no power on earth that can protect him from being bombed," the former prime minister intoned. "Whatever people may tell him, the bomber will always get through."

The last phrase would echo throughout the country; its reverberations lingered a half century later. Continuing his speech, Baldwin painted a grisly picture — an early version of mutually assured destruction. "The only defence is in offence, which means that you will have to kill more women and children more quickly than the enemy if you want to save yourselves."

A few years after the speech, nothing had happened to change this gloomy outlook. During the 1934 summer air exercises, the Air Ministry itself was "destroyed" in the first mock raid. The Houses of Parliament soon followed. But as the threat of Hitler's army grew, a few outspoken leaders could no longer tolerate the government's seemingly fatalistic view. Winston Churchill led the general cry for rearmament and railed against the "cursed, hellish invention and development of war from the air . . . " His good friend Frederick Lindemann, an Oxford professor, publicly challenged the government's position on air defense. In a widely read letter to the *Times* on August 8, 1934, he noted: "That there is at present no means of preventing hostile bombers from depositing their loads of explosives, incendiary materials, gases, or bacteria upon their objectives I believe to be true; that no method can be devised to safeguard great centres of population from such a fate appears to me to be profoundly improbable . . . To adopt a defeatist attitude in the face of such a threat is inexcusable until it has definitely been shown that all the resources of science and invention have been exhausted."

In June 1934, in the eye of this rising tempest, a young civil servant named A. P. Rowe stirred in his relatively obscure cubbyhole inside the British Air Ministry's labyrinth of directorates. A small, bespectacled, pipe-smoking man on the headquarters staff of the ministry's director of scientific research, Rowe mirrored the quintessential civil servant: concerned about propriety, anticipating problems, meticulous. He took it upon himself to survey his country's plans for air defense and rounded up every available file on the subject, fifty-three in all.

He found nothing at all promising. While a great deal of effort clearly had gone into making planes faster and boosting fighter armament, it was just as plain that developing air defense technology had attracted little of Britain's scientific resources. Rowe prepared a memo for his immediate superior, Henry Wimperis, in which he stated that unless science corrected the situation, the country would likely lose the next war.

Coming amidst the surge of public discontent, Rowe's memo helped galvanize Wimperis. On November 12, 1934, he proposed to the secretary of state for air that a committee be established "to consider how far recent advances in scientific and technical knowledge can be used to strengthen the present methods of defence against hostile aircraft." As the natural leader for the new body, Wimperis suggested Henry Tizard, who had served with the Department of Scientific and Industrial Research during the German military buildup of the early 1930s, and was then chairman of Britain's Aeronautical Research Committee. Wimperis also put up as possible members the names of A. V. Hill and Cambridge University physicist Patrick M.S. Blackett, both Great War veterans interested in military technology.

The idea was quickly accepted by the Air Ministry's highest echelons, and the Committee for the Scientific Survey of Air Defence was formed within a few weeks. Tizard, Hill, and Blackett all became members. Wimperis himself rounded out the committee, with Rowe serving as secretary.

Even as the government moved to attack the problem of air defense from a scientific standpoint, Wimperis began pursuing a private line of inquiry. For a decade, officials had received oddball proposals for the ultimate weapon, a ubiquitous death ray to slay enemy soldiers — or pilots in the case of air attack. The Air Ministry offered a standing reward of £1000 for anyone who developed a ray that could kill a sheep at a hundred yards. No one had claimed the purse, but Wimperis wanted to cover every option. For help, he turned to Robert Watson Watt, superintendent of the Radio Department of the National Physical Laboratory, who operated what was known as Radio Research Station, a respected facility for radio studies of the ionosphere near the Berkshire town of Slough.

The two met to discuss the death ray on January 18, 1935. Wimperis told the radioman he had come outside any official capacity to examine the feasibility of developing a system of what he called "damaging radiation." Watson Watt immediately answered that the prospects didn't seem good. But he promised to study the problem more scientifically.

Robert Alexander Watson Watt was at the time perched on the outskirts of the highest circles of British scientific policy making and looking for a way in. An engineer by training, the charismatic and ambitious Scotsman had volunteered for the Army when World War I broke out, but instead got called to the Meteorological Office at the Royal Aircraft Establishment. There, in 1915, he had begun the atmospherics work that determined the next two decades of his life, initially studying thunderstorms and moving on to the general problem of radio static. Throughout his career, he had risen steadily through the ranks, finally reaching his present position atop a world-class center of ionospheric research. But he had not cracked the inner circle.

A great many of Watson Watt's problems stemmed from his effusive personality. He stood just five feet six inches tall, and was chubby to the point of outright fatness. Words, said one colleague, "bubbled out of him like a fountain." He talked in long-winded, often convoluted sentences, full of homilies and novel metaphors that mesmerized listeners. This unceasing zeal struck more than a few in the scientific establishment as self-centered and unrefined: one high-minded acquaintance described the Scot as a man with no hobby and a failure to appreciate good food.

But while Watson Watt could be off-putting, the force of his character often brought quick and impressive results when focused on a problem. He prepared a formal statement of the death ray inquiry and passed it on to a trusted junior scientific officer, Arnold F. "Skip" Wilkins. In keeping with his strain of Scottish frugality, the superintendent scrawled the note on an old calendar leaf, which he left on his subordinate's desk. It read something like: "Please calculate the amount of radio frequency power which should be radiated to raise the temperature of eight pints of water from 98 degrees F to 105 F at a distance of five km and a height of 1 km."

Despite the Scotsman's attempt to camouflage the military nature of the assignment, Wilkins saw right away that eight pints of water somehow levitated in space represented an aviator's blood, and thought immediately of the death ray. His calculations showed that given the technology of the day, generating enough power to raise a man's temperature enough to kill him could not be even remotely possible. He reported the conclusion to his superior, who replied, "Well, then if the death ray is not possible how can we help them?"

The young scientific officer thought for a moment. He had recently heard that engineers at the government post office, whose dominion extended to shortwave communications, had noticed disturbances when airplanes flew near their receivers. Perhaps this phenomenon could somehow form the basis for an aircraft detection scheme, he mused. Watson Watt seized on the idea, pressing for calculations to establish the effect's magnitude. Would it be possible, he wondered, to detect planes as far as five or six miles away with the energy from a one-kilowatt transmitter?

Wilkins considered the size of a typical airplane and calculated roughly, to his surprise, that the energy reflected from its surfaces should be detectable. Watson Watt quickly checked the figures. Finding no obvious errors, he included the idea in a memo for Wimperis. This was the first of two historic memos the Scotsman would write. It arrived in time to be presented at the inaugural meeting of Tizard's CSSAD on January 28. In his own unique terms, the Scotsman discounted the death ray idea but held forth the alternative: "Meanwhile attention is being turned to the still difficult but less unpromising problem of radio-detection as opposed to radio-destruction, and numerical considerations of the method of detection by reflected radio waves will be submitted when required."

The committee pounced on Watson Watt's slightly cryptic offer. On February 6, after confirming the idea's basic feasibility, Rowe wrote the Scotsman asking for the promised facts and figures. The superintendent again turned to Wilkins for more specific calculations of the current needed to detect an aircraft up to ten miles away. When the answer came back, it seemed too good to be true. Even allowing generously for losses, Wilkins concluded, at a wavelength of fifty meters, a transmitter sending fifteen amperes of current through a simple aerial should easily produce a detectable echo from planes ten miles distant and flying at an altitude of 20,000 feet. Unable to contain his excitement, Watson Watt hurriedly drafted his second landmark memo, "Detection of Aircraft by Radio Methods," on February 12, 1935. "It turns out so favourably that I am still nervous as to whether we have not got a power of ten wrong, but even that would not be fatal," he wrote in a cover note. "I have therefore thought it desirable to send you the memorandum immediately rather than to wait for close rechecking."

The twenty-paragraph epistle, a hallmark of off-the-cuff scientific foresight, formed the basis of Watson Watt's legacy as the father of radar, though the term itself, coined by U.S. Navy Lieutenant Commanders Samuel M. Tucker and F. R. Furth as an acronym for RAdio Detection And Ranging and adopted by their service in November 1940, was not accepted by the British until July 1943. The memo took into account all the day's relevant technologies, from the cathode-ray tube display screens so familiar to ionospheric researchers to the latest means of generating radio pulses. The pulse technique was crucial. In the split second, actually on the order of microseconds, it took for the pulse to be reflected from an aircraft and return to a ground-based receiver, the transmitter would remain off. This would allow the interval between transmission and detection to be timed. Since radio waves move at the speed of light, the distance, or range, to the target then could be easily computed. For example, light travels at about 186,000 miles per second, meaning it takes a radio wave pulse one microsecond to journey .186 mile, or 328 yards. An echo coming in ten microseconds after an emitted pulse would have traveled 3280 yards, or just under two miles. The target itself would lie half this distance away, since the pulse had to venture out from the transmitter and back to the receiver.

Watson Watt saw, too, that range could be shown on a basic oscilloscope calibrated with a linear distance scale. A spot of light starts at the left side of the screen when the pulse is transmitted and moves across to the right, leaving a bright trace in its wake. The echo signal is amplified and used to deflect the trace upward in a momentary spike. Even in the mid-1930s, as the memo noted, it was relatively easy to perform such measurements, "the whole technique already being worked out for ionospheric work at Radio Research Station."

The Scotsman proposed that zones of radio waves be established to guard against enemy bombers, stressing that any such detection scheme must include a way to tell friendly planes from the enemy, or friend from foe. Present radio technology made aircraft detection easiest on a wavelength of fifty meters. But Watson Watt acknowledged the need to push to shorter wavelengths less vulnerable to ionospheric interference. He also estimated that range might be extended as far as 300 kilometers, or nearly 190 miles — and pointed out the importance of adding elevation and azimuth finding to any distance-measuring capabilities.

The importance of this far-reaching vision did not escape the CSSAD members. As the nation's primary sounding board for air defense technologies, the Tizard Committee had spent its first few sessions getting the lay of the land, considering old schemes like barrage balloons along with newer but remote possibilities such as infrared detection. The somewhat jaded members quickly realized that Watson Watt had dropped into their laps the best idea of all, one possible with existing technology. Following a luncheon at London's highbrow Athenæum club with Watson Watt and Tizard a few days after the second memo was written, Wimperis professed to like the idea enough to recommend immediately allocating £10,000 toward its development.

It would not be quite that simple. In order to shake loose such funding, the Tizard Committee needed the support of Air Marshal Sir Hugh Dowding, then air member for research and development. And "Stuffy," as Dowding was known for his lack of humor and conviviality, wanted more than a scrap of paper. He had already witnessed one of Britain's high-tech air defense schemes in action, a concave wall two hundred feet long and twenty-five feet tall built as a prototype for the sound mirror system, and had not come away particularly impressed when a horse-drawn milk cart interfered with the sensitive mirror's operation. Before he got behind any large expenditure, Dowding wanted a test.

The showdown was set for February 26. Wilkins, again delegated to the detail work, did not even have time to ready a pulse transmitter. So he suggested "borrowing" a British Broadcasting Corporation station at Daventry, in Northamptonshire. The continuous signal would make it difficult to gauge distance. Still, it was easily strong enough to bathe a plane in radio waves, and Wilkins could jury-rig a receiver and cathode-ray tube display in a nearby field to pick up any echoes.

Wilkins and a Radio Research Station handyman, Dyer, worked by matchlight until midnight the eve of the test to set up the crude apparatus. The next morning dawned crisp and clear, with a strong southwesterly breeze. Watson Watt and Rowe arrived from London in the Scot's treasured Daimler. In the skies above Daventry, Squadron Leader R. S. Blucke piloted the Heyford bomber that served as the guinea pig. The pilot never fully got his bearings. He made four passes in the general vicinity of the radio station's beam at about 10,000 feet. In the field below, the observers gathered around the tiny screen. All but a minute fraction of the Daventry signal had been balanced out, so the indicator showed just a tiny glowing green vertical line, about an eighth of an inch thick, that flared up to an inch or so with what the Scotsman called "atmospherics."

The Heyford trailed a long communications aerial — not mentioned in any published account of the test — that undoubtedly enhanced its signal and may have been a Watson Watt ploy to help stack the deck in his favor. The first pass took the twin-engined biplane far to the east of the transmitter beacon, so nothing showed up on the cathode ray tube. When the craft returned, however, the beam painted it clearly. The glowing stub started to throb and swell, reaching more than an inch in height before dwindling away as the hum of the plane's engine faded into the distance. The signal stayed strong on the remaining runs, and rough calculations based on the hundred mile per hour airspeed and the duration of the signal indicated the Heyford had been tracked to a distance of eight miles. "Britain," intoned Watson Watt, "has become an island once more."

Although the test fell far short of actually locating an airplane in space, it heralded another era. Nothing coming down the pike was even remotely as promising as RDF, as radar was soon called in order to fool enemy agents into thinking it was radio direction finding, a common technique of homing in on radio signals. Once back in London, Rowe quickly spread the word. "It was demonstrated beyond doubt that electromagnetic energy is reflected from the metal components of an aircraft's structure and that it can be detected," he reported. "Whether aircraft can be accurately located remains to be shown. No one seeing the demonstration could fail to be hopeful of detecting the existence and approximate bearing of aircraft at ranges far in excess of those given by the 200-foot sound mirrors."

Dowding needed no further proof. Wimperis gleefully recorded in his diary that the air marshal had agreed to bankroll the new technology with "all the money I want, within reason." A scant six weeks after the test, the

Treasury sanctioned an impressive £12,300 — more than $60,000 at the time — to set up a secret program to develop RDF. The money would cover expenses for the project's first year — after which more would be allocated if things panned out. The official decision came on April 13, Watson Watt's forty-third birthday. The jovial Scotsman threw himself into the task with typical fervor, feeling a sense of urgency but never dreaming he wasn't the only one in the world hot on the same trail.

•

In the often-mystical process of discovery, a few innovations arise solely from original thinking, flying in the face of convention and challenging the status quo. Einstein's special theory of relativity, advanced in 1905, languished for nearly two decades until the physics mainstream finally caught up. Many inventions stem from a hybrid case, whereby the originator brings a unique point of view and set of experiences to known facts. Still other advances can be traced ultimately to the steady march of technology: given the state of the art, they must occur, and often take root in several places at once. So it was for radar — a classic case of simultaneous invention. Within a few months of the Daventry experiment, researchers in at least six other nations were pursuing systems with pulse transmissions, cathode ray displays, and distance, elevation, and direction finding. One country, the United States, even enjoyed several months' head start on Watson Watt's crew.

The emerging era drew upon more than a half century of advances, starting with the pathbreaking work of James Clerk Maxwell and Heinrich Hertz. Maxwell first unified the equations of electromagnetism, culminating in the 1873 publication of his landmark *Treatise on Electricity and Magnetism*. Nearly fifteen years later, Hertz brought things to reality. In a classic series of experiments, the German investigator generated and detected the first known radio waves, then proceeded to show that the waves traveled with the velocity of light and fell subject to, among other things, reflection, refraction, diffraction, and polarization. In short, they exhibited all the properties of visible light, just as Maxwell had predicted.

The discovery set off a race to manipulate and employ the newfound phenomenon. In 1896, young Guglielmo Marconi, his ideas unappreciated by the Italian government, moved to Britain. With the backing of the government post office, he soon established radio links nearly two miles across Salisbury Plain and then over the Bristol Channel. Some five years later Marconi successfully sent radio signals across the Atlantic, opening the door for a worldwide communications industry by demonstrating the practicality of long-range transmissions.

The evolution of radio carried with it the ingredients necessary for

ELECTROMAGNETIC SPECTRUM AND RADAR WAVELENGTHS

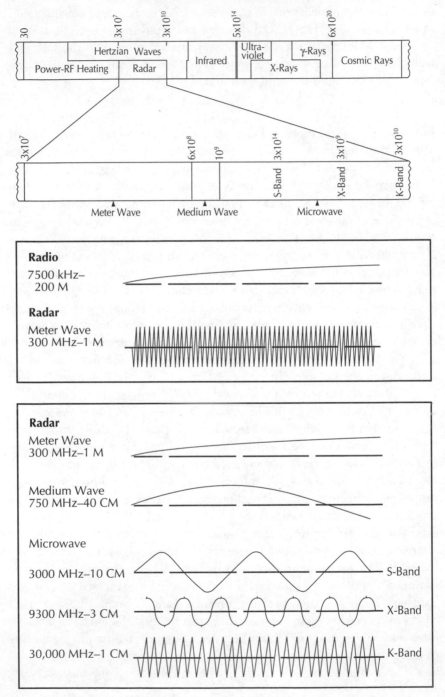

Source: Guerlac, Radar in World War II, pp. 7 and 9.

radar. At first, long-distance communications were thought to depend on waves of at least two hundred meters. The American Arthur Edwin Kennelly and Britain's Oliver Heaviside independently suggested that the transmissions ricocheted between the Earth and some sort of conducting surface or high-altitude boundary, later called the ionosphere, that kept them from vanishing into space. Shorter waves would also bounce along this atmospheric and geological pinball machine but were largely ignored because researchers generally believed the signals would attenuate over long distances.

Before radar and the ability to detect and pinpoint the position of some object could come into play, however, much shorter wavelengths on the order of fifty meters or less were a virtual necessity. Just as waves of water ripple out from their point of origin, radio signals spread over a broader area while propagating through the air. Unless transmitted pulses were tightly focused, too little of their energy would return to the receiver after traveling to a target and back. It wasn't until World War I brought the need for tightly directed transmissions secure from enemy ears that this criterion began to be satisfied. Over the course of the war, physicists and engineers at places like AT&T's Western Electric arm pushed the art of wireless transmission ahead by as much as a decade. Researchers soon realized that shortwaves did not attenuate nearly as much as previously believed. The finding suddenly made higher frequency signals much more attractive. Because shortwaves started out more concentrated, a greater portion of the signal actually arrived at its destination than did a long wave broadcast. That not only enabled transmissions between specific locations, it reduced power demands and relieved transmitters and receivers of a host of bothersome vagaries. In the early and middle 1920s, led by talented amateur hams, a shortwave bridge sprang up across the Atlantic. On its heels came industrial concerns — RCA, Westinghouse, and Germany's Telefunken among them — racing to establish a network of commercial shortwave radio stations.

The last major piece of the puzzle fell into place in concert with the shortwave revolution. The theoretical boundary layer envisioned by Kennelly and Heaviside not only attracted immense interest from radio enthusiasts, it also drew in atmospheric researchers seeking to understand the earth's mysteries. For more than two decades, scientists poked and pried at the hypothetical invisible shield, trying to prove its existence. Among those participating in the hunt were the American investigators Merle A. Tuve, then a graduate student at Johns Hopkins University, and Gregory Breit, a physicist with the Carnegie Institution's Department of Terrestrial Magnetism. In July 1925 the pair set up an experiment that measured the height of the ionosphere by sending up radio pulses rather than a continuous transmission of energy.

In one purely scientific sense, the innovation came too late. About six months earlier Britain's E. V. Appleton and colleague M.A.F. Barnett had used a different radio technique to prove the existence of the ionosphere and measure its height — work that won Appleton a Nobel Prize and robbed Tuve and Breit of the same honor. But the Americans' pulse method was quickly adopted worldwide as the simplest and most direct. It did not take long before important improvements showed up. A cathode ray tube was attached to the receiver to provide a live image of incoming echoes. Moreover, in the initial Tuve-Breit experiment, the pulses and the rest time between transmissions were roughly equal. Various researchers, however, soon added a multivibrator circuit to the transmitter that compressed pulses to around a thousandth of a second, separated by much longer rest intervals on the order of a few hundredths of a second. Shorter pulses helped resolve nearby targets since transmission was less likely to be in progress when the echo arrived, while the extended rest interval provided time for the signal to travel farther and reveal the presence of more distant objects.

Although both the pulse transmission method and the shortwaves necessary for radar existed by the mid-twenties, it took at least five years for the pieces to coalesce. Industry had to gear up to mass-produce the components, and researchers needed to put them to use. The awakening was spurred by the immense consumer market for home and car radios that opened in the late twenties, a surge that brought with it the triode, a much-improved transmitting and receiving tube. A few years later, the infant television industry, led by such pioneers as RCA, the Marconi Company, and the British giant Electrical and Musical Industries, contributed crucial refinements in the cathode ray tube, transforming it from an obscure laboratory tool to a compact, lightweight device of widespread utility.

As the thirties got under way, radio waves permeated the air. It became simply a matter of time before someone — or several people — hit on the idea of radar. In the nascent years of radio, a few had come close. As far back as 1904, an obscure Düsseldorf engineer named Christian Hülsmeyer filed patents in Germany and Britain for the *Telemobiloskøp*, a collision prevention device for ships that worked much like a searchlight, only with radio waves in place of visible rays, and even possessed a crude range finder. Nearly two decades later, in September 1922, the wooden steamer *Dorchester* plied up the Potomac River and passed between a transmitter and receiver being used for experimental U.S. Navy high-frequency radio communications. The two researchers conducting the tests, Albert Hoyt Taylor and Leo C. Young, had sailed on ships and knew of the difficulty in guarding against enemy vessels seeking to penetrate harbors and fleet formations under darkness. Quickly putting the serendipitous finding together, the men proposed using radio waves like a burglar alarm, stringing up an electromagnetic curtain across harbor entrances and between ships. But receiving no response to the suggestion, and with many demands on their time, the investigators let the idea wither on the vine.

By the time Watson Watt set out to investigate the death ray idea in 1935, the technological landscape for radar had been prepared. The early thirties brought a spate of reports of radio signals being interfered with by planes, ships, and even cars. The government post office report Wilkins mentioned to Watson Watt served as just one example. From there it was not a great leap, as Taylor and Young's 1922 report made clear, to envisioning the immense defense implications of a novel technology. In the mid-1930s numerous radar-like inventions began appearing before patent examiners, though most lacked some key element — the pulse system, money, or just the time to pursue a still-dubious venture.

Ironically, the first system to incorporate the pulses synonymous with modern radar was cobbled together in late 1934 by Robert Morris Page, working under the direction of Albert Taylor and Leo Young at the Naval Research Laboratory at Anacostia in the District of Columbia. Taylor and Young had resuscitated their original suggestion a few years earlier, after high-frequency radio signals began being interrupted by aircraft. The low-priority radio detection project had struggled through the Great Depression, until early in 1934 Young convinced his boss, Taylor, to give pulses a try.

At Taylor's direction, Robert Page built the first makeshift system the following December — transmitter in one building and receiver in another, both linked to rooftop antennas. As a small airplane flew up and down the Potomac through the transmitter beam, the outgoing signal so saturated the receiver that an echo could only be detected through the wavering or beating of the mass of clutter, and nothing was seen at all unless the target passed close enough that a visual sighting would have served much better. Nevertheless, the test proved the feasibility of a pulsed system a full month before Robert Alexander Watson Watt even dreamed of radar.

This was only the tip of the iceberg. In March 1934, the same time Page got started building his pulsed set, a primitive continuous wave radar detected echoes from the German warship *Hessen* in Kiel harbor. Back in Italy a bit earlier, while testing a telephone link between Vatican City and the pope's summer retreat at Castel Gandolfo, Guglielmo Marconi had noticed a rhythmic disturbance in his test signal that turned out to be caused by a moving steamroller. This had sparked the inventor's interest in radar and led to his lab manufacturing prototype sets for the Italian Army by August 1935. By that same watershed year, the Leningrad Electrophysics Institute had fashioned its own crude radar system that detected airplanes nearly two miles off; the newly built French liner *Normandie* carried on its deck a microwave obstacle detector; and inklings of radar could be found in Japan and Holland.

Like the British work, most of these projects took nourishment from military exigencies and existed under a veil of secrecy. What set Watson Watt's effort apart from its unknown competitors was largely the scope of the vision — a defensive network, accepted in principle at the highest governmental levels and backed by a leading member of the military. Moreover, none of the rival endeavors exhibited the British urgency. The most advanced, in Germany and the United States, potentially the two greatest military powers on Earth, simply found no strong reason to push ahead. Hitler's military command, already thinking of panzer blitzes and Luftwaffe thunder strikes, did not see the pressing need for a seemingly defensive weapon. America, separated by vast seas from its potential enemies and swept with an isolationist fervor, could not yet be bothered to stir from its slumber.

•

The seafarer's village of Orford glistens like a slightly dusty jewel on the dank Suffolk coast about ninety miles northeast of London. A Norman keep, an outpost to guard against marauders from the sea, looks over a broad expanse of windswept flatlands from a knoll on the town's southern edge. The village itself, with its handful of shops and pubs, stretches for just a few hundred yards along a single main road that slopes down to the River Alde, the fisherman's conduit to the North Sea hunting grounds. A hundred yards or so across the fast-flowing river lies a desolate fifteen-mile-long spit of salt marsh and shingle called Orfordness.

Often considered an island but thinly connected to the mainland several miles upriver, the narrow isthmus runs along the coast — parallel to both river and sea — well after the Alde merges into the River Ore a few hundred yards beyond the village. Orfordness hosts a treeless, windswept landscape crisscrossed by dykes and scattered pea bush. Watson Watt considered the "island," far from the public eye, an ideal spot to set up his infant radar program in the cold, wet spring of 1935.

Eddie Bowen motored into Orford as part of a small convoy — six people, two cars, and two Royal Air Force lorries loaded with equipment — the evening of May 13. Hail and sleet greeted the entourage as it assembled on the quay the next morning. During World War I the island had become home to an airfield on one end and a bombing range on the other. Gunnery practice kept up sporadically into the early 1930s. But by the time Bowen's group arrived, Orfordness lay largely derelict. A government-contracted motorcraft, driven by a taciturn boatman, carried the men across. On the opposite shore, another close-mouthed local operated a strange vehicle built on the combined chassis of a Ford Model T and an ancient fire truck. This contraption carted men and equipment across the empty aerodrome to a confluence of wooden huts. The buildings had recently been repaired, and power cables laid to the village. But notices from the station adjutant, going back to 1918, still clung to the walls, and the site never lost its abandoned feeling. "It was a desolate, forbidding place whose only redeeming features were the birds," an early worker recalled. A dusty, unopened wine bottle was found tucked on a chimney ledge in one of the buildings.

Only four members of the entourage would be staying. Skip Wilkins fell naturally into the role of leader. Shy and modest, he had a penchant for inventing nicknames and proved himself a shrewd judge of character. Then there was Bowen, who would emerge as Wilkins's right hand. Rounding out the crew were L. H. Bainbridge-Bell, a highly talented but moody circuit designer, and his technical assistant, George Willis.

Bowen had been cast into this group by one of those early twists of fate that shape the course of a lifetime. After earning an honors degree at University College in his south Wales hometown of Swansea, he had undertaken the pilgrimage to King's College in London to study with the renowned Appleton. While pursuing his doctorate, the Welshman had often journeyed to Watson Watt's radio research center in Slough to utilize its facilities, coming under what he termed the superintendent's "benign influence." When Bowen completed his Ph.D. and went job hunting, the enthusiastic Scot asked him to apply to a secret project just under way. This turned out to be radar. Watson Watt's top assistant, Jimmy Herd, briefed Bowen on the endeavor one day in April, stressing that the work fell under the Official Secrets Act and reading with gusto the penalty for leaking information: hanged by the neck until life is extinct.

At Orfordness just a few weeks after that sobering indoctrination, it fell chiefly to Bowen to develop the transmitter, while his comrades concentrated on getting the receiver and cathode ray indicator up to snuff. Orfordness contained no proper workshop and few tools. Nevertheless, the team plunged ahead, starting work at 8:30 each morning and frequently staying past ferry hours; the men either rowed themselves back to the village or slept in one of the huts. "Many were the nights I spent on a camp bed alongside the transmitter, with a piece of cake and a bottle of beer for supper, and as likely as not, the same for breakfast," Bowen later recalled.

Watson Watt soon arrived to hold the first of what became ritualistic discussions at the Crown and Castle Hotel, an inn and tavern across a small gully from the picturesque keep. In those early days, the Scotsman visited almost every weekend to check on progress and plot the future over drinks in the hotel's comfortable lounge.

Despite the often-dire Suffolk weather, Bowen counted the days at Orfordness as among the happiest of his life. He lived with an attentive spinster and kept tabs on gossip at the Jolly Sailor, a pub just up the street from the quay. He delighted in the colorful village, where one man served as hairdresser, barber, insurance agent, and secretary-treasurer of both the soccer and cricket clubs.

The work also proved satisfying. Though still lacking tools and finely engineered components, Bowen quickly succeeded in stringing together a hodgepodge of scrounged parts — including two metal cans, one fitted inside the other, that served as a condenser — into what he called "a marvel of crudity." Within just a few weeks he had boosted transmitter power to a respectable seventy-five kilowatts. On the other side of the fence, Wilkins and Bainbridge-Bell coaxed the receiver along, feeding its signals into a cathode ray oscilloscope that automatically showed range along a horizontal scale.

Behind these improvements, the station sprang to life. Two seventy-five-foot lattice towers rose alongside Bowen's hut, transmitting antenna wires strung between them. A stone's throw away, four more towers, corner-stones of a rectangle about thirty meters wide and one hundred meters long, formed the receiving corps: two separate sets of aerials were strung in parallel between towers on the short sides of the configuration. The entire system was up and working by mid-June, in time for the station's grand opening.

Tizard Committee members ventured over from London for the occasion. They didn't expect a live demonstration, but the ever-optimistic Watson Watt promised one anyway. Prospects for an impressive showing *did* look good: while warming up the system the day before the big event, the "Islanders," as the townfolk began calling the radar men, followed a Singapore flying boat cruising off Orfordness out to about fifteen miles and back. However, on the official test day, Saturday, June 15, their luck did not hold. Amidst a thunderstorm and heavy ionospheric clutter, a Valencia biplane lumbered through the antenna beam, but only the Scot claimed to see a slight flicker on the screen. Early the next morning, the old troop transport roared aloft for a second go-around; this time no one saw a thing.

The setbacks did not seem to perturb the visitors, who expressed satisfaction with progress to date. Still, Watson Watt stayed on after the committee departed to investigate the failure. On the Monday, with no modifications to the set, the tiny group unexpectedly picked up a magnificently strong signal. A twin-engined Scapa flying boat, practicing bombing runs at sea, came in clearly on the range scope at seventeen miles. The

Islanders tracked it up and down the coast, as far out as twenty-nine miles, or forty-six kilometers. "We were all wildly excited," Bowen recounted. Watson Watt telephoned the airfield at nearby Felixstowe, imploring the commanding officer to send the plane up again on the same flight path.

After that things sailed along for a while. In late July the group passed another milestone. Watching an echo with an unusually shaped amplitude, Wilkins correctly guessed that the strange blip held not one, but three planes, hinting at a time when fighters not only reacted to a threat but deployed according to the magnitude of the impending attack. The group pressed on throughout the summer, clearing technical obstacles. Pulse time was compressed, and the signal's rest interval stretched to improve sensitivity and range, and draw a bead on their working goal of locating planes at least fifty miles out. The researchers steadily drove down the operating wavelength — first to twenty-six meters and then to thirteen meters — to avoid interference from commercial signals. Ultimately, to counter deliberate jamming, the system would operate on four interchangeable wavelengths between ten meters and thirteen meters.

Steering fighter pilots close enough to see enemy planes posed a far more difficult challenge than extending range. The radar system also had to determine an aircraft's height and direction, or azimuth. Repeated measurements of range and azimuth would reveal the enemy's heading. Early on, Wilkins solved the height-finding problem by adapting a method he had developed at Slough for measuring the downcoming angle of transatlantic telephone communications. The technique involved comparing signals received at two parallel horizontal antennas in order to find the elevation angle of the incoming echo. However, before the war this system was modified to employ a pair of aerials set at different heights: a comparison of signal phase determined the target's angle of elevation.

Shortly after clearing the initial height-finding hurdle, Watson Watt announced a solution to the formidable direction-finding problem. Originally, the group intended to rely on simple triangulation — using two or more stations — to locate a target. But the nightmare involved in trying to track several planes at once, a real-life necessity, forced them to abandon the plan. The Scotsman's idea hinged on placing two sets of aerials in a crossed dipole configuration, one slightly above and perpendicular to the other. The nonparallel position of the two antennas in effect simulated two stations, with a goniometer sorting through the incoming signals to fix the target's azimuth.

Taken as a whole, these advances catapulted the infant technology into adolescence. Within a few months, the Islanders could track planes forty miles out and gauge the height of an aircraft at seven thousand feet to within a thousand feet—good enough for daytime interception. The men set about replacing the seventy-five-foot aerial arrays with gigantic lattice towers about 250 feet tall. The increased height provided greater ranges— even without boosting transmitter power, another priority—by allowing for better propagation over the Earth's curvature. At the same time, efforts continued to improve the receivers and expand the range of operating frequencies.

By September 1935, the project had evolved so far that an enthusiastic British government approved in principle the construction of a chain of five RDF stations—the first stage of the radio wave shield envisioned by Watson Watt. The stations would be erected at twenty-five-mile intervals up and down the east coast from the River Tyne to Southampton, flanking the Thames estuary, gateway to the great nerve center of London with its vast dockyards, railway hubs, and government halls. The first stations, as it turned out, aimed eastward at the Belgian port of Ostend, taking their cue from Air Ministry tactical data that omitted the possibility the Germans might occupy the French side of the English Channel.

Work on the early warning chain started that fall, but it soon became apparent the rapidly swelling RDF project required a larger staff and a more hospitable base than Orfordness. One day, probably still in the early autumn, Bowen, Wilkins, and Watson Watt drove along the southern Suffolk coast in the Scotsman's new Daimler on a leisurely scouting expedition to find another home.

As the trio cruised over meandering two-lane roads, Wilkins mentioned seeing a sprawling estate with buildings that could serve as laboratories and living quarters just ten miles or so south of Orford, near the confluence of the River Deben and the North Sea. Located on a small bluff, and with no high ground between it and the water to interfere with signals, the manor could be ideal for the radar group.

Watson Watt insisted on reconnoitering the property that day. Soon Rowe and Wimperis drove out. Looking for local gossip, the duo approached the purveyor of a ginger-beer stall near the winding lane that led up to the coveted site. Bawdsey Manor, they learned, was a country estate built by Sir W. Cuthbert Quilter, a wealthy stockbroker and businessman. The family might be willing to sell. Encouraged, the men pressed on. By year's end, the government had approved a £24,000 outlay for the 168-acre estate.

In contrast to the bleakness surrounding windswept Orfordness, Bawdsey Manor brought a dreamy quality to the business of defense. The stately structure, its Gothic front connecting two minaret-capped towers—one red brick, the other white stone—overlooked the rough-hewn coast with an air of elegant benevolence.

A brisk stroll along the bluff, the walkway populated by purple emperors, painted ladies, and other exotic butterflies, found a stairway leading to the pebbly beach and the sounds and smells of the North Sea. An eight-acre lawn sprawled off the east wing, separated from the sea by dunes and a fringe of tamarisk. Although in need of some attention, the estate held all the other accoutrements of the landed gentry: peach trees, bougainvillea, swimming pool, orangery, lily pond, topiary, rose garden.

Wood-paneled rooms peered out over this panorama. A central dining hall contained a musician's gallery—down to the pipe organ. The Quilters left a majestic billiards room, filled by the roar of the sea. The government wanted the table removed. But the former owners offered it up for the special price of £25: Eddie Bowen paid for it out of his own pocket.

The move to what became known formally as Bawdsey Research Station wasn't complete until May 1936. Labs sprang up in the White Tower and the stables. The hills behind the manor, overlooking marshland and the Deben, played host to imposing wooden receiving towers.

Watson Watt, forsaking the long weekend commute, took palatial quarters overlooking the water and began scouring the nation's physics laboratories for young talent. By August, when the station formally severed its National Physical Laboratory ties and became part of the Air Ministry, the staff totaled close to twenty. Since single men predominated, the superintendent ordered a bachelor's quarters established in the manor house. The few married members either lived in cottages dotting the estate or, more likely, commuted by ferry from across the River Deben in Felixstowe.

The close quarters and easy camaraderie gave the place the air of a college dormitory. One of the new recruits, a sharp-minded Imperial College student named Robert Hanbury Brown, arrived just two weeks before his twentieth birthday with an oversized Dalmatian dubbed Gay Domino. Before the manor got too crowded, the pet merited a suite of its own. Working hours nominally ran from nine to five, but in truth the men stopped when the mood struck to enjoy a swim, keep up the gardens, or play cricket and soccer on the lawn. They made up for their time off by burning the midnight oil. Technical discussions often took place in a timbered hall before a roaring fire. With the exception of Watson Watt, the entire entourage worked on a first-name basis.

As fall approached, two research groups developed—the main Chain Home effort at Bawdsey and a smaller program at Orfordness aimed chiefly at putting together the transmitter, receiver, and other components necessary to move to shorter wavelengths. Eddie Bowen's assignment fell a bit outside both efforts. After the move to Bawdsey, the young physicist continued to bear chief responsibility for the chain's transmitter development. But he devoted more and more time to RDF Project 2, a radar system small enough to fit into a plane. Almost from the beginning such an airborne interception, or AI, set was considered a vital supplement to the early warning network. The Chain typically could fix aircraft within a mile in range from a distance of seventy miles, but elevation-finding was notoriously poor and bearing could be off by as much as twelve degrees. In broad daylight, eagle-eyed aviators could make up the difference. But nightfighters needed to be steered to within roughly a thousand feet of their quarry before pilots could see well enough to carry out an attack. Lording over a staff of one—himself—the Welshman formed the country's first airborne radar group.

The odds didn't look good. The Orfordness transmitter alone took up a full room and weighed several tons. Receiver equipment filled another small hut. A practical airborne radar, on the other hand, could take up only about 8 cubic feet and weigh no more than 200 pounds. The meager 500-watt power supply on existing aircraft, already fully occupied serving other systems, highlighted the need for a bigger generator. Given the problems of aerodynamic drag, the antenna must be shrunk to a veritable stub compared to the ground towers, dictating an operating wavelength of close to one meter, much shorter than had so far proven feasible. Finally, any airborne radar needed to be so simple that a pilot or special operator could use it during the rush of combat.

As the second half of 1936 progressed, Bawdsey entered a frustrating period. Unable to move directly to anything close to one meter, Bowen settled for building a prototype of a scaled-down Orfordness system on 6.8 meters. Meanwhile, that September the manor group suffered through a failed demonstration before a host of dignitaries that included Sir Hugh Dowding, who by then had taken over as commander in chief of Fighter Command. Dowding, watching the trial in the receiver room, heard the planes before anything appeared on a radar screen. A few days later Watson Watt received a harsh letter from Tizard, which he read to a downcast staff: "To say that I was disappointed on Thursday is to put it very mildly. As you put it yourself, you have to face the fact that very little progress in achievement has been made for a year. Unless very different results are obtained soon, I shall have to dissuade the Air Ministry from putting up other stations."

Watson Watt told his charges only a few months remained to turn things around. The chief problem turned out to rest with the new transmitter, which was only partially complete in time for the trial. Launching a push to improve the system — the superintendent himself climbed a tower in a screaming gale to work on the antenna — the radar pioneers quickly recovered. The following April, the manor station successfully detected planes a hundred miles away. By August, the first official Chain Home stations looked out to sea from Bawdsey, Canewdon, and Dover.

Eddie Bowen had temporarily put his one-man airborne interception project on the back burner to help resurrect the main transmitter. Back at work in the first few months of 1937, he made rapid progress. Working out of a small lab in the manor's White Tower, he combined a cathode ray tube with a tuned radio frequency receiver designed for television into a unit that weighed less than twenty pounds and drew little power. Designing an AI transmitter proved a more difficult hurdle. But by that summer, bolstered by the addition of a few assistants and the acquisition of powerful "doorknob" transmitting tubes just out from Western Electric, he pieced together a serviceable system on the respectably short wavelength of 1.25 meters. The first test flight of a fully airborne radar took place on August 17, 1937. Bowen took leave to visit his parents in Wales and missed the historic event — to his infinite regret.

Gerald Touch and Keith Wood took the radar aloft. It did not locate another plane. Instead, it detected ships a few miles off the Felixstowe coast. The news sparked a transformation in the airborne project, shifting the emphasis from airborne interception to what was called ASV, for air-to-surface-vessel radar. Watson Watt quickly volunteered the airborne unit for upcoming air-sea exercises, in which Coastal Command aircraft would defend the island against an "enemy" armada by trying to locate portions of the British fleet sailing from the Strait of Dover into the North Sea.

The games were scheduled to begin on Saturday, September 4. The day before the event, having just completed a final checkout of his airborne radar set, Bowen sipped a beer in the sergeant's mess at Martlesham Heath airfield with Wood and Sergeant Kenneth Edward Naish. The threesome were still on the first round, musing about the upcoming exercise, when the aviator predicted the naval task force would be charting a course up the English Channel at that very moment. As Wood and Bowen digested this tidbit, Naish added: "Let's go down and take a look."

It sounded like a good way to glean some vital intelligence on the fleet's whereabouts for the next day. The men downed their beers and piled into an Anson put at their disposal. Naish found the fleet without trouble. The battleship *Rodney*, carrier *Courageous*, and cruiser *Southampton*, attended by six destroyers, had not yet made it out of the Channel. The radar checked out perfectly. Satisfied it would give a good accounting the next day, the men returned home.

As dawn broke the following morning, the trio waited on the flight perimeter for the last wisps of fog to clear. When the trees at one end of the field could be seen, the Anson roared off, crossing the coast at Felixstowe and heading due east to where its crew calculated the naval force should be. The men spotted six flying boats — rivals in the hunt for the enemy flotilla — taking off from Felixstowe and heading north of their own position. Cruising at three thousand feet in a square search pattern, Bowen's team noticed nothing but an occasional echo from small boats until eight o'clock. Suddenly, a large return showed on the screen at a distance of five or six miles. As the little plane closed on the blips, the entire task force with the exception of the *Rodney* came into view.

Sharp-eyed lookouts must have spotted the solo airplane, for all hell broke loose. "Signal lights flashed in all directions, guns were fired — no doubt firing blanks — and aircraft started to take off from the *Courageous*," Bowen remembered. Although the radar group didn't know it at the time, all the carrier's fifteen Swordfish took off, their progress flaring across the Anson's radar screen in the world's first air-to-air radar detection. To avoid the slow-flying swarm, Naish brought the plane up to nine thousand feet and moved off northward to try to find the missing battleship.

After more than an hour of fruitless reconnaissance, fuel running low, the three decided to give up. The hunt had been so intense that no one had paid much attention to the deteriorating weather or their position. Naish brought the craft down low through a thick haze, until, a few hundred feet above the sea, the glassy surface showed itself. Maneuvering to just twenty feet over the waves, he tried to divine their location. Finally, someone spotted a large black-and-white marker designating the Dutch coast, giving the pilot the bearing he needed to turn back toward Martlesham Heath. The squadron leader climbed to a safer height above the weather and brought the Anson down again when the English coastline showed on the radar screen. The tiny craft landed with about fifteen minutes fuel to spare.

On the ground, Bowen dug out the phone number of an Air Ministry official Watson Watt had asked him to call with the test results. Dutifully, the Welshman reported finding the naval force some two hours earlier, naming the vessels and their position. "Yes, that is correct. An aircraft did intercept the Fleet in that position and at that time," the officer agreed. "But what squadron is this reporting?" When Bowen replied he did not belong to any squadron, puzzlement turned to astonishment. Not only was the man unaware of the radar group's participation, he told Bowen the foul weather had forced exercise organizers to call back all planes.

The significance of this news did not escape Bowen. "We had found the Fleet under conditions which had grounded Coastal Command, we had detected other aircraft for the first time with a self-contained airborne radar and, simply by returning home in one piece, had demonstrated some of its navigational capabilities," he once summarized. It had been a banner day for radar.

Its star on the rise, Bawdsey Manor came ever more under the watchful eye of government and military officials. Since the Air Ministry ran the station, the RAF took an active interest in the radar project, opening a training and operation school at the manor. Its graduates took over the Chain's operation. The War Office attached Army scientists and technicians to Bawdsey, where they developed mobile gun-control and searchlight radars for coastal defense. The Admiralty, too, placed a naval liaison officer at the manor to aid its own separate effort to install radar on ships. Both service arms — once content to study radio communications, gunnery, and sonar — also set up RDF projects at their own research establishments.

As radar became incorporated into the mainstream of military life, the firebrand days slipped away and the manor gradually took on a more formal character. The transition to officialdom was cemented in late 1937, when Watson Watt climbed a bit closer to the inner circle by stepping up to director of communications development at the Air Ministry. A. P. Rowe, the prickly Tizard Commitee secretary, took over as Bawdsey's superintendent.

Rowe's officious manner did not endear him to the freewheeling staff, who preferred Watson Watt's Scottish familiarity. Hanbury Brown enraged the incoming administrator by shooting a rabbit from the window of a White Tower lab: the young recruit claimed to need the meat to supplement his £214 annual salary. Hearing the shotgun blast, Rowe rushed in to point out a station order prohibiting killing game on the grounds. "Rabbits," countered Brown with contempt, "are not game, they are vermin."

Rowe forced the men to sign in and out and issued a stream of infamous Station Orders that further disenfranchised the Bawdsey workers. The first called for the "standard of living and general conduct to conform to what is expected of professional civil servants. . . ." Another instructed the men to save razor blades by sharpening them on glass.

Still, a somewhat begrudging respect formed between the two sides. The superintendent gave his charges credit for bringing a fresh outlook, even genius, to the enterprise. And despite Rowe's adverse affect on overall cohesion, he took a broad view of the Bawdsey work's importance that was not lost on his staff.

Partly because of Rowe and partly in spite of him, the station continued to flourish. The Army group worked so quickly that by the end of 1938 the government approved orders for a thousand anti-aircraft gun and searchlight radars and forty Coastal Defence sets designed to peer over the water and provide gun batteries with the range and bearing data needed to engage enemy ships. The all-important Chain Home outposts also thrived. The five originally contracted stations went on twenty-four-hour alert in time for the Munich crisis of September 1938: operators followed Neville Chamberlain's aircraft for a hundred miles as the prime minister winged off to meet with Hitler and thrash out a way for the Reich to seize control of Czechoslovakia's Sudetenland. As the international situation worsened, the government pressed ahead with plans to establish a twenty-station network up and down the east and south coasts, and Bawdsey switched to seven-day workweeks. On Good Friday 1939, stations stretched from the Isle of Wight in the English Channel to the Firth of Tay halfway up Scotland. The gritty island nation was drawing a line in the ether.

Parallel with the Chain Home development, Bowen's small airborne radar band struggled to resolve its remaining technical hang-ups. Planes carried no oxygen for the crew. Testing equipment as high up as 19,000 feet, the men sometimes returned with splitting headaches, and a few showed signs of disorientation. Still, by the spring of 1939, the team had cleared several important hurdles. The rise of the twin-engined aircraft meant planes carried an extra driveshaft. Bowen's contingent seized on the fact, installing an engine-driven alternator that cranked out an additional eight hundred watts and solved the on-board power problem. The ASV antenna configuration was also refined, though it would be upgraded several more times. The exact setup depended on the plane. However, in one popular scheme a Yagi transmitting aerial was fitted to the nose, with receiving aerials on either side. A second transmitting array crowned the midsection, with two other receiving antennas looking port and starboard. Planes relied on these eyes for maximum coverage while in search mode. Upon detecting a ship, the pilot banked toward the signal and switched to the forward-looking system, in which overlapping lobes fanned down ahead of the plane. By balancing the returns from these two lobes, he could home in on the target.

A steady stream of dignitaries, including all the top Royal Air Force commanders, trooped to Bawdsey for demonstrations. On the heels of the strong showing in the 1937 air-sea exercises, the original priority for an airborne interception radar waned in favor of air-to-surface-vessel (ASV) systems. After the Munich crisis, when the reality of impending war heightened the threat of air attacks, the focus shifted back to AI. Fighter Command's top officer, Hugh Dowding, showed up to view the system one morning in mid-June 1939. He and Bowen climbed into the back of a Fairey

Battle. Without seat belts, the two sat side by side on a plank of wood, their heads tucked under the black cloth that shielded the cathode ray screens. Thinking of night-fighting difficulties, Dowding showed particular interest in the minimum range feature — how close the AI set could steer an interceptor to an enemy plane before the echo was lost in the clutter of the radar pulse. The air marshal asked to be notified when the goal had been reached, so that he could see for himself their relation to the target. As Bowen relates it, when the pilot finally gave the cue, Dowding looked around. "Where is it? I can't see it." The radar man pointed straight up: the Battle flew almost underneath the other plane. "My God," ordered Dowding, "tell him to move away, we are too close."

A few days later Winston Churchill arrived to make his own assessments. Although not part of the government, the formidable statesman strove to rally the nation with his constant calls for rearmament and stood in line for a top post if war broke out. Unlike the Fighter Command leader, Churchill stayed on the ground for his demonstration. In ceremonious fashion he removed his black homburg and handed a partially smoked cigar to the Martlesham commanding officer. Replacing the hat with a forage cap borrowed from an onlooking airman, the visitor squeezed his portly frame into the Battle, equipped for the memorable occasion with wooden stairs to facilitate boarding.

As a target plane droned around the airfield, Hanbury Brown treated the eminent guest to a successful demonstration of airborne radar — on the ground. Afterwards, Churchill emerged from the plane, reclaimed cigar and hat, and retired to the officers' mess. Tea had been arranged, and as the statesman sat down, someone asked him how he would like his served. "Tea, tea," the visitor proclaimed. "Fetch me a brandy — a big one."

On August 1, 1939, probably as a direct consequence of Dowding's test flight, Bowen's group received a crash order for thirty sets to be installed in Blenheim nightfighters within a month. The small crew could not possibly meet the deadline, but as the summer came to an end an ominous tension rose in the world that heightened the urgency surrounding their task.

Over the past eighteen months, with barely a whimper of international opposition, Hitler had fortified the swelling Third Reich with the annexations of Austria and Czechoslovakia. The previous April, Italian forces had seized tiny Albania, a perfect springboard for assaults on Greece and Yugoslavia. Finally, in the waning days of August, the Soviet news agency Tass stunned the West by announcing the imminent conclusion of a non-aggression pact between Germany and the USSR.

At Bawdsey Manor, Eddie Bowen and his compatriots daily passed by the watchwords, carved in French in the arched stone entrance: *Plutôt mourir que changer* — better to die than change. The message harkened back to another time. Around them, as the world seethed, the radar men could not afford to play a waiting game.

4 • A Line in the Ether

"The odds were great; our margins small; the stakes infinite."
WINSTON CHURCHILL

THE EVACUATION OF BAWDSEY MANOR was ordered the evening of September 2, 1939, a day after panzer tank divisions and Stuka dive-bombers stormed Poland and secured Hitler's eastern frontier. In early August a Zeppelin outfitted with an array of electronic listening devices had mysteriously appeared off Bawdsey and moved up the east coast near other Chain Home stations. It seemed likely that the research facility's purpose had been divined, and perched so close to the Reich it made a vulnerable target. Everyone felt certain war would soon be declared. A. P. Rowe, for one, was further convinced the manor formed a prime enemy target. When the evacuation order came, he rose and issued a panicky prediction that the entire installation soon would be leveled by bombs. The exodus, it was announced, would start on the morrow.

Contingency plans for a rush move far north to Scotland had already been drawn up. Packing cases lay in wait. Eddie Bowen had pleaded against moving on the hectic opening day of conflict, but no one paid heed. Arriving at Martlesham Heath at dawn on September 3 to supervise the loading of some thirty planes with airborne radar equipment, the Welshman found a crazy scene. Ten-ton trucks spread soot to obscure the airfield: he figured the resulting black cloud fixed the position better than any enemy reconnaissance. On the radio late that morning, an exasperated Prime Minister Neville Chamberlain made the official announcement that Britain and Germany had entered a state of war. As some three million Londoners stampeded from the city, the Bawdsey staffers continued their work. Face and clothing bathed in soot, Bowen didn't see the last aircraft lift off until dusk.

The move from Bawdsey Research Station closed the book on the early days of British radar. At the time, A. P. Rowe's staff numbered around 250, including approximately a hundred scientists. For the most part, the men had worked in close harmony in a tranquil seaside setting. Neither the camaraderie nor the pleasant conditions survived the onset of war.

Robert Watson Watt had arranged for the enterprise, renamed the Air Ministry Research Establishment, to open a new base at University College in Dundee, his alma mater. Rowe's contingent arrived to find no one waiting. It seemed that after extracting an agreement-in-principle, the Scotsman had failed to nail down the details, and school officials had forgotten the whole episode. With no allocated space, the radar researchers had to dump their equipment in a parking lot and crowd temporarily into two rooms.

Rowe nurtured the venture back onto its feet as fast as possible. He faced a formidable task. With enemy bombing raids likely imminent, a top priority lay in improving the Chain Home system, upgrading components and boosting transmitting power and receiver sensitivity. The chain's glaring weakness lay in its inability to spot planes at altitudes below an angle of elevation of about two degrees, or roughly twenty-eight thousand feet at a distance of one hundred miles in normal conditions. Another rush project focused on adapting the Army's coastal defense set to a different type of radar that scanned closer to the horizon and filled in gaps in the coverage; once deployed later that year, these sets would be interspersed with the regular chain stations to form the Chain Home Low network. At the same time, the establishment started work on the Plan Position Indicator, forerunner of modern radar's easy-to-read cathode ray tube. In the PPI, a pencil-thin sweep rotates rapidly around the screen, painting in all aircraft within its coverage zone in their exact relationship to the station, dramatically simplifying picture interpretation and obviating the need to convert separate range and azimuth data to a distinct coordinate, steps that avoided confusion and saved precious time directing fighter interceptors toward the enemy.

To staff his greatly expanded program, Rowe began recruiting from previously prepared lists of the nation's engineers and physicists. AMRE almost doubled in the fall of 1939, as personnel poured into the university.

Bowen's airborne unit, numbering perhaps two dozen, set up shop some twenty miles west of Dundee in Scone, home to the nearest airfield. Its leader arrived to find a situation as bleak as the one back at headquarters. The airport manager also remembered a planning session with Watson Watt, but once again, in the absence of a follow-up, no preparations had been made. As a compromise, Bowen's forces were awarded one of the two existing hangars and a share of available office space. Since Scone continued to function as a civilian airfield, the arrangement meant that top-secret radar work would be conducted uncomfortably close to public view.

British industry would soon gear up to mass-produce airborne interception sets. In the meantime, the top priority lay in completing the Royal Air Force order for thirty pre-production models that had come down a month earlier; before the move Bowen's men had managed to deliver just six. At

CHAIN HOME STATIONS—RADAR COVER
September 1939 (at 15,000 ft.)

Shetland
Is.

Orkney
Is.

NETHERBUTTON

SCHOOL
HILL

DOUGLAS
WOOD

140 Miles

Edinburgh

DRONE
HILL

OTTERCOPS
MOSS

Belfast

DANBY
BEACON

STAXTON
WOLD

Liverpool

STENIGOT

WEST
BECKHAM

STOKE HOLY CROSS

HIGH STREET

GREAT BROMLEY

BAWDSEY

CANEWDON

London

DUNKIRK

170 Miles

Ostend

Portsmouth

POLING

DOVER

PEVENSEY

RYE

VENTNOR

110 Miles

Cherbourg

■ = RADAR SITES

Sources: Swords, "The Beginnings of Radar," p. 296, and Bowen, Radar Days, p. 27.

the same time, with the war on, Coastal Command urgently wanted radars modified for air-to-surface-vessel (ASV) search. The German submarine had come extremely close to winning the First World War by destroying merchant shipping and severing the flow of supplies to England. Anything that increased the chances of spotting enemy subs was a hot commodity.

Bowen's crew soon included a talented core of scientists, among them Robert Hanbury Brown and the young physicist Bernard Lovell. But the operation floundered trying to satiate the radar hunger of the two often-competing service arms. Research on increasingly powerful and accurate models ground to a halt, and morale slipped as the men grew increasingly isolated from the main nerve center at Dundee and found themselves functioning more as mechanics than scientists, installing sets one after the other in the stream of aircraft descending on the tiny field. The feeling of being cut out of the loop only heightened in mid-October, when the unit was ordered to the RAF base at St. Athan, far to the south in Wales about thirty miles from Bowen's birthplace.

St. Athan played host to No. 32 Maintenance Unit, originally established to service aircraft engines and perform other overhauling duties. With airborne radar beginning to roll off industry production lines, A. P. Rowe wanted the maintenance group to take responsibility for all fitting and testing. Bowen's men would show them how to do it, then get back to research and development. The plan was sound. But as it turned out, the Air Force base teemed with war recruits, leaving little room for the latest arrivals. Bowen and his coterie were shunted into a large hangar unshielded from the wind and wintry air, with no heat or electricity. The men spent a week putting in a power supply, laying plywood and duckboard over the cold concrete floor, and spreading a series of camouflage-green canvas partitions over a framework of battens to serve as laboratory walls. "The sight of a future University Professor, in coat, gloves and cap, working on a radar chassis with a soldering iron will stay with me for the rest of my life," Bowen once bitterly recalled.

Never a fan of A. P. Rowe and his civil servant style, the Welshman could not forgive the AMRE superintendent for subjecting the airborne group to such appalling conditions. By early 1940, when Bowen learned Rowe had launched a second airborne radar effort independent of the St. Athan group, the estrangement between the two men had widened into a bitter chasm. The upstart project centered on improving the airborne set's minimum range — how close it could track an enemy plane before the outgoing pulse obscured any echoes from the target. This was a critical issue for night fighting. Back in the early days of Bawdsey Manor, Air Force officials had insisted that fighters be guided to within a thousand feet of the enemy to ensure pilots could see their objective. The set being developed by Bowen's group in early 1940—the AI Mark III, signifying their third version—came within a few hundred feet of the desired performance. But Rowe, after reviewing reports of poor performance and listening to recent Fighter Command demands, became convinced the minimum range figure had to be cut at least in half, to five hundred feet. Wishing to move quickly, he set up the separate project without consulting Bowen, placing a contract with the talent-rich British electronics giant Electrical and Musical Industries.

An infuriated Bowen learned of the EMI deal sometime that February, not long after King George VI and Queen Elizabeth visited the base and praised the assembled airborne group. He felt convinced that AI's minimum range troubles started and ended with the inadequate training and servicing entailed by its rush into active duty and that no further modifications were necessary.

Bowen was so incensed he insolently stopped filing status reports to Rowe and carried on his work largely alone. Progress was swift, but he paid a price for his revolt. By early spring, with airborne radar models installed in close to a hundred planes, and production orders for three thousand sets each of the AI and ASV units placed with British industry, he was ready to return to his original research and development task. Rowe, though, had other plans. After watching the cohesion of his own headquarters group deteriorate in the dank and isolated Scottish environs, he decided to move AMRE to a more amenable setting closer to London and the centers of British government and industry. A site was found for the organization, to be renamed the Telecommunications Research Establishment, along the Dorset coast in south England, near Swanage and the quaint village of Worth Matravers.

The move took place during the first week of May. As part of his effort to decrease strife, Rowe disbanded Bowen's airborne unit. He first stripped away the Welshman's talented research staff—namely Bernard Lovell and Alan Hodgkin, a future Nobelist in physiology—then dispersed the rest of the members. Some were assigned to maintain radars at operational airfields. Others, including Bowen, were transferred to a lonely airfield at Christchurch, not far from Worth Matravers, to continue the lowly job of installing radar prototypes in aircraft. Eddie Bowen, virtually confined to quarters at a local flying club with one grass runway and a single wooden hut, had reached the low point of his career.

Although war had been declared in September 1939, a long period of quiet had descended over the European landscape, engendering the name Phony War. On May 9 and 10, just as the drama between Bowen and Rowe played to its end, Hitler's armies shattered the calm and steamrolled through Western Europe. Within a fortnight the Germans had severed the link between the main French Army and the British Expeditionary Force sent to bolster continental defenses, sparking the seemingly miraculous evacuation of some 340,000 trapped French and British personnel from the port of Dunkirk. Winston Churchill had replaced Neville Chamberlain as prime minister almost at the moment of the German attack. In a June 4 parliamentary address, the resurgent leader passed on news of the successful Dunkirk sealift but reminded his countrymen that wars were not won by evacuations: ". . . we shall defend our Island, whatever the cost may be. We shall fight on the beaches, we shall fight on the landing-grounds, we shall fight in the fields and in the streets, we shall fight in the hills; we shall never surrender . . ."

At the Telecommuncations Research Establishment, the German blitz carried a heavy irony. The removal of the original radar group from Bawdsey Manor just eight months earlier had been designed to keep the effort away from the battle zone. Rowe had moved south still figuring to be out of harm's way. But as an armada of vessels — sloops, fishing trawlers, tugs, ferries, fireboats, destroyers — hauled troops back across the Strait of Dover, the Nazi army had unexpectedly taken control of the Cherbourg peninsula just sixty miles away. The superintendent canceled all leave as the establishment flew into a frenzy to wrap up any projects close to completion and rush them into operational service. Protective earthworks were erected alongside laboratory huts to absorb the shock of bombs.

Heading into the summer, as German air raids against England picked up steam, opening what came to be called the Battle of Britain, a dejected Eddie Bowen lingered on the fringes of TRE's frantic activities. In 1938, before the move from Bawdsey, the Welshman had married Enid Vesta Williams, a schoolteacher he had known since his undergraduate days. Vesta remembered the time as a dark period, when her husband would come home muttering about Rowe. A few projects kept him somewhat occupied. But while his colleagues labored to meet the threat of invasion, the father of airborne radar sometimes wandered around with nothing to do — until the cavity magnetron burst onto the scene and cast him back into the limelight.

•

The magnetron had surged to life only a few months earlier, on February 21, 1940. Just as Bowen's personal situation was unraveling, the physicists John Randall and Henry Boot huddled around a disembodied car headlight in a makeshift laboratory at the University of Birmingham. Suddenly, at the flick of a switch, power shot from a strange piece of electronic gadgetry that resembled a pint-sized engine rotor with a series of cylindrical holes cut into its body. This was the cavity magnetron. As a small blue arc sizzled off the wires connecting the device to the headlamp, the six-watt car bulb shone brightly for an instant and burned out, overwhelmed by the energy coursing into it.

It had taken almost no time to glimpse the immense ramifications of what had occurred. Testing their invention over the next few days, burning out progressively higher-rated bulbs and moving on to more powerful neon lights, the men realized they had discovered a revolutionary way to generate microwaves. A tight veil of secrecy had descended over the project, as a team led by Eric Megaw of the GEC Research Labs at Wembley was brought in to transform the crude amalgamation of equipment, literally held together in places with sealing wax, into a workable design for mass production. Staff at the Telecommunications Research Establishment, striving to incorporate the magnetron into radar systems, did not appear to have even heard of the work for weeks; and a full accounting of the discovery did not trickle out until long after the first devices arrived at TRE that summer.

Like so many other technological innovations, the cavity magnetron was invented accidentally on purpose. From the onset of the British radar project, the advantages of microwaves had been fully appreciated: their greater immunity to jamming, greater range and directional accuracies, reduced ground clutter, and ability to discriminate between closely bunched targets promised to easily outshadow longwave systems. Watson Watt's second memorandum in early 1935 had drawn attention to the need for shorter wavelengths. That October, while still at Orfordness, Eddie Bowen had tried in vain to construct a fifty-centimeter radar set. A few years later, again unsuccessfully, he attempted to develop a centimeter airborne radar set. The chief problem lay in the lack of an adequate power source, one that could produce a strong enough beam of energy to provide the ranges needed in a military radar.

As 1938 closed, and with war looking ever more likely, the Admiralty, representing all three services, had established a special committee to push the development of microwave systems at two of the nation's top research universities. Oxford's Clarendon Laboratory strove primarily to design microwave receivers, while in parallel scientists at the University of Birmingham physics department were asked to concentrate on developing ten-centimeter radar transmitters.

The Birmingham laboratory was run by Marcus Oliphant, an energetic New Zealand–born physicist brought in from the Cavendish Laboratory a few years earlier to turn around a physics department that had become a science backwater. He had done it, planning the school's first cyclotron and assembling a talented staff. Although the highly respected Oliphant would be made privy to many of the nation's defense secrets — going on to help kick-start the American atomic bomb program — he did not set aside the lab's nuclear studies to concentrate on war-related work until the summer of 1939. At that point, he took lab members to the Ventnor Chain Home station on the Isle of Wight for a briefing on general radar technology, setting in motion the serendipitous chain of events that led to the cavity magnetron's invention.

Most of the visitors went home after a few weeks. But Randall and Boot had stayed on after the war started on September 3. By the time they returned to Birmingham, Oliphant had already assigned other workers to pursue the primary avenues likely to lead to the desired ten-centimeter radar transmitter. Randall and Boot were stuck with the dregs.

At the time only a few feasible ways existed to generate microwaves, none of them yielding the substantial powers needed for a radar system. One was a scaled-down version of the old-fashioned spark gap pioneered by Heinrich Hertz in 1887 and used commonly in radio transmissions. Some success had also been demonstrated with the Barkhausen-Kurz valve, a German vacuum tube. Another possibility lay in the magnetron, a special form of diode invented around 1920 by Albert W. Hull at the General Electric Research Laboratory in Schenectady, New York. Although originally conceived as a low-frequency alternative to the vacuum tube triode, the magnetron had been adapted for very high frequency power output independently by reseachers in Japan and Europe.

A promising newcomer, the klystron, outshone all these possibilities. The brainchild of brothers Sigurd and Russell Varian at Stanford University, the klystron had been developed partly in an attempt to find a microwave power source for a blind landing system: As a Pan Am pilot, Sigurd Varian had flown in the heavy rains and fogs of South America. The ingenious device drew its name from the Greek verb *klyzo*, for the breaking of waves on a beach, signifying the sorting of an electron stream into bunches that converted part of the energy carried by a direct current input into high-frequency alternating current power.

It was the klystron that attracted the bulk of Oliphant's resources, with the lab concentrating on coaxing more power from the device. When Boot and Randall returned from Ventnor, Oliphant assigned the latecomers to ancillary tasks, including a search for ways to employ Barkhausen-Kurz tubes as microwave detectors, presumably for use in radar receivers.

Although the two researchers knew each other and got on well enough, this seemed to have been their first collaboration. John Turton Randall was a few months shy of his thirty-fifth birthday. He had come to the university less than three years earlier after a long stint at the GEC's Wembley labs, where he had established himself as an expert in solid-state physics. Small and erect, he burned with get-up-and-go and once modestly described his only special attribute as a "substantial capacity for hard work." Henry Albert Howard Boot, twelve years Randall's junior and still working toward a graduate degree, was more shoot-from-the-hip and dreamed of glory. Although neither man had been happy with being put on second-tier tasks, Boot was especially miffed.

In fulfilling their assignments, the men built a number of Barkhausen-Kurz detectors. However, they had no way to generate microwaves for the tubes to detect. A conventional magnetron seemed the simplest transmitter to build, so the pair set to work designing the special electromagnet the device required. While waiting for it to be constructed, they took the opportunity to study magnetrons in more detail.

What they found piqued their interest. The device consisted of a vacuum tube placed in a magnetic field so that electrons followed curved paths while traveling from one side of the tube to the other. As known then, the magnetron could not produce much more than thirty or forty watts at centimetric wavelengths. But when Randall and Boot stacked it against the klystron, they began to see a way around this obstacle by combining the best features of each device.

The klystron's chief problem lay in the lack of a way to funnel large amounts of power into the tightly focused electron beam: the small cathode restricted the number of electrons that could be passed through. Randall and Boot realized, though, that the magnetron's large cylindrical cathode and anode would be free from this drawback. At the same time, the klystron employed internal resonators to convert direct current to high-frequency radio power. Fashioned of solid copper, the resonators gave low losses and offered the possibility of large heat dissipation, both advantages lacking in magnetrons. The challenge rested in adapting the klystron's doughnut-shaped cavities to the cathode and anode structure of the magnetron, which depended on cylindrical symmetry. In short, the men needed a different kind of resonator, around which a new type of magnetron could be fashioned.

The previous summer Randall had gone on vacation with his wife and son to the Welsh seaside town of Aberystwyth, stopping at a favorite used-book store. Rummaging through the selections, the young scientist had come across an English translation of Heinrich Hertz's *Electric Waves.* Confronted with the magnetron-klystron problem, he remembered reading about the German master's famous spark gap experiment, in which a wire loop resonator generated high-frequency radio oscillations. Randall realized that if the wire loops were simply extended into three dimensions they would form cylindrical resonators. Moreover, Hertz had shown that the wavelength of the electromagnetic power emitted from such a resonator would be equal to 7.94 times the loop's diameter. As Randall shared his insight with Boot, the men assumed that diameter would remain the critical factor even with a three-dimensional resonator and ignored the height of the cylindrical column. Since the goal in Oliphant's lab was to produce a ten-centimeter transmitter, they rounded the 7.94 multiplier up to 8, and with a simple bit of algebra calculated that the cavity's diameter should be about 1.2 centimeters.

Randall's epiphany came one day in November 1939. He and Boot plowed into the problem that afternoon, with the senior scientist literally scratching out a diagram on an envelope while Boot did the calculations. Their initial design called for six resonant cavities symmetrically grouped around a central cathode hole, although the early model carried to America by the Tizard mission, and manufactured in the United States, featured eight of the cylinders.

By teatime, the first rough cut had been worked out. As I. I. Rabi would correctly intuit a year later, the Randall and Boot creation was a kind of whistle — what engineers would call an oscillating circuit — operating under the combined influence of electric and magnetic fields instead of a burst of air. In broad terms, after a voltage was applied, electrons were attracted by the anode's potential. Simultaneously, the magnetron's magnetic field acted to force them around the cathode, the combined effect being to circulate the particles in spoke-like bunches, or clouds, across the mouth of the cavity slots. Not unlike the way air flowing in front of a whistle hole causes a tone to be emitted — the frequency of which is largely dependent on the instrument's size and shape — the electrons oscillated at a specific radio frequency determined by the cavity dimensions. Each bunch of current induced a facsimile in the cavity walls, and the power generated there could be conducted to an antenna in standard fashion.

The physical laws behind the magnetron's operations were well understood, but the theoretical details of how the electrons traveled remained largely a mystery. Much of what Randall and Boot drew up, therefore, was empirical, almost guesswork. While Oliphant displayed little enthusiasm for the idea, he nevertheless agreed to give his researchers a free hand. The men begged a couple of transformers from the government and built a pair of mercury rectifiers to change the alternating current in the lab to direct, drawing heavily on Boot's workshop skills. Slowly, and crudely, the cavity magnetron came together: When a metal disk was needed to plug one end, Boot produced a halfpenny. Joints were sealed with wax.

A few days after Christmas, Oliphant's staff began moving to a newly

CAVITY MAGNETRON

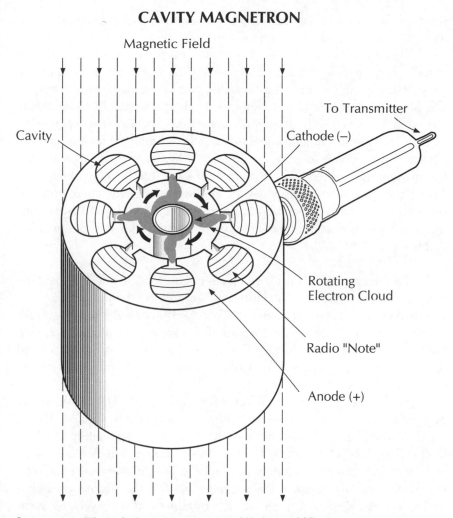

Sources: anon., "The Multi-Cavity Magnetron," pp. 219–21, and Miller, "Secret Weapon," p. 10.

built facility that became known as the Nuffield Research Laboratory. The move disrupted the magnetron's development, but within two months the device was ready for its first trial. The car headlamp was chosen for want of something better to demonstrate the power output. When the magnetron lit up neon tubes a few days after the initial February 21 test, it became clear the device was putting out close to half a kilowatt. This kind of power seemed too great for a ten-centimeter transmitter, so Boot and Randall set about carefully measuring the wavelength with Lecher wires. On the twenty-fourth, Boot confirmed the wavelength at 9.5 centimeters and duly scrawled the measurement into his black laboratory notebook. Power output was soon established as a robust four hundred watts.

Oliphant had been called in to view the magnetron within a half hour of the original test. By April, GEC was designing versions with proper seals, a pulsed power supply, and a higher anode current to enable greater output. On May 9, just as Paris was falling, French researcher Maurice Pomte arrived at Wembley with samples of magnetrons developed independently by colleague Henri Gutton, whose oxide cathodes enabled far higher pulse power outputs. Before long, as the devices trickled into production and the Birmingham lab looked into variations with eight, fourteen, and thirty slots — and even shorter wavelengths — the original six-slotted model churned out twelve to fifteen kilowatts, some thirty times the output of its crude forefather and at least a thousandfold better than the klystron or any other competitor.

The resonant cavity magnetron fell straight into the midst of the Telecommunications Research Establishment turmoil. By the time word of the remarkable device reached Swanage, the organization was struggling to get its own floundering microwave program off the ground in collaboration with GEC and Electrical and Musical Industries. In one fell swoop, Boot and Randall's invention breathed fresh life into the project and refocused the entire British radar program.

The first magnetrons arrived at TRE on July 19. The revolutionary devices were turned over to a new centimeter group led by Philip I. Dee, a standout member of Ernest Rutherford's Cavendish Laboratory at Cambridge University. Dee's talented clique had absorbed both Bernard Lovell and Alan Hodgkin after their detachment from Bowen's airborne radar project. It did not take long for the group to harness the magnetron into a workable radar system. On August 12, mounted on a swivel, the radar tracked a lone aircraft buzzing down the coast a few miles away, the first time a system based on the cavity magnetron followed an aircraft in flight.

Watson Watt and Rowe hurried over for a demonstration. The next afternoon one of the team's assistants, Reg Batt, held up a tin sheet while riding a bicycle in front of a nearby cliff face. With the current longwave systems, Batt's signal would have been lost amidst a swell of ground returns. But, records Lovell, "As we swivelled the paraboloids to follow Batt and the tin sheet a strong echo appeared on the cathode-ray tube." The dream of employing microwaves to pinpoint even low-flying enemy fighters had moved a step closer to reality.

On the heels of this success, TRE came alive with microwave activity. The main need was for an airborne interception radar working on about ten centimeters to counter expected German night raids. Dee's group threw itself into the task. Other projects focused on improving the magnetron's power output, building better microwave detectors for receivers, and beefing up all the other gadgetry from antennas to tubes that had long characterized general development. At the same time TRE joined specialized microwave radar programs getting under way in both the Army and Navy. The Army's Air Defence Research and Development Establishment focused on building more accurate radar-steered anti-aircraft guns, while Admiralty engineers sought to design shipboard radars.

Even as the age of microwave radar dawned, it was evident that war-strained British electronics makers would need help developing and producing magnetrons. Largely for this reason, Britain's greatest technological secret became part of the Tizard Mission's black box. A few days after an August 1 meeting with Churchill, Sir Henry received official clearance for the scientific exchange with the United States. He immediately started making the preparations, which included finalizing the list of technological items to be shared and deciding on mission members. He had already selected John Cockcroft as his deputy, and the two men tapped Eddie Bowen as the team's scientific member. Tizard, familiar with Bowen's airborne radar work since the Bawdsey days, had closely followed the unit's subsequent trials and tribulations. While neither he nor Bowen left a clear record as to how the Welshman came to join the mission, it seems likely that Sir Henry needed a radar expert, and knowing Bowen was one of the best, decided to rescue him from oblivion.

In any event, Bowen became an official part of the Tizard Mission sometime in early August, joining fellow members collecting the manuals, blueprints, and other items bound for the black box. On the seventh he visited Wembley for a briefing on how the cavity magnetron worked. Four days later the Welshman returned to the labs to pick up a working model, the same copy he demonstrated less than two months later to stunned electronics experts assembled for a secret Sunday session at Bell Labs.

•

The fall and rise of Eddie Bowen took place as Hitler's forces across the English Channel prepared for Operation Seelöwe, or Sea Lion—the invasion of Britain. Any miracles of microwave radar would never be ready to help stem the angry German tide. Fighter Command knew an invasion loomed but did not have the fuel or planes to maintain standing patrols in anticipation of enemy raids. Nor did the country have the time to breed another crop of brave and intelligent young men if German bombers surprised planes and pilots on the ground. The nation's best hope for hanging on rested on being able to spot the Luftwaffe far out over the English Channel and then deploying its thin resources to meet the threat at hand. On this vital front, everything depended on the Chain Home radar network.

To pave the way for invasion, *Reichsmarschall* Hermann Göring planned to pound Britain into submission with his bombers. When the attacks opened in earnest the afternoon of July 10, 1940, twenty-one operational Chain Home stations spanned the British east and south coasts from Aberdeen to Southampton, opposite the Cherbourg peninsula. The fragile-looking structures stood tall against the sky—wooden receiving masts typically measured 250 feet in height, while steel transmitting towers rose 350 feet—and were sometimes able to extend their gaze across the Channel to spot enemy formations assembling over Belgian and French airfields. Another thirty Chain Home Low units had been rushed into service late in 1939 to cover the airspace under the main net, reaching out some fifty miles with their more tightly concentrated 1.5-meter beams.

A series of air games had helped fold the radar stations into a tight-knit interception scheme that in daylight, at least, regularly guided fighters close enough to see their targets. At the heart of the system lay the critical job of sorting through the often-contradictory mass of sightings from various outposts. Without a good "filtering" system, for example, two stations might put the same bomber formation in slightly different positions, resulting in an overestimation of the enemy force.

The Royal Air Force set up the first experimental filter room at Bawdsey Manor in August 1937. In fall of the next year, with the first five Chain Home stations on twenty-four-hour alert, the center moved to Fighter Command headquarters at Bentley Priory, built on the site of a twelfth-century monastery in the affluent London suburb of Stanmore. By the July 1940 attacks, the filter center took up much of the priory's concrete basement. Radar plots streamed in via dedicated military phone lines to a crescent-shaped table that featured an outline map of the British Isles and a large swath of the occupied European continent. Filter officers sorted through the data by position, bearing, height, and enemy strength, cross-referencing the reports to eliminate duplication and assigning each formation a coded battle number. Landlines and teleprinters carried the filtered radar information to the operations rooms of the four Fighter Command groups, numbered 10, 11, 12, and 13. Each group covered a specific portion of the country, and the operations area functioned as a central beehive for parceling out fighters to meet any Luftwaffe raid in that area.

It was not a perfect scheme. As the Battle of Britain opened, the entire system—from Chain Home stations to filter center to operations rooms and fighters — ticked and hiccuped like a piece of slightly clinky clockwork. The information got through, the gears just needed constant adjustment. Radar stations often underestimated the height of enemy aircraft by several thousand feet, resulting in pilots arriving below the bomber stream and out of position to attack. Filter room personnel tried to compensate by arbitrarily adding a few thousand feet to a plot, a move compounded by the pilots' own tendency to mentally tack on elevation. Down below in the operations rooms, airmen or members of the Women's Auxiliary Air Force listened on headphones to filter reports, constantly making adjustments to markers placed on gigantic maps of the battle scene while a loudspeaker broadcast the controller's decisions of which squadrons to scramble. Since the radar web did not extend inland, phoned-in Observer Corps reports formed an integral part of the system. The dauntless volunteers proved surprisingly efficient in picking up where the Chain Home stations left off. The biggest problems came in their dependency on good visibility, and in estimating aircraft altitudes, a difficult prospect even in the best conditions. But then, the Chain wasn't too adept at height-finding, either.

Had Göring fully appreciated the radar net's capabilities, he could have tailored his tactics to exploit its weaknesses; after all, he had mustered close to 2400 front-line fighters and bombers against Britain, roughly four times the defender's numbers. But in failing to study closely the British radar chain, Göring's planners had made a massive blunder that at least evened the odds — an error that glared all the more brilliantly in the cold light of history since through a remarkable series of bad luck and miscalculations intelligence officials squandered one opportunity after the other to unravel the mysteries of the spindly towers.

At least two years before the war, unarmed German planes had crisscrossed Britain, purportedly gathering weather data for Lufthansa airline. In reality, the flights served also as photographic reconnaissance missions, and they did not fail to spot the gigantic Chain Home towers springing up at Bawdsey Manor and along the eastern coast. Initially, the Abwehr, one of Hitler's intelligence arms, identified the aerials as radio station arrays. But Major General Wolfgang Martini, chief of the Luftwaffe signals organization, wasn't so certain. It was his insight into the towers' true intent that had led to the Zeppelin flyby of the radar chain in the summer of 1939.

The voyage of LZ 130 that August marked the world's first electronic intelligence gathering flight. But nothing had been learned, probably because the Zeppelin's receivers were focused on much shorter wavelengths, closer to the operating frequencies of the Reich's own radars. The end result was that Hitler's military planners remained ignorant of the Chain's true purposes. If Martini even passed on word of his suspicions to the Luftwaffe, the message was lost or ignored. As the main bombing campaign opened in mid-July 1940, a seminal survey of British air capabilities issued by Colonel Josef "Beppo" Schmid, chief of Luftwaffe intelligence, failed to mention the enemy radar net.

Germany paid a heavy price for its intelligence letdown. As Göring's incursions shattered the summer calm, the British air defense system, for all its fits and starts, provided the country's outnumbered pilots with just enough of an edge. As the 1940 summer wore on, the Luftwaffe sallied forth daily against Channel shipping, accentuating the strikes with raids on ports and nighttime minelaying forays. Time and time again, radar spotted the intruders, and fighters scrambled to the challenge. Early warning did not guarantee success; defenders might arrive on the scene only to find themselves outgunned. Almost invariably, however, because controllers knew the approximate numbers they faced, the RAF deployed enough Hurricanes and Spitfires to shoot down more planes than were lost. Although most attacks could not be thwarted completely, British fighters often made enough of a dent in the German ranks to spare their country the full weight of the onslaught, and even when outdueled, pilots could parachute over friendly territory and live to try again.

As August opened, British forces had downed 270 enemy planes, against a loss of 145 of their own aircraft. Despite the fierce opposition Luftwaffe commanders continued to exhibit a seeming lack of concern about their foe's air defense network, encouraged probably by a dramatic overestimation of British losses. Toward the end of summer Göring's staff did become aware of the Chain Home's function, though its central role in the entire British defense scheme was certainly not appreciated. The network finally came under fire the morning of August 12. At nine o'clock, as a prelude to concerted attacks on airfields and other targets that lasted throughout the day, bombers struck several radar stations on the Kent promontory, leveling a few huts but inflicting little lasting damage. The most serious attack took place later that morning against Ventnor on the Isle of Wight, the outpost visited by Oliphant's Birmingham lab members. Fifteen Junkers 88s dive-bombed the station, sparking a fire and knocking it off the air.

Three days of furious fighting followed, as the Luftwaffe switched to a policy of massive raids designed to overwhelm the defenders at a given point. On the thirteenth, waves of attackers stormed the flanks of Fighter Command's No. 11 Group, which covered the London area and all of England's exposed southeast region, and bore the brunt of the Battle of Britain. A slew of Junkers 88s got through to Southampton, raining ruin on the docks and warehouses and setting off widespread fires. On the fourteenth, heavy German strikes pounded fighter bases. The fifteenth dawned warm and fair, as a ridge of high pressure settled over England. Seizing the advantage, Göring cast all three major *Luftflotten*, or air fleets, into the fray and extended the raids into the north. The attackers blasted the Chain Home stations at Rye and Dover off the air, pummeled airfields, flattened part of a Stirling bomber factory, and followed up with fierce night attacks on ten cities, including Birmingham, Southampton, and Harwich.

Guided by the radar net, defenders challenged each wave, outshooting the stream of marauders two to one. Although heavy raids continued deep into August, the Luftwaffe never came close to achieving the knockout blows Göring intended. A few sporadic air strikes continued to target the radar chain: a follow-up assault against Ventnor kept the station out of service until the twenty-third. But the towers were hard to hit from the air and the radar buildings were protected by blast walls; in any case, the *Reichsmarschall* still didn't appreciate radar's fundamental importance to the opposition. After the raids of the fifteenth, unaware of Ventnor's plight and apparently heeding erroneous technical reports that concluded all key British radar equipment must be safely underground, Göring ill-advisedly told his senior staff and *Luftflotten* commanders: "It is doubtful whether there is any point in continuing the attacks on radar sites, in view of the fact that not one of those attacked has so far been put out of action."

Even without stepping up forays against the Chain Home stations, Göring sent Fighter Command reeling during the last week of August, as his bombers relentlessly intensified their massive assaults on airfields and major cities. On the last day of the month, Air Chief Marshal Sir Hugh Dowding suffered his heaviest losses of the battle — 39 planes down and 14 pilots killed. The tally mirrored the German losses of 41 aircraft — a parity the island could ill afford. September brought no end to the onslaught. In the two weeks that ended on the sixth, Dowding's forces had lost 295 planes, with another 171 significantly disabled, outstripping replacements by some 200 aircraft. Far more serious than the dismal airplane figures, 231 pilots were killed, missing, or severely injured.

The Luftwaffe also suffered mightily, and since each bomber carried a crew of four or five, its manpower was being depleted at a faster rate than the defenders'. Still, instead of driving home his attacks, Hitler made a huge strategic error that brought Fighter Command some respite. The August 24 raid on London had been accidental: bombers looking for Rochester and oil storage tanks straddling the Thames flew past the targets and dropped their payloads on the great city. An incensed Churchill ordered a retaliatory strike against Berlin the next night and three more in the coming week. None of the raids claimed much damage. But Hitler, unable to bear seeing the cradle of his empire so scarred, vowed to strike back on London. His fury overrode the objections of Beppo Schmid, among others, who argued that attacks on RAF air installations should be continued until more life was sapped from Fighter Command. The Luftwaffe launched its first large-scale attacks on the British capital shortly before five on the afternoon of Sepember 7, the same day the Tizard Mission left Halifax for Washington. The initial wave of three hundred bombers set the stage for twelve hours of continuous assaults that killed 306 persons and filled the air with the acrid smoke of a thousand fires. The droning instruments of Hitler's revenge returned the next night and the next, leaving nearly another eight hundred dead in their wake.

Dowding later confessed that his thinly stretched squadrons might not have been able to go on opposing the German raids had direct attacks on RAF airfields continued to interrupt normal operations and deplete both pilots and planes. As it was, Hitler vented his wrath on London over the next week. By the time Göring was set to resume his concentrated daytime attacks and airfield assaults, Fighter Command had regrouped and was ready for a climactic battle.

A light morning mist lingered over the English countryside on Sunday, the fifteenth of September 1940. As the sun rose, the haze burned away to reveal a fine day marred only by patchy clouds. Winston Churchill was at Chequers, the prime ministerial estate west of London. Surveying the weather, the stalwart leader decided to drive to the nearby Royal Air Force base at Uxbridge. The day looked right for flying and he wanted to see an air battle.

No better place existed in all of Britain to witness a Luftwaffe raid than Uxbridge, home to No. 11 Group of Fighter Command. Air Vice-Marshal Keith Park commanded the defenders. He joined Churchill at the entrance to a concrete bunker that housed the operations room, where radar reports were tabulated and fighter aircraft directed to the attack.

Two narrow flights of stairs, slow going for a man of Churchill's girth, led to the subterranean theater. "I don't know whether anything will happen to-day," Park advised as the pair descended. "At present all is quiet."

At the bottom of the staircase, a long corridor pointed the way to the plotting room, where airmen and WAAFs wielded wooden rakes like casino croupiers, pushing markers representing the converging attackers and defenders across a blown-up map of the United Kingdom, northern France, and the Belgian and Dutch coasts. The plotting floor at No. 11 Group, like its three counterparts, functioned around the clock in four shifts. On a busy day, as this would become, the room filled with up to seventy people.

Churchill took a second corridor, to the left of the main hallway and up a few steps, that led to a bank of three viewing cabins. The Dress Circle, he liked to call it. In the first box, men traced the German attack routes, building up an operational picture of the enemy's strategy. On the opposite end sat members of Anti-Aircraft Command, alerting their superiors when fighters scrambled and trying to keep flak barrages directed toward the enemy and away from friendlies.

The guest of honor took a seat in the center compartment—the battle control station. Phone lines linked controllers to the various airfields, which communicated with individual planes by high-frequency radio. A special red hotline went directly to Fighter Command headquarters at Bentley Priory. The cabin looked out through curved glass windows designed to cut noise and minimize glare and reflection, affording the prime minister a clear view of the action. Plotters hovered around the situation map—one end was tilted slightly for better viewing—and its shadowy outlines of the war zone. Royal Air Force units were represented by wooden blocks that resembled miniature A-frame houses. Numerals slipped into tracks on their sides revealed altitude in thousands of feet—"angels" to the RAF, so that ten angels meant 10,000 feet—as well as the number of friendly fighters in the formation. A small flag attached to the markers showed the squadron number in bright red figures. Disks showcasing German forces bore a group code number and the radar-derived estimated attack size: 30 + or 40 +.

A vast electric tableau, glowing in a bewildering array of colored lights and numbers, spanned the wall opposite the viewing cabin like a movie curtain. On this totalizator, or tote board, controllers could see at a glance the pertinent operational details—latest weather, height of the balloon barrage layer guarding key cities, and, most especially, fighter status.

Aircraft flew from a handful of sectors, each of which controlled several airfields and squadrons. A series of boxes running down the board provided the latest word on all aircraft, listing such factors as whether fighters were on the ground, in the air, in need of refueling, or engaging the enemy. Switchboard operators on the floor below flicked levers to keep the electric grid up to date as reports came in. But in a rapidly shifting air battle, controllers had to know the freshness of the data before them. The key lay in stacks of four electric bars—white, blue, yellow, and red—that spanned the tote board's middle. The bars were linked to a special clock mounted on the wall under the totalizator. The outer edge of every five-minute swath on the clockface had been painted alternatingly in red, yellow, or blue. Whenever a squadron's status was updated, the switchboard staff illuminated one of the bars, its color chosen to match the swath containing the clock's minute hand when the news arrived.

Like grandmasters in some deadly chess game, controllers quickly became adept at reading the constantly changing colors. If a blue bar shone brightly and the clock's hand had moved into a red zone, the information was approximately five minutes old. If the hand rested in a yellow zone, some ten minutes had passed without word from the squadron. A glowing white bar signified news more than ten minutes old; that rendered it unreliable.

Around 11 A.M., not long after Churchill arrived, plotters began stirring. Radar showed massive enemy formations building up over Boulogne and Calais. On this day No. 11 Group comprised twenty-six squadrons, mostly Spitfires and Hurricanes. Soon the attack's growing size and scale glared from the German markers on the situation map: 20+, 40+, 80+. In concert, the tote board's kaleidoscope of data flickered to life. By 11:30, two squadrons lit up as Detailed to Raid. Nine others showed In Position, meaning planes were aloft and in their assigned area but not yet vectored to an attack: it took Hurricanes about ten minutes to climb to attack position at 20,000 feet.

As the prime minister watched, reports kept streaming in. Park strode up and down the cabin, barking out directions, reinforcing some areas, moving fighters to defend new zones. Soon, the tote board showed every available squadron committed. Park spoke over the hotline to Dowding, seeking three squadrons from No. 12 Group in case another wave of attacks arrived while his fighters refueled. It was done.

"What other reserves have we?" the prime minister ventured.

"There are none," came the reply.

Churchill looked grave. It was of this moment that he later wrote: "The odds were great; our margins small; the stakes infinite."

Within five minutes or so, most planes began returning to base to refuel. At this critical juncture, the Germans disengaged. Chain Home stations tracked a stream of enemy planes heading back to the continent. The relieved prime minister climbed the stairs to the surface and emerged almost in sync with the All Clear.

It had been a glorious day for Britain. The next morning's headline in the *Daily Herald* screamed in exaltation: "175 NAZI PLANES DOWN, RAF TRIUMPHS IN BIGGEST AIR BATTLES OF WAR." Subsequent analysis put the actual German losses at sixty. Nevertheless, the tally still exceeded any day except August 18 and dwarfed the Royal Air Force losses of twenty-six. More critically, Hitler had been categorically denied the daytime air supremacy necessary to support an invasion. Göring was not done. Changing tactics, he concentrated on trying to demoralize the enemy population by continuing the bombing raids against London and other major cities that had begun a week earlier and marked the onset of the Blitz. But on September 17, the Führer postponed Operation Sea Lion indefinitely. Over in America, as the cavity magnetron was demonstrated and the Radiation Laboratory formed, the line in the ether still held.

5 • The Rooftop Gang

"In many ways it was more like a scientific convention than a research labora-
tory, except that it was a convention which kept running year after year. Here
was the cream of American scientists, hell-bent on doing all they could for the
war effort — some fourteen months before America itself actually entered
the war."

EDDIE BOWEN
on the Radiation Laboratory

ACROSS THE ATLANTIC in the months following the Battle of Britain's dramatic climax, Eddie Bowen continued to regale members of the freshly
minted Radiation Laboratory with stories of radar's triumphs — and failings.
He fielded questions about Watson Watt and Bawdsey Manor, explained the
extensive British filter system, and documented his accounts by showing
riveting combat footage of the Royal Air Force in action. The bedazzled
American radar novices saw before them someone roughly their own age
who had already spent years in the thick of the action, someone who had
even drawn the specifications for the centimetric radar systems that formed
the basis of their lab's existence. In no time, the effervescent Welshman
became a sort of living legend.

The bond forged between Bowen and his hosts was cemented on Friday
nights, in the bar of the Commander Hotel, where many Rad Lab staffers
stayed before finding apartments. Drinking sessions became known as Project Four, after the laboratory's three charter goals of developing an airborne
interception radar, an anti-aircraft gunlaying set, and a navigation system.
Murals lining the bar's walls depicted the Boston Tea Party, Paul Revere's
ride, and the Minute Men, scenes from America's Revolutionary War triumph over Britain that Bowen found amusing. Proceedings usually opened
with a good-natured toast: "The hell with the Limeys." It was in America,
possibly at these very get-togethers, that Bowen was christened with the
nickname he cherished the rest of his life: Taffy, the traditional epithet for
Welshmen reputedly taken from the River Taff that runs near Cardiff in
South Wales.

Bowen's life had been nothing short of frantic since the Radiation
Laboratory's founding in November 1940. He had spent much of the year's

last weeks arranging for the latest AI and ASV longwave sets to be demonstrated to American military brass, a move that ultimately led to widespread production of the air-to-surface-vessel radar by the Philco Corporation and Canada's Research Enterprises Ltd. He had launched a tour of top American university and industry radio research labs to assess the state of microwave electronics, uncovering an astonishing level of technical expertise that showed the United States to be substantially ahead of his countrymen in receiver design, crystal mixers, and waveguides. To bring British compatriots up to speed, Bowen had filed a series of reports, among the first of some 140 he would author while in America, and packed klystrons, acorn tubes, and other components off to Washington, where embassy officials shipped the booty overseas by diplomatic pouch.

When the existing U.S. technology was combined with the cavity magnetron transmitter Bowen had escorted across the ocean, America plainly harbored all the ingredients for making powerful microwave radars. All the same, the Welshman saw only a lackadaisical attempt to develop the technological wellspring. His rounds included stops at RCA, General Electric, and Bell Labs. Either to advance the state of their own technology or in conjunction with military ventures, all three had started small microwave radar programs several years earlier. By the end of 1940, however, these systems still constituted little more than ill-funded sideshows. Only AT&T held a production contract — a limited agreement to produce microwave gunlaying, or fire-control, sets for Navy ships — and it was not going exactly gangbusters. "There appears to be widespread appreciation of the military importance of detection methods, and a large amount of work is being done," Bowen reported to Tizard and Cockcroft back in England. "The urgency of the situation has not yet been realized, however, and progress is about half as fast as in England."

The Radiation Laboratory promised to change all that, bringing a vital sense of urgency to the task of developing centimeter radar. As 1940 wound down, Bowen kept squeezing in trips to check on progress. He took a break for Christmas, spending the holiday at Alfred Loomis's private playground on Hilton Head Island. But after the New Year, his Cambridge visits became more and more frequent until he became a virtual consultant-in-residence, setting up a formal liaison office and taking his own apartment at the Commander. With his assignment in America looking more permanent, he wanted to bring Vesta over but had been unable to find her passage in the wartime travel crunch. The Rad Lab's unexpected warmth and camaraderie eased some of the loneliness he felt.

In those early days, though, the pleasant aura of newfound friends banding together for a common purpose was tempered by changing conditions back in Britain. Over the final months of 1940, and ultimately extending into the following spring, the increasingly isolated country suffered through the Blitz. As had been foreseen, the Chain Home stations and elaborate filter center operations that had performed so magnificently against the massed daylight raids of the Battle of Britain were failing miserably after sundown. Not only did the radar net often lack the accuracy to guide pilots close enough for night fighting, the early airborne radars intended to ease the situation also suffered a series of drawbacks. The sets possessed little ability to accurately determine an invader's direction. At the same time, a large portion of the pulsed energy headed straight down and reflected off the ground back to the receiver. Any echoes from other aircraft at distances greater than the height of the fighter were lost in the reflected signal, restricting range. Exacerbating such failings, radar-equipped Blenheim nightfighters simply did not have the speed or weaponry to readily down enemy bombers. By early 1941, after going up against thousands of invaders, the RAF counted less than ten confirmed scores in night action.

With hardly a major city being spared the ravaging effects of Göring's latest strategy, Bowen rallied his Rad Lab comrades around their prime task of developing a microwave airborne interception set. The Americans responded to the challenge. Luie Alvarez considered the Welshman the infant lab's most important personage. Yale physicist Ernest Pollard took it almost as a matter of faith that the Americans could build the set on the schedule laid out the previous December: a rooftop system by January 6, a test model in a bomber by February 1, and a working airborne radar in a fighter a month after that. "We accepted it because we had E. G. Bowen in our midst, who knew that it could be done, or at least pretended to know, and we weren't to be outdone by any Britisher," Pollard recollects. The Yale man, himself a former British subject, had received his American citizenship less than a year earlier.

The Yanks had gotten off to a fast start. Staff members spent the initial eight weeks of the lab's existence in an all-out blitz to meet the January 6 target. Recruits streaming into Cambridge were granted only a moment to greet old friends before receiving cursory indoctrinations and being thrown into the fray, as the lab spread over parts of two floors in Building 4 and spilled down the famous Infinite Corridor linking MIT's central classrooms to Building 6 and the rooftop experimental station. Out of every nook and cranny emerged the steady sounds of technical discussions and electronic tinkering as men labored to get their first test radar up and running.

On Saturday, January 4, two days ahead of schedule, everything seemed ready. In contrast to the rain and snow that had closed out the previous year, the morning opened crisp and bright as several Rad Lab members assembled in the Building 6 shack and gazed at the compact Boston skyline across the Charles River. In the cold winter air, sunlight reflects off the river's icy crust, and the golden dome of the State House sparkles with an unusual clarity.

The Rad Lab's inaugural radar system, built with the cavity magnetron, spanned the entire rooftop. An unwieldy transmitting antenna was mounted on one end, with the receiving aerial set up on the other and shielded from its counterpart by a loose screen cage. Like two wide-set eyes on some giant's forehead, the parabolas gazed stoically at the panoply of buildings across the Charles. Perusing the rickety contraption, a few of the men must have harbored their doubts. They were not radar engineers, and the whole thing was expected to work on the almost unthinkably short wavelength of ten centimeters.

Within minutes of being turned on, though, the clunky system registered its first echoes from somewhere in the man-made tangle opposite. True or not, laboratory lore credited the historic signal to the dome of the Christian Science Mother Church. Few staffers seem to have witnessed the weekend display. But by Monday, the official deadline for getting a system into operation, the primitive radar was the talk of the lab. Dale Corson, a Berkeley-trained physicist who had recently started teaching at the University of Missouri, was reporting for his first day of work. He remembered a column of men climbing up to the penthouse and crowding behind makeshift curtains that shielded the display from glare to catch a glimpse of the echoes. That same day Director Lee DuBridge telegraphed Ernest Lawrence on the West Coast, merrily predicting that the next goal of installing a radar set in a bomber by February 1 would also be met: "ROOF OUTFIT IN FALL [sic] SWING LOOMIS IS JUBILANT . . . FEBRUARY FIRST DATE LOOKS EASY IF SHIP COMES IN HOPE YOU ARE AS PROUD OF YOUR BABY AS WE ARE OF OURS = LEE"

Much of what DuBridge brought to the lab was embodied in that early communiqué. The soft-spoken Indianan approached his job almost as a benevolent father. He believed wholeheartedly in the cyclotroneers, as particle physicists often called themselves in their field's infancy. He knew most personally; in fact, he was one of them. And from the beginning, while keeping a watchful eye on his flock, he stayed far in the background on technical issues, cultivating a hands-off management style that forever warmed him to the hearts of the men he supervised.

Under DuBridge's calm guidance, the lab attacked its next hurdle. Before staff members could hope to downsize their giant rooftop radar into a system small enough to fit in a bomber, let alone a fighter, they needed to find a way of combining the widely separated transmitter and receiver aerials into a single antenna. This duplexing, or transmit-receive, problem had been a top priority almost since the beginning. In a single-antenna system, the transmitter and receiver had to share the same circuitry. Unless a way could be found to shield the receiver for a fraction of a microsecond or so while the outgoing radar pulse was generated, the energy jolt from the cavity magnetron would overwhelm the sensitive crystal detector.

One widely considered possibility was to create a spark gap, a gas-filled volume between the transmitter and receiver. Energy spilling over from the pulse would enter the gap, causing a spark that momentarily ionized the gas and shorted out the receiver. Such duplexers had been built for some long-wave radars, where wires and lines allowed convenient places to set up spark gaps. But working with microwaves and airborne systems nearly sent the problem back to square one. For one thing, early warning radars looked out over great distances, which allowed ample time for the gap's ionization cloud to vanish so the receiver could once again pick up echoes. Airborne radars needed to follow craft at extremely short distances, hopefully as close as a few hundred yards. That cut allowable recovery time down to a few microseconds. In the Rad Lab's early days, project leaders would diagram the key radar components on a blackboard — antenna, receiver, pulser. When it came to the vital transmitter-receiver, however, they simply drew an empty box labeled TR. No one knew what belonged inside.

Jim Lawson, one of the few lab members with a strong background in amateur radio, jumped on the problem. "If we had been paid in proportion to our contributions to the success of the first microwave radar program, Jim Lawson would have earned more than half the monthly payroll," Luie Alvarez once asserted. Unusually gaunt, with a skull-like face marked by deep-set eyes and an extremely high forehead, the University of Michigan Ph.D. was a topflight experimentalist who burned like a fire to solve problems. And the duplexing question struck him as irresistible.

Even before the lab's first rooftop success, Lawson gathered a small rooftop crew to address the issue. Within a few weeks, the team had managed to fashion a TR Box by using a klystron as a buffer between the crystal and the transmitter. On January 10 a slightly less cumbersome rooftop apparatus employing a single paraboloid sprang to life, once again detecting echoes from the Boston skyline. DuBridge, in Washington for a Microwave Committee meeting, happily received a cryptic telegram announcing the event: "HAVE SUCCEEDED WITH ONE EYE."

After this second important advance, however, the lab's early momentum came to a crashing halt. The contraption the physicists had slapped together was crude, to say the least, a fact followed with some satisfaction in MIT's vaunted electrical engineering department, whose professors found themselves largely excluded from the venture. The plumbing that ferried signals between the transmitter-receiver and the antenna amounted to little

HOW RADAR WORKS

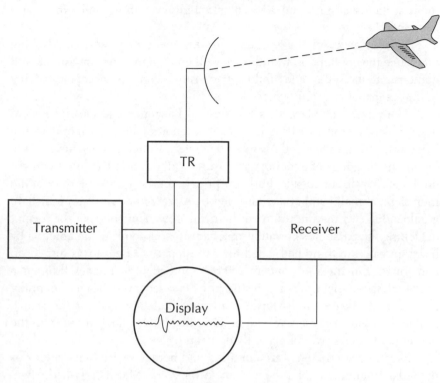

Simplified radar set schematic: A transmitted pulse reflects off a target and is detected in the receiver, where it is converted to a lower frequency and amplified for display on a cathode ray tube. The challenge in late 1940 and early 1941 was to devise a TR box that could shield the delicate crystal detector from the outgoing radar pulse, then recover in time to let in echoes.

more than homemade beaded coaxial lines, albeit with a touch of Lawson elegance. The radar itself was riddled with tuning stubs, each of which had to be hand-adjusted to ensure the signal stayed on the proper frequency. The whole thing fairly crackled with inherent electronic noise — the radar equivalent of snow on a TV set — that made it difficult to pick out even a strong signal such as a skyscraper. Detecting a much fainter aircraft echo from miles off seemed a dim prospect. On the rooftop shack, a newspaper quote cut out and tacked to a bulletin board somewhere around this time said it all: "What we know here is very little; but what we are ignorant of is immense."

As the February 1 target for putting a prototype into a B-18 Bolo passed with little visible progress, even Microwave Committee members began to doubt the lab's chances. When E. O. Lawrence visited the rooftop lab four days after the deadline, he took one look at the unappealing mass of equipment scattered around and advised scrapping everything and exploring a fresh tack.

Luie Alvarez, on the rooftop with his mentor, grew upset. He bet Lawrence the contraption would work well enough to detect an aircraft and demonstrate the system's feasibility. The visiting dignitary accepted, adding he was prepared to eat his words.

The wager took place on a Wednesday. Lawrence and DuBridge were due in Washington two days later to give another progress report to the Microwave Committee, and Alvarez desperately wanted to track his airplane before the meeting. The rooftop gang worked all day and late into the night on Thursday with no results. Early on Friday, February 7, while most of the men slept, a small band of die-hards led by Alvarez and Lauriston Marshall, a fellow Berkeley recruit, resumed their tinkering. Someone got the idea of detaching the parabola's elevation mechanism and pointing the antenna by hand, in case the thing had failed to scan properly. The set was turned on and tuned. On the roof, one researcher scouted for airplanes through a crude telescopic sight, feeding instructions to a second colleague manning the antenna. Alvarez and Marshall, meanwhile, stared idly at the scope. Suddenly, a blip appeared. A bit incredulously, the two looked out from the penthouse to see a civilian plane flying nearby.

Skeptical co-workers were corralled and herded over to take a look as the radar continued to plot the aircraft. Alvarez and Marshall dashed downstairs to try to telephone DuBridge. Their call caught the Rad Lab director at a gloomy meeting of the Microwave Committee, convening as usual in Vannevar Bush's digs at the Carnegie Institution, where confidence in the AI system was waning. As Alvarez and Marshall took turns detailing how they had followed the airplane out to two and a half miles, DuBridge attracted E. O. Lawrence's attention and held up two fingers, signifying the range. "Ernest caught it right away," DuBridge remembered. To the rest of the committee, the Rad Lab director reported with some glee: "We've done it, boys."

The eleventh-hour telephone call breathed fresh life into the Rad Lab. Microwave Committee members voted on the spot to reaffirm their confidence in the facility; and shortly afterwards Alvarez received a telegram from E. O. Lawrence: "I HAD MY WORDS FOR LUNCH."

Edwin McMillan, the Berkeley physicist who had shared the train ride east with Alvarez to join the lab, was immediately placed in charge of modifying the penthouse mock-up into a prototype able to fit into a bomber. A small group under his direction labored intensively over the next four weeks, finally rising around four the morning of March 7 to move equipment to East Boston Airport so that the installation could begin as soon as a newly consigned B-18 arrived. Three days later the team was ready. America's first airborne centimetric radar rode in a special Plexiglas nose cone transparent to microwave energy: at this stage, the single parabolic antenna was not powered, so it had to be hand-pointed.

What transpired remains unclear. Dale Corson, who had known McMillan at Berkeley, took off with the plane. He vividly recalled the crude radar, with a range-only display, picking up echoes from another aircraft a mile or two distant. However, the official history, pieced together perhaps two years after the event, relates only that "the equipment was flown on March 10 with poor results. . . ." Which is correct matters, as does the timing of the trial, since on that same day, the Telecommunications Research Establishment recorded its initial air-to-air detection with microwave radar. One of the trials marked a world first.

After its maiden foray, the U.S. group spent two weeks refining the radar, building a rotating aerial and proper display, before McMillan declared everything ready. On the morning of March 27, 1941, the weather-beaten B-18 again lumbered off the runway with a roar of its twin engines. On board were four of the Radiation Laboratory's top brain trust: Ed McMillan, Luie Alvarez, Ernie Pollard, and Taffy Bowen. This time the test was more definitive.

The Bolo steered southeast for the open skies over Cape Cod. Within minutes the hopes of those aboard were realized. At a distance of two or three miles, the set picked up a single-engined target plane borrowed from the National Guard, marking the first confirmed air-to-air detection in U.S. microwave radar history. Several additional runs reinforced the success. The foursome might have called it a day had not one of the men spotted several merchant ships plying the waters below. The radar's elevation scan was disconnected, leaving the set in air-to-surface-vessel mode, providing azimuth and range signals only. With the plane cruising some two thousand feet above the graceful crescent of Cape Cod Bay, a large vessel showed up clearly on the cathode ray tube from some nine miles away.

Stacked up against British meter wave sets already in production, the ship-detection showing could not be called exceptional. Nevertheless, it marked another first for American centimetric radar. To make certain what appeared was not some screen artifact or a stationary object, Bowen cautiously asked the pilot to fly a zigzag course. When the signal proved real and continued to change position, the passengers could not contain their enthusiasm. Bowen decided to press their luck and root about for an even smaller target. The previous month he had presented specifications for a microwave sub-hunting radar to the Rad Lab, and just four days before the March test he had discussed the need for such a radar set with Alfred Loomis at Tuxedo Park — so subs were on his mind. "I remember the moment precisely," the Welshman told an interviewer years later. "I said, 'My God, let's go and see, with that range we can detect a submarine at three to five miles.' "

The others hardly needed convincing. Caught up in the enthusiasm and overriding the crew chief's concerns about their rapidly dwindling fuel supply, the pilot turned the craft toward the Navy's submarine yards at New London, Connecticut. As the B-18 roared in low across Long Island Sound, the radar picked up several subs cruising the surface some three miles distant. The men shared another current of excitement. Back on March 10, the British had apparently beaten the Americans to the punch by recording the first successful air-to-air detection with microwave radar, but no one had yet detected a submarine with an airborne centimeter system. Bowen regarded the submarine-finding flight as the start of an epoch. In his memoirs he writes: "We returned in triumph and the news spread around the Laboratory like wildfire."

•

By early March 1941, the Rad Lab could no longer be considered a provisional facility. It was clearly meeting most expectations, and with the coming of spring projects sailed out of the "if" stage to become a question of "when." The ranks swelled toward two hundred, more than half scientists and engineers. New recruits were drawn from fields as disparate as physiology, political science, architecture, music, optics, mathematics, anthropology, and astronomy. Often, these outsiders made significant contributions: University of Pennsylvania physical chemist Britton Chance, badge #156, proved a special genius at circuit design. However, noted one observer, the lab remained "a physicist's world, run for, and as completely as possible by, physicists."

Lee DuBridge, who had initially understood that the organization would never grow beyond thirty or forty people, proved remarkably adept at maintaining the freewheeling atmosphere and blending the talents of his diverse group. Sometime in those early days he hosted a blowout party for staff members and spouses. A limerick contest highlighted the evening; limericks soon became a lab tradition. While someone measured audience response with a decibel meter, contestants bellowed out their entries. DuBridge writes: "I think the good will and good fun at that party set the tone for the good will that permeated the Rad Lab group all during the war."

On the professional end, the director still kept an arm's length between himself and his often more accomplished staff, acting as the head of a scientific republic rather than a military systems development autocracy. He would step in to resolve policy disputes. But mostly, DuBridge relied on a steering committee that met weekly to review general tasks and left implementation to individual research teams.

As it ramped up to meet the demands placed on it, the Radiation Laboratory quickly evolved into a full-fledged research and development organization. Its wildness and lack of structure had helped get things off to a running start. But within a few months that time was past. DuBridge recognized the need for a falconer, someone who could keep the assembled prima donnas focused on the main objective of building radar systems while paying bills, sorting out sticky personnel problems, and handling the other vagaries of day-to-day operations.

Eventually, the key management role was divided between two associate directors. Tying down the administrative side was F. Wheeler Loomis. The respected University of Illinois spectroscopist had been singled out on one of DuBridge's initial recruiting trips, but could not join the lab before January 1941. Wizened, already in his mid-fifties and the enterprise's senior statesman by a decade, Loomis brought the same amiable yet autocratic hand that had fashioned the Illinois physics department into a national force. Although a physicist, he claimed to be unversed in electronics and immediately distanced himself from technical matters. Instead, he did the recruiting, massaged egos, and, if need be, dressed a man down. One Rad Lab member called the Illinois man just what the doctor ordered: "a son of a bitch." Another insider put it more subtly, noting that there was a division of labor at the Rad Lab: DuBridge said "Yes." Loomis said "No."

I. I. Rabi handled the research front. His core group labored in the original Room 4-133 next to DuBridge's office. The development of power-measurement techniques and a special frequency analyzer, combined with old-fashioned trial and error, enabled the team to extract greater power from the magnetron. Rabi also broke off a separate group to try to adapt the original ten-centimeter device to transmit at three centimeters, where the benefits of microwaves would be even more pronounced. On a much broader scale, the Columbia experimentalist and teacher functioned as an oracle any staffer could approach with technical dilemmas, though he saw his role as much narrower than in a university setting. In Rabi's own words, the lab's function was "to develop something which would do as much harm as possible to the enemy." His typical reply when considering any new idea became almost a lab mantra: "How many Germans will it kill?"

Against the backdrop of the lab's more clearly delineated framework, airborne interception remained the overriding concern. The German bombing attacks against Britain, curtailed by bad weather during the first two months of 1941, escalated into all-out assaults. Glasgow's shipyards were knocked out of action for months. Two nights of London bombings claimed 2300 lives. On April 29, in the midst of Göring's assault, Sir Hugh "Stuffy" Dowding stopped at the Rad Lab during an inspection tour of U.S. military installations. Although dry and stiff in keeping with his nickname, the architect of the Battle of Britain victory showed 16-mm movies of the Royal Air Force in action and impressed his hosts with the need to develop a much-improved nightfighter radar.

Bowen, DuBridge, and McMillan took the visitor aloft in the B-18 to demonstrate the lab's progress. By that time, the radar housed a much-improved duplexer that enhanced signal-to-noise ratio and thereby improved range. For the Welshman, the flight brought back vivid memories of the foray he and Dowding had taken in a Fairey Battle two years earlier at Martlesham Heath. After the Rad Lab's set tracked a target plane at a very satisfactory three miles, Dowding once again asked to witness firsthand the minimum range capability. The air marshal sat before the radar screen in the plane's nose compartment, just below the cockpit. When the B-18 closed to minimum range, he climbed into the main cabin and found the target practically alongside. Dowding was clearly impressed, and Bowen hoped his country's Ministry of Aircraft Production would soon put in an order for America's first centimeter airborne radar.

The lab's two remaining original projects — long-range navigation and anti-aircraft gun direction — also progressed nicely. As far back as the second Tuxedo Park meeting between Alfred Loomis and members of the Tizard Mission in October 1940, Taffy Bowen had informally sketched for his host the broad outlines of a British navigation system called Gee, in which ground stations transmitted signals that enabled aircraft to fix their locations. The disclosure had set Loomis's mind racing. The next day he announced that while taking a shower he had been inspired to conceive a long-range navigation system, hinging on a highly synchronized crisscrossed grid of pulsed radiations broadcast across the sky, through which navigators on planes and ships could determine their position to an accuracy of one percent from a thousand miles off. This became the basis for the Rad Lab's own project, ultimately dubbed Loran, for LOng-RAnge Navigation.

A small body of lab staffers began developing the system early in 1941. The first tests transmitted signals on thirty megahertz, or ten meters. By spring, however, the group began to see that longer waves bounced off the ionosphere and spread farther around the Earth than microwaves, a realization that soon led to the adoption of a lower transmission frequency of around two megahertz. Loran therefore became the only Rad Lab product that didn't employ microwaves.

Easily the lab's most innovative early work, and among its best of the entire war, took place in the anti-aircraft project that got under way toward the end of January. The value of a microwave radar feeding position and bearing data to gun batteries was starkly evident. Existing British meter wave fire-control radars amounted to little more than a prayer: a plane's approximate position was detected, searchlights were directed in the intruder's general direction, and guns let loose static barrages in the hope a bomber might run into the flak. Britain placed such a premium on a microwave gunlaying radar that it launched its own effort and recruited both the Rad Lab and the Canadian government to undertake parallel ventures.

Beyond the move to more accurate microwaves, neither the British nor the Canadian program looked to add much to the state of the art. Once a target was acquired, human hands would still need to steer antennas and aim anti-aircraft batteries. The Rad Lab's physicists sought to do what the rival programs didn't consider feasible: build a gunlaying radar that automatically locked onto and followed enemy planes through evasive maneuvers, then piped the data to computers that ordered up precisely aimed fire.

Louis Ridenour, a razor-tongued University of Pennsylvania physicist, pushed the decision to go automatic. He was backed by a talented band of fellow believers, including baby-faced former Rhodes scholar Ivan Getting and Lee Davenport, a Schenectady-born electronics wizard who had interrupted his doctoral work at the University of Pittsburgh to join the lab.

The cornerstone of the approach the group adopted was called conical scanning, another concept attributed to Alfred Loomis but also developed in Britain and Germany. The technique drew its name from the cone of radiation traced in space by a tiny microwave dipole antenna rapidly nutating — rotating but without spinning, so that the top always remained the top — around the focal point of a parabolic dish. A target on axis inside the cone would return a constant signal, indicating the radar was on the mark. If the target strayed off center, the variation in the echo's amplitude could be analyzed instantly and converted to a direct current voltage that was fed to servomotors rigged to realign the radar on the moving target. It was the single beam equivalent of lobe-switching on the early British airborne interception sets, where operators tracked targets by balancing the signals from receiving antennas mounted on opposite sides of an aircraft. With conical scanning, though, angular tracking could become automatic. As the target moved, the antenna followed.

Such a system tracked aircraft in elevation and azimuth — up and down, left and right. Ridenour's boys worked to fill in the third dimension: slant range, or distance to the aircraft. Ironically, measuring range was the easy part of any conventional radar. Since radio signals travel at the speed of light, determining distance was a simple matter of measuring the time interval between the outgoing pulse and the returning echo, and doing some division. But for first-rate fire control, targets needed to be tracked with at least ten times the accuracy of any existing radar counterpart, to within about ten yards. Even more challenging than achieving this precision, the ranging system needed to be so smart the radar would not shift its attention to unwanted returns from trees, buildings, or other aircraft. Recalled Lee Davenport, it was easy to imagine a tracker that "kept bouncing among different signals, or got hung up between two echoes."

Here the choice of physicists for the lab paid off. These were the infant days of particle physics, when experimenters at virtually every major university had rigged cosmic ray traps to detect exotic particles. A radiation quantum — a particle — would hit a Geiger counter–like detector and cause a short electrical signal, on the order of microseconds. To record such a fleeting glimpse into the universe required electronics unlike those used in communications or other conventional engineering systems. A chief pitfall lay in confusing the signals from two independent particles that hit the detector virtually at the same instant. So researchers would line up two or more traps, designing highly synchronized or coincident circuitry that would be triggered only by particles passing through the detectors sequentially.

The brief electronic signal sparked by cosmic rays amounted to nothing more than a pulse. Rad Lab investigators applied the same basic coincident techniques to handling radar signals. A highly precise electronic quartz clock, cued to the moment a signal was transmitted, orchestrated the movements of electronic doormen — range gates — designed to open and close in such a way that only an echo from a precise distance could get through to be tracked. That way, the radar stayed locked on the target.

The whole design was a work of art for its day. Integral to the scheme were two screens, mounted side by side on an indicator panel. Instead of the usual cathode ray range tubes that showed distance along a horizontal scale, Ridenour's crew adopted ones with a circularized trace — J-scopes — so that the sweep went around the screen rather than across it. On the first scope, the trace completed one pass in about a five-thousandth of a second, giving it a maximum range of 32,000 yards, or eighteen miles. Although the radar itself could detect planes much farther out, this was the point at which accurate, close-in automatic tracking could begin. The second J-scope was similar — only its trace painted in sixteen times faster, affording it a maximum coverage of 2000 yards per revolution. Linked by the quartz clock, or crystal oscillator, the two sweeps began their circular journeys the instant a radar pulse was transmitted. On the 32,000-yard scope, an echo appeared as a spike so nice and sharp it seemed drawn in ink. But on the second screen, the sweep moved so quickly that the same signal showed up as a long and wide hill. It was like having one clock with only minute hands and a companion clock to fill in the seconds, except in this case the second timepiece enabled measurements shorter than one hundred millionth of a second. Instead of gauging range to within a hundred yards, the men could get as close as ten feet.

From that point, tracking was virtually automatic. The radar range operator would turn a handwheel that positioned a hairline cursor squarely on the front edge of the second scope's elongated echo. The hand crank itself was coupled to a potentiometer that drove a motor designed to keep the hairline on the blip. If the cursor fell behind, the operator adjusted the control, speeding the motor up in the process so that it was less likely to lag again.

The entire system was conceived within a few weeks of the first January 1941 meeting. Encouraged by their ideas, Ridenour's technical virtuosos dived into the project. In order to provide all-around coverage, it was decided to mount the radar on a servo-driven, 50-caliber machine-gun turret conscripted from General Electric's B-29 bomber program. When the turret arrived that spring, the men began building their first real prototype in the Building 6 rooftop laboratory. Local air traffic was too infrequent to provide a reliable supply of test targets. But Getting's friend Dave Griggs, a Harvard geologist whose legs had been crushed in an automobile accident, had gotten enough money from the ensuing insurance settlement to buy an all-metal Luscombe. For $10 an hour, he agreed to fly around simulating an enemy aircraft. Range circuits were still under development. But as Griggs buzzed Cambridge with Lee Davenport in back chatting over a radio to their excited colleagues, Ridenour's team achieved the first automatic tracking of an aircraft by elevation and azimuth on May 31, 1941.

To measure the accuracy of the tracking and to record the event, the men attached a 16-mm movie camera with telephoto lens to the antenna mount. The film showed the jittery image of the Luscombe moving across the sky. The herky-jerky picture resulted partly from fluctuations in the reflected signal and partly from the fact that the dominant returns continually shifted between different parts of the target's body at close range. For accurate fire control, something would have to be done about that. But in the meantime, the men watched plane and radar maneuver almost in tandem. Griggs might duck behind a cloud, obscuring vision, but the radar signal continued undaunted. Then the aircraft emerged, virtually on the scope's crosshair. Marveled Getting, "It was just like magic."

That whirlwind first spring soon found the Rad Lab at another critical juncture. With all three of the charter projects off the ground and running, Lee DuBridge launched a handful of endeavors outside the original mission. As a direct result of the improvised submarine-fishing expedition off New London on March 27, a small team began developing an air-to-surface-vessel radar. Virtually simultaneously, a crew that at various times included Jerrold Zacharias, Dale Corson, and Ernie Pollard also installed a prototype microwave search radar on the USS *Semmes*, a World War I–vintage four-stacker destroyer.

Despite such promising activity, though, the Rad Lab was entering a frustrating period. Taffy Bowen had hoped Hugh Dowding's visit would prompt a large order for airborne interception equipment. But well into May the British had yet to make a request. At home in the United States, the message from the military was equally ambiguous. Although Army representatives had also witnessed an AI demonstration, that service branch had not moved past the window-shopping stage. Nor was there much interest from the Navy beyond allowing the *Semmes* installation. Bowen wrote home to Cockcroft, discouraged: "All the M.I.T. equipment is now in a well-engineered form, ready for manufacture — certainly in a far better state than the experimental AI's and ASV's ever were in England. But the darned U.S. is not going for it. There has been a terrible pause while the Army try to make up their minds to buy. . . . The same applies to the Navy, who, instead of asking for all their destroyers to be fitted with 10 cm. gear, have asked for one only. If they always wait for England to do a job first, I am beginning to think the U.S. will never catch up. Here is their heaven sent opportunity to go right ahead on 10 cms. air and ship gear and do the old country some good, but they are not seizing on to it."

Bowen suspected the lack of British interest was influencing American opinion of the new microwave radars. He hoped to break the logjam by urging his countrymen to order a thousand U.S. airborne interception sets under Lend-Lease. "There is nothing to lose by making this request, and everything to gain," he argued to Cockcroft, "since the U.S. Army would act on it at once."

Adding to the paradox, even with military acceptance hanging in the balance, the Rad Lab continued its meteoric expansion, gobbling up space at the Massachusetts Institute of Technology. Costs far outstripped allotted capital, and the enterprise still lacked a permanent source of funding. To keep the lab afloat until September, MIT's steering committee voted to underwrite up to $500,000 in expenses. In light of Karl Compton's private entreaties, John D. Rockefeller Jr. agreed to cover, anonymously, the technical staff's salaries from September through June 1942.

Both the school and the financier trusted that more enduring solutions would be forthcoming. But as the summer of 1941 approached, Compton wondered what, if anything, the future held. While much of the country's power elite conspired to back Britain's cause, deep in its heartlands America wavered far from war.

6 • The Radar Bridge

"Our impudent assumption of superiority, and a failure to appreciate the easy terms on which closer American collaboration could be secured, may help to lose us the war."

ARCHIBALD VIVIAN HILL,
Secretary of the British Royal Society,
confidential memo to government authorities,
June 18, 1940

"The essential point of this is that we should offer any information desired, without condition, since we realise that America is fundamentally engaged in the same struggle for civilization as we are. The exchange would follow naturally."

HILL, June 18, 1940,
letter to Brigadier Charles Lindemann

THE FUTURE CAME more quickly than anyone guessed. With every bomb that fell on Britain in the frenzied spring of 1941, the United States spiraled closer to war, and with astonishing speed the basic question marks surrounding the Radiation Laboratory evaporated. Money woes disappeared with a stroke of the presidential pen. The military shook off its lethargy and began, in fits and starts, to embrace the emerging technology of microwave radar.

Vannevar Bush stood squarely behind the financial solution; he had resolved everything by pulling off a masterful Washington coup. Viewing war as inevitable, the country's needs already outstripping the emergency measures enacted less than a year earlier to mobilize scientists for war, the National Defense Research Committee chairman had petitioned Franklin Roosevelt to establish a federal agency to coordinate civilian science. Such a body would look much like the NDRC but with the power to go straight to Congress for real money, rather than existing merely as an Oval Office offshoot limited to special coffers. Besides, more than dollars were at stake. The far-thinking engineer also sought to free scientists, at the Rad Lab and elsewhere, to take ideas beyond the research stage to the development of pre-production prototypes, something excluded by the NDRC charter. This kind of freedom, he ardently believed, was essential to keeping scientists independent of the military's tendency to pursue only weapons and devices that struck the fancy of admirals and generals. It could make or break a war.

Bush laid out his proposal that May in a report to the Bureau of the Budget. The idea seemed to mesh perfectly with FDR's carefully crafted moves — including the Lend-Lease bill signed the previous March — to prepare America for war. The President chewed things over only briefly before abruptly concurring, announcing plans to create an Office of Scientific Research and Development to be run by Vannevar Bush. The executive order was signed the next month.

The President's landmark decision effectively brought all civilian war-related research and development, from radar to fighting malaria and developing the atomic bomb, under the skinny, tough-minded scientist's control. Since the Radiation Laboratory constituted the single biggest enterprise in Bush's empire, overnight money became almost a nonissue; and John Rockefeller Jr. never turned over a dime.

Barely had DuBridge taken stock of this financing revolution than the military on both sides of the Atlantic came knocking. On June 9, Britain requested ten Rad Lab airborne interception sets, tentatively ordering two hundred more subject to further review. Within a few days the U.S. Navy placed its own order for a shipboard system based on trials aboard the *Semmes*. Earlier in the month Ernie Pollard had joined the destroyer for exercises, when a heavy fog descended around the formation. Standard operating procedure called for dropping anchor until the haze lifted, but the seasick Rad Lab man convinced the captain he could spot harbor buoys on the Plan Position Indicator. The *Semmes* had docked safely, three submarines in her wake. The forthcoming production contract marked the first U.S. order for ten-centimeter radar: when the SG set arrived from the Raytheon Corporation, it proved able to track ships nearly to the horizon.

With money in the bank and the military finally moving beyond window shopping, Lee DuBridge moved to strengthen ties with the Telecommunications Research Establishment. Facing the fury of the Blitz, British scientists had motivation and a nearly six-month head start developing radar systems around the cavity magnetron; but the Rad Lab's wunderkinds were catching up at a dizzying rate. Staff members quickly built five improved airborne interception prototypes. On the Building 6 rooftop, a next-generation three-centimeter system picked up echoes from ground objects six miles away. There was a lot to share. As in the Great War, Germany would be defeated in World War II partly by the Bridge of Ships steaming supplies from America to Britain. DuBridge was instrumental in establishing a much smaller convoy of radar experts that may have been equally as important.

The foundations of the radar bridge, dating back to the Tizard Mission, had been fortified the previous winter. Harvard University president James Conant led an expedition to the United Kingdom to establish an NDRC liaison office; and shortly after Conant's trip, the British opened a reciprocal Washington bureau run by the respected mathematical physicist Charles G. Darwin, a grandson of the great evolutionist. These measures set the stage for regular scientific exchanges, not just at the policy level, but among the researchers who actually developed the technologies of war, including radar and everything else that had sprung out of the Tizard Mission's black box. The first Rad Lab emissary to go across was Ken Bainbridge, the Harvard cyclotron pioneer. On March 3 he left on a three-month mission to brief British officials on the Cambridge doings. In the summer, as the lab emerged from its haze of uncertainty, DuBridge sent a swarm of other staffers to step up the interchange. Less than a year after Tizard's arrival, technical expertise began to flow back Britain's way.

The next Rad Lab member to venture overseas likely was Norman Ramsey, leader of the Advance Development Group that had constructed the experimental three-centimeter system atop Building 6. In early or middle June, the lanky twenty-five-year-old physicist boarded a Pan Am Clipper in New York and hopscotched across the Atlantic to Bermuda, the Azores, and into Lisbon, where a British plane waited to ferry passengers the rest of the way. The trip marked a homecoming of sorts. After being awarded a bachelor of arts with a concentration in mathematics by Columbia University at the age of nineteen, and before rejoining his alma mater for doctoral studies under I. I. Rabi, Ramsey had spent two years at Cambridge University earning a second undergraduate degree in physics. Back in England as a Rad Lab ambassador, he would visit Marcus Oliphant's lab at Birmingham and spend time watching nightfighters in action from an east coast Chain Home station before heading to TRE to discuss three-centimeter systems.

Close on Ramsey's heels, anxious to show off the lab's prototype airborne interception system, came Taffy Bowen, Dale Corson, and Fred Heath. The resourceful Heath, an engineer borrowed from Canada's National Research Council, rode shotgun on six hundred pounds worth of radar equipment flown up to Montreal by Eastern Airlines, then transferred to a Liberator for the transatlantic haul. Billy Bishop, the World War I fighter pilot, saw him off from the Quebec metropolis. Heath's job was to reinstall the radar in a Boeing 247-D shipped over earlier in June as deck cargo. On the twenty-eighth Bowen and Corson flew across to operate the set. With Corson running the project, Bowen was not deemed essential. But Charles Darwin granted the Welshman a trip home to see Vesta as thanks for a job well done in getting the American radar project up to speed.

A fifth Rad Lab staffer in England at this time was Jim Lawson, the cadaverous TR box architect. All the men had different missions or responsibilities. But all five converged on Swanage in the first days of July. P. I. Dee's centimeter team worked apart from the main group, at Leeson House, a girl's school with an unobstructed view of the Isle of Wight. The war, the cavity magnetron, and Taffy Bowen's stories had instilled a deep respect of the British in the Americans' minds. But the Welshman was piqued to find that Dee, at least, resented the intrusion into TRE affairs. As he remembered, "They had forgotten, or perhaps did not appreciate, that all the essential transmitting and receiving valves in our early . . . radars had come from America, and they did not want to believe that the United States was anywhere in the picture—a curious denial of Tizard's thesis that America had much to offer."

Any tension was dispelled within a fortnight. Before joining the Boeing in Liverpool, where the wings would be reattached and the plane checked out before moving nearer to TRE for flight tests, Heath had set up the American radar on a benchtop at Leeson House. On July 15 the two groups began conducting side-by-side comparison tests of their competing models —rooting out planes, ships, and ground objects on the Isle of Wight some forty miles away. In terms of overall performance, both sets gave comparable results, although the edge went to the British. Tests of individual components, however, revealed important differences. The Rad Lab set was better engineered, with the souped-up magnetron provided by Rabi's group, combined with new Eimac pulser tubes, enabling the transmitter to put out significantly more power than its rival. More than offsetting that advantage, though, the British receiver picked up signals several times fainter than what the Rad Lab unit could detect.

For a short while Bowen and Corson were mystified. When the Welshman had toured the United States as part of the Tizard Mission in late 1940, American microwave receivers outperformed the best Britain could muster. Lawson and Ramsey joined in to try to explain the turnabout. At first the American contingent attempted vainly to coax more range from its set, poking and prying and checking connections. Finally, they borrowed a British receiver and cabled it to the American set. "Immediately we were picking up airplanes three times further away," Ramsey recalled.

From that point it did not take long to identify the British receiver's silicon crystal detectors as the main factor behind its superior showing. The Rad Lab radar employed a grounded grid triode. The finding stunned the Americans, since they had started with crystals but abandoned them earlier in the year after tests showed vacuum tubes to be superior. Only later, Ramsey related, did it become apparent that the crystal used for the American tests had partially burned out. The British, by contrast, had gone with crystals from the start. By spring 1941, Oliphant's lab had designed a better capsule that lessened susceptibility to shock and vibration, and rendered crystals superior to anything in the United States; these had gone into mass production at British Thomson-Houston. Around the same time the duplexing, or TR, problem was solved by the ingenious application of a reflex klystron, itself an adaptation of the Varian brothers' original creation. The so-called soft Sutton tube was filled with a low-pressure gas that was rapidly ionized by the transmitter pulse, providing a short circuit that protected the crystal from burnout. Once the pulse ended, the gas recovered, allowing received signals passage to the detector.

The face-off between the TRE and Rad Lab airborne interception radars had far-ranging consequences for the war effort and beyond. Norman Ramsey, the first to return stateside, drew the assignment of bringing the rediscovered technology back to the United States. With him on the Pan Am Clipper, he carried three British crystals and three duplexers. "I guess I had in my possession on that trip one hundred percent of the good solid-state receivers that worked in this country," he related. Once back in Cambridge, Ramsey summed up his trip at one of the Rad Lab's standard colloquia. Enjoying his moment in the sun, the young physicist first covered three-centimeter developments, proudly reporting that the lab led the Telecommunications Research Establishment on this front. He concluded the presentation with a discussion of the ten-centimeter comparison test and revelations about the British crystals and TR box. With something of a flourish, he handed the samples over to DuBridge.

Ramsey hoped the United States would rush to put the British crystal technology into immediate use. However, he remembered, although Bell Telephone Laboratories carried out a study of crystal mixers, at the time no extensive crystal research program existed in the country. Embarking on what amounted to a silicon public relations tour, Ramsey traveled the East Coast and told his tale to industrial R&D arms working on microwave radar technology. The effect was profound. Within a few months Bell Labs, Westinghouse Research Laboratory, and Sylvania Electric Products Inc. undertook important crystal research or production programs. The Rad Lab strove to improve crystal reception by adding minute traces of the right impurities, moving systematically through the periodic table, trying out different elements and combinations of elements that might enhance conductivity. Purdue and the University of Pennsylvania started companion studies.

For most of the war years, crystal making remained a black art. Some detectors worked beautifully, others burned out. But over the next decade the reemergence of crystal studies as a major area of research arguably would change the American landscape more than any war-related technology. More immediately to the point, as Hitler's forces continued their quest for European domination, the TRE and Rad Lab teams agreed to combine the best of both their worlds, a move that dramatically enhanced the effectiveness of forthcoming Allied microwave radars. This scientific interplay, a result of the radar bridge, was something Germany and its allies never shared.

Just as Norman Ramsey went home, TRE sent a second representative to the Rad Lab as an adjunct to Taffy Bowen. The new man, Denis Robinson, was a thirty-three-year-old electrical engineer lured from commercial television research into the British radar organization in the Dundee days. Robinson was almost singularly responsible for his country's use of crystal receivers. A. P. Rowe had dispatched the recruit to find a suitable detector for the establishment's struggling centimeter program. Visiting the station's small technical library, Robinson had stumbled across a series of books from Springer Verlag, Germany's great scientific publisher. Married to an Austrian and consequently fluent in German, he had been drawn to one title. A later investigation by the physicist Frederick Seitz concluded it was assuredly *Physik und Technik der Ultrakurzwellen*, "Physics and Technology of Ultra Short Waves," by microwave pioneer Hans E. Hollmann. The author had stated categorically that crystal rectifiers, much like the old-fashioned cat's whiskers favored by radio buffs, made the best microwave detectors. Robinson had rushed the book to his immediate superior, W. B. Lewis, a Cavendish man who could read German only haltingly. "You see," he cried triumphantly, and the matter was decided.

For his mission to America, Robinson had one high-priority assignment that explained some of the stiffness Taffy Bowen had seen that summer as his old colleagues greeted their Rad Lab counterparts. He was to convince DuBridge's group to forget about airborne interception radar, once so desperately sought by Britain, and switch its efforts to building air-to-surface-vessel systems for hunting enemy submarines.

As Robinson later explained matters, the astonishing British volte-face could be traced to three central concerns. First, P. I. Dee and others at TRE felt certain the Americans could not develop a workable system without direct feedback from combat pilots. Far more importantly, the need for any microwave set had lessened dramatically. After a monstrous May 10 attack on London in which three thousand died and the House of Commons was leveled, Hitler had called off the massive night raids that characterized the Blitz, secretly mustering Luftwaffe resources for Operation Barbarossa, the surprise assault against Russia on June 22.

Even before the switch, British nightfighters had started performing far better than most had dreamed possible. A vastly improved training program and better maintenance had substantially increased the effectiveness of the Mark IV longwave AI radar. At the same time, the sets were now carried in high-performance Beaufighters equipped with 20-mm Hispano cannons and six machine guns. As the coup de grâce, the RAF had introduced a radar system called Ground Controlled Interception. Built around modified Chain Home Low sets, each GCI station came complete with its own plotting board and switchboard. So instead of relying on elaborate filter and operations rooms, controllers could see enemy formations on a Plan Position Indicator and talk to pilots continually over radio links, guiding aircraft into combat with unprecedented precision. The result of the moves was that after shooting down only two night bombers the previous December, the RAF rose to destroy fifty-two planes in April. The next month, the figure nearly doubled to 102 confirmed kills.

The final factor behind Denis Robinson's mission was the *Unterseeboot* itself. As the air menace subsided, the U-boat danger rose dramatically. In the First World War the German submarine had nearly brought Britain to its knees. Yet between the wars the Royal Navy, convinced that Asdic, or sonar, could handle the threat, had failed to conduct a single exercise in protecting merchant convoys against submarine attack.

This shortsightedness had lingered well into 1940. While the British necessarily attached the highest priority to beating back the air threat and thwarting an invasion, officials had seriously deluded themselves about the U-boat. From pens along the Bay of Biscay on France's west coast, *Führer der Unterseeboote* Karl Dönitz could order his submarines well into the mid-Atlantic gap beyond the reach of British and Canadian escorts. His commanders also learned to attack on the surface at night, rendering Asdic useless, since the device only tracked underwater. By spring 1941, Dönitz's *Rudel*, or wolf packs, terrorized shipping in the North Atlantic, around the Azores, and off West Africa. As shipping losses exceeded a hundred month after month, the period formed the core of *Die glückliche Zeit* — the Happy Time — prompting Winston Churchill to declare the Battle of the Atlantic.

By the time of Robinson's trip, the Royal Navy had taken a few compensatory steps. New facilities in Iceland extended the air umbrella and provided a safe haven for hunted ships. Escorts were beefed up and extended across the Atlantic as destroyers became available. That same spring, after salvaging an Enigma encoding machine from a captured U-boat, the British also began cracking German naval codes. Reluctant to compromise the windfall on limited tactical gains, however, Churchill set out a policy that often avoided outright attacks and favored instead rerouting convoys around U-boats whose positions the intercepted traffic revealed.

Radar, therefore, took on greater importance as Britain struggled to gain the upper hand. Longwave shipboard systems helped fill in Asdic's surface gap and provided a measure of relief at night and in poor weather. But air-to-surface-vessel sets attracted the bulk of the attention, since a single plane could cover large ocean tracts and offered the only viable way of taking the hunt to the U-boat. By late spring, the ASV Mark II developed by Taffy Bowen's group had been hurriedly installed in more than a hundred British aircraft.

The sets played a vital role in the cat-and-mouse game that led to the destruction of the *Bismarck* on May 27, 1941. Early on the twenty-sixth a Catalina flying boat fitted with the radar spotted the warship after a thirty-hour hunt. Although the public was told the plane came out of a cloud close enough to the *Bismarck* that the crew could see their prey steaming toward France, an analysis of the Catalina's search pattern by naval historian Tony Devereux indicates the Mark II may have been the crucial factor. In any case, the carrier *Ark Royal*, part of a large force converging on the German battleship, dispatched two waves of Swordfish torpedo bombers that were guided by the Mark II to their quarry through heavy overcast. During the second assault the battleship took a hit under the stern that left her bereft of steering and virtually adrift.

U-boats, though, presented a much tinier target for radio waves than a battleship. In optimum conditions the Mark II could detect surfaced submarines at twenty miles from a cruising altitude of two thousand feet. In practice, clutter from turbulent seas, coupled with problems in maintenance, training, and aerial gain, cut markedly into detection ranges. Air crews found the sets a great help warning of looming coastlines and aiding rendezvous with convoys in bad visibility. Against the submarine, however, the Mark II's record was dismal. Throughout the first months of 1941, exacerbated by the limited number, range, and armament of sub-chasing planes, the radar played a role in only one partially successful attack on a U-boat; early in the war, shrapnel from anti-submarine bombs bouncing off the sea surface and exploding brought down at least three British planes.

To Denis Robinson and many others in Britain, it was painfully clear that success against the U-boat virtually dictated the enhanced resolution of microwaves. With the airborne interception problem well in hand, officials impressed by Taffy Bowen's stream of Rad Lab accolades decided American talents might be put to better use in an all-out push to develop air-to-surface-vessel radar.

In later years Robinson delighted in the story of how he was tapped to go overseas. One day early in 1941, Dee had come around contemptuously waving the government notice ordering another liaison man to the Rad Lab to work alongside Bowen. It couldn't be anyone too good, he stressed, because the British needed all their best talent. On the other hand, Whitehall would never accept a flunky. "It's all wrong because the Americans will never do anything useful," Dee concluded in disgust.

Robinson's heart jumped. After a German bomb had nearly leveled his house in Swanage a few months earlier, his wife, Alix, and their two sons had moved to America to escape the war. They lived in Belmont, next door to Cambridge, with Alix's brother, a civil engineering professor at Harvard. If that weren't enough, after earning his doctorate from King's College at the University of London, Robinson had spent two fellowship years at MIT studying under Vannevar Bush. He felt at home in New England.

Although he desperately wanted the assignment, Robinson held his tongue while Dee made a list of candidates, then eliminated one name after another for various reasons. Finally, he spoke his mind. "After all, I think I ought to be on the list." Dee couldn't understand at first why anyone would want to leave the center of action, but he crumpled up the Whitehall paper and tossed it in a wastebasket. "You're it." Robinson fired off a letter to Alix, telling her in guarded language that he might show up one day out of the blue.

A nonplussed Denis Robinson arrived in America sometime near mid-July. The Americans visiting England had been struck by the grim truths of war, from bomb craters to the technologically impoverished conditions at Swanage, where workers often made do with cramped quarters and inadequate supplies, to the way the fighting touched ordinary lives. Dale Corson never shook the memory of holding one of P. I. Dee's daughters on his lap and listening to the girl, maybe five years old, tell of Luftwaffe bombing raids in graphic detail.

Robinson, on the other hand, found himself nearly overwhelmed by the relative opulence and tranquility on the western side of the Atlantic. Lots of little things impressed him. He remembered being dazzled by the bright red and yellow of the Shell fuel trucks on the tarmac at Gander, Newfoundland, where his plane touched down to refuel, so unlike the drab camouflage colors around British airfields. Later, safely landed at the U.S. Army's Bolling Field near Washington, he had been ushered into an officer's club equipped with pinball machines, an unthinkable frivolity at bases back home. On the heels of that display came the spectacle of the wives or girlfriends waiting to greet American officers on board the flight. As the men lined up for required medical exams, each with a thermometer sticking from his mouth, Robinson watched the women, decked to the nines. "And these girls looked so smart," he would say. "I hadn't seen anything so smart for years."

After checking in at the British embassy, and orienting himself for a few days, he went straight to Boston, where Alix met his plane. The next working day, it was over to the Rad Lab to outline the British submarine-hunting proposal. Without committing himself, Lee DuBridge took Robinson back to Washington to confer with Alfred Loomis, chairman of the NDRC's Microwave Committee.

The critical rendezvous took place one evening after dinner in Loomis's suite at the Wardman Park Hotel. Robinson expected a formal meeting and knocked on the door anticipating a servant to open it. Instead, the financier answered himself, clad only in boxer shorts.

"Come along in, Denis," he said gaily. "I want to talk to you. Do you like bourbon or scotch?"

Robinson was flabbergasted. He accepted a bourbon, his first. He savored the smooth taste while gathering his wits, then explained the general situation, half-expecting the Microwave Committee chairman to take offense at the notion that the Rad Lab adopt a new priority project. But Loomis, sipping his scotch, reacted cheerfully. "Anything you want," he told his guest. "Tell DuBridge I said so."

It occurred to Robinson that the statement was probably disingenuous, for DuBridge and Loomis must have already worked out the U.S. response. Whatever the case, back in Cambridge a few days later, the newcomer found the Rad Lab head entirely accommodating. Before undertaking his overseas trip, Robinson had spent about a month inspecting the various American planes coming into Britain under the Lend-Lease bill signed by FDR the previous March. As a result, he explained to DuBridge, his countrymen had decided on the B-24 Liberator as the most suitable craft to carry an anti-submarine radar. DuBridge immediately granted the Englishman carte blanche to set up and manage a project to develop the radar, assigning him to share an office with I. I. Rabi. The director laughed, though, at the idea of putting aside the airborne interception effort in order to concentrate on ASV systems. "We'll do it all," he told his guest.

Robinson couldn't get over the American willingness to cooperate, especially since his hosts weren't even at war. Before long two Liberators were diverted to East Boston Airport and placed at his disposal. Moreover, he found the Rad Lab much further along than expected: airborne systems employed Plan Position Indicators, antennas scanned 360 degrees, and the lab's modulators and transmitters could be used without modification on his

ASV configuration. Unlike the war-stretched British, the Americans enjoyed plenty of everything. "I had a glorious time," Robinson remembered. "I'd never seen so much technology available — there wasn't any shortage."

Over the course of the summer, the Radiation Laboratory pressed on with its rapid expansion. Many staff members rightfully came to consider themselves radar experts, rather than physicists impersonating electrical engineers, and the facility continued to gain ground on the British. Bowen returned from England that fall reporting flight trials of the lab's airborne interception prototype had gone so well that his countrymen had decided to equip their aircraft with equal numbers of British and American sets.

Although it wasn't evident then, the news heralded one of the lab's greatest successes. The American set exhibited superiority in the important area of antenna scanning and display. Dee's crew relied on a conical scan similar to that under development with the Rad Lab's anti-aircraft radar, but without any lock-on features. It located targets — left, right, up, down — relative to the plane. However, explains Fred Heath, "The problem with the British one was that if it was off center, you knew it was off center, but you couldn't tell how far off it was." The American system, by contrast, employed a thirty-inch-diameter parabolic antenna, mounted in the Boeing's plywood nose, that scouted for aircraft much like a lighthouse beacon. It came with several displays. But the main screen for airborne interception plotted an echo's azimuth, or bearing, horizontally, so a target to the left of the search plane appeared on the screen's left side; the farther away the target, the farther out the blip. Elevation was plotted similarly along the vertical. If the interceptor flew straight at a target and at the same elevation, the echo lay directly in the screen's center. The bottom line: the Rad Lab system made it far easier to home in on an enemy aircraft.

When it came time to return to the States that September, Taffy Bowen worried that the British would not service the American radar properly, causing performance to degrade and endangering their hard-won edge. He ordered Heath to sabotage the set to prevent such a fate; and the Canadian bent the hard copper coaxial line connecting the magnetron to the antenna so that the system short-circuited.

With Bowen's report, Lee DuBridge started phasing out the original Project I, airborne interception, raising the fledgling ASV program's priority in the process. It had taken only a few months to fulfill the British desire for a major push in anti-submarine warfare. Each aircraft type required its own specialized system. Since no plane had yet been designed with radar in mind, this was often dependent on where space could be found to house the components. Reflecting this constraint, by the fall of 1941 the Radiation

Laboratory carried at least five air-to-surface-vessel projects on its books for craft as different as PBM-1 flying boats and blimps.

The most sophisticated was Norman Ramsey's experimental three-centimeter system. Though it still lingered in the embryonic stage, the Navy showed an acute interest in the project: given the same-sized antenna and pulse output as a ten-centimeter counterpart, a three-centimeter radar would bring roughly eleven times greater energy to bear on a target, enough of an improvement to make a periscope visible amidst somewhat turbulent seas.

Another hot project, likely to see action far sooner, rested in Denis Robinson's hands. Progress had come fast and furious in the few months the Englishman had been on board. The Liberators sprouted bulbous radar domes under their noses, earning them the nicknames Dumbo I and Dumbo II. In November, even before Robinson managed a flight test, the British ordered a dozen sets from a special Rad Lab offshoot called the Research Construction Company, Inc., for which the contractor was the New York–based Research Corporation, a nonprofit supporter of scientific studies. This enterprise, in essence a model shop set up on Albany Street near the MIT campus, evolved into a center for the crash production of limited numbers of the lab's radars. Since industry could not generally spare the manpower and facilities for such small-scale engineering, the RCC filled the need to get a few copies of each new radar set into service hands as quickly as possible for evaluation and special combat situations. In parallel with such efforts, development followed a more standard course. The idea, anyway, was that the Army or Navy would evaluate a breadboard model, and if it looked good, the design would be turned over to a large manufacturer, which took full responsibility for meeting military specs for ruggedness and tropicalization and generally seeing the set into the field.

As December winds carried a harsh cold snap to New England, Robinson pressed to get Dumbo I's radar ready for an initial test flight. He continued to marvel at the Rad Lab's keen sense of cooperation, despite America's state of nonbelligerence. The erstwhile TRE man still lived with his wife and children at his brother-in-law's big house in Belmont. On the first Sunday of the month, he had stepped out for a brisk afternoon walk when a passing car stopped alongside. "Did you hear the news?" the driver called out. "The Japanese have bombed Pearl Harbor."

•

The attack had begun in the predawn hours of December 7, late morning on the East Coast. Radar might have prevented much of the devastation. At that moment, as it soon became known, the island of Oahu, third most westerly in the Hawaiian chain and home to the United States Pacific Fleet, lay enveloped in a radar curtain. The five sets providing the coverage, all U.S. Army mobile SCR-270s delivered the previous August and positioned around the island, penetrated the Pacific mist and extended their radio eyes some 150 miles out to sea. The stations, on the air since four that morning, were due to stay on alert until seven.

Radar should have picked up the initial 183-plane Japanese strike force that took off shortly after six that morning from aircraft carriers positioned about 220 miles north of the island. And it did. As early as 6:45, three of the sets — at Opana on the northern tip of Oahu, Kawailoa farther west, and Kaaawa in the east — detected the first faint glimmers of the invaders but no action was taken. Just before seven, with an alarm yet to be raised, the order to shut down for the morning came through from the Information Center at Fort Shafter, where reconnaissance reports were charted.

All five stations began to sign off. At Opana, though, an excellent radar observation point 230 feet above sea level, Private George E. Elliott asked the more experienced Private Joseph L. Lockard for additional instruction time. Suddenly, the cathode ray screen came alive with so much clutter Lockard thought it had gone haywire. After a system check indicated all was well, the enlisted man decided that just beyond 130 miles northeast of Opana lay the biggest aircraft formation he had ever seen on radar. At Elliott's insistence, Lockard raised the Information Center's switchboard operator, who passed them on to Lieutenant Kermit Tyler, the only other man still on duty. As Tyler listened to Lockard's report, a bell went off in his head. The Army paid a local radio station to broadcast music throughout the night whenever aircraft were headed for the island base as a way of furnishing the planes with a friendly homing signal. The lieutenant remembered hearing the Hawaiian melodies while driving to work that morning and concluded a flight of bombers must be approaching. He told Lockard to forget the radar signal. As a consequence, the two Opana-based privates silently plotted the first attack in to about twenty miles, when it disappeared in clutter from hills behind the station.

Beyond the green-capped crests, roughly an hour after the first detection, all hell broke loose. At 7:49 A.M., Lieutenant Commander Mitsuo Fuchida, chills innervating his spine, gave the attack signal — "To, To, To," the first syllable of the word totsugekiseyo, or charge. Fuchida's plane swung around Barbers Point on the island's southwest rim: Pearl Harbor lay almost due east, with the sprawl of Honolulu and the rising peak of Diamond Head beyond. The initial bombs had already dropped four minutes later when Fuchida cried out the famous Tora! Tora! Tora! — "Tiger" three times — that told the Japanese Navy the U.S. Pacific Fleet had been caught unawares. An hour after the first strike, another wave came to finish the job. Seven battle-ships had floated peacefully in the harbor as dawn broke, with another in dry dock. All eight were lost or damaged, along with three destroyers, as many cruisers, and a handful of other vessels. The shards of some three hundred aircraft glittered in nearby airfields. All told, 2403 civilians and service personnel had been killed. More than a third of the deaths came on the battleship *Arizona*, where a bomb set off the forward magazine and blew the doomed vessel briefly out of the water before she found a final resting place on the harbor's shallow bottom.

The Pearl Harbor assault not only brought America into the war, it also assured radar's place in the conflict. Aware the Opana station had provided early warning of the attack, the military looked hard at its radar programs. By early to middle December, the Navy had taken delivery of 132 longwave radar sets — six models in all — for shipboard early warning and fire control. On the eighteenth, officials reported that seventy-nine sets were operation-ally ready, and the service moved quickly to increase its stock. The first RCA production model of the CXAM search set, sunk at Pearl Harbor with the USS *California*, was salvaged and put into service at an Oahu radar training school whose cadre of young teachers included Ensign Henry Loomis, youn-gest son of Alfred Loomis. The Army also scurried into action. By Pearl Harbor, several hundred early-warning sets, both the long-range SCR-270 type deployed on Hawaii and shorter-range SCR-268 sets for more accurate aircraft detection, had been delivered to Iceland, Panama, and other sites. On the day of the attack, Ivan Getting and Lee Davenport happened to be at Fort Hancock, New Jersey, filling the service in on the latest developments with their fire-control radar. By then, an experimental truck-based version of the set — the XT-1 — was well along. The two were lunching in Harlem when people started shouting the news. "Lee and I dashed back to Fort Hancock, but no one was allowed anywhere near the fort," writes Getting. "We could see one SCR-268 after another leaving under armed convoys."

Pearl Harbor marked perhaps the most dramatic turning point in the Radiation Laboratory's history. Until the attack, Lee DuBridge still half-expected the enterprise to fold sometime in 1942. With America in the war, he sold his house in Rochester, shipped up the furniture, and moved into a Belmont rental. The Navy quickly set up a liaison office at the lab, both to stay abreast of technical developments and to bring radar developers up to snuff on suddenly exploding service requirements. The Signal Corps and Army Air Corps followed suit. The lab's relationship with industry also grew tighter. For one thing, the idea of simply turning over a design to a manufacturer and forgetting about it proved infeasible. Many companies did not have the facilities or experience to construct novel microwave systems.

For instance, a lock manufacturer built waveguide elements. In the weeks after Pearl Harbor, the Rad Lab opened a Transition Office designed to train and educate manufacturers in centimeter radar matters. In spring 1942, after director Armand Herb died in a commercial airline crash, the office was run by H. Rowan Gaither Jr., a prominent San Francisco attorney. As scores of electronics concerns, subcontractors, and vendors proliferated into the hundreds, Gaither compiled a detailed list of which were suitable for specific tasks. Companies sent engineers to the Rad Lab for indoctrination, and lab staff often followed through into the field.

Both industry operations and service needs had to mesh before a radar system could be approved and mass-produced, and the various parties did not always see eye to eye. Ernie Pollard still endured "endless trips to queer rooms in the Pentagon where odd people had to be met and where all kinds of prejudices had to be overcome." In one case, Bell Labs resisted changes in magnetron design, prompting the Radiation Lab to recruit Raytheon to produce revamped transmitters. Nevertheless, what Jerrold Zacharias termed a "wholesome jealousy" built up between the Cambridge group and Bell, and the same held true for the lab's relations with other companies. Largely through Lee DuBridge's operation, the United States was on its way to enjoying an unprecedented partnership between civilian science and the services, exactly the kind of relationship missing in World War I, and which Vannevar Bush had hoped to engender when he spawned the National Defense Research Committee in June 1940.

The demands of war raised a host of additional issues. DuBridge needed to scare up more room for engineering design, small-scale production, installation and maintenance of the lab's units, and training of service personnel. Already a small secret community with almost five hundred employees, the facility had spilled out of its allocated space in Buildings 4 and 6 into MIT's mechanical engineering department. A nearby structure had been purchased to accommodate more overflow. Another edifice, Building 24, had been built from scratch on the northwest side of the campus. Yet with a war on, the director estimated the lab still needed to swell to six times its current size. He won quick NDRC approval to add four floors and a penthouse to Building 24. A supposedly temporary three-story wooden frame structure, Building 22, went up next door, connected to its neighbor by an overpass.

In concert with the physical expansion, DuBridge launched another recruitment drive. Staffers splayed out across the country to locate talent. Wheeler Loomis wrote directly to university officials, seeking their personal recommendations. A whole new breed, even younger than the first, came into the fold. The average staff salary was about $350 a month, down $50 from the early 1941 figure. Among the prize catches was Robert Dicke, a highly original thinker from the University of Rochester: DuBridge himself had recruited Dicke the previous summer, but the budding physicist needed to finish his doctorate and fulfill some teaching duties and didn't arrive until that September. Another gold-plated arrival was Jerome B. Wiesner, an engineering graduate of the University of Michigan. Early in 1942, Wiesner worked in the Library of Congress propaganda arm, putting together radio transmitters for what became the Voice of America. His old professor George Uhlenbeck, leader of the Rad Lab's theory group, asked Wiesner to leave Washington to conduct war research. The student wrote back, seeking more details. Uhlenbeck's response: "We are doing important war research." With that, Wiesner signed on. He would eventually become science advisor to John F. Kennedy and president of MIT.

As these and other talented conscripts arrived in Cambridge, uncertain even of the nature of the work ahead, DuBridge put in motion a massive reorganization that separated the laboratory into ten divisions; ultimately there would be twelve, with a separate patent office and security staff, but this was the final form. Security was stepped up. Fences already protected the main areas. Civilian guards, many retired Cambridge policemen, stood sentry at building entrances and patrolled hallways, checking staff badges and issuing laggards a dollar fine and oversized replacements stamped "Forgotten." Now, though, with America at war, a contingent of Military Police with riot guns arrived to cover the grounds. No documented case of espionage occurred at the Rad Lab. However, shortly after the MPs arrived, A. J. "Ajax" Allen, head of the building and facilities division, as well as security in the early days, drove to the main gate and was challenged by one of the new guards, who asked for identification. Allen opened his jacket to reach for his wallet, exposing his own revolver, kept in a shoulder holster. The startled sentry didn't wait for an explanation. A bullet grazed the Rad Lab man in the chest or stomach, luckily inflicting only a flesh wound. People smiled over the incident for years.

One thing that didn't change, according to later interviews and written accounts, was the laboratory's special flavor of a tight-knit community bound by an important secret purpose. The carnage at Pearl Harbor, while a national outrage, engendered a particular sense of shared pride in the year's head start members had given their country. As the secret few became a small city, that aura persisted.

Work usually began around 8:30. Long lines of staffers jammed the cafeteria for coffee and doughnuts. A typical day ended by five or six, though many stayed late into the dimmed-out night. Lab members naturally tended to spend time with their fellows, before and after hours. Colleagues shared lunch, often at their desks or in the cafeteria, but sometimes walking the few blocks into Central Square for a sandwich, over to the MIT Graduate House, near the Charles River, or even across the river into Boston for good Chinese fare. They carpooled to save gasoline rations. Families played and even vacationed together on the windswept beaches of Cape Cod and Maine. Parties were rampant. Taffy Bowen demonstrated how to put footprints on the ceiling: discard shoes and socks, stand back-to-back with a friend on a table, lock elbows and swing up the legs. Romance flourished, too. The lab employed very few female scientists. But as administrative needs grew women came as secretaries, mail room staff, bookkeepers, or model shop technicians, so the final ratio of men to women was about 65 to 35: "It is true," reads what is basically the Rad Lab's yearbook, "that when early radar sets began spraying high-frequency energy about, there was some concern felt as to the possible effects on the reproductive organs; and it is further true that this concern was allayed when three people in the Propagation Group began having babies at once."

Predominant in the midst of all the comings and goings, the eating and drinking and courting, was the rush to develop effective, often deadly, microwave radar systems. Here again, DuBridge and the lab hierarchy created a climate of university-like openness that in many senses became a model for postwar corporate research. The lab sponsored group meetings, panel discussions, and weekly seminars, and deliberately left operations more decentralized than in most contemporary industrial organizations. Wheeler Loomis took particular pride in the authority of group leaders to negotiate directly with vendors and even select subcontractors without higher approval. At a lower level, while formal projects were arrived at in consensus with the services and industry, the atmosphere encouraged men to explore their own brainstorms. Someone might float an idea among his peers and develop it all the way to the breadboard stage, which entailed mounting a mock-up system on plywood, without regard to scale or engineering parameters. Only then would the group leader take it to a higher level, where there was always I. I. Rabi asking, "How many Germans will it kill?"

By the spring of 1942, Lee DuBridge's charges were almost ready to send their creations to war. The airborne interception effort had been turned over to industry, and the remaining original projects were rapidly coming together. A minor glitch occurred in Alfred Loomis's Loran network when the frequency chosen, 2.2 megahertz, turned out to be the channel used for a ship telephone link, and a test of the Rad Lab's high-powered transmitter caused phones to ring around the Great Lakes. Systems had quickly been pushed down to 1.95 megahertz; and the first five stations, spanning the Atlantic coast from Greenland to Nova Scotia, would be ready by fall.

Gunlaying was shaping up as a potentially war-changing success. The Rad Lab's theory group found a way of instantaneously averaging the changing signal strength as radar pulses struck different parts of an aircraft, eliminating the herky-jerky tracking manifested during initial trials. Smoother data also helped a computer, or predictor, calculate where the aircraft would be by the time a shell arrived, a necessary step to determining the aiming point and precise moment to fire. About the time of Pearl Harbor, tremendous strides were made on this vital front through a collaboration with Bell Labs.

During the war years Bell mounted a microwave radar research and development effort second in size and scope only to the Rad Lab. Not only did the company's expertise in high-frequency communications lend itself to the design and manufacture of crystal detectors, antennas, and other components, its Whippany Radio Laboratory thirty miles west of New York City had been developing Navy centimetric fire-control radars with manual tracking since late 1937 as well.

Bell's radar team had stumbled onto a gem. A twenty-nine-year-old company engineer, David "Parky" Parkinson, had invented what he called the automatic level recorder, in which a potentiometer automatically drove an ink pen across a strip of paper to chart voltages for the phone system. Parkinson worked outside Bell's radar program. However, one night shortly after Hitler's blitzkrieg into Western Europe in May 1940, he experienced a particularly vivid dream. He was in an anti-aircraft revetment. A gun fired occasionally, and Parkinson had the impression that every burst brought down a plane. After a few shots a soldier beckoned him closer. There, unmistakably, sat his potentiometer. Parkinson awoke with a remarkably clear memory of the details. As he noted, "It didn't take long to make the necessary translation — if the potentiometer could control the high-speed motion of a recording pen with great accuracy, why couldn't a suitably engineered device do the same thing for an anti-aircraft gun!"

Immediately after the dream, Parkinson and his boss, Clarence A. Lovell, spent two exhausting weeks hurriedly drawing up the specifications to turn the recorder into a quick-thinking analog computer. Existing gun predictors were mechanical devices equipped with a cumbersome array of gears, differentials, and cams. In most cases, operators tracked the target in range, elevation, and azimuth through special telescopes. The act of moving the scopes onto the target caused the contraption to align gun barrels in the proper position. Although accurate in many instances, such predictors were unable to follow low-flying, fast-moving targets. The design Parkinson and Lovell came up with performed almost every mechanical step electronically.

Originally, their goal was to build a predictor that worked with existing radars or optical sights. But over the course of the war, much of Bell Labs' work, including the predictor development, was sponsored by the National Defense Research Committee. Given the close relationship developed between its staff and the MIT group—beginning in May 1941, the Rad Lab's Jerrold Zacharias and another staffer had spent eight months at Whippany striving to adapt the initial airborne interception design for P-61 Black Widow nightfighters, and men from each facility constantly circulated project updates—it didn't take long for word of the new predictor to reach Ridenour. Beginning shortly after Pearl Harbor, the XT-1 and what eventually became the M-9 electronic analog computer were designed concurrently.

In its final form, the Bell predictor consisted of a series of input potentiometers—the loose equivalent of a modern computer keyboard—incorporated into the radar set. As the XT-1 adjusted itself to track airplanes, mechanical connections rotated the potentiometer shafts, so that voltages proportional to the target's elevation, azimuth, and distance—the coordinates needed to plot a plane in space—were passed on to the main vacuum tube processing units, which used the information to divine bearing and speed. Other potentiometers, in essence a computer memory, stored gun tables that provided the time of flight needed for an artillery shell to cross the enemy plane's path at a given instant. These, in turn, transmitted voltages corresponding to where guns needed to point. Signals were relayed through an array of synchros, establishing fuse settings and orchestrating the movement of hydraulic-powered gun mounts to the proper angle.

Lee Davenport loved the predictor. With it, the pace of the XT-1's development picked up dramatically in the post–Pearl Harbor frenzy. Under the Rad Lab's reorganization that March, the gunlaying program became a separate entity, Group 81. Louis Ridenour moved up the ladder to become a division head. Ivan Getting, just turned thirty, took overall charge of the work, with Davenport handling day-to-day affairs. By April Fool's Day in 1942, the prototype was ready for competitive trials at Army anti-aircraft command headquarters at Fort Monroe, on the Virginia coast. Once the set locked onto a signal, only two humans were needed to determine a target's fate—a Rad Lab staffer to monitor and adjust the range-tracking handwheel and an Army man to physically set the timer on the shell fuse. The rest of the process, aside from the gun crew loading shells and the battery commander who ordered the firing, came automatically.

CONICAL SCANNING

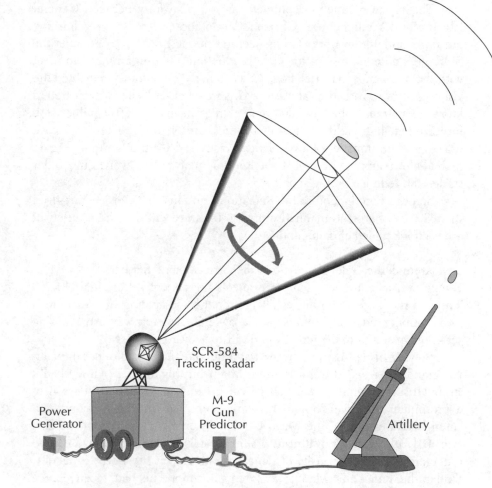

The radar locks onto a target within the overlapping part of a revolving microwave beam. The M-9 then predicts the target's future position and angles the guns accordingly. *Source: Buderi, "The V-1 Menace," p. 30.*

For the tests, airplanes flew over the sea towing cloth banners laced with copper wire threads to simulate a target's radar reflection. Although the banners trailed well behind the towing craft, Davenport worried that the whole project might be blown if the guns accidentally shot down a plane. He climbed onto the truck roof and watched through a sighting telescope mounted on the antenna to make sure the eye was properly pointed. In the end, the set performed magnificently. Targets were hit after only a few rounds, without gunners needing visual contact. The next day, the Army ordered 1256 units, formally designating the radar the SCR-584.

The prime production contract was split sixty-forty between General Electric and Westinghouse. Chrysler Corporation, already busy churning out tanks and 40mm Borfors guns, seemed the best bet for redesigning the SCR-584's antenna mount for mass production. Getting arranged to meet with the company's maverick boss, K. T. Keller. The automaker greeted the visitor gruffly and asked what was up. Unsure whether Keller was cleared to know about radar, the physicist replied that he had a $100 million job for Chrysler, but could not reveal specifics, only hand over representative drawings. The impatient Chrysler head dismissed Getting abruptly. Afraid he had blown an important deal, the Rad Lab man spilled everything. Keller started the redesign on the spot.

So began production of the Radiation Lab's first large order. But given the diversity of projects that had surged to life in recent months, a variety of other efforts trailed close behind.

Around the time of the SCR-584's triumphant field trials, Lee Du-Bridge oversaw close to a score of projects not part of his original charter. Three of the boldest and biggest — an aircraft blind-landing system, a precision bombing radar called Eagle, and a far-seeing Microwave Early Warning set — had arisen from the fertile mind of Luie Alvarez.

Belying his last name, Alvarez was a tall, fair-skinned blond. His paternal grandfather hailed from northern Spain but his mother's family had a strong Irish background. He had grown up in San Francisco, the son of a prominent physician, but in his early teens had moved to Minnesota when his father joined the Mayo Clinic. When he signed on at the Rad Lab in 1940, Alvarez was already a shining star in physics. He had earned a doctorate at the University of Chicago and was on the Berkeley faculty. Under the Rad Lab's March 1942 reorganization, he had taken charge of about twenty workers and created Division 7, Special Systems, almost solely to develop his inventions. The group operated out of a lab in Building 22, the wood-frame low-rise hastily thrown up on the northwest side of the campus. The rough-hewn quarters consisted of a vast open bay, with several desks and a drafting table in the middle, and workbenches and a small conference room lining the perimeter. Everybody could see what everybody else was doing. Alvarez loved the place. ". . . It reminded me of the Radiation Laboratory at Berkeley, which was an old broken down wooden building, where one could drill a hole into the wall anytime he felt like it," he reminisced.

The longest-running project was the blind-landing system. Alvarez had conceived it the previous August, shortly after his return from gall bladder surgery at the Mayo Clinic, where his father still conducted research. The operation, at the hands of William Mayo's son-in-law Waltman Walters, could be counted a success. But Alvarez had suffered for weeks from nauseous aftereffects of the ether anesthetic. Bedridden and unable to stretch his limbs, he developed phlebitis in both legs. The condition proved so limiting that he initially felt unable to take up regular work, so he had busied himself by getting up to speed on the various lab endeavors.

One of the first things Alvarez witnessed was a rooftop demonstration of the gunlaying radar. As the set followed Dave Griggs flying overhead, Alvarez had suddenly realized that if a radar tracked planes accurately enough to direct guns to shoot them down, it should be able to guide pilots to safe landings in bad weather. He envisioned a system in which radar signals would be converted to positive and negative voltages that indicated an airplane's departure from an ideal glide path. Controllers would then "talk" the pilot back on track. Such a system, he knew as a pilot himself, would fill a critical void. One of the biggest problems facing the military was its inability to carry out bombing and air reconnaissance missions because planes couldn't land in fog and other conditions of low visibility.

Almost immediately, the Rad Lab had launched a low-level blind-landing project called Ground Controlled Landing, or GCL. Joining Alvarez as project engineer was twenty-three-year-old Lawrence Johnston, who had interrupted his doctoral work in physics at Cal Berkeley to follow his mentor east. The group rigged a crude prototype of the system, with optical sights instead of radars, and in late March 1942 began conducting trials at the Quonset Point Naval Air Station outside Providence. Although accompanied by a co-pilot whose vision was not restricted, Navy Chief Machinist Mate Bruce Griffin became the first to fly under the hood—that is, with a hood literally placed over his head to prevent cheating—all the way to touchdown. Griffin's feat did not qualify as a radar landing because of the optical sights. Yet it seemed a simple matter to unplug the sights' selsyn wires and plug them into corresponding selsyns on the XT-1. As soon as Ridenour's group could spare one, the men borrowed a gunlaying prototype and began converting it for blind landing.

Alvarez's other two big projects—the Eagle high-resolution bombing radar and the Microwave Early Warning system, or MEW—were linked closely. The core idea behind both traced back to a spirited conversation between Alvarez and Taffy Bowen on November 19, 1941. The pair wanted to build a bombing radar with a resolution comparable to that reputed for the optical Norden bombsight—something able to clearly pick up factories and bridges instead of showing indiscriminate blobs on the screen.

The men had resolved little beyond establishing the need to build the system on three centimeters instead of the more basic ten-centimeter wavelength. After Bowen left, though, Alvarez suddenly got the idea of mounting a twenty-foot-long antenna along the edge of an aircraft's wings. To focus the pulsed energy into the narrow, high-resolution beam he needed, the waveguide would be built with a series of slots, a leaky pipe in engineering jargon. From each slot, a bit of radiation would escape. As an expert in interference phenomena, Alvarez knew, just as children simultaneously dropping a series of pebbles in water could see, that the potpourri of waves would cancel each other out in some places and reinforce themselves in others. Properly designed, such a linear array would result in a tall and narrow swath of radiation shaped like a beaver tail turned to one side — just what the radar doctor ordered.

From there, Alvarez had quickly brought the bombing project to life. Early in 1942, before acquiring the more politically correct name Eagle, the radar was dubbed EHIB, for Every House in Berlin. Early that May, Alvarez adapted the idea of using a leaky pipe array to produce narrow, high-resolution beams as the basis for the Microwave Early Warning system. The Berkeley physicist first envisioned a reflector a hundred feet long and ten feet high that would scan by being mounted on a railroad car and be driven slowly around a circular track. Later, after some recalculations, he adopted a more manageable fixed array, twenty-four feet wide and eight feet tall.

Alvarez revealed his MEW idea over dinner one evening to Lauriston Marshall, head of the Rad Lab's Roof Group and a member of the steering committee. Marshall was impressed enough to suggest taking the short walk to DuBridge's house. The director gave Alvarez the green light on the spot. Work got started quickly, with the Berkeley man replacing his leaky pipe slots with dipole antennas — a move probably intended to give him more precise control over the amount of energy coupled out of the waveguide. However, both MEW and Eagle were plagued by unwanted antenna beams known as side lobes that angled off to either side of the main beam and robbed it of power. The lobes, which developed from competing wave fronts caused by the way energy radiated from the dipoles, posed a formidable challenge.

Alvarez's ingenious solution not only carried his creations over the hump into war, it was also destined to have a major impact on antenna design. He delighted in solving problems, and he periodically shouted aloud when an insight struck his fancy. The men of Division 7 winced at such brainstorms; they usually meant making complicated changes. On May 19, though, Alvarez triumphantly announced the answer to the side lobe problem: adding another set of dipoles between each existing dipole. Such a layout by itself created a condition where alternate dipoles would be 180 degrees out of phase with their neighbors, preventing any beam of radiation from being generated. But Alvarez saw that reversing the orientation of every other dipole — by twisting them 180 degrees about their coaxial axis — put the new set back in phase with the original set. Outside the waveguide, the tighter dipole spacing eliminated the conditions for side lobes.

It was classic Alvarez — a beautifully simple arrangement that made others wonder why they hadn't thought of it. Reverse dipoles soon became a widely used antenna feature, standard on the search radars at any modern airport. It solved the chief problem with MEW right away and rid Eagle of one of its main hang-ups.

There was an added bonus. That same May, around the time Alvarez conceived MEW, the blind-landing system flopped in tests at the Oceana Naval Air Station near Norfolk. All through the trial the XT-1's antenna kept bobbing wildly, alternately locking onto Griffin's aircraft and its mirror image three degrees below the runway surface. It soon became evident that another type of radar system was needed, one that delivered greater angular accuracy than the XT-1's conical scanning. The men had gone home feeling defeated.

Of all Division 7's projects, GCL ranked nearest and dearest to Luie Alvarez's heart. Griffin had just taught him how to fly by instruments, and he almost desperately wanted to fill the need for a blind-landing system. Not long after the trial, the California transplant confessed his frustration to Alfred Loomis when the financier came to town and invited him to dinner at the Ritz Carlton in Boston. Dining in Loomis's suite, probably on June 1, the two men reconstructed the entire sequence of events at Oceana. Loomis felt strongly that GCL offered the only blind-landing solution that feasibly could be developed in time to help the war effort. "I don't want you to go home tonight until we're both satisfied that you've come up with a design that will do the job," the Microwave Committee chairman told his guest.

Alvarez found the answer just before midnight. "I'm convinced that without Loomis's ultimatum that night World War II would have seen no effective blind-landing system," he confides. "I would have immersed myself in other interesting projects to forget my disappointment and embarrassment. Many lives would have been lost unnecessarily." The system he devised, with crucial help from Loomis, was a complicated array that depended on four radars. Two were general purpose, for search and range finding. The other pair had to be extremely precise, one to guide planes in elevation angle, the other for tracking left and right of the approach path.

The design achieved some of its precision by shifting to three centimeters from ten, but hinged on reverse dipoles to produce the narrow beams needed by the two precision antennas.

The revitalized GCL team soon worked up a prototype called the Mark I, with a patent filed in the names of Alvarez and Larry Johnston. The Division 7 leader insisted the system be self-contained and fully mobile so that it could be carted to different runways as the wind shifted. The men crammed everything into two trucks, which would take a position a hundred feet to one side of a landing strip's upwind end. Built onto the roof of the larger truck were three radars and their antennas. A diesel generator powered the units, with just one engineer needed to keep things running.

The second, more comfortably outfitted vehicle contained a radar ranger and the director — a mechanical analog computer designed by Johnston. He had programmed an entire library of large Masonite cams with the most likely glide paths and horizontal variances. Depending on airport conditions and runway widths, he would select the appropriate cams and, like a disc jockey playing a record, lock them into his apparatus. The cam followers in the director were linked to selsyn transducers, whose shaft angles represented the ideal glide path. These selsyns were linked electrically to counterparts in the radar in such a way that any deviation of the airplane's signal from the selected glide path showed up as a voltage output on error meters read by the controller. Three other men — human servos — moved cursors on their screens to actuate the radar's selsyns and generate the range, azimuth, and elevation information fed to the error meters.

About this time a high-ranking Air Forces officer informed Alvarez that instrument flying all the way to touchdown was not only impossible, it was forbidden. The men bristled, but their leader cheerfully changed the name from Ground Controlled Landing to Ground Controlled Approach. No matter what it was called, the landing project had taken off.

•

One last vital piece of unfinished business, the German submarine, occupied the Rad Lab in the early months of 1942. For a few weeks after Pearl Harbor, the country had concentrated on the Pacific. Japanese forces stormed onto the offensive in Malaya, the Philippines, and Hong Kong, and worried U.S. officials warned of possible attacks on the American mainland. But much of the focus shifted back to the Atlantic on January 12, when Karl Dönitz inaugurated Operation *Paukenschlag*, Drum Beat.

The strategy was elegantly simple. Once mobilized, the United States would become the principal British arms and supply merchant. Therefore, striking East Coast sea lanes afforded the best chance to stop that transformation in its tracks. Many of the raw materials that fed the country's industrial appetite steamed up from South and Central America: oil from Venezuela and the Dutch West Indies; bauxite, the basic ingredient for aluminum aircraft fuselages, out of Brazilian and Guianan mines. In addition, U.S.-produced coal, iron, concrete, lumber, and oil were often shipped around Florida to the great cities of New York, Philadelphia, and Boston.

Dönitz's drum ushered in the second Happy Time. Outside of planting some mines near Chesapeake Bay and erecting anti-submarine nets at a few East Coast harbors, the United States mustered almost no defense against the U-boat. No convoy system existed along the Atlantic seaboard. The country possessed no search-and-destroy vessels, and, as of December 1941, could count only nine Army Air Forces planes for offshore patrols. Three weeks into *Paukenschlag*, a handful of U-boats had sunk at least thirty-five ships in the lightly protected North American coastal waters.

The United States pitifully tried to counter by dispatching a few destroyers from transatlantic convoy duty, and ended up losing one of its warships. In January and February, a slightly beefed up Air Forces — without radar — managed attacks against only four U-boats in some 8000 flying hours. Losses mounted throughout the quarter to a staggering 216 vessels in the North Atlantic alone, the vast majority in waters patrolled by the once-respected, now almost disgraced, United States Navy.

At the Rad Lab, reflecting the bleak situation, Lee DuBridge pushed ASV as his top priority, and a number of broad changes took place in the laboratory's anti-submarine efforts. Experimental projects to install ten-centimeter radars in a Lockheed transport and a Navy PBM-1 flying boat were abandoned, with staff members quickly shifting their expertise to ongoing efforts. One surviving program was George, the model designed for blimps, which was already in pre-production at Philco. It had come so far so fast that the Navy was demanding a similar radar for patrol bombers. The Army, too, had grown interested in a variant for its Liberators. Around the same time ten rather ragtag B-18s arrived at East Boston Airport as part of an Air Forces request for the crash installation of air-to-surface-vessel radars, intended at various stages of the war for sentry duty in the Panama Canal, the West Coast, and the Aleutian Islands. This effort, too, drew heavily on the phased-out programs.

Norman Ramsey's experimental three-centimeter system, arbitrarily called ASD, later Dog, was finally ready for its first test run that spring. Ramsey climbed into the aircraft with a few other men and flew over Boston harbor to look for floating tin cans or barrels that might serve as a suitable model for a periscope. As the pilot cruised over the water, nothing appeared on the Plan Position Indicator.

"Find us a boat and we'll go over that," one of the radar men called to the cockpit over the hand-held microphone.

"Fine," the pilot related, soon adding: "There, we went over one."

Still, the radar screen remained empty. Ramsey grabbed the mike. "Well, find a bigger boat," he requested. The process was repeated a few times, until the pilot finally called back in a disgusted tone: "That was the *Queen Mary* we just flew over."

It took only a little fiddling with the waveguide to coax the radar to life. Back over the water about a week later, to the radar team's astonishment, the prototype spotted a German submarine just off the coast. The pilot, a Navy officer already upset at having to cart around civilians during wartime, grew furious over the dearth of weaponry on board, and disgustedly hurled a monkey wrench at the U-boat. "This slowed up work at the Radiation Lab for the next several months," Ramsey recalled, "because then the pilots insisted on loading up with depth bombs before they ever took off on our experimental flight. . . . " As far as he knew, none were dropped.

Denis Robinson's late-starting campaign to outfit B-24 Liberators cata-pulted past all its predecessors. When Robinson arrived at MIT the day after Pearl Harbor he had met a hail of well-wishers. "They all came and shook my hand and said, 'Now we are allies, I will do anything you want,' " Robinson remembered. "I thought they were working pretty well before, but boy did they go to work!"

Dumbo I's radar was ready for its maiden flight just four days after Pearl Harbor. The successful test took place the day Germany and Italy declared war on the United States. Shortly thereafter, the elephantine craft went through its paces before a contingent of U.S. and visiting British officials. In March a Royal Air Force crew came over to fly both Robinson and the plane across the Atlantic for a series of trials against a British submarine at Loch Neagh in Northern Ireland. In head-to-head competition with the British meter wave Mark II ASV, the Liberator's centimeter waves demonstrated their clear superiority.

An elated Robinson flew to London and discussed the tests with Air Marshal Sir Philip Joubert de la Ferté, commander in chief of Coastal Command. Until the meeting took place, airborne radar's record against the U-boat had proven nothing less than dismal. In all of 1941 the three hundred or so planes equipped with longwave ASV sets could thank the technology for only two kills. Joubert desperately sought to step up the pressure on Dönitz's marauders, especially as boats sallied to and from their concrete pens at St. Nazaire and Lorient on the Bay of Biscay. Microwave radar promised salvation. "Robinson, what the hell are you sitting there for?" the air marshal commanded. "Get back across to America and get more of these!"

But neither in the United States nor the United Kingdom would things come quite so easily.

7 • Battle for the Atlantic

"The U-boat at sea must be hunted, the U-boat in the building yard or in dock must be bombed."
> Winston Churchill, declaring the Battle of the Atlantic in secret parliamentary session, June 25, 1941

"Admiral King had a terrible blind spot for new things—and about as rugged a case of stubbornness as has been cultivated by a human being."
> Vannevar Bush on Ernest J. King, U.S. naval chief of staff during World War II

Surveying the Washington landscape in the first gray months of 1942, Vannevar Bush was not pleased. America's entry into the war, the Radiation Laboratory's new respectability, even the rising armed service awareness of radar's virtues—none of it went far enough for his tastes. Beginning some eighteen months before Pearl Harbor, Bush had mustered the country's rich scientific talents to help win the war he felt certain would involve the United States. With microwave radar and other National Defense Research Committee technologies nearing production, he foresaw an even tougher fight to see the bounty properly employed.

The steely civilian scientist, a student of the effect of technology on warfare dating back to his World War I days developing submarine detectors, had two major problems with U.S. military thinking. On one level, despite the fact that Army and Navy representatives swarmed to the Rad Lab in greater numbers, top brass still had not really embraced microwave radar. In the First World War military planners had erected an insulating wall between their own shops and civilian science. Bush was convinced that in the present fast-paced, increasingly technological conflict, the Allies could not afford to repeat that mistake.

On a deeper level, even if military leaders came around to integrating the latest NDRC developments—radar, rockets, and everything else—into their operations, Bush wanted his charges to have a stronger hand in formulating the strategies to exploit the weaponry. Bush did not intend that scientists should plot out broad military campaigns. But he was convinced that sharp technical men were needed to help plan key operations—with anti-submarine warfare the prime example—that depended on a sound knowledge of the latest technology.

It was an extremely tall order. Throughout 1941, Bush had hammered for closer cooperation between his organization and the various service branches. But while he could point to some success—witness the Rad Lab's swelling Army and Navy liaison offices—the research czar could not crack the military's closely guarded inner ranks. Although sometimes willing to listen, the two main Army commanders, Chief of Staff George C. Marshall and Air Forces boss Henry H. "Hap" Arnold, suspected the cagey engineer of worrying more about his own power base than the war. With the Navy, relations were far worse. From the early NDRC days, Rear Admiral Harold G. Bowen, then directing the Naval Research Laboratory, had resented the Rad Lab's dominant role in microwave radar, which left his own shop to concentrate on longwave systems. At the same time, stubborn Navy leader Ernest J. King seemed to put technical people and their gadgetry in the same forgotten mold as bean counters. The previous spring King, then Atlantic Fleet commander, had brushed aside the pioneering centimetric system installed on the USS *Semmes*, declaring, "We want something for this war, not the next one."

Dismayed at the ravaging effects of Operation *Paukenschlag* on U.S. shipping, Bush opened an ambitious campaign to sell microwave radar—and what he liked to call the scientific method—to the American military. At the time Bush considered stopping the German submarine the most crucial aspect of the war effort, and nothing seemed better suited to bringing about the downfall of Dönitz's fleet than the Rad Lab's microwave air-to-surface-vessel systems. Yet all the technology in the world did no good if it wasn't used properly, and he meant to erect a bridge between science and the armed services that ensured the fruits of his outfit's labors were fully appreciated. Such a desire, carried into action, added a political and ulti-mately personal dimension to the technological arms race. And like the Radiation Laboratory, the NDRC, and its parent organization, the Office of Scientific Research and Development, Vannevar Bush's latest ambition marked another crucial difference between the Allied and Axis war efforts.

The Army took the friendliest view of civilian science. Bush started with its top civilian official, War Secretary Henry Stimson, the respected Republican lawyer who had served as William Howard Taft's war secretary from 1911 to 1913 and then as secretary of state under Herbert Hoover. The eminent statesman, born in the shadow of the Civil War, counted himself an old friend of Roosevelt—Teddy Roosevelt. But he had been coaxed back to public life by Democrat Franklin, who not only wanted the veteran warhorse's experience but also shrewdly wished to head off opposition dissent during a time of turmoil.

Bush's assault on the secretary was aided by Alfred Loomis, chairman of the Radiation Laboratory's overseeing body and a Stimson first cousin. Bush also engaged the ear of Harvey H. Bundy, a sympathetic Boston lawyer who worked as Stimson's chief personal assistant. As it turned out, the war secretary didn't take much convincing. He, too, found military officers overly stodgy when it came to adopting new weapons. At Bush's urging, by April 1942 he had set up the Joint Committee on New Weapons and Equipment under the Joint Chiefs of Staff. The three-person group, consisting of a brigadier general, a rear admiral, and chairman Bush, aimed at duplicating the Rad Lab's liaison offices on a higher level — namely, keeping the Joint Chiefs up to speed on all technical developments while at the same time informing civilian researchers of military needs. One of its key subcommittees, under Karl Compton and including Lee DuBridge, handled radar research and development.

Bush never saw the weapons committee as the answer to his prayers, but things were moving in the right direction. It also helped that Stimson exhibited a particularly keen interest in radar. "The use of radar in the Battle of Britain had set him on fire," Bush later recounted. Stimson was so unsatisfied with the country's use of the technology that he decided to hire a personal radar advisor. He wanted a man who could keep him abreast of the latest technical advances, as well as problems with radar's deployment and use not likely to reach his ears. The war secretary turned to Harvey Bundy for a nominee. Bundy went back to Bush. The pair first considered Alfred Loomis, but could see no way around nepotism charges. So Bush suggested Edward Bowles, his old MIT student and the current Microwave Committee secretary. Putting forth Bowles's name proved to be one of his most incisive acts of the war.

Together, Vannevar Bush and Ed Bowles served up a one-two punch needed to help break down the military's barriers to microwave radar. Both men worked tirelessly at getting the technology to the front lines, especially in the U-boat war, and placing scientists who knew how to use it in the War Room. Bush carried the torch to the highest Army and Navy echelons, adeptly wielding his presidential access as a trump card. On a daily nuts-and-bolts level, Bowles, who quickly became an intimate of Stimson, Marshall, and Arnold, nailed down the Army and Air Forces. He also saw more of the combat situation than his onetime mentor and influenced naval decisions by playing upon the heated interservice rivalry.

Bush and Bowles maintained a complicated relationship. They often worked closely together, drawing on each other's resources as friends. But at the same time, the pair had different styles, and an undercurrent of animosity could not be concealed. Such a dynamic can arise anytime one powerful and respected man—in this case, Bowles—always rests slightly in the shadow of another. "Bush has got to be dictator. He's got to be boss. He's got to be 'it,'" Bowles once complained. But the underpinnings of the rocky relationship ran deeper, all the way back to their MIT days.

The critical incident had occurred in the early thirties. After completing his master's, Bowles had joined the electrical engineering faculty. Institute president Karl Compton wanted to promote Bush from professor to vice president and dean of engineering and sounded Bowles out on the idea. By that time, the younger man had come to look askance at his former advisor's ethics, seeing potential conflicts in outside consultancies and never forgetting Bush had not paid him for some research back in his student days. "And I, like a stupid idiot, told Karl that I had nothing but admiration for Bush's ability and not one damn bit of use for his methods," the engineer later remarked. A short while later the telephone rang in Bowles's office. It was Bush, wondering if the two could chat. Trapped but unrepentant, Bowles hurried over and listened as Bush laid things on the table.

"I understand you had a talk with Dr. Compton and that you told him that you—I think I can put it exactly: You had nothing but admiration for my ability and not any use for my methods."

Bowles hit back. "No, Van, I said, 'not one damn bit of use' for your methods."

Bush prided himself on straight talking. He asked for an explanation. The two men hunkered down for an airing out. After an hour or so Bush called his secretary and canceled all remaining appointments; the meeting lasted another two and a half hours. Bowles pulled no punches. Some of the tenseness between the men was dispelled. But their relationship was never the same. As Bowles related, "he was like a person who knew I knew plenty, if not too much. And we always treated each other pretty well. But there was always, after that, a kind of feeling among two crooks."

For Bowles, the job in Stimson's camp represented nothing short of a godsend. At the time, besides his secretarial duties, the young engineer served as the Microwave Committee's representative at MIT. He was feeling increasingly unwanted among what he disparagingly called the Rad Lab's "fraternity of physicists." Staff members resented his presence and sought to diminish his role in their affairs. By the time of DuBridge's reorganization early in 1942, Bowles found his workload reduced to escorting distinguished visitors around the lab. "I am a 'secretary' to write letters and do the bidding of others," he complained to Karl Compton.

Bush himself left no clear record as to why he sprang to Bowles's rescue.

But he did not miss the backroom tactics conspiring against his old graduate student. He later told an interviewer that Bowles tried to bring engineering methods into the Rad Lab and that the physicists "pasted the hell out of him." Any personal animosity aside, Bush knew Bowles as a very good electrical engineer well acquainted with microwave radar. He would have considered it criminal to let a man of Bowles's talents waste away during a war, especially when microwave radar needed a champion inside the military.

Bowles arrived in Washington on April 1. No one had yet told him of Stimson's plan, and the engineer thought he was about to sign on as an innocuous consultant to the Army Air Forces—the last twist of the knife in the Rad Lab power play that had left him no choice but to find another job. He felt sold down the river: "I was shocked, humiliated, and discouraged."

In the capital, however, a perplexed Bowles found himself ushered into the war secretary's office in the old munitions building on Constitution Avenue. He shook hands with Stimson nervously. The old man handed over a scathing indictment of the radar network around the Panama Canal. The report had been authored by Robert Watson Watt as part of a special mission to survey the U.S. air defenses. Stimson asked Bowles to read it and state whether he agreed or disagreed. Bowles got through about four points, taking issue with some of Watt's conclusions, when the Army boss stopped him. "I've been receiving a lot of gratuitous advice on radar, and I want someone in my office to tell me what to do and I'll do it." The war secretary added that his sources had "suggested a man, name of Bowles."

The job offered a golden opportunity to rescue his wartime career, but Bowles hesitated slightly. Never a particular fan of Alfred Loomis, and still stinging from his Rad Lab experience, he accepted the position only with the assurance that Stimson's cousin would stay out of his hair. The next Monday, that issue settled, he started work as expert consultant to the secretary of war. Bowles jumped into the role by asking for various Army regulations in order to learn about the organization. As he sat there, an aide kept bringing in books and papers until it seemed the whole war could be spent reading. Bowles resolved to find more effective ways of getting things done.

Outfoxing the opposition, cutting corners, finding devious ways around problems—it was all in the nature of Edward Lindley Bowles. Growing up in the foothills of the Missouri Ozarks, son of a lawyer turned country doctor, he had spent many hours in the backwoods learning to trap wild animals. One winter he beat the local professional three minks to none after he learned to put bait up inside a rotted-out tree core. The shrewd minks avoided the trap going in, but the animals had no reverse sense: all three were snared by the hind legs as they backed out.

The young man carried his hunting experience into the bureaucratic wilderness. After Stimson's move to the recently opened Pentagon in November 1942, Bowles set up shop in Room 4-E-936, on the floor above the secretary's own office. "I spent a good part of my young life outwitting the lower animal and facetiously speaking, later life outwitting the upper ones," Bowles once said. The radar advisor strove to endear himself to Miss Neary, Stimson's secretary, relying on her discretion about when to see the boss. On particularly important matters, he learned to approach Stimson only after an afternoon game of deck tennis. They played at Woodley, where the war secretary had a court next to his house. Stimson would make up rules as they went along, always free to call a violation, play points over, or require that younger opponents hold the racquet with their off hand. Afterwards, they would retire to the living room — in fall and winter before a roaring fire — and drink tea. Bowles used to marvel at how refreshed and sharp-minded his aging boss could be at those times.

Bowles devised many other stratagems. He buttonholed Arnold and Marshall in the exclusive officer's mess. Even though as a civilian he lay outside their jurisdiction, the advisor always passed reports by the generals before sending them on to Stimson. And he sidled up to the Pentagon barber, who provided the skinny on officers up and down the ranks.

From the start, Bowles's chief concern was bringing radar to bear against the U-boat menace. As he took office, the carnage unleashed three months earlier continued unabated. A ragtag group of lightly armed private yachts, tugs, and other small craft — the Bucket Brigade — constituted a first feeble stab at an East Coast convoy system. Dönitz's fleet operated so close to the American shoreline that beachgoers could witness what the German admiral boastfully described as "the red glorioles of blazing tankers." The rampage threatened the entire Allied war effort. Churchill and Roosevelt plotted to defeat Germany before facing the Pacific threat. But if enough oil tankers and other supply ships couldn't get through to England, the planned invasion of the European continent would have to be called off. At least that was how Stimson saw it, and his advisor agreed.

A 1920 convention gave the Army control of land-based aircraft, even those flying over water. Bowles immediately plotted to get the Rad Lab's microwave ASV sets into Air Forces planes. Like Vannevar Bush, he was well aware that despite Dumbo I's successful demonstrations in the United States and Britain, its radar had failed to generate much interest from high-ranking American officers. After both Marshall and Arnold ignored invitations to see the set in action, Bowles coaxed the war secretary up in Dumbo II to check things out for himself.

Their foray took place on April 22. A surface ship simulated an enemy sub, but only Stimson knew where the prey would be. The pilot had orders to fly blind, according to the radar operator's instructions, and at least some of the plane's windows were covered to prevent peeking. When the radar showed the aircraft in position, the operater alerted Stimson. Bowles remembered the event with pleasure. "Lo and behold, when he got the signal, he pulled up the curtain and there was the ship underneath them."

"That's good enough for me," Stimson said. "Let's go home." On their desks the next day, Marshall and Arnold found carbon-copy notes from the war secretary, tantamount to orders: "I've seen the new radar equipment. Why haven't you?"

Stimson's total conversion to microwave radar—he quickly became convinced it would revolutionize anti-submarine warfare—fueled Army interest in the technology. Back at the Rad Lab, Lee DuBridge launched a crash program to build seventeen sets to be divvied up between British-bound Liberators and American planes. For Bowles, however, these first few steps to get centimeter radar into the services didn't begin to approach the problem. Too many people still saw the technology as an end in itself: get the thing out there, it will do good. Or they envisioned radar solely as an adjunct to existing military dogma—a way of providing support for the tried-and-true anti-submarine tactic of the convoy.

Bowles's trapper instincts steered away from the defensive posturing inherent in convoys. Ever since his first day on the job, the Army consultant had quietly developed a plan for placing microwave radar at the core of a U-boat purge, with land-based Army bombers dogging enemy submarines up and down the American seaboard. He spent weeks firming up the details, spreading research materials over the bed in his cramped wartime digs in the Marlin apartments.

What distinguished his report from a mere white paper urging the adoption of the latest technological gizmo was its vision of placing microwave radar inside a broadened military machinery. The special bombardment unit Bowles wanted to form would continually study the latest techniques and equipment with an eye toward determining how they could be assimilated into ongoing operations. That meant thinking of radar as something that worked hand in glove with other systems such as the emerging Magnetic Airborne Detector, designed to track submerged U-boats through their extremely slight distortion of the earth's magnetic field.

The nascent field of Operations Research—Operational Research in Britain—figured prominently in Bowles's plans. The roots of "OR" ran at least back to England during the First World War, but the field had come into its own only during the present conflict. The central tenet lay in applying sophisticated mathematical and statistical techniques to warfare. Scientists would take in raw data—meteorological conditions, combat charts, numbers of rounds fired versus enemy hit—and put them through the number cruncher to try to improve the effectiveness of both weapons and tactics.

OR had already paid off in Britain. Early in 1942 a group led by Patrick M.S. Blackett had studied all airborne attacks on U-boats in the North Atlantic for the previous year and concluded that a Whitley patrol plane's black paint rendered the craft visible to submarine lookouts twelve seconds earlier than if the same airplane was painted white to blend in with background skies. The Whitleys got a fresh coat of white paint, and the ratio of sinkings to sightings, or so one chronicler claimed, promptly rose by nearly a third.

At the time of Bowles's appointment, the only American OR program was the infant Anti-Submarine Warfare Operations Research Group. Although under naval control, ASWORG's civilian staff was paid by Vannevar Bush's NDRC. Its scientific leader was Philip Morse, an MIT acoustics expert who had worked briefly at the Rad Lab. Under his guidance, ASWORG grew quickly into the record-keeping center for the entire U.S. anti-submarine effort, with a state-of-the-art IBM data-processing system to help analyze and track the rapidly enlarging body of information about the U-boat.

One of ASWORG's first reports, issued in early May 1942, made a profound impression on Bowles. Examining optimal patterns for finding submarines, the OR group showed that a radar-fitted destroyer could cover 75 square miles of ocean each hour. Planes using meter wave radar could patrol 1000 square miles in the same time. But aircraft employing the Rad Lab's microwave sets, with their vastly improved range, boosted the patrol capability to 3000 square miles an hour. The report added statistical fodder to Bowles's drive to form hunter-killer groups with bombers carrying microwave radar.

Although such mathematics was hard to contest, Bowles harbored few illusions. Bringing his vision off would take all his backwoods skill and acumen. Not only would it entail unraveling decades of anti-submarine doctrine rooted in convoy tactics, his idea also meant flying Army planes well beyond normal coastal patrol routes and far out over the ocean—Navy territory. By late spring 1942, the omnipresent rivalry between the two service branches had flared into a titanic clash of wills between colorful Air Forces boss Hap Arnold and Navy top dog Ernest King. Arnold maneuvered to build a bona fide strategic force of long-range bombers to carry the battle to enemy territory. King resisted violently, seeking to protect the Navy's sole jurisdiction in foreign lands. Bowles felt certain that if he presented his submarine-fighting plan formally to both services, it would only add fuel to the fire — and meet with a King roadblock.

A frontal assault out of the question, Bowles poked around for a back door. He found it in the fledgling Radiation Laboratory effort to furnish Air Forces B-18s with microwave radar for Panama Canal duty. The mission had since been changed several times: the planes were assigned to the Aleutian Islands off Alaska, then diverted to Florida, and finally sent up to Langley Field in Virginia to work out kinks. On April 1, its first night of patrol, one of the bombers had detected echoes from three enemy submarines in succession, claiming to sink the last boat. Another B-18 had chalked up a second kill on May 1, apparently duplicating in thirty days several months of effort by planes without microwave radar though neither claim was verified by postwar records.

Encouraged by the twin successes, Bowles sought to shanghai the B-18s and make them the core of his East Coast hunter-killer squadron. He first secured the approval of the maverick B-18 project leader, Colonel William C. "Bid" Dolan, then took the idea through back channels to Arnold, who gave him an informal green light. Not until May 20, with all the pieces in place, did he make a formal pitch to Stimson. The war secretary approved the plan the same day. Ten days later Arnold ordered the activation of the Sea Search Attack and Development Unit, SADU, which operated from Langley Field under Dolan. A few weeks after that, following trials in Florida, the microwave-equipped B-18s, joined by Dumbos I and II, were renamed the First Sea Search Attack Group, although the name seems to have been interchangeable for a time with SADU. The move infuriated the Navy brass, who had always considered Army sub-hunters a temporary expedient.

The First Sea Search Attack Group served chiefly as a development and training unit, both to test emerging weapons and to evolve a tactical submarine-chasing doctrine. With its creation, Langley became a sorely needed link between research facilities such as the Rad Lab and the front lines. When scientists came down to visit, they could see operational problems firsthand. Explained Bowles, "We merely put them in an airplane and let them go out and search for a day, and after they had flown out over the water in a B-24 or a B-18, 800 feet above the water perhaps, and having nothing but a horizon to look at, they began to realize the nature of the problem, the human aspects, that is the psychological aspects, also the practical aspects because their life was at stake in the damned airplane. . . ."

Actual submarine chasing, however, was not nearly as successful. The main Army sub-hunters at the time were in the First Bomber Command, which had been activated the previous December and partially fitted with longwave air-to-surface-vessel radars modeled after Taffy Bowen's Mark II system. Sent out occasionally on detached service to help the First, Dolan's smattering of microwave-stocked planes would fly 209 missions, log 1274 hours of patrols, and spot only eighteen subs over the next year. Of those, just four were sunk. Dale Corson, one of the Rad Lab men detailed to help Dolan's outfit, remembered the grim situation. Radar operators sat on the flight deck, relaying enemy positions by intercom to the bombardier down in the nose. Even with these constant updates, however, payloads could not be dropped accurately until the man had the sub in his sights. In the black of night, when U-boats were most likely to surface, the visual part of the process was reduced to total guesswork. "You couldn't hit anything that way," Corson noted. He and Byron Havens, a highly talented engineer just arrived from undergraduate studies at Caltech, talked things over with an imaginative bombardier, Lieutenant Ned Estes Jr., then set about designing a system with a duplicate radar scope so the nose man could control the action.

Even if the cobwebs could be shaken loose, nothing in the Sea Search group's orders addressed the fundamental split between the services. To implement Ed Bowles's hunter-killer approach, planes needed to communicate with ships and put them on the trail whenever the aircraft lost sight of a quarry, missed with its payload, or had to turn home for lack of fuel. That entailed Navy cooperation. And Admiral King remained staunchly opposed to taking the offensive, which he felt would deplete resources from the main job of protecting convoys. "Escort is not just one way of handling the submarine menace," he asserted, "it is the only way that gives any promise of success."

The U-boat situation deteriorated as the summer wore on. Dönitz proved a cagey and elusive foe, ordering his fleet to the Gulf of Mexico and the Caribbean just as the United States finally put together the first formal East Coast convoys and began beefing up traditional air coverage. Over the war's first seven months, U.S. naval forces, destroyers and aircraft combined, chalked up only seven U-boat kills — roughly as many submarines as German factories churned out every ten days. Meanwhile, Allied shipping losses reached devastating proportions. The gloomy month of June closed with a global record of 834,196 tons lost to enemy attacks, the vast majority to U-boats prowling Atlantic waters. Half the sinkings took place within three hundred miles of the North and South American coasts. That summer the U.S. merchant marine's losses surpassed its World War I total.

Optimists could find a few rays of light penetrating the gloom. While only a handful of enemy submarines were destroyed, the hodgepodge of uncoordinated actions — convoys and airborne patrols, sometimes with long-wave radar and occasionally with the Rad Lab's microwave ASV — slowly but surely drove the U-boats farther from American shores. On July 19, Dönitz formally ended *Paukenschlag*, seeking safer waters off Brazil and Trinidad, in his old mid-Atlantic haunts southeast of Greenland, and the Barents Sea near the northern tip of Norway and Finland.

While welcoming such developments, however, realists like Bowles foresaw that merely driving the U-boats farther offshore proved little. Even if the dangers were slight, Dönitz wouldn't risk his crews when he could do as much damage in less hazardous areas. The Army's radar advisor kept pushing for the total, coordinated offensive effort to defeat the U-boat that his original report envisioned. Radar, he had argued, could not be considered "simply as a magic gadget" that once placed in an armed aircraft "ipso facto creates an effective anti-submarine weapon." Just how much remained to be done became crystal clear late that spring when he rode in a B-18 to observe a test of the Magnetic Airborne Detector. In the midst of the trial, the plane flew near a tanker that had just been hit by U-boat torpedoes off Barnegat Light, in New Jersey coastal waters north of Atlantic City. A billow of brown smoke rose from the stricken vessel. Bowles could make out a few crew members bobbing in a lifeboat as debris and oil fanned out across the calm seas. The B-18 passed low over the doomed ship, dropping another life raft. Soon, a potpourri of other airplanes, blimps, and ships arrived on the scene, some to pick up survivors and others ostensibly to chase down the submarine. Watching the hunt, Bowles was struck by the lack of any unified command. Ships and planes fanned out with no systematic search pattern, "each craft for himself so to speak and no leadership," he reported to Secretary Stimson on June 1. Even in his own plane confusion reigned. The magnetic detector made more than twenty contacts with an object believed to be the submarine. Because he was the only one who could whistle loudly, Bowles became the link between the pilot and those in the rear, signaling for the bombardier to drop depth bombs and flares. Meanwhile, a Morse key tie-in with the shore served as the only means of passing on messages to Navy ships in the area. After the B-18's only two bombs failed to detonate, the pilot frantically hunted for the message book, scrawled out communiqués, and passed them back to the operator.

A shaken Bowles concluded far more sweeping action than he originally envisioned was needed to bring cohesion to anti-submarine warfare. He spelled out his ideas in a succinct two-part memo to Stimson that August, noting tersely that ". . . the factual record of sinkings are ample evidence that drastic steps must be taken to destroy the submarine." Bowles blamed both the Navy and the Army for the poor showing. He railed against Admiral King's total devotion to the convoy. At the same time he condemned the slipshod manner in which the Army's First Bomber Command had recently taken over air patrols in the Gulf Sea Frontier, with commanders often assigning aircraft without enough range to pursue the U-boat out to sea.

In almost pleading terms Bowles called for a uniform land-based anti-submarine command under one service, the Army. "Scientific aids are help-less or at most only partly effective without a command and operational framework expressly and objectively set up to make them 'click,' " he wrote. The veteran trapper anticipated driving the U-boat to even deeper waters and noted that the sprinkling of aircraft at hand would be inadequate for pursuing them farther offshore. He wanted 20 percent of the long-range B-24 Liberator production diverted to ASV use. That would bring approxi-mately five hundred heavy bombers carrying microwave radar into the U-boat war by the end of June 1943.

On August 10, three days after submitting the report, the radar advisor met with Stimson and Hap Arnold to consider the bomber allocation. Ar-nold, already planning the North African landings that would take place that November, cut Bowles's request for B-24s in half, but that still meant some 250 submarine hunters. Around mid-September, Army Chief of Staff George Marshall, with King's concurrence, decided to transfer all subma-rine-hunting elements of the First Bomber Command into the new Army Air Forces Antisubmarine Command, which at least held out the hope for the integrated fighting force Bowles envisioned. The command formed the next month, and many of its planes were fitted with microwave radar, ini-tially the Bell Labs' SCR-517. This set, the first mass-produced centimeter air-to-surface-vessel model, was a descendant of the Rad Lab's pioneering airborne interception radar that Jerrold Zacharias had gone down to Bell Laboratories the year before to help develop for the Army's Black Widow nightfighter. Around the time of Pearl Harbor, as the set proved too cumber-some for the P-61, the core transmitter-receiver was adapted for ASV use. Development was completed in March 1942, and the trailblazing radar was rushed into production to meet the mounting U-boat peril. Whereas only about one hundred AI units were built by Western Electric, something closer to two thousand of the various SCR-517 models saw war service. At least in the beginning, air crews picked up the Liberators coming off produc-tion lines, then went to Wright-Patterson Field in Dayton to have the radars installed.

But that was an Army effort. Nothing and no one seemed able to secure the Navy's complete cooperation. King would not place naval forces under his rival service's control, nor could he envision molding his own planes and ships into hunter-killer squads when so many resources were tied up protecting convoys. As a result, the newly hatched Air Forces Antisubmarine Command met yet another roadblock. From the American end, at least, the noose around Dönitz's neck may have been tightening, but it was being cinched at an agonizingly slow pace. Unfortunately for the Allied war effort, these same sorts of clashes and false starts were hampering the U-boat fight in British strategic waters on the other side of the Atlantic.

The vast waterway known as the Bay of Biscay stretches some three hundred miles from Brest in northwest France to Spain's Cape Ortegal. During World War II three-quarters of the German and Italian submarine fleets passed through the bay on their way to and from Atlantic patrols. It was imperative for the Allies to sever this crucial artery, and much of the United Kingdom's submarine-hunting activities focused on the area. As 1942 drew to a close, however, the efforts had met with only limited success, conjuring an ominous sense of foreboding in the highest government circles.

As it was, Coastal Command's first positive strides in the region had not come until midyear. To avoid British air patrols, German submarines usually remained submerged during the daytime, surfacing at night to recharge their batteries. For every twenty-four hours of operation, U-boats needed about four hours of recharging. The nighttime forays rendered Air Marshal Joubert increasingly dependent on radar, and an improved maintenance and training program had overcome many of the meter wave ASV Mark II's liabilities. But there was a catch. Echoes disappeared in clutter as the craft closed on its prey, and at night the human eye wasn't sharp enough to close the gap.

After months of struggle the British had adapted to the situation by combining the Mark II with a powerful naval searchlight, the Leigh Light. Mounted in a retractable turret or cupola beneath an aircraft's fuselage, the light could be controlled remotely by an operator squatted in the nose. Guided by the radar, the man would wait until the plane closed to within a mile of the target, then turn on the beam, hopefully catching the prey unawares and without time to dive.

The marriage of the new searchlight with the old radar in June 1942 initially proved so successful that a late-starting microwave ASV system under development at TRE limped through the summer with an extremely low priority. Several enemy submarines were damaged that first month and two others sunk during July: The first Leigh Light kill, on the fifth, went to Pilot Officer W. Howell, an American who had joined the RAF before the war. Far more important than the sinkings, however, the omnipresent pressure forced U-boats to increase their time underwater, adding to crew strain, depleting batteries, and shortening effective operational time.

Once again, though, Dönitz wriggled out of the closing vise. Probably through listening posts on the English Channel, and perhaps a captured set, German engineers gathered enough information to outfit U-boats with radar detectors. Such devices had always been a great British fear, for a radar set's very emissions can form its Achilles' heel. Because the transmitted pulse had to travel out to a target and *back*, an enemy who knew its wavelength could build a listening device that would let him know he was being detected before the radar operator saw anything on his screen.

Metox, the German listening receiver, came into use that August, just two months after the Leigh Light's introduction, and was fitted to most U-boats by the next month; boats that didn't carry the device formed a convoy behind one that did. U-boat crews called the antenna frame *Biscayakreuz* — Biscay cross. Originally it consisted of nothing more than a wooden cross rigged with wires and clipped onto a submarine conning tower, but even with its crude "ear" *Metox* provided ample time for a U-boat to escape. The general situation soon reverted to pre–Leigh Light conditions. In November 1942, as Allied resources were diverted to support the invasion of North Africa, U-boats claimed a wartime record 117 ships.

The deteriorating outlook brought centimeter radar back into the limelight. Not only could microwave sets theoretically spot surfaced submarines at four times the range of the Mark II, the high-frequency signals were also invisible to *Metox*. At TRE that fall, about the time the Army Air Forces Antisubmarine Command formed in the United States, the existing air-to-surface-vessel program was reorganized and placed under Taffy Bowen's old colleague Bernard Lovell, who was already leading efforts to develop a ten-centimeter bombing radar. To beef up the waning Biscay Bay air offensive, the government ordered the first forty bombing sets converted to ASV and installed in Wellingtons operated by Coastal Command. But it was too little, too late. The radars were not designed for Wellingtons and had to be adapted to work with the Leigh Light, slowing deployment. Moreover, Bomber Command fought any hint of Joubert taking its resources and was able to retain overall priority: future radars would go to it.

In terms of deploying microwave radar against the U-boat, then, that left the Americans. A few of Denis Robinson's Liberators were almost ready. But Churchill suddenly wanted more. On November 20 the prime minister wrote Roosevelt: "The German U-boats have recently been fitted with a device enabling them to listen to our 1½-metre A.S.V. equipment, and thus dive to safety before our aircraft can appear on the scene. As a result our day patrols in the Bay have become largely ineffective in bad weather, and our night patrols, with searchlight aircraft, have been rendered almost entirely useless. . . . No improvement can be expected until aircraft fitted with a type of A.S.V. to which they cannot at present listen, called 'centimetre A.S.V.,' become available." He asked FDR "to consider the allocation of some thirty Liberators with centimetre A.S.V. equipment from the supplies which I understand are now available in the United States."

Whether the President took direct action is unclear. Through one forum or another, however, the message got through and ultimately highlighted the wisdom that created the radar bridge. Many hurdles remained to be overcome, many battles to be fought—at sea and on land. But the seeds planted two years earlier by Henry Tizard, and cultivated more recently by Denis Robinson to help fight the U-boat, were almost ready for harvesting.

•

Karl Dönitz's forces came very close to winning the war in the early months of 1943. As an increasingly exasperated Vannevar Bush examined the Atlantic balance sheet for the year's first three months, the fact became drearily evident.

Dönitz became grand admiral of the German Navy in late January and wielded his added clout to bolster the U-boat fleet. The submariner's latest campaign had gotten off to a slow start. Tempestuous January weather drastically hindered operations. But with calmer seas, Allied losses shot up in February. March brought outright disaster. Wolf packs savaged four successive heavily laden convoys, culminating in a double battle around the last two convoys that raged for days as more than 140 merchantmen, destroyers, frigates, corvettes, and submarines flayed out across the Atlantic. When the waters cleared, at least twenty-one Allied vessels had been sunk for the price of a single U-boat. The losses helped bring the staggering total for the first twenty days of March to ninety-five ships—more than half a million tons. Against that, the Allies managed only twelve kills, barely half the German monthly production.

Bush was stunned. He saw the entire hard-fought Allied war effort as a giant house of cards, on the verge of toppling. "As the depredations of the U-boats mounted early in 1943 there was no doubt in my mind that we were headed for catastrophe," he writes. "It was clear enough that, if U-boat success continued to climb, England could be starved out, the United States could mount no overseas attack on the Nazi power, Russia certainly could not resist alone. We would be forced into a situation where only the advent of an A-bomb could alter the trend toward world conquest by the dictatorships of Germany and Japan, and as far as we then knew, the development of that weapon might go either way."

For months the reports streaming across Bush's desk had foretold of a wealth of powerful submarine-hunting weapons—homing torpedoes and the Magnetic Airborne Detector among them—coming down the pike. Microwave radar topped the list. Lee DuBridge's operation had doubled again, to nearly two thousand workers tackling fifty-odd projects. And while only a handful of shipboard and airborne systems had yet trickled into service, the technology was ready to explode. Rad Lab–developed sets such as the SCR-584 and various bombing radars would account for more than a third of America's 1943 total radar production—meter wave and microwave—and more like two-thirds of U.S. output by war's end. Befitting this impending renaissance, Bush had recently completed a massive reorganization that saw the Microwave Committee elevated to full divisional status inside the NDRC.

Microwave radar, in Bush's mind, was now primed for the U-boat war. The first of Denis Robinson's Liberators had arrived in England that January to bolster Coastal Command patrols in the Bay of Biscay. Fourteen radars, designated the DMS-1000, had been hurriedly built by the Research Construction Corporation, with Robinson scurrying between Cambridge and Fort Worth, Texas, where the sets were installed in planes rolling off a Consolidated Vultee Aircraft Corporation assembly line. These had been supplemented by two squadrons of U.S. Army Liberators—nineteen planes equipped with the early Bell Labs SCR-517—which had begun arriving in southern England even before Churchill's request: the first American operational mission had taken place November 16, but not all the planes trickled into service until early January. By late March, as Bush contemplated the worsening situation, these squadrons had taken up Mediterranean duties in Morocco and were being replaced by British planes fitted with the Philco-built George, of which some 5547 models would be produced. Besides its easy-to-read Plan Position Indicator, George picked up U-boats from an unprecedented average distance of thirteen nautical miles. In addition, at least thirty-two Wellingtons were fitted with Leigh Lights and British microwave systems.

Yet for all the rush to get the sets into service, the effort to really use the improved technology seemed halfhearted, even contrary to war policy. At January's Casablanca Conference, the Allies had declared victory over the submarine their top priority. The United States committed itself to heavily reinforcing the Atlantic, and the Combined Chiefs of Staff recommended filling the Greenland air gap with eighty very long range bombers. For the most part these were specially rigged Liberators, with armor and gun turrets stripped out and gas tanks added. Fitted with microwave radar, the B-24s could take up a patrol station a thousand miles from their base and cover it for four hours, double an unmodified plane's normal tour of duty. At the end of March, however, the Allied very long range bomber fleet in the region numbered just twenty, with no planes based west of Iceland. A couple squadrons of unmodified Liberators based in Northern Ireland housed the only centimetric ASV operating in the North Atlantic.

Perhaps worse, those microwave sets in operation were performing dismally. The Bell Labs' radars carried by the two U.S. squadrons briefly assigned to England ran satisfactorily barely 60 percent of the time. Operators were poorly trained, and maintenance was slipshod. A British Coastal Command study had rated them only marginally better than longwave sets. True enough, an SCR-517 had contributed to First Lieutenant William L. Sanford's sinking of U.519 on February 10, the sole kill by American forces in the period. But the set was big and heavy, taking up scarce space and slowing aircraft performance, and radar had been involved in just two of the eight U.S. attacks on German submarines; an airman's eyes did better.

It was just as Bush had feared: merely coming up with a deadly technology did not ensure its effective application. The engineer was an astute observer of people, power, and processes. He knew that a period of ineffective use always accompanied the introduction of a new device; it took time to overcome training difficulties and superstitions. But he also knew that the problems ran much deeper. The quagmire of military conservatism, interservice rivalry, and lack of political leadership on both sides of the Atlantic had so gummed up the works that even the deployment of microwave radar for purely defensive purposes was being dramatically slowed. In Britain, operational research experts had calculated that a Liberator crew based in Iceland for escort duties would save six ships over its effective lifetime, while the same plane operating as a bomber would drop a hundred tons worth of payload and kill no more than two dozen Germans. Yet a host of influential officials insisted that Germany could be defeated by bombing alone and had prevented many aircraft from being fitted with centimetric ASV radar.

Bush could do little about Britain. Similar divisions, however, permeated the U.S. command. Ed Bowles had helped win the Army over to the idea of a scientific offensive based on microwave radar. But the Navy remained a formidable stumbling block. Not only did Admiral King view hunter-killer groups with disdain, he also did not rank the Battle of the Atlantic on a par with the Pacific campaign, where American forces pressed forward after their impressive showings in the Coral Sea and Midway the previous spring and had only just recently declared total victory in the fierce battle for Guadalcanal. Vannevar Bush decided he might be able to do something about King.

Not many people, from the President on down, knew what to make of Ernest Joseph King. As both Chief of Naval Operations and Commander in Chief of the U.S. Fleet, he held more power than any other American naval officer in history. No one questioned his inherent abilities. Over his forty-year career King had served aboard everything from submarines to aircraft carriers and knew the Navy from boiler room to bridge. Moreover, his private studies of war history took an unusually broad view that covered land and air combat. Nevertheless, King's unshakable belief in the powers of his own intellect ran to the point of extreme cussedness. He trusted almost no one — and even questioned an American political system that permitted civilian intrusion into military affairs. The wry joke among insiders was that King shaved every morning with a blowtorch.

Even when forced to face the mounting German submarine threat in early 1943, King steadfastly refused to come off the defensive. On March 1 about a hundred participants, including twenty Allied general and staff officers, gathered in Washington for the Atlantic Convoy Conference. The Navy boss told the attendees pointedly: "I see no profit in searching the ocean or even any but a limited area such as a focal area — all else puts to shame the proverbial 'search for a needle in a haystack.' Let me say again, by way of emphasis, that anti-submarine warfare — for the remainder of 1943 at least — must concern itself primarily with escort of convoys."

King could dig up a strong argument to support his view. For the last six months of 1942, only thirty-nine of nine thousand ships had been lost in East Coast and Caribbean convoys. Still, to people like Bush, the admiral was behind the curve. He failed to see that with the new technologies and options becoming available — microwave radar, depth charges, homing torpedoes, long-range bombers, escort carriers, and all the rest — convoys could be protected all the way across the Atlantic even while hunter-killer forces sought out U-boats more aggressively. As losses to German submarines mounted early in 1943, a growing number of King's own officers waxed impatient and pleaded for formal seek-and-kill orders that would step up the pressure on Dönitz, exhaust his crews, and make it more difficult for packs to form and mount effective attacks.

Several powerful people already had sought to crack King's shell. The most cogent arguments for an alternative approach had been spelled out in a March 1 report, *The Acute Problem of Ocean Borne Transport and Supply*, delivered by Ed Bowles as a counterblast to the Navy boss's convoy conference address. Microwave radar, Bowles contended, made it relatively easy to find the needles in the Atlantic haystack. At the time, the Army had deployed around a hundred centimetric ASV sets, most installed in B-18s divided between the Army Air Forces Antisubmarine Command and the Caribbean Defense Command. Critically to Bowles, of the 250 heavy B-24s Hap Arnold had promised would also be fitted with microwave radar, only 25 had arrived —well below the 220 slated to be ready by the end of February 1943. The Army advisor laid out a plan whereby these long-range hunter-killers would fix their location via Loran, which was already transmitting across the northwest Atlantic and preparing to go on the air in Labrador and Greenland. That way, if aircraft needed help chasing submarines, surface vessels would know precisely where to converge. "Were it not for delay, indecision and, I believe, lack of appreciation of the problem (exemplified by a failure to make available B-24 aircraft) we would today have some forty of [sic] fifty more microwave ASV installations in heavy bombers for antisubmarine warfare," he railed.

King had been unmoved. The admiral remained unshakable when an impressed Stimson backed Bowles's report before the President. He refused to waver when Stimson and the War Department sought to persuade Navy Secretary Franklin Knox to help establish an autonomous offensive air task force to fight the U-boat. Nor did he concede an inch of ground when Marshall urged the creation of a comprehensive anti-submarine command under the Joint Chiefs of Staff. King's unwavering obstinacy brought Stimson to the verge of raising hell with FDR. A worried Marshall had headed him off at the pass, pleading, "Mr. Secretary, I must work with King."

The President also had been monitoring the situation. After the first stunning U-boat victories of March, he sent a sharp note of inquiry to both Marshall and King. Because of the admiral's Pacific orientation, not a single U.S. Navy Liberator was yet operating in the Atlantic, with radar or without. Roosevelt's note shook loose a promise of sixty planes, to be supplemented by another seventy-five Army aircraft and 120 Royal Air Force counterparts. These aircraft trickled into operational duty. By the end of the month, only twenty very long range bombers were assigned to the Atlantic.

Vannevar Bush did not consider the presidential arm twisting enough. There remained the question of whether to use aircraft for more offensive patrols or mere convoy support. Not only that, Bush adamantly believed scientists needed to be incorporated into the command process. It kept coming back to this central bugaboo. "The usual way is for the military man to use the scientist as an aid and auxiliary," Bush once told a congressional committee. "It is essential in this type of war that the matter go further than that: that there be a partnership of professional men in approaching problems which in this type of war are always military and scientific."

Like so many others, the research czar simply could not fathom Ernest King's refusal to consider the benefits of technology and all that went with it. Bush had once listened in stunned silence as King went over the designs for a new cruiser with a group of young officers. The ship fairly sparkled with the latest radar, but King complained: "There's too much radar on this ship. We've got to be able to fight a ship with or without radar." Determined to shake things loose, the engineer considered, as had Stimson, going straight to FDR to challenge King's competence. That might or might not work, he decided, but it would almost certainly spoil overall relations with the Navy. Instead, after making a personal study of the admiral, Bush concluded King was at once a man's man and an iron horse who constituted a living example of the fallacy of leaving technical and scientific innovations entirely in military hands. The best course, the hard-nosed Yankee scientist resolved, was to tackle the stubborn Navy leader head-on.

The opening gambit came on April 12. Bush sent the Navy leader a long letter detailing his view of the central problem with current anti-submarine warfare efforts. Apparently laboring to avoid giving offense, the OSRD director repeatedly stressed that he had no intention of intruding on military affairs. At the same time, the letter described the close relations Bush held with the Army hierarchy — Stimson, Marshall, Arnold. "I have not been similarly called in or consulted by the High Command in the Navy," the engineer stated. ". . . The condition which obtains, and which warrants study, is that the scientists of the country are not brought into planning of antisubmarine activities at the top level. I quite frankly, think that this is a mistake, and I will state my reason for this belief."

He got to the meat a few paragraphs later:

> Antisubmarine warfare is notably a struggle between rapidly advancing techniques. It involves, to a greater extent than any other problem I could name, a combination of military aspects on the one hand, and scientific and technical aspects on the other. It is certainly true that no scientist could hope to grasp fully the military phases of the problem. This can be attained only as a result of a life spent in close association with the sea, with naval tradition, and with the responsibilities of command. Yet it is equally true that no naval officer can be expected to grasp fully the implications and trends of modern science and its applications. This requires, equally forcefully, a lifetime spent in science, and in the personal utilization of the scientific method.

Bush told King he expected the submarine problem to worsen, running down the list of technologies intelligence experts believed the Germans would soon bring to bear: means to avoid sonar and radio direction finding, radar detectors, better torpedoes, and liquid oxygen boosters to help submersibles shake off pursuers with high-speed underwater bursts. The Allies could counter with their own systems, he noted, but fast action was needed. Bush went on, hinting strongly that microwave radar and OSRD's other weaponry had rendered King's convoy-only strategy obsolete:

> Recent advances in airborne weapons greatly increase the potential probability of a kill of a submarine after aircraft sighting. These are of such striking nature as to warrant a review of our plans in regard to the balance between seaborne and airborne attack, and a re-orientation of our strategy. . . .
>
> I regard these matters as a part of the very core of the problem. I know that the detailed aspects of the several weapons are being carefully studied by naval groups, and men of my organization participate in these discussions. I do not know, however, that the full significance of the modern technical trends is being weighed in the councils where the strategic planning occurs, and that scientific and technical matters and trends are there examined by those who have made a lifetime career in the application of the scientific method. Because I do not know this, and because I feel that the problem is very serious, and the outcome not in sight, I am exceedingly troubled.

Bush offered no outright solutions. He merely promised all possible future cooperation and left the ball in King's court. Still, he felt good about the letter and wrote Bell Labs president Frank Jewett two days later: "I feel more hopeful than I have for some time that we may be getting somewhere."

Around the same time King received similar exhortations from Admiral Julius A. Furer, a former Annapolis classmate who served as coordinator of Naval Research and Development and as an OSRD advisor. Furer considered Bush an opportunist, but he nevertheless stressed to his boss that it would be worthwhile to listen. So far, he advised, the Army derived all the benefit from people like Bush and Bowles, and that might run counter to the Navy's best interests.

In light of these orchestrations, King called in Bush for a wide-ranging talk on April 19, probably in the admiral's expansive Navy Department office on Constitution Avenue. The men spent nearly two hours in vigorous discussion, breaking for lunch with King's staff. Bush found the encounter extremely stimulating. When the Navy leader pressed for specific recommendations, the scientist shrewdly avoided the issue, telling King he preferred to sketch objectives and methods and leave the Navy to run its own ship. Otherwise, Bush explained, his main objective might be lost merely because as an outsider he had failed to fully take into account the inner workings of the service.

Mulling things over, King broached his own idea. A week or two before the meeting he had named Rear Admiral Francis S. "Frog" Low, a submariner and cruiser commander who had once served on the Atlantic Fleet staff, as his right hand in anti-submarine warfare. "Wouldn't our problem be met if one of your scientists sat with Admiral Low and participated in planning with him . . . ?" he now asked.

Bush was close to getting his wish. He praised the suggestion, but held out for three scientists. When King acquiesced, the engineer pressed to get the details on the table. Advisors must have full access to all naval information pertinent to the U-boat war, he argued. Their recommendations on tactics and even strategy had to go straight to Low and King. After the admiral agreed, Bush put everything down in writing, then dispatched his synopsis to King the next day, promising to follow up quickly with nominees.

The final phase of the Navy transformation in anti-submarine warfare began in May, when King created the Tenth Fleet. With Low as its chief of staff and King himself as overall commander, the fleet contained no actual ships. But it consolidated all Atlantic anti-submarine warfare arms — air and sea patrols, research and development, and anything else — under a single umbrella. Bush's influence glared out from the formal statement King issued to the Joint Chiefs of Staff: "The headquarters of the Tenth Fleet will consist of all existing anti-submarine activities of U.S. Fleet headquarters, which will be transferred intact to the Commander Tenth Fleet. Such additional officers will be assigned to the Tenth Fleet as are necessary for its function, in the same manner as any other major command. In addition, a research-statistical analysis group will be set up composed of civilian scientists, headed by Dr. Vannevar Bush." This became the three-person Civilian Scientific Council, chaired by John Tate, head of the OSRD division on submarine warfare. Later that summer the ASWORG operations research staff was formally incorporated into the Tenth Fleet.

In conjunction with the fleet's creation, King approved more aggressive offensive tactics against the U-boat. The message to the Joint Chiefs noted that anti-submarine aircraft, carriers, escort ships, and submarines would be allocated "to reinforce task forces which need help, or to employment as 'killer groups,' " as King deemed necessary. The admiral would soon become a dutiful fan of the rising hunter-killer breed.

•

By early 1943, as Allied forces jockeyed for a coherent U-boat strategy, Karl Dönitz had twice shifted his headquarters, first from near Calais to Paris, a house on the Avenue Maréchal Maunoury, then to Berlin-Charlottenburg after he became grand admiral. The Navy leader and his select *Stab ohne*

Bäuche, the staff without potbellies, worried increasingly about microwave radar's ability to disrupt their campaign.

That March, Dönitz had become especially concerned about the cause of a falloff in convoy sightings reported by his packs. His operations research officers assembled a detailed intelligence mosaic using all pieces of information known to have been in enemy possession. The investigation concluded that enemy aircraft were "almost certainly" locating submarines through microwave radar. This view was reinforced by news trickling in that a British bomber shot down over Rotterdam in early February seemed to contain a centimeter wave radar transmitter: the cavity magnetron.

The German deductions benefited the Allies in a sense, because they concealed what was probably the central cause of the U-boats' failure to find convoys in late 1942: code breaking. Early in the year Dönitz had added a fourth wheel to the encryption machine's standard three, making codes far more difficult to crack. It had taken some ten months of effort, but that December British cryptanalysts had finally penetrated the extra wheel and begun the old ploy of routing convoys around enemy submarine positions.

On the other hand, Dönitz's intelligence experts had been right — the Allies did have microwave radar. As spring neared more evidence trickled in, intensifying the grand admiral's apprehensions. On March 5, Lieutenant Werner Schwaff signaled that U.333 had been attacked in the Bay of Biscay without the usual *Metox* warning. Luckily, he reported, the boat suffered only minor damage, and the crew had managed to shoot down the enemy aircraft. The next day word of a similar undetected attack arrived from Lieutenant Commander Werner Hartenstein, commanding U.156 near Trinidad.

Throughout the rest of March and into April, somewhat overshadowed by the great wave of U-boat successes in the North Atlantic that Vannevar Bush so fretted over, German submarines passing through the Bay of Biscay continued to report air attacks coming out of the blue. By this time, a dozen Wellingtons equipped with Leigh Lights and the Telecommunications Research Establishment's own ten-centimeter radar were flying night patrols out of the Coastal Command base at Chivenor. Combined with the American sets already pulling daylight duty, the area was cloaked in centimeter waves around the clock, though the bulk of the patrols still depended on meter wave sets. Only three U-boats were sunk over the two-month period, but the pressure was unrelenting. After a series of undetected night attacks Dönitz changed tactics again, ordering sub commanders to recharge batteries in daylight and shoot it out with attackers if escape seemed unlikely. He boosted anti-aircraft gunnery aboard the fleet. Soon small bands of U-boats crossed the expanse in groups, to concentrate firepower.

The strategy backfired. From the Coastal Command perspective, even if the U-boats saw aircraft coming, the wondrous opportunity to attack by day only raised the odds of a kill. During the first week of May alone, three U-boats were sunk in Biscay Bay and another trio damaged. Allied forces claimed four more later in the month. Radar robbed U-boat commanders of almost any chance to traverse the bay undetected. Out of 120 U-boat crossings in May, British and American patrols picked up ninety-eight, mounting attacks on two-thirds of those. Besides the vessels sunk, at least seven enemy submarines suffered damage.

The North Atlantic campaign also escalated. Increasing numbers of Liberators, outfitted with extra gas tanks and microwave radar, and navigating by Loran as Bowles had envisioned, left almost no area uncovered. As the spring of 1943 wore on, any remaining air gaps were filled by a new breed of radar-rich escort carriers and destroyer escorts: seventeen carriers by midyear, compared to just two a year earlier. Many warships also carried centimeter radar, along with newer and highly accurate radio-direction-finding equipment, Huff-Duff, to home in on German communications sources. After its long dormancy the Allies' latent technological power surged to life. Coupled with a more aggressive search-and-destroy policy and better communication between aircraft and ships at sea, U-boats could count almost no area safe. For the month of April, Allied losses plummeted to a mere thirty-nine ships in the North Atlantic. The Reich lost a dozen submarines.

Dönitz could blame bad weather for part of the calamitous performance. But by May, as more British and American long-range aircraft arrived in the Atlantic theater and King's Tenth Fleet picked up steam, he had run out of excuses. Wolf packs that did manage to locate a convoy were unlikely to arrive undetected. When commanders mounted an attack, it was not likely to succeed. When hastily fired torpedoes did somehow bring down a merchantman or two, just as many U-boats were likely to die in the assault.

A strike against the forty-three-ship convoy ONS5 in early May turned into a typical disaster for the German submariners. Some thirty U-boats of the *Fink*, or Finch, wolf pack closed in on their prey south of Greenland. Gale winds scattered the convoy, as at least one destroyer escort turned back to refuel. But for good measure, Dönitz threw in *Amsel*, Blackbird, giving him another eleven submarines on the scene.

A patrolling Catalina aircraft sank one U-boat the first day, May 4. But that night Dönitz's forces picked off a few stragglers and managed to fire volleys into the crowded main pack, destroying seven ships at a cost of several submarines damaged. The battle continued the next day and night, as thick fogs off Newfoundland swallowed up the westward-bound vessels.

Steadily, Allied forces began to flex their technological muscles. Microwave radar provided an inestimable advantage in the fog: U-boats were forced to attack blind. Following a strong centimetric radar contact, the destroyer *Oribi* rammed U.125. Crippled and forced to remain on the surface, the submarine radioed for help. Converging on her signals via Huff-Duff, the British corvette *Snowflake* finished the job. That same night, still in dense fog but armed with radar and sonar, the *Vidette* sank U.531 using Hedgehog, a relatively new multiple mortar that fired impact fused bombs ahead of a ship instead of astern—before sonar contact was broken.

By morning light on the sixth, Dönitz counted six submarines lost and four others badly damaged against just twelve enemy ships destroyed. He called off the attack in the face of clear tactical defeat. The grand admiral blamed the gamut of Allied measures from mortars to escorts. But most of all he blamed centimetric radar. His war diary recorded: "Enemy radar devices operated both from the air and from surface vessels very greatly hampered the operations of individual U-boats, and in addition they provide the enemy with an opportunity, which he is obviously not slow to use, of discovering the preparatory positions of the U-boats and then seeking to evade them. The U-boat is now threatened with the loss of its most important advantage: the difficulty of being located."

As the month progressed, the U-boats could do nothing to stem the tide. A series of actions resulted in losses that nearly matched the number of Allied ships sunk. Gone were the days when Dönitz's aces would sink 80,000 tons per submarine; the figure hovered near a woefully unacceptable 10,000 tons. In rising despair, Dönitz on May 15 broadcast a message to his forces at sea: "The enemy in his efforts to deprive the U-boat of its invisibility has developed a system of radio-location which puts him several lengths ahead of us. I am fully aware of all the difficulties you encounter in fighting the convoy defence forces. Rest assured that as your Commander-in-Chief I have done, and will continue to do, everything to change this situation as soon as possible. . . ."

Only a few days later, Dönitz sent the seventeen-boat *Donau* pack after SC130, sailing with thirty-eight ships out of Halifax. Shortly after midnight on the calm, moonlit night of May 19, the first U-boats spotted the convoy. By morning seven had assembled, smelling the kill. But a Royal Air Force Liberator flying out of Northern Ireland burst on the scene at first light, bracketing U.954 in a hail of bombs and sinking her within minutes. One of Dönitz's sons went down with the sub. Flying on, the Liberator located five more submarines. Bombs expended, the pilot radioed for assistance. Allied forces now kept a series of support groups, usually a small carrier and a handful of destroyers, at sea at all times. If a convoy fell under fire, these roving brigades would rush to the rescue. Support vessels steaming to aid convoy SC130 sank another U-boat with a mortar attack. Meanwhile, a second Liberator arrived and forced six U-boats to dive. Two additional long-range bombers took up the slack that afternoon and attacked three more submarines. Over the whole battle only one attacker got into position to fire, missing with all torpedoes. A blanket of air cover arrived the next day and forced fourteen additional U-boat dives. The convoy didn't lose a ship.

Dönitz could take it no longer. In the first three weeks of May his forces had lost a record thirty-one U-boats in the Atlantic alone, with two more missing. On the twenty-fourth the grand admiral admitted failure in his war diary: "Thus our losses so far in May have reached an intolerable level. The enemy air force played a decisive role in inflicting these high losses. This was due to the increased use of land-based and carrier-borne planes, together with the opportunities for surprise afforded to the enemy both day and night by his improved location devices."

That same day Dönitz ordered most of his North Atlantic boats into southern waters. Allied forces sank thirty-eight submarines in May — more than twice the February total of eighteen, the war's previous high. Radar accounted for more sightings than all other sources combined. Meanwhile, only fifty Allied ships were sunk worldwide, thirty-four in the North Atlantic. It was the second straight month of sharp decline.

In the South Atlantic and the Bay of Biscay, the beleaguered U-boats fared no better that summer. In June and July another fifty-four went down — all but ten to aircraft, many of which were based on escort carriers — while Allied shipping losses dwindled to almost nil. Patrolling aircraft developed sophisticated attack procedures. In Biscay Bay, to counter German packs, Coastal Command aircraft flew in formations of their own — seven planes strung out in parallel. If one spotted a U-boat, the pilot hovered overhead and radioed for surface assistance. American forces in the Azores pioneered a two-plane assault tactic. Upon spotting a sub, an F4F Wildcat would begin a strafing run, forcing the boat to dive. Then a TBF Avenger would drop an acoustic homing torpedo into the diving swirl.

That summer the United States finally managed a significant presence on the eastern side of the Atlantic. The 479th Antisubmarine Group arrived in Britain during early July to replace the two squadrons that had gone to Morocco four months earlier. In the first four weeks of service, the planes launched six attacks and shared credit for sinking three U-boats.

Meanwhile, the 479th's predecessors, reconstituted as the 480th Antisubmarine Group, had found new life in North Africa. By July, Lieutenant Colonel Jack Roberts, headquartered in Port Lyautey, had supplemented his forces with several extended-range bombers. Some planes also carried fresh-from-the-factory Bell Labs SCR-717 microwave radars, which were smaller and lighter than the 517s and came with a Plan Position Indicator, but even the old models performed better as maintenance improved and operators grew accustomed to the technology. That summer the group's sets ran at better than 90 percent efficiency, and average detection ranges on submarines jumped from eight miles to around sixteen. Largely as a result, kill rates also improved. Over a ten-day stretch beginning July 5, the group launched twelve attacks on enemy submarines — eight after radar detection. Three U-boats were sunk, two seriously damaged, and two slightly damaged.

An attack the afternoon of July 12 typified the changing situation. American Second Lieutenant Ernst Salm piloted his B-24 through complete cloud cover some two hundred miles north of Lisbon. It was Salm's first operational flight as aircraft commander. The plane was cruising at five thousand feet when radar operator Sergeant Llewellyn A. Williams noted an echo on his SCR-517 screen: the range indicator read twenty-three miles. Guided by his radar man, Salm steered blind toward the blip. Bringing the craft down to two hundred feet, he emerged from the haze to see a surfaced U-boat off the starboard bow about a mile away. As Salm closed on his prey, the bombardier loosed seven Mark XI 250-pound depth charges. Five hit just short, but the sixth exploded forward of the conning tower, and the seventh went off on the afterdeck, catapulting men on the bridge into the sea. Salm banked sharply and returned for a second run. It wasn't necessary. The sub's conning tower was missing — its stern down, bow knifing into the air. Then the U-boat disappeared. Salm dropped a rubber dinghy for the dozen or so survivors and let loose some smoke flares. Six of the enemy were picked up a few days later. It turned out his victim was U.506, commanded by *Kapitänleutnant* Erich Würdemann, a Knight's Cross recipient. Würdemann went down with his ship. Salm had never even spotted a U-boat before and thanked Williams for the kill. "He deserves the credit. I told him to take us to it, and, by golly, he took us."

And so it went. On August 2, discouraged over losing four U-boats in the Bay of Biscay in two days, Karl Dönitz temporarily confined all U-boats to their French bases. His forces sank only four ships that month in the Atlantic, against twenty-six submarines lost. In the four months that began May 18, not one of the 3546 merchant ships sailing in convoys between America and Britain fell to enemy torpedoes, a testament both to convoys and the more aggressive philosophy of attacking with a bucketful of science. Meanwhile, the ratio of U-boats sunk to those spotted shrank from just 1:40 before 1943 to 1:4, with aircraft accounting for roughly two-thirds of the scores.

No one figured the submarine to disappear from the war. In Washington, Vannevar Bush knew from intelligence reports — interviews with war prisoners and technical leaks — that the Germans were readying technologies designed to give U-boats back their edge. One was *Aphrodite*, strings of decoy buoys designed to fool airborne radar by simulating a submarine. The Germans also possessed their own acoustic torpedo, code-named *Zaunkönig* (Wren), that homed in on the sound of ship propellers to enable submarines to attack from farther off. In July 1943, aware the devices would make their appearance sometime that fall, Bush traveled to England to discuss countermeasures. He was Churchill's guest at a War Cabinet committee meeting on anti-submarine warfare.

Aphrodite never worked against airborne radar and drew relatively little attention. The homing torpedo made its first appearance in September. By then, the antidote, a noisemaker called Foxer towed astern of a ship, was ready for deployment throughout the Tenth Fleet. As soon as headquarters received confirmed reports of the torpedo's characteristics — speed, approximate range, and the like — Tate's scientific council and operational research specialists devised a set of procedures to go with Foxer. Within forty-eight hours the countermeasures package sped seabound. "I think it was the fastest thing I ever saw done," Bush later told Congress.

Dönitz's counterattack that fall failed miserably. Throughout September and October, U-boats managed to sink just nine merchant ships out of nearly 2500 that sailed in convoy. Twenty-five of their own number were lost. The official Allied scoreboard credits air patrols with 19½ kills, one being shared with a surface escort.

It was impossible to single out any one factor in Germany's defeat in the Battle of the Atlantic. Even without airborne microwave radar, the U-boat certainly would have been contained, probably overwhelmed, by the sum of Enigma code breaking, escort carriers, and other weaponry: shipboard radars, Hedgehog, Huff-Duff, homing torpedoes, and magnetic airborne detection, which proved especially effective patrolling the deep and narrow Straits of Gibraltar.

Yet despite its shaky start, the Radiation Laboratory's air-to-surface-vessel radars took the dominant place in the story. That was the view of Captain S. W. Roskill, the official British naval historian: "the centimetric radar set stands out above all other achievements because it enabled us to attack at night and in poor visibility." This sentiment is echoed by Samuel Eliot Morison, Roskill's American counterpart: "Microwave radar made possible the large number of kills by aircraft in the spring and early summer of 1943." Dönitz concurred, although he remained ignorant of Enigma. It was also

Vannevar Bush's conclusion, and he knew the full story of Allied code cracking. Of the myriad anti-submarine weapons that came into play during World War II, he writes, "The greatest of all was centimeter radar."

After the severe blows dealt to the U-boat in late summer and fall of 1943, many Allied observers elatedly declared the U-boat a dinosaur — made extinct by superior technology. More astute warriors knew the determined Dönitz was not likely to give up. Still, the German naval commander could do nothing to change the essential tactical situation. In addition to the ominous Allied anti-submarine war chest, the massive American shipbuilding program was churning out vessels far faster than submarines could hope to sink them. If by some miracle the U-boat returned to glory, Dönitz still could not bring down enough ships to stem the stream of supplies and soldiers crossing to England in preparation for the Allies' long-delayed continental invasion.

Not even the cautious Ernest J. King denied the facts. During a Navy Day speech on October 27, 1943, the crusty admiral downgraded the U-boat "from menace to problem." The Atlantic had been secured. The main battles in the European campaign would be fought in the air and on the ground — and that bloody work was well under way.

8 • The Will to War

"It is the sovereign people who will war today, and it is their nerve and morale that must be broken. The great cities are convenient assemblages of the sovereign people; therefore smash the cities and you smash the will to war."

JAMES MOLONY SPAIGHT,
British Air Ministry analyst, 1924

"The bombardment of cities, towns, villages, dwellings, or buildings not in the immediate neighbourhood of the operations of land forces is prohibited."
From Rules of Aerial Warfare — draft,
Commission of Jurists, The Hague, 1923

"It is doubtful whether such rules for air bombardment as those drawn up by the jurists at the Hague . . . will save the world's great cities."
JAMES MOLONY SPAIGHT

EVEN AT THE HEIGHT of the U-boats' fury, the real German menace was the imposing German war machine dug in on the European continent — a line of soldiers, guns, tanks, and airplanes stretching from Italy to Norway. Virtually from the beginning of the conflict, as submarines threatened the slender lifeline between Britain and the United States, Allied political and military commanders resolved to break Hitler's stranglehold on Europe. Smash Germany first was the overall plan. Then deal with Japan.

Until an invasion could be mounted, bombing provided the only means of striking the Reich. Enraged by the bombardment of London, Churchill had ordered the first raids on Berlin in August 1940. But for the next eighteen months, military dogma and a lingering sense of fair play, even after the German strikes on English cities during the Battle of Britain and the Blitz, led British strategists to base their campaign on the pinpoint bombing of enemy nerve centers: submarine yards, railways, synthetic oil plants, factories. The strategy only slowly collapsed as it became clear that pinpoint bombing was a myth. The European continent lay shrouded in inclement weather nearly three days out of five, making it no simple matter even to find the right city, let alone drop bombs on a specific factory. With British Bomber Command preferring the cover of night operations, the operational net result was that good bombing meant getting within a mile or two of the target. And that was *good* bombing.

By spring 1942, policy was shifting dramatically toward the goal of winning the war through the massive bombardment of cities and the obliteration of the German people's will to fight. To keep up such an offensive, aircraft needed to find their objectives in any sort of weather. Since radio waves penetrate cloud cover when visible light cannot, radar and radar-like aids to navigation suddenly became crucial to the Allied area-bombing campaign. Centimeter bombing radars took top priority at both the Telecommunications Research Establishment and the Radiation Laboratory. The resulting generation of offensively geared microwave systems, while not as accurate as hoped, would form the cornerstone of a massive bombing crusade that surpassed anything the world had ever witnessed.

If there existed a single driving force behind radar bombing, and indeed the British bombing strategy, it was Winston Churchill's science advisor Frederick Alexander Lindemann. Known formally as Lord Cherwell after being raised to the peerage in 1942, and familiarly called the Prof, Lindemann had been active on the outskirts of governmental policy before the war, but soon came to wield immense power. He had been entrenched as Churchill's friend and close advisor since not long after World War I. The prime minister depended on Lindemann's ability to translate complex scientific issues into plain language. And the scientist, who eventually joined the War Cabinet, did not hesitate to use his close ties to the British leader to advance policies he considered vital to the country's cause.

Unlike his American counterpart Vannevar Bush, however, Lindemann knew little about the art of delegation and empire-building. Instead, he seemed to create a froth of discontent around him. He was a strong, if not gifted, scientist who had studied in Germany under the Nobelist Walther Nernst. There, he had become close friends with fellow graduate student Henry Tizard, who had later secured for the Prof the directorship of Oxford's Clarendon Laboratory. Lindemann had gradually built up the Clarendon — a poor second to Rutherford's Cavendish Lab at Cambridge — by rescuing from Hitler's yoke a string of top Jewish scientists that included the Hungarian theoretical physicist Leo Szilard and preeminent German chemist Francis Simon. At the same time, though, Lindemann was an oddity who remained an outsider in Britian's scientific circles. A strict vegetarian who loved egg whites, Port Salut cheese, and olive oil, he was independently wealthy and delighted in high society. Scientifically, he trusted mainly in his own ideas and viewpoints, and was often autocratic and unbending. All these factors conspired to make him unpopular beyond Downing Street.

Through Churchill's rising influence, Lindemann had secured a spot on Tizard's Committee for the Scientific Survey of Air Defence shortly after it formed in early 1935 and launched the British radar effort. At the time Lindemann and Tizard were still friends. But the Prof constantly challenged Sir Henry's leadership, urging repeatedly, and ahead of his time, that a higher priority be given to detecting aircraft by their infrared emissions — and to adopting a fanciful scheme that involved dropping bombs suspended from parachutes in front of enemy planes. The disruptions undermined committee unity until an internal coup resulted in the CSSAD being disbanded, then re-formed with the radio wave expert E. V. Appleton in Lindemann's place.

The Prof had come back with a vengeance when Churchill took the government's reins in May 1940, pushing Tizard farther and farther out on the fringes of science policy. Sir Henry suspected the mission to America that September was a way of getting him out of the picture. And despite the venture's success, when Tizard returned from the United States he found himself reduced to little more than a freelance advisor to the Ministry of Aircraft Production.

Alone at the top of scientific policy matters, Lindemann had turned his mind to the bombing campaign. The British had launched precision strikes against targets in occupied Europe almost from the start of the war. Initially, Bomber Command claimed a high success rate. In reality, however, the raids proved little more than wishful thinking. Moreover, the antiquated fleet of Hampdens, Blenheims, Whitleys, Wellingtons, and even single-engined Battles could not operate in daylight without unacceptable losses — virtually dictating night attacks that reduced accuracy even further.

Lindemann exposed the earlier Bomber Command claims of pinpoint accuracy as so much hot air. Early in 1941 he visited Medmenham, headquarters of the Central Photographic Interpretation Unit. After examining photos that at least partially confirmed intelligence reports of payloads falling far from targets, he ordered cameras installed in bombers and launched an independent study of Bomber Command's effectiveness, putting D.M.B. Butt of the Cabinet secretariat in charge of the job. The Butt report, issued that August and taking into account some seven hundred photos from the summer's raids, dropped a bombshell of its own. It concluded that less than a quarter of the bomber fleet came within five miles of its objectives. Missions launched against the Ruhr's heavily defended industrial belt fared even worse: only seven aircraft out of a hundred dropped payloads anywhere near the target.

The report provided Lindemann with a devastating indictment of the folly of precision night bombing. The Royal Air Force resisted the conclusions feebly, arguing that the analysis covered an extraordinary spate of overcast that lowered performance. But its credibility had been battered. Fortunately for the command, Lindemann didn't dispute the need for bombing; he just questioned the means. The science advisor pushed for new blind-bombing techniques, turning to the Telecommunications Research Establishment for help. On Sunday, October 26, 1941, in response to the Prof's call, A. P. Rowe and Watson Watt convened a meeting to discuss the matter.

Meetings on the sabbath, called Sunday Soviets, formed a TRE ritual dating back to May 1940. In the heat of war Rowe had chosen Saturdays as his staff's lone day of rest. Unlike on the traditional holy day, shops, barbers, and such were open, making it easy for the men to handle essentials. Sunday sessions also afforded the opportunity for senior military and government officials to escape London for discussions with TRE technicians. The term "soviet" reflected the freedom of the exchange. Sometimes as many as forty radar men and officials would gather to banter ideas back and forth. "The rules were quite straightforward," remembered Denis Robinson, who took part in several encounters before leaving for America. "No notes were kept, no names, and it wasn't necessary for us civilians to say, 'Yes, air marshal, no air marshal.' We just said: 'No, it isn't right.' " The next morning, the radar developers returned to work armed with a better understanding of operational needs and performance, and feeling more connected with those who actually employed their technology.

During the October 26 soviet Lindemann and the other attendees covered every feasible idea for blind bombing, down to somehow locking onto the magnetic fields emanating from power grids. The most promising techniques were code-named Gee and Oboe. Neither relied on reflected signals to detect targets and could not be considered classic radar. However, both employed radio pulses to try to accurately fix a bomber's position over enemy territory.

Gee, the brainchild of TRE investigator Robert J. Dippy, was furthest along, just a few months from full operational use. It employed three widely spaced ground transmitters — a "master" in the center, "slaves" on either side — that sent out carefully timed pulses. Bombers carried a receiver that measured the difference in time of receipt of the three sets of pulses. The locus of all points where this difference remained the same for one slave and its master formed a hyperbola. By consulting special charts, a navigator could find the aircraft's position and compare it with the known coordinates of enemy targets.

Although a nice concept, Gee did not have the accuracy needed for bombing. From 350 miles off, the transmitters could only put an airplane somewhere within an elliptical area six miles wide and one mile long.

Moreover, Gee depended on straight-line radio transmissions between ground stations and aircraft, and so was limited in range by the Earth's curvature. It would prove useful in the war as a close-in navigational aid. But perhaps its main contribution came in helping inspire Alfred Loomis's Loran, which was basically a longwave equivalent where signals bounced off the ionosphere to achieve a greatly extended range.

Oboe relied on two radar stations, cat and mouse. An aircraft housed a beacon transmitter through which it could be guided to fly at a set distance from the cat, therefore tracing out the arc of a circle with the station at the center — and with the radius selected so that the bomber passed directly over the target. The mouse, meanwhile, also tracked the plane's range via the beacon, transmitting the bomb release signal at a precalculated instant. Oboe was deadly accurate. Planes flying at 30,000 feet some 250 miles from England could be placed so precisely their bombs would fall within 120 yards of a target. Like Gee, however, it was limited to line-of-sight transmissions and could not reach beyond the Ruhr.

Lindemann refused to accept either alternative as a final answer. He wanted to reach deep inside Germany and so insisted on a blind-bombing device that could be carried inside a bomber. The Sunday Soviet broke up without a solution being put forth.

Shortly after the session, however, Philip Dee recalled the inaugural test of the cavity magnetron in a prototype airborne interception system back in August 1940. The stunning feature of that experiment had been the way Reg Batt, holding a tin sheet while riding a bicycle in front of a cliff face, stood out on the radar screen instead of being lost in ground clutter. Perhaps, postulated Dee, centimeter radar could similarly pick out towns from surrounding terrain. On November 1, 1941, less than a week after the soviet, two of Dee's crew tested a modified system in a Blenheim. Flying at seven thousand feet, the radar clearly picked out the towns of Salisbury and Warminster, as well as various military encampments.

The idea of a bombing radar had been described by Taffy Bowen in December 1937, and again in a secret and largely forgotten report the Welshman submitted to Rowe in February 1940. This time around, the TRE head seized on the concept. Dee's men photographed the cathode ray tube as proof a bombing radar could work, and their leader excitedly carried the still-wet pictures to Rowe's office. The superintendent could not contain himself, exclaiming, "This is the turning point of the war."

On New Year's Day 1942, with the concept proved and a small production contract being placed with Electrical and Musical Industries, Rowe named Bernard Lovell head of a program to develop the bombing radar. At the time Lovell was busy with a lock-and-follow set to allow fighters to track enemy aircraft. He did not want the new job — even found it objectionable — but Rowe insisted. Lovell became a convert later that spring after massive German raids pounded his adopted hometown of Bath, killing four hundred and reducing large tracts to rubble. With communications to the city severed, he and his wife, Joyce, were cut off from their families. The couple drove back roads to avoid Army checkpoints, finally making it to Joyce's family home, where their second child had been born just a few weeks earlier. The house stood abandoned, its windows, doors, and part of the roof blown out. Happily, both families were safe. Recalled the scientist, "At last, I needed no further urging to do my utmost to help our own night bombers."

Initially, the new radar was called BN, for blind navigation. But by that spring it was known only as H_2S. One story holds that Cherwell said it was a stinking matter the set hadn't been produced earlier and gave it the chemical symbol for hydrogen sulfide, the stench from rotten eggs. The most common belief, however, was that H_2S stood for Home Sweet Home, because it enabled a bomber to home in on a target.

Almost from its moment of conception, H_2S was king. "Nothing was too good for it — as the Prof pushed the prime minister to give it his highest priority," writes Rowe. Lovell's crew quickly developed an all-around scanning antenna, which was positioned under an aircraft's fuselage and encased in a Perspex dome some eight feet long and four feet wide. The cupola lessened the payload and even threatened bomber performance, but Churchill overrode RAF objections. In early May, a few weeks after the first prototype picked up towns four or five miles off from eight thousand feet, the prime minister stepped up the pressure. "I hope that a really large order for H_2S has been placed," he urged the secretary of state for air, "and that nothing will be allowed to stand in the way of getting this apparatus punctually."

Churchill's wish was not fulfilled, as a series of disruptions forced developmental delays. In spring 1942, amidst rumors that a German parachute division had assembled on the Cherbourg peninsula to attack TRE, the site took on the aspect of a beleaguered military outpost. Barbed wire went up. The Army and Home Guard patrolled the area. Preparations were made to put demolition charges on secret equipment. Every night all the magnetrons and many secret files were driven inland to greater safety. Finally, it became too much. At the end of May, in the heat of the H_2S buildup, the entire establishment packed its bags once again.

The home this time was Malvern College, a hundred miles north. There was another chaotic upheaval, and the local population resented the intrusion. Lovell arrived on a rainy day to discover equipment stacked in mud. Beds were scarce and the lines depressingly long at a converted canteen where the thousand-odd TRE workers trooped for meals. No sooner had Lovell gotten back on track when a greater disaster struck. EMI had assigned the H_2S job to its best engineers, a group headed by Alan Blumlein, the television and stereophonic pioneer who had developed the modulator for the successful Mark IV longwave airborne interception radar used during the Blitz. On Sunday, June 7, Blumlein was among eleven engineers and military personnel testing an H_2S unit aboard a refitted Halifax bomber. The plane crashed in the Wye Valley. No one survived.

The loss of Blumlein, which Lovell considered a national disaster, by itself set radar development on its heels. At the same time a political squabble surrounding the cavity magnetron almost paralyzed the program. A magnetron transmitter lay at the heart of the radar, but the idea that a plane might be shot down and the enemy learn about the top-secret device sent shivers into the highest reaches of government. The entire war effort could be set back if the Germans fitted their own nightfighters with microwave radar or started jamming Allied centimeter systems. To make matters worse, the magnetron's main copper block proved virtually indestructible. Engineers tried fitting a destruct device to the transmitter. But in a trial on a captured German Junkers 88, the demolition device blew a ten-foot hole in the fuselage, while the magnetron remained salvageable.

Cherwell, as Lindemann was called by then, had at first insisted on replacing the magnetron with the lower-powered klystron, which was known to the Germans. The Prof reasoned that if planes could navigate to within a few miles of a target by conventional means, a klystron-powered bombing radar could get them the rest of the way. To Lovell's group, losing the magnetron meant losing the system. Only it had the power to guide planes to targets outside the range of Gee. Besides, the TRE contingent argued, it would take the Germans at least two years to learn the magnetron's secrets and develop their own systems. Cherwell conceded only partly, insisting that both systems be pursued.

In the face of all this, the official machinery churned on. On July 3, Churchill called a meeting to discuss the H_2S situation. Lovell arrived with Dee and found a number of men awaiting the prime minister at 10 Downing Street. Lovell recognized Cherwell, Watson Watt, and a few others, but could not put names to many faces. Churchill finally appeared in a blue boiler suit zipped up the front to his wide neck. He asked them into a meeting room and sat down, taking one side of the table for himself, his back to a fireplace. Like Lovell, the prime minister did not know everyone. So he pointed to each in turn while the men identified themselves. Churchill then proceeded to announce that he needed two hundred H_2S sets by October 15. The radar men explained that production models would never be ready in time—that the only working system had gone down in flames a few weeks earlier. As Lovell records the event, Churchill tossed the first of several chewed but unsmoked cigars over his shoulder toward the fireplace, and stated: "We don't have objections in this room. I must have 200 sets by October."

After several minutes, with no encouragement forthcoming, the prime minister finally turned to Cherwell. "What does the Professor think?"

"They can be built on breadboards," the Prof responded.

Lovell was astonished. After all the delays and disaster that had plagued the program, the idea that TRE itself could produce two hundred sets in a few months seemed incredible. But both Bomber Command and Churchill wanted those sets. The prime minister ended the two-hour meeting by calling H_2S the sole means of striking back hard at the enemy. He dispatched the group to an adjoining room to figure out how to equip two squadrons by October.

A crash program to develop the sets at TRE began at once. Until mass production was under way, these were earmarked for Pathfinder squadrons, planes that would use the radar to illuminate targets with flares and other devices for the main bomber force. Much of Malvern College was turned upside down to meet the order. "We had had crash programmes before," recalled Rowe, "but never one like this."

Among those present at Downing Street for the July 3 meeting was a stern-looking man with a round face and short-clipped mustache that made him look like an English Hitler. This was Bomber Command leader Arthur Travers Harris, known to his contemporaries as Bert, and his crews as Bomber, Butcher, or just Butch. During World War II the Butcher raced his Bentley between the Air Ministry and Bomber Command headquarters at High Wycombe. A policeman stopped Harris late one night and reportedly told him, "You might have killed somebody, sir." Harris shot back: "Young man, I kill thousands of people every night!"

Harris had started World War I as a bugler boy and ended it as an RAF major with the Air Force Cross, and then kept on rising. He took over Bomber Command toward the end of February 1942, barely a week after a blunt Whitehall directive changed the nature of the bombing offensive: "It has been decided that the primary objective of your operations should now be focussed on the morale of the enemy civil population and in particular, of the industrial workers."

In effect, the order marked the beginning of area bombing, which aimed at breaking an enemy's morale through massive bombardment of its soldiers, factories, and civilians. Until this policy came through, and Harris arrived to implement it, military commanders had resisted such a change. However, the Butt report had shattered the last delusions that anything close to pinpoint bombing was possible. Moreover, the Blitz had convinced Harris and others, as the directive made clear, that the right kind of bombs and scale of attack could break a country's yen for war.

If any misgivings existed about this "fact," the Prof soon put the matter to rest. Extrapolating wildly from a study of the damage done by Luftwaffe raids on British cities, he issued a highly influential memo examining the fruits of a greatly intensified bombing campaign coldly aimed at razing Germany's largest cities. As he memoed Churchill on March 30:

> In 1938 over 22 million Germans lived in fifty-eight towns of over 100,000 inhabitants, which, with modern equipment, should be easy to find and hit . . . If even half the total load of 10,000 bombers were dropped on the built-up areas of these fifty-eight German towns the great majority of their inhabitants (about one-third of the German population) would be turned out of house and home.
>
> Investigation seems to show that having one's house demolished is most damaging to morale. People seem to mind it more than having their friends or relatives killed. . . . There seems little doubt that this would break the spirit of the people.

Tizard and Blackett saw the memo and argued that Cherwell had vastly overestimated potential damage to the enemy. The untimely counter proved Sir Henry's final undoing. Bombing had evolved into such a matter of faith that Tizard was labeled a defeatist and ostracized. Crushed, he would withdraw at the end of the year to an honorable safe haven, becoming president of Oxford's Magdalen College. From there, apart from serving the government in a perfunctory advisory capacity, the man who had done so much to hone Britain's system of air defense and entice America into the radar game sat out the war.

For Arthur Harris and Bomber Command, however, Cherwell had provided the final rationalization for area bombing. On March 28, just before the memo was disseminated, Harris had sent 234 bombers to level historic Lübeck. Thereafter, he stepped up the heat. In April, Harris dispatched aircraft to the coastal city of Rostock for four straight nights. Like Lübeck before it, the old port went up like a tinderbox. The master stroke — a public relations ploy designed to cement his policies — was saved for Cologne. On May 30 a total of 1046 planes — the Thousand Bomber Raid — took off for the Rhineland city. In less than three hours the aircraft saturated the area with their deadly rain; the typical technique was to drop conventional explosives first, reducing structures to kindling, then follow up with incendiaries. The city burned so brightly that some pilots believed their own planes were afire. When it was over, more than 450 Germans — the vast majority civilians — lay dead, and nearly 5000 required medical treatment. Some 45,000 people were left homeless. "We are going to scourge the Third Reich from end to end," Harris proclaimed. "We are bombing Germany city by city and ever more terribly in order to make it impossible for her to go on with the war. That is our object; we shall pursue it relentlessly."

And he did. When Harris took it over, Bomber Command listed 7448 air crew killed or missing. When the dust settled on World War II, the figure would be nearly ten times higher. Almost all the casualties were officers and noncommissioned officers, many among the empire's best-educated and most highly trained young men.

On the German side, the numbers reached staggering proportions. Entire cities were obliterated, and with them, cathedrals, civic centers, museums, and hospitals. Five million were left homeless. Somewhere between 300,000 and 600,000 were dead, mostly civilians. Bomber Command's leader was right about something — he could unleash a reign of terror. Fundamentally, though, he and Cherwell were wrong: the enemy's morale never broke.

On the heels of the July meeting with Churchill, Bernard Lovell did all he could to quickly deliver H₂S to Pathfinder squadrons. Within a week he finally persuaded the Prof to abandon the klystron program. On July 15 the secretary of state concurred, with the proviso that the magnetron would not fly unless the beleaguered Russians held the line at the Volga River.

While the decision helped focus the TRE effort, Lovell's team could not meet the prime minister's October deadline. The men staggered through what was left of summer, hurriedly fitting experimental systems without proper trials. Antenna scanner problems plagued the new radar: although towns initially showed up from fifteen or twenty miles out, the echo would mysteriously fade out as the plane neared the target, only to reappear later. In September, in the face of the deteriorating U-boat situation, Lovell was told to adapt some H₂S sets for anti-submarine work. This order helped spark the battle over whether top priority went to Bomber Command or Coastal Command, and probably spurred Churchill to ask Roosevelt to allocate some microwave ASV sets to the British.

Meanwhile, pressure to deliver something — anything — for the bombing campaign kept mounting. The night of October 14, Lovell was at Lakenheath, where he sat in on the debriefing that followed a conventional raid on Cologne and heard pilot after pilot admit he had never even seen the target. In a hail of resolution, Lovell threw his crew back into the H₂S effort and managed to equip twenty-four planes—half Halifaxes, half Stirlings—by New Year's Eve.

The plan called for Pathfinders to navigate far into Germany by radar and mark a target with flares, enabling the main bomber stream to find it. Most of the next month was spent maintaining the H_2S equipment and training navigators in its use. Meanwhile, the Russians did their part, cutting off tens of thousands of Hitler's stunned troops near Stalingrad and driving others back to the west. Finally, on Saturday, January 30, 1943, a few days before the dramatic surrender of the surrounded German Sixth Army to Russian forces, the new radar was cleared to fly over enemy territory. Lovell, passing a weekend in the country, received a phone call telling him to hurry to the Wyton airfield near the town of Huntingdon. Berlin was the original target. But after a bad weather report, the raid was shifted to Hamburg. Thirteen Pathfinders took off at midnight.

Lovell spent a sleepless night. The Pathfinders weren't due back until early the next morning. But by six, things looked bad: many bombers had turned around in the face of nasty weather, among them seven H_2S planes. As dawn rose, however, news came in that the remaining six Pathfinders had marked the target, and that a hundred Lancasters had flown perfectly to their marks.

A quick debriefing by one of Lovell's colleagues showed that the crews had come back elated, reporting no difficulty in picking out key landmarks from an average distance of more than twenty miles: Den Helder on the Dutch west coast, Cuxhaven and the Frisian Islands near the mouth of the Elbe, then finally Hamburg itself. Over the next few days, the Pathfinders marked Cologne, returned again to Hamburg, and dipped down to hit Turin in western Italy. An enthusiastic Donald Bennett, the Pathfinder force's maverick Australian leader, cabled Lovell: "Heartiest congratulations from myself and the users to you and your collaborators on the development of the outstanding contribution to the war effort which has just been brought into action."

As events developed, Bennett was not the only one entranced by the radar. During the Cologne run the night of February 2—just the second H_2S attack of the war—a Pathfinder was shot down near Rotterdam. German technicians undertaking a routine inspection of the wreckage the next day did not fail to discover the cavity magnetron, and engineers got right to work unraveling the secrets of the *Rotterdam-gerät*, or Rotterdam apparatus.

Countermeasures would eventually appear. But by then both TRE and the Rad Lab were well on their way to developing higher-frequency systems. Staying one step ahead depended on vision, talent, and a bit of luck. The first two ingredients had been supplied years ago with the establishment of the British and American radar efforts. Just a few months before the Germans discovered their *Rotterdam-gerät*, the Allies' good fortune had also fallen out of the sky.

•

Air raid sirens catapulted George Valley out of a sound slumber. Curious, more than frightened, he threw back the covers and walked to the window, peering out over Oxford Street and, off to the right, Hyde Park. Moving a few feet out onto the tiny hotel balcony, Valley stood in amazement. Ack-ack guns, he guessed more than a hundred, blazed away from positions around the park. Flares, almost beautiful in the surrounding fury, lit up the night sky. The young man could make out a half-dozen or so German bombers amidst the bursts, droning eastward toward the Thames River. The aircraft looked somehow defiant in their disregard for the puffs of flak swirling around them.

Valley, an American come over to England from the Radiation Laboratory to discuss anti-aircraft gunnery, observed the spectacle from the Cumberland Hotel—the same grand old edifice adjacent to Marble Arch where Taffy Bowen had holed up with the cavity magnetron before embarking on the Tizard Mission. It was around two in the morning on a clear night in autumn 1942, probably September. He watched a few more minutes, dimly aware of the tinkling sound of debris crashing to earth in the distance. Suddenly, the tinkling grew louder and the balcony shook under the weight of a small, jagged piece of metal that landed a few feet from where Valley stood. "I better get the hell in," he thought, ducking for cover.

Even as he moved Valley registered that what had clunked against the balcony was not a piece of enemy aircraft but an anti-aircraft shell. The thought struck a strong note of discord. In that instant, the tall, solidly built physicist of barely twenty-eight years hurried back inside the hotel room with a fundamental change in thinking about the war. Valley worked on the Rad Lab's Project II—gunlaying. But now he keenly felt the ultimate futility of anti-aircraft defense in the face of an assault like the one he had just witnessed.

As he walked around London during the remainder of his trip, this view was reinforced. Images stuck in Valley's mind—brownstone houses reduced to brownstone walls, boarded-up windows, bomb craters. Through all the air raids, he believed, the city's anti-aircraft guns contributed only a rain of their own steel to the debris and destruction. Although one point of anti-aircraft fire was simply to disrupt enemy bombing patterns, for the life of him, Valley could not see an ounce of evidence the batteries did any good whatsoever. The damned bombers, he figured, just dropped their loads and flew on home. If they were brought down, it was thanks to interceptors, not A-A guns.

George Valley was no different from almost any man or woman drafted into wartime service — in uniform or out. He wanted to help win the war, in the way that made the best use of his talents. But whatever combination of nature and nurture fashioned him, the result was an unusually hardheaded and determined individual. After experiencing the epiphany in London, the native of Flushing, New York, put his mind to figuring out what he could do besides anti-aircraft. His thoughts turned back to the German bombing run. One of the great problems in retaliating against Hitler lay in the fact that Germany enjoyed one of the best air defenses ever created — Mother Nature — and Valley knew that the cloudy European skies would keep Allied bombers grounded through much of the coming winter. Then a second realization hit him: microwave radar provided a way to break down those defenses.

Even before returning to the Rad Lab, Valley had made up his mind about what to do. Another unavoidable facet of his character was that he simply never mastered the art of being subtle. As he put it, slightly mellowed in later years, "I didn't call a spade a spade, I called a spade a fucking God damn shovel, you son of a bitch." Almost upon landing, he remembered, "I came back and announced to one and all that anti-aircraft was for the birds." George Valley Jr. wanted into the bombing business.

Valley knew his abrupt condemnation of one of the lab's three original projects might not sit well with some colleagues. But, he reasoned, "It didn't make me more enemies than I cared about, because the enemies you have to worry about are smart enemies, and smart people didn't get mad at me unless they had a good reason to."

Fortunately for Valley, one of his friends was Lee DuBridge. After completing a bachelor's degree at MIT in 1935, Valley had earned a doctorate under the director at the University of Rochester, then followed DuBridge to the Rad Lab in August 1941 after two years at Harvard as a research associate and National Research Council fellow. Once back from England, Valley met with his former advisor and pressed the case for switching to bombing. DuBridge listened and agreed: no one could dispute the need for a good bombing radar.

The only such projects on the Rad Lab books in late fall of 1942 were Eagle and NAB, for Navigation And Bombing. The former, Luie Alvarez's extremely sophisticated creation, was enmeshed in technical difficulties. NAB, begun the previous June, was a far simpler ten-centimeter radar modeled after H_2S, but it wasn't much. Try as they might, the small NAB staff could not match the British performance. After examining the early system

—and personally tweaking the controls—Denis Robinson had found it unable to even distinguish between cities and countryside. Not long before Valley confronted DuBridge about switching to bombing, Robinson had written P. I. Dee back in England, gloomily advising: "H_2S did not work in America."

DuBridge put Valley on the NAB project. By December, the erstwhile anti-aircraft man was running the show. The failure to reproduce British performance had strained relations with TRE. A few months earlier I. I. Rabi and Edward Purcell, the lab's Fundamental Developments leader, had visited Worth Matravers as part of a high-level mission arranged by Watson Watt. Purcell remembered that they considered H_2S "a pretty lousy radar" and made no bones about it. In fact, the visitors had pointedly advised their hosts, continuing development would only result in the Germans getting a magnetron. It wasn't just the prospect of Hitler's forces finding countermeasures to the bombing radar that weighed on the Americans' minds. The lab's ten-centimeter ASV systems were gearing up for deployment against the U-boat, and all might be lost if the enemy navy also learned of the cavity magnetron. "What we were afraid of," recounted Purcell, "was that the submarines would know just what to defend against."

From the British viewpoint these objections couldn't have come at a worse time. Rabi and Purcell had arrived at the height of the magnetron controversy—before the device had been officially approved for fitting in bombers and the klystron effort phased out. The American opposition registered on British forces skeptical of H_2S and aggravated TRE's difficulties, a fact that rankled Bernard Lovell years later, even though the go-ahead for the magnetron's use came within days of the Americans' visit.

The matter rested there, with tension lingering between the two radar houses, as George Valley took up the NAB reins. At first, the young physicist found nothing to dispel his more experienced colleagues' gloomy assessment of H_2S. Even in the latest trials, cities showed up only in areas with a sharp land-water contrast: the water, which reflected microwaves away from the radar set, appeared as a completely dark area on the cathode ray tube. On inland flights over Springfield, Massachusetts, and Hartford, Connecticut, nothing distinctive showed up. It was difficult to escape the conclusion that the set would be of almost no help in dropping bombs over Germany—until, Valley related, a model malfunctioned during a test at East Boston Airport.

Rather than lugging the device back to the Rad Lab shops, the impatient physicist tried to fix it on the spot. While examining the radar closely, he saw that in the final amplifier stage the Rad Lab's receiver group had incorporated a feature called a limiter, which cut off incoming pulses at a fixed amplitude. It was a simple trick of the trade that Valley figured reduced the tendency of the PPI screen to bloom in the face of strong signals. That might ease the shock to an operator's eyes if he patrolled over water and suddenly a submarine echo flared out on the radar screen. But in the case of a bombing radar used mainly over land, the addition prevented the set from distinguishing the relative intensity of ground and building echoes — so that a radar man could fly past a city and never see it. Once the feature was removed, the radar not only matched British claims, it exceeded them. From that point on, recalled Valley, "I stopped suggesting that the British H_2S was a bucket of crap — and said look, we could do it better."

The discovery placed the Rad Lab at a crossroads. Valley could have proceeded with an improved version of H_2S. But a series of events, coupled with the twist of fate that had led him to bombing in the first place, conspired to change the American course. Around this time, early in 1943, the NAB leader suddenly found himself visited by two Air Forces officers expressing a keen interest in bombing radar. Major William Cowart Jr. and Captain George Schmidt were helping develop a radar bombing program for the Eighth Air Force, by then under Brigadier General Ira C. Eaker, which had set up shop in England early in 1942 and flown its first mission that summer. Its bombing arm, the VIII Bomber Command, was charged with carrying the offensive to Germany. But bad weather had grounded the planes virtually all of that December and January 1943, with Air Forces meteorologists estimating that European conditions would permit only 189 operational days of visual bombing per year.

Rumors soon reached Valley that President Roosevelt was turning up the heat, insisting that Eaker get going with the offensive or pack up and head for home in the next six months. This had the Eighth brass almost desperate for an alternative to the Norden optical bombsight. By the middle of February, Eaker had already deemed Gee unsuitable for bombing. At the same time Oboe was available only in limited quantities, and H_2S had yet to be fitted into U.S. planes.

In the face of all this, Valley concluded the best course would be to abandon plans for an H_2S-like ten-centimeter radar and move straight to a shorter wavelength. Three-centimeter technology seemed ready, given that the lab's ASV prototypes had produced strong images in trials off Nantucket and other New England areas. And the factor of three increase in resolution could prove vital. Valley imagined a radar so good it could clearly illuminate broad avenues like Berlin's majestic Unter den Linden.

He started work right away, but made his formal pitch at a radar bombing conference held at the Rad Lab on February 16. DuBridge called the meeting to coordinate the facility's efforts in light of increasing Army and

Navy interest in radar bombing aids. All told, twenty-six people gathered for the day-long session, including some of the lab's top guns: the director, Rabi, and Luie Alvarez. Valley took the minutes. The Navy and the Army Signal Corps each detailed an officer. Ed Bowles sent along Dave Griggs, who had become one of his top aides in the war secretary's office. Also present were Norman Ramsey and Dale Corson, both moved to the Pentagon as radar advisors. They, too, at least in the beginning, worked for Bowles.

Much of the morning session concerned adding a synchronized blind-bombing attachment to the standard ten-centimeter anti-submarine radar under development: at the time of the meeting, the U-boat was nearing the climax of its depredations. Valley followed with a presentation of the NAB program, announcing he had already converted the system to three centimeters and was about to test it against the original wavelength.

He did not get a warm reception. In the afternoon talk centered around Eagle, Alvarez's precision-bombing radar. It also worked on three centimeters. But the mammoth antenna, ultimately fitted with 250 dipoles and mounted parallel to the wing under the fuselage, was expected to provide twenty times the angular resolving power and one to two times the range resolution of a typical ten-centimeter system. The figures made Valley's threefold improvement pale by comparison.

The mere thought of Eagle warmed Air Forces hearts. In summing up the service arm's official position toward the close of the meeting, Ramsey stressed the American distaste for the British area offensive and noted his bosses would only entertain precision-bombing techniques like Eagle. First, the Air Forces wanted anti-submarine radars, Ramsey related. After that came Eagle — period.

Unofficially, however, Ramsey added a caveat. Personally, he could envision policy changing enough that Valley's three-centimeter radar might enter the picture. A shift from precision bombing to area attacks would herald a fundamental turnabout, he admitted. But it should not be considered out of the question, and in that case Valley's system might serve just fine. On the heels of this disclosure DuBridge ended the meeting with the vague suggestion that the Rad Lab might develop a three-centimeter version of NAB if the British did not.

Valley viewed DuBridge's statement as a hint to go ahead with his project. He decided to call the new bombing radar H_2X. Not terribly original, he knew, given the British code name H_2S. But it made sense. In radio, wavelengths around ten centimeters comprised the S-band region of the spectrum. The three-centimeter area was called X-band.

The effort immediately drew in some of the laboratory's best talents. As one of his first steps, Valley charged the antenna group with redesigning the aerial for Norman Ramsey's three-centimeter air-to-surface-vessel system so that it cast a vertical, fan-shaped beam downward and ahead of the plane. Arden H. Frederick and Britton Chance, the physical chemist–turned–radar man extraordinaire, developed the precision-ranging circuits. Huddled over the dining room table at his tiny Frost Street apartment in North Cambridge, Valley himself had already invented the bombsight computer. It consisted of a small drum the size of a soup can. Around the drum the physicist had wrapped a hand-drawn chart that showed at a glance how far in front of the target bombs should be dropped from a given altitude. Bombardiers or radar operators would turn a tiny knob to input their altitude, then consult the chart to find this slant range. Crosshairs marked the spot on the chart and caused a bombing circle to appear on the Plan Position Indicator. In general terms, depending on wind, or drift, the bombardier would release his pay-load when the radar signal coincided with the circle.

At the February conference, Ramsey, Corson, and Griggs had agreed that the Army sought an accuracy comparable to that of the Norden bomb-sight, with its romanticized ability to drop a bomb in a pickle barrel from 20,000 feet. Valley soon grew discouraged, however, trying to bring H_2X anywhere close to such a performance and even thought about quitting. At first, DuBridge's blessings alone kept him going. But as the spring wore on, the physicist came to realize the Norden's capabilities were vastly overrated: the only way it could come close to any pickle barrel was if the pilot held a steady course on a windless day, both virtual operational impossibilities. In fact, he decided, the Norden's sole unusual aspect was its detachable, football-shaped sighthead, which contained a telescope and an analog com-puter — and which a bombardier could disconnect from the stabilizer and carry out in a specially made leather bag manacled to his wrist. Valley brushed that off: "Wowie, zowie."

Besides, he somewhat coldly considered one day, American bombers flew in vast and intricate daylight formations — hundreds of aircraft staggered in altitudes to form a three-dimensional rhomboid — designed to cover the approach of German fighters from any direction with maximum firepower. The formation could often extend over a square mile of airspace, so what was the point of worrying about pinpoint accuracy? "When I saw this, then I saw the solution to the bombing problem," he related. "It was to forget about pickle barrels and to realize that what they were aiming at was at least a mile in diameter, and that that was the area you should worry about." He gave up any pretensions of accuracy and thought only about being able to blanket a target.

In late April 1943, as head of the United States Special Mission on Radar, Karl Compton went to England to discuss radar policy at the highest levels to date. The mission, reflecting the technology's growing importance to the war effort, was authorized by the Joint Chiefs of Staff. Lee DuBridge traveled at Compton's side. Over the next month the contingent visited all the top British radar research and service establishments. The two Allies agreed the United States should generally take the lead in longer-range projects, such as pushing the development of one-centimeter radar components, an area where both TRE and the Rad Lab already nurtured small programs. Meanwhile, both countries would continue to pursure overall radar systems separately.

Toward the end of May, apparently after Compton and the rest of the mission had gone home, DuBridge traveled to TRE to bring Dee's group up to date about H_2X. By this time, the American system was advancing rapidly, with Eagle still mired in development difficulties. DuBridge proposed the widespread adoption of H_2X. The very idea stunned the British, especially Bernard Lovell. A year earlier, when Rabi and Purcell had visited TRE, the Americans had eschewed the entire concept of radar bombing. Now, they were offering to supplant the pioneering H_2S before it even got into the main bomber force. Lovell found the notion hypocritical and fought the proposal instinctively, setting off a policy battle that quickly extended up to Cherwell.

Earlier in the year Lovell had started his own three-centimeter program. It was a poor man's effort already significantly behind the Rad Lab's venture. He could spare only one staffer, and a test flight in March had shown only moderate success. But he wanted at least a share of the action. After a week of stirring up the debate, the two sides gathered on June 7 at the Air Ministry. Sir Robert Alexander Watson-Watt — once knighted in 1942, the Scotsman formally hyphenated his name — oversaw the proceedings. On hand were Dee, Lovell, and Cherwell on the British end, and DuBridge and several of his staffers representing the Americans. After two tense sessions that took up most of the afternoon, it was finally agreed to share the burden: the Americans would press on with H_2X, while the British would attempt to outfit three squadrons with their own X-band version of H_2S by the coming Christmas.

Suddenly, everything seemed to go into high gear. Almost in sync with DuBridge's return from the Compton mission, Dave Griggs came back from a separate radar fact-finding expedition led by Robert A. Lovett, assistant secretary of the army for air. The Lovett mission had discovered H_2S production beset by troubles. With the Eighth Air Force still having trouble getting its hands on the British sets, Griggs told DuBridge, Eaker urgently wanted twenty H_2X radars for his own Pathfinders. To his superiors, he reported: "The success of the 8th Air Force as a strategic weapon depends on its ability to carry out continual raids against its assigned targets. Clouds over Germany this winter may be expected to make visual bombing from high altitude impossible ninety per cent of the time. Radar bombing *must* be introduced if we are to reap results from this Air Force commensurate with its size."

An impressed DuBridge shot into action. In mid-June he declared a crash program to fulfill the request. Western Electric won the Army contract to mass-produce H₂X, while Philco secured rights to build a Navy version. From almost complete obscurity less than six months earlier, George Valley's bombing radar came to dominate the Rad Lab's agenda. Other projects slowed or ground to a halt, while components were rushed through the shops and assembled in a special hangar at East Boston Airport. Norman Ramsey's predicted change of heart had come true: the United States was about to go in for area bombing. As with H₂S in Britain, nothing was to be spared for H₂X. DuBridge told his forces: "the radar war is on."

•

Shortly after the first H₂S mission on January 30, 1943, a new bombing directive reached Allied commanders. It arose from the just-ended Casablanca Conference between Churchill, Roosevelt, and their top generals, where FDR had announced the Allied war effort would be waged under the doctrine of "unconditional surrender." The corresponding bombing mandate called for "the progressive destruction and dislocation of the German military, industrial, and economic system and the undermining of the morale of the German people to a point where their capacity for armed resistance is fatally weakened."

The directive applied equally to British Bomber Command's mainly nighttime forays and the massive daytime formation raids to be conducted by the American VIII Bomber Command. The U.S. force had launched its first token sortie over Germany on January 27 and was gearing up for more full-hearted measures.

While the Americans still professed their aversion to area bombing, Arthur "Bomber" Harris apparently saw the Casablanca directive as a virtual carte blanche to continue his offensives. The edict repeated almost verbatim a British Chiefs of Staff report from the previous December that embraced Harris's own ambitions of bringing Germany to its knees through bombing alone. In chillingly matter-of-fact terms, the report had set out the awful price for victory: the destruction of six million German homes and the deaths of nearly one million people.

The clamor for H₂S continued during the year's opening months, even as Bomber Command concentrated on targets within Oboe's range. Starting with a devastating Oboe-led Essen raid on the moonless night of March 5 and continuing through July 12, Harris ordered forty-three major attacks, most against heavily defended objectives in the Ruhr. All told, his aircraft poured 58,000 tons of incendiary and conventional bombs on Germany, more than the Luftwaffe dropped on Britain in 1940 and 1941 combined.

Harris had never given up his desire to strike deep into Germany, however. As he moved against inland cities later that summer, radar bombing rose to the fore. He chose Hamburg as the first big test of H_2S. Rabi and Purcell had been right: it was a pretty lousy radar. After the initial euphoria surrounding the set's introduction, the reality had become apparent. Gaps and fades kept appearing in the Plan Position Indicator picture, a failing that was aggravated as bombers increased their operational height from 15,000 to 20,000 feet to avoid fierce enemy flak. Lovell's crew pushed to make adjustments, but the production models that began appearing in May showed little improvement over crash-program units. Hamburg, though, marked a special opportunity for H_2S. At the confluence of the North Elbe and Alster, the old shipbuilding center formed a perfect radar target: the rivers should show up as unmistakable snaky, dark areas on a cathode ray screen.

Operation Gomorrah, the sweeping assault on Hamburg, opened with a 791-plane raid on July 24, 1943, a clear Saturday night. A force of twenty Pathfinders arrived over the target two minutes before zero hour at 1 A.M. and dropped their yellow Target Indicators blind on H_2S. In perfect harmony, eight other Pathfinders followed in their wake, visually dropping red TIs. Then came fifty-three more aircraft to add green indicators — each stage in the rainbow increasing accuracy and making it easier to tell true markers from German decoys.

The main bomber stream began arriving at Zero + 2 minutes and continued for three-quarters of an hour. In previous raids the tendency of bombers to release their payloads early and turn to avoid flak had led to the phenomenon of creep back, in which bombs unrolled progressively farther from the target along the line of approach. For Hamburg, to reduce creep back, Harris ordered the main force to overshoot the markers by two seconds; but the bombs still unfurled some four miles back to the north, saturating sprawling residential areas. Over the next two days the Americans attacked in daylight, stoking fires lit the first night. The British followed with a nuisance raid the night of the twenty-fifth, intending to tire out already stretched fire and air raid services. Two nights later Harris's forces came back with a vengeance — a raid of nearly eight hundred planes that dropped a fresh wave of incendiaries onto the fires already ravaging the city.

Bomber Command sent another 777 planes on the twenty-ninth and 740 more two nights after that. But these last needn't have come. It was the third night raid, on July 27, that set off the *Feuersturm* — firestorm. Lancasters, Halifaxes, Stirlings, and Wellingtons came in from the northeast this time, so that the creep back extended over a different set of workers' apartments. As more and more planes flew over the scene, it became apparent to pilots something different was under way. Instead of many fires, they saw only one — a gigantic inferno that sucked up air and suffocated thousands, while creating winds of fire that turned air raid shelters into crematoriums. The large stores of coal and coke in the cellars of nearly every house joined the conflagration. Sparks the size of coins rained down on those caught in the open. Human beings cooked in their own fat. Streets melted, trapping some residents temporarily alive in the asphalt.

The raids destroyed an estimated three-quarters of the city — an area half the size of Manhattan. At least 42,000 German citizens perished in the fire and rubble, most of them women, children, and the elderly.

9 • Tangled Web

"A delusion, a mockery, and a snare."

LORD DENMAN

"It is double pleasure to deceive the deceiver."

LA FONTAINE

THE MORNING AFTER the first Hamburg raid, residents of a vast stretch of northern Germany awoke to find the landscape littered with a glittery tangle of aluminum-coated strips, each nearly a foot long and about half an inch wide. It was as if the local population had decided to decorate the countryside — trees, hedges, houses, cars, telegraph wires — for Christmas five months early. Instead, the strips, cut to precise lengths designed to throw German fire-control radars into disarray, had streamed out of British bombers. Packet after packet of the foil had been pushed overboard as planes approached the target area, with an effect almost greater than imagined. As bombers closed on Hamburg, the once-feared German searchlights, linked to the radar net, wandered drunkenly across the sky, and the city's ninety-two flak batteries fired blind barrages. Only twelve planes failed to come home. Total losses for the whole series of attacks numbered just 86 of 3095 bombers, a 2.8 percent attrition rate significantly under the barely acceptable 6.1 percent figure suffered in six previous raids against the city.

The dropping of Window during Operation Gomorrah marked the formal declaration of a phase of World War II later labeled the Wizard War. From the Hamburg attacks on, the ether would become a vast stew of electronic conjuring in which measure was followed by countermeasure — and often counter-countermeasure. The battles ahead would prove critical to the victory over Hitler. But the lines on this dark front had been drawn almost since the opening of war, only slowly building to their crescendo.

As much as any conflict can claim easily recognizable catalysts, the Wizard War arose from the convergence of two men — Hans Ferdinand Mayer, a German physicist and electronics expert opposed to Hitler's regime, and a young and exceptionally keen-minded British scientist named Reginald Victor Jones. The two were brought together one evening in November 1939. As Jones sat working at his desk in the British Air Ministry, a top official of the country's intelligence service, MI6, strode into his office and plunked down a package. "Here's a present for you!"

The package had been forwarded through the British Legation in Oslo, where it had been received by mail a few days earlier. It contained seven pages of typewritten German text, to which the Oslo legation had appended an English translation. It took Jones just a glance to see that here was the stuff that won wars: detailed technical overviews of nearly a dozen purported German weapons or scientific developments, from the Junkers 88 aircraft to torpedoes. The paper claimed that rocket-driven gliders were being built at a secret site called Peenemünde, which turned out to be on the Baltic island of Usedom. And it mentioned both a radio distance-measuring beam for guiding bombers to targets and two versions of an "Aircraft warning device" that appeared to be radar. The document was unsigned. Jones didn't learn the author's identity until years after the war.

The Oslo Report, as the mysterious message became known, held several items of dubious validity. It claimed that Junkers 88 production was five thousand a month, when British intelligence estimates put it at three hundred tops. Largely for this reason, but partly because it also seemed too good to be believed, the report was discounted in Whitehall as a German plant. To Jones, however, its overall air of authority and the author's obvious command of technical details rang true. He decided not to neglect the document.

R. V. Jones was an oddity inside the Air Ministry. Just twenty-eight years old, he was a research staff member attached to Air Intelligence — and within a year would hold the formal title of assistant director of intelligence (science), which made him the highest-ranking civilian on the Air Staff. A tall, broad-shouldered man with a prodigious memory, a bent for practical jokes, and a doctorate in physics from Oxford, he was also the only air force official investigating scientific intelligence in the early stages of the war.

Since he worked in the Air Ministry, Jones was most concerned with the Oslo Report's mention of rockets and radio beams — such beams would be navigational aids like Gee and Loran and not classical radar — that provided previously unrecognized means for Germany to attack Britain. At the same time he thought about German radar, which might help the enemy shoot down RAF planes. Among the report's other troubling assertions was its claim that a disastrous RAF attack on the port city on Wilhelmshaven in early September had been detected seventy-five miles off by a string of stations that emitted pulses of radio energy on an unknown frequency. It also mentioned a second radar-like system under develoment on what was then the incredibly short wavelength of fifty centimeters.

At the time Jones received the report, Hitler's forces still sat massed on France's eastern border. England and Germany were at war, but it was that waiting game called the Phony War, and British intelligence had received barely a trickle of other information. Nothing the Oslo document claimed about German beams and radar could be confirmed. Still, the strange communiqué put Jones on guard. If the Germans did have radar and radio bombing techniques, he wanted to be ready with an appropriate technical response. British worries about German jamming of the Chain Home system stretched back to the Orfordness and Bawdsey days, and the early radar pioneers had consequently built the network to operate on any one of four different wavelengths between ten and thirteen meters. The country had given almost no thought, however, to the flip side of the coin: German radar. Nothing close to a radio countermeasures program existed.

Through good luck, inspired deduction, and dogged detective work, R. V. Jones deftly used the head start provided by the Oslo Report to spur the creation of the British countermeasures effort. The next break didn't come until the following March, when secretly recorded conversations between German prisoners, interrogations, and a paper fragment recovered from a downed German bomber revealed the existence of two enemy radio beam systems. Though details were sketchy, and neither description exactly matched anything in the Oslo Report, both devices seemed to involve transmitting signals that could guide bombers to their targets.

As the war heated up, more snippets of information arrived from prisoner reports, intercepted Enigma traffic, and various orts salvaged from other German aircraft. Jones met with Tizard, Watson Watt, and Lindemann before persuading authorities to use a few Chain Home stations to try to determine the source and frequency of any suspicious German transmissions. Finally, in mid-June 1940, amidst Hitler's blitz into France and the first sporadic bombing attacks on Britain, newly obtained enemy papers provided strong evidence that very high frequency radio signals were being transmitted across the country's airspace from stations in Kleve, along Germany's Dutch border, and Schleswig-Holstein, near Denmark.

Jones was digesting the latest piece of the puzzle when he arrived for work the morning of the twenty-first and found a message on his desk saying he was wanted in the Cabinet Room. Because the Scotsman himself loved a good practical joke, he didn't hail a taxi until first checking the note's validity. Arriving at 10 Downing Street, Jones was shown through double doors into a large meeting room dominated by a long, cloth-covered table. Churchill sat imperially on the left side, fireplace behind him, flanked by Lindemann and Lord Beaverbrook, the minister of aircraft production. Everyone else, including Tizard, Watson Watt, and then Bomber Command chief Sir Charles Portal, sat opposite. They had been discussing the German beams — and the possibility the signals might provide precise navigational data for bombers seeking blacked-out British targets — for nearly half an hour.

The newcomer quietly took a seat at one end. No one seemed to grasp the essentials, and debate still raged over whether a beam system really existed. Then Churchill asked Jones a question. Rather than answering directly, He countered: "Would it help, sir, if I told you the story right from the start?"

The prime minister seemed nonplussed. After a pause he replied, "Well, yes it would!"

Jones then went over his whole tale, laying out the evidence for German transmissions being aimed at Britain. He spoke for some twenty minutes, sensing he was making a deep impression on Churchill. He sensed correctly, for the prime minister later confided that hearing the young physicist confidently assert that the Germans might well use the beams to bomb accurately at night when the country already had its hands full holding off daytime attacks formed one of his blackest moments of the war. Almost in the same breath, however, Jones suggested that countermeasures could be developed to disrupt the German bombers. The prime minister loved the idea. He banged the table emphatically. "All I get from the Air Ministry is files, files, files!"

R. V. Jones left Downing Street invigorated by Churchill, whom he called "a source of living power." Tizard and other scientists remained skeptical of the beams: it was commonly held that meter-wave radio signals like those identified could not negotiate the Earth's curvature and travel the three hundred–odd miles from Germany to England. Jones was able to continue his efforts to counter the enemy signals only by issuing pointed reminders that the prime minister himself backed the idea. He dispatched special receiver-fitted aircraft that quickly removed all doubts of the beams' existence. By fall, his litany of prisoner interrogations, recovered equipment, and deciphered Enigma traffic had identified three distinct radio-navigational networks. The systems were different and varying in complexity. Two employed an approach beam that aircraft equipped with special receivers could "ride" to the target and one or more crossing beams that signaled the bomb release point. The more sophisticated third system relied on a solitary ground station that radiated a lobe-switching signal for direction finding and an audio tone for ranging: the craft would pick up and reradiate the tone on another frequency, with the ground station determining range from the phase shift in what came back. This last system, code-named Y-Gerät, or Y apparatus, matched perfectly the description in the Oslo Report.

As the unmasking of the German webs progressed, a dozen or so members of the Telecommunications Research Establishment began to study the beams in detail and design countermeasures. Under the leadership of thirty-one-year-old Robert Cockburn, a former municipal college science master, the TRE team produced its first jammers — simple noisemakers that transmitted an earful of mush on the beams' frequencies and made it difficult for a bomber to pick up the intended signals. As other beams were discovered, Cockburn's crew worked night and day developing additional countermeasures. They also subtly doctored some beams, superimposing signals on enemy transmissions that led bombers astray. In the case of Y-Gerät, Cockburn commandeered the dormant British Broadcasting Corporation television transmitter at Alexandra Palace in London and used its powerful signal to send false ranging information to the German ground stations.

The Blitz would continue several more months, but the Battle of the Beams was effectively won by February 1941. Hitler's radio specialists dabbled in counter-countermeasures, later developing a variant of one beam by concealing a special supersonic frequency transmission above the limit of human hearing into the signal. But the British learned the basics from Enigma traffic and quickly worked out the important details. As the spring wore on, German bombs found a smaller percentage of their targets. Coupled with the rapidly improving British nightfighter radar and Ground Controlled Interception system taking a heavy toll on the Luftwaffe, the outlook for the defenders brightened considerably. The next month Hitler largely abandoned the Western Front to launch his attack on Russia.

Through most of the Battle of the Beams, R. V. Jones struggled to confirm the existence of German radar. Despite their foe's highly advanced radio-navigational systems, many government scientists and advisors still felt convinced that radar was exclusively an Allied prize. In 1937, some two years before the war, Watson Watt had spent a holiday in Germany secretly scouting for evidence of the technology. Unable to find anything resembling Chain Home towers, he had concluded that Hitler's engineers did not know the secret. Lacking anything more from wartime air reconnaissance photos or other intelligence, the conclusion had remained the overriding view.

Jones, though, could not believe that the absence of towers translated to the absence of German radar. For one thing, he had the evidence of the Oslo Report, which indicated the enemy used wavelengths so short that large aerials would not be necessary. For another, back in July 1940, one of his sources had passed on word that German fighters had intercepted some British planes thanks to Freya-Meldung-Freya, or Freya warning. Aware Freya was a Norse goddess, Jones looked for clues to her real-life purpose in mythology. The Nordic deity of Beauty, Love, and Fertility, he learned, possessed a prized necklace that was guarded by Heimdall, watchman of the gods, who could see a hundred miles in all directions, night and day. That sounded like radar.

As more references to *Freya* trickled in, Jones grew almost certain that the name referred to an early warning radar. He just could not pin it down. His intelligence, including text of a smuggled German report, even gave approximate *Freya* locations. But repeated air reconnaissance missions that summer and early fall drew a blank.

That October, as Jones continued the air search, a TRE radar expert named Derek Garrard corralled a few colleagues and set up his own unauthorized effort to identify German radar signals. Loading his car with commandeered receivers, Garrard drove to the Dover coast and set up a watch. Almost immediately, his crew noticed previously undetected signals on a wavelength of about eighty centimeters. Garrard was able to link the broadcasts to enemy shelling of British ships in the Strait of Dover and correctly concluded he had found a German radar that helped direct artillery. The system turned out to be *Seetakt*, widely used on naval vessels and for ship watching.

The British military and scientific establishment had only recently accepted the existence of enemy radio-bombing beams. This identification of German radar, apparently used to locate and shoot up the vaunted British Navy, threw officials into a mild state of uproar. Not only did Germany seem to possess the technology, Garrard was arguing that Hitler's forces were using a system on a much shorter wavelength, and therefore probably far more advanced, than anything yet in service in Britain. Some senior staff balked at Garrard's conclusions. For R. V. Jones, the TRE man's finding meant he had a lot more work to do, for whatever *Freya* was, it did not appear to be a naval radar.

Jones didn't find *Freya* until early 1941, in the midst of the Battle of the Beams. Sometime in January he got a call from Claude Wavell, a colleague in photo reconnaissance who said he had various "curiosities" to exhibit. When Jones rushed over, Wavell threw down photos of The Hague peninsula. On the edge of a field cleared sometime since the last pictures of the area had been taken a few months earlier, Jones could see a pair of circular objects that seemed to run twenty feet in diameter. A bell clanged in his mind. Scattered intelligence reports about *Freya* had indicated a station might exist in this general area. Could it be that the goddess — and her watchman — were betrayed at last?

Jones took the photos to Charles Frank, a physicist recently assigned to his small staff. Frank carefully viewed the pictures through a stereoscope and found that a shadow thrown by a thin, tall object extending from one of the "circles" had changed by a tenth of a millimeter on two successive frames — an indication that whatever was there had rotated, as if tracking an object in the sky. Jones immediately ordered a special low-level reconnaissance run, but the pressures of war delayed the flight. A mission was not attempted until February 16, but the Spitfire sped by the site too quickly and failed to get good photos. The circles were finally photographed six days later. By the twenty-third, staring at magnificent images showing steerable aerials atop each circle, Jones knew he had found *Freya*.

The enemy radar was soon established to operate on a wavelength of 2.4 meters. Detectors were set up to identify the signals and home in on other transmission sources. As it became clear the *Freya* network was as extensive as the Chain Home web, and capable of tracking RAF planes well inland over Britain, jammers would be developed. Meanwhile, Jones hatched a plan to photograph and map every station. He explained the idea one day to A. P. Rowe, the TRE superintendent.

"What good is it?" Rowe queried.

Jones reminded him that the Allies would have to retake Western Europe to win the war. "Some day we're going back," he asserted, "and we shall need to deal with those stations if we are going to land successfully."

•

Freya was almost assuredly the aircraft warning device mentioned in the Oslo Report. But what about the separate fifty-centimeter system referred to in the same communiqué? Just as Jones was unraveling the last clues to *Freya*, an Enigma decrypt indicated that a set was being sent to German-occupied Romania, a German ally, with something called a *Würzburg*, and that two more of these companion units were heading to Bulgaria. Jones quickly widened his search for German radar. Within about a month listening posts detected the first *Würzburg* transmissions — not exactly on fifty centimeters, as the Oslo Report had indicated, but on fifty-three centimeters. Good enough.

Special receiving stations traced the newly discovered radar signals to general areas. As in the early hunt for *Freya*, however, air reconnaissance could not locate any actual equipment. Gradually, Jones turned his attention back to the *Freyas*, ordering numerous photos of every site and hoping for a *Würzburg* in the bargain. His diligence paid off toward the end of 1941.

Once again the key clue came when Charles Frank examined a curiosity — this one in shots taken November 22 of a well-known *Freya* station at

Cap d'Antifer, a chalk headlands about ten miles north of Le Havre near the French village of Bruneval. A series of frames showed two *Freyas* atop a four hundred–foot cliff. Some sort of track led from them to a large villa several hundred yards away. To an untrained eye, the path seemed to go all the way to the house, as if providing an access road for radar operators bunked inside. But Frank had noticed that the track stopped just short of the villa, ending in a small loop a bit too far away to serve as a convenient driveway. Inside the loop was a tiny dot, so obscure Jones and Frank consulted several photos before convincing themselves a speck of dust hadn't somehow collected on the negative. Could this be the long-sought *Würzburg?* Jones ordered another low-level photo run.

The photographs were taken on December 5. Two frames came back with the object of interest beautifully centered: Instead of a nondescript dot, Jones and Wavell could see a parabolic reflector about ten feet in diameter. It had to be the *Würzburg*. Once the British knew what to look for, reconnaissance flights began turning up additional sets. One photo of a site not far from a known German nightfighter airfield east of Brussels showed a larger version of the newly uncloaked radar, a model they learned was called *Würzburg Riese* — Giant *Würzburg*. Near it was a *Freya* and three searchlight emplacements.

With these details, Jones began unraveling the German air defense system. For months, bomber pilots had been reporting a great searchlight belt that seemed too accurate to be based on visual sighting. He deduced that *Freya* provided long-range warning, while *Würzburgs* must be more accurate short-range radars that somehow told the lights where to point. Another batch of photos taken early in 1942 showed a German radar outpost on the island of Walcheren with a slightly different configuration than usual — one *Freya*, two Giant *Würzburgs*, and no searchlights. While some British authorities argued that the second *Riese* functioned as a backup, Jones maintained that the enemy didn't have that much technology to spare: instead of directing beacons, one must be tracking Allied bombers while the other controlled German nightfighters.

Whoever was right, the *Würzburgs* obviously held the key to the German nightfighter system. Neutralize them, and the enemy's defenses might be compromised severely. Jones favored a raid to steal one — or at least have an expert study the device close up. TRE officials backed the plan, and not far into 1942 the proposal went to the Combined Operations headquarters under Lord Louis Mountbatten, who was already plotting an Allied return to Western Europe. Two French secret agents — code names Pol and Charlemagne — reconnoitered the original Bruneval site's defenses: about a hundred soldiers were garrisoned in a compound four hundred yards from the villa, with some fifteen machine-gun emplacements scattered about the chalk cliff. The beach below was not mined.

The hundred-strong enemy force precluded a frontal assault, so officials drew up a plan that called for 119 Scottish paratroopers to be dropped on headlands near the target. They would split into three roughly equal groups — one to secure the villa and beachhead, a second to storm the *Würzburg* site, and the last to form a protective line to prevent the garrison from coming to the rescue. Meanwhile, a small naval force would wait offshore to cover the withdrawal and bring troops off the beach.

Under the operational scheme, a technician would drop with the second group and try to dismantle the *Würzburg*. A second radar man, a scientist, was to wait with the naval force. If the military situation made his capture unlikely, he would also come ashore. A number of radar specialists volunteered to jump with the paratroopers. The peculiar choice was C.W.H. Cox, a Royal Air Force flight sergeant and radar mechanic who had been a cinema projectionist before the war. He had never been in a ship or a plane, and he had a wife and newborn baby. Jones briefed him on the raid, noting that the Germans did not seem to be torturing their prisoners. But, he warned, watch out for interrogators who seemed extra nice and asked seemingly innocent questions. Cox considered the advice and replied: "I can stand a lot of kindness, sir."

Jones himself volunteered for the second assignment. But he knew too much to risk on such an adventure. In the end, the man chosen was Donald Priest, a Bawdsey Manor veteran employed at TRE.

The raiding party practiced on a mock-up of the site based on reconnaissance photographs. On February 20 the team was pronounced ready. Night after night, violent weather forced the men to stand down. Finally, the twenty-seventh dawned bright and frosty. The previous days' bitter wind had dropped. The raid was a go.

A dozen Whitley bombers carried the troops aloft. Two planes were forced to fly off course by anti-aircraft fire and dropped twenty of the forty soldiers assigned to capture the beach a mile and a half south of the intended touchdown. The main force of fifty men under Major John D. Frost, which included Cox and a radar dismantling crew, could not wait for their comrades to catch up. Shortly after midnight, part of Frost's group charged the villa, cutting down the only enemy soldier present as he fired an automatic weapon out a window. The rest stormed the *Würzburg* site, killing a few of the enemy and forcing the rest to flee. One radar operator fell over the cliff and barely managed to grab onto a bush. He was taken prisoner.

Cox made his way over a low ring of barbed wire to the *Würzburg*. A lieutenant started snapping photographs, but the flashes drew fire from the German garrison down the road. Forced to shield himself behind the antenna, Cox helped attack the set by hand. A sapper sawed off a chunk of the aerial, while other men ripped out components with crowbars. The plan allowed a half hour to dismantle the set, but after about ten minutes, mortar shells falling around them, Frost signaled the retreat. Cox and a group of men loaded the equipment onto a special cart and began maneuvering their plunder down a path to the beach. The few Germans patrolling the shore below had not yet been challenged and began raking the cliffside with gunfire. At the same time, the ousted German garrison broke down the second group's defensive perimeter and recaptured the villa. Adding to the confusion, shouting could be heard: "Cabar Feidh!," the Scot Seaforth Highlanders' battle cry. The missing twenty paratroopers, followed by their own German pursuers, had finally arrived and launched the beach attack. Hand grenades and machine-gun fire punctuated the night.

The beach was finally taken, and all three British sub-groups made their way down to wait for the still-absent landing craft. With German reinforcements bearing down hard, Frost thought he had been abandoned. Suddenly, from close offshore a fusillade sprayed bullets at enemy soldiers approaching the clifftop. As landing craft began plowing onto the beach, Cox and the paratroopers scrambled aboard.

By around 2:30 A.M., the operation was over. Two British soldiers had been killed and six captured: all those taken prisoner survived the war. On the German side, five were reported killed, while Frost and his men took two prisoners of their own, including a radar technician.

Cox had gotten almost every important *Würzburg* component — transmitter, modulator, receiver, and amplifier. Although the display screen was missing, British experts easily had enough to divine the range of wavelengths the set could cover and evaluate the radar's capacity to resist countermeasures.

Beyond its importance to the Wizard War, the Bruneval raid guaranteed the future of British paratroopers. Major Frost received the Military Cross. Cox won the Military Medal. Bruneval is the first battle honor inscribed on the drums of the First Parachute Regiment.

•

For at least a century leading up to World War II, as R. V. Jones well knew, the rich German scientific tradition had spawned and nurtured some of the world's leading engineers and physicists — from Heinrich Hertz to Max Planck. As he was chasing down a *Würzburg*, Jones felt certain it was not the last German radar. He was right.

Unknown to any of its enemies, by the time war erupted Germany had developed the world's best engineered, most accurate, and in many ways most versatile radar systems. It had shipboard radar, fire-control models like *Seetakt*, and the *Freya* medium-range early warning set. By the time Jones got wind of German radar, the country had deployed a sophisticated ground-based nightfighter control system that depended on the *Würzburg*. The country was not close to developing a cavity magnetron. But any problems Germany had in the radar war stemmed not so much from its technical sophistication as from its military mindset. Hitler and his war planners thought in terms of lightning offensives, and so did not push the development of mainly defensive radars. This one-dimensional thinking, accentuated by the lack of civilian scientists with the clout of Vannevar Bush or Frederick Lindemann, had a devastating effect on Germany's war effort.

Germans knew radar as *Funkmessgerät* — radio measuring device. The modern pioneer was Rudolf Kühnold, who in the early 1930s served as chief of the German Navy's Signals Research division. A veteran of sonar development, he applied the same basic principles to aboveground detection with radio waves. To develop the technology, Kühnold and his assistant worked closely with a newly formed company, *Gesellschaft für Elektroakustische und Mechanische Apparate*: GEMA.

In March 1934, GEMA's crude continuous-wave radar sat on a balcony and detected the battleship *Hessen* lying at anchor in Kiel harbor about a half mile away. The first German pulsed radar made its debut in May 1935, nearly a half year behind the American set developed by Robert Page, and three months after Watson Watt's Daventry test. Early the next year GEMA moved to a more sophisticated design on 2.4 meters. This was *Freya*'s progenitor.

The Navy took delivery of the first production *Freya* in early 1938. The sets had a maximum range of seventy-five miles and could not measure an aircraft's altitude, rendering both its range and height-finding capabilities inferior to Britain's Chain Home stations. Offsetting these failings, it was fully mobile and could look in all directions. Meanwhile, GEMA also pushed development of the first *Seetakt*, a shipborne precursor to the shore-based model of the same name. It was a beautiful piece of work used for sea search, and the full production unit was ready at war's outset. Two years passed before Royal Navy ships carried its equivalent.

While GEMA worked to perfect its radars, German corporate giant Telefunken designed its own systems. The elusive *Würzburg* that R. V. Jones strove so hard to find in 1941 was its creation. The set worked on the uncommonly short wavelength of fifty-three centimeters. Its maximum range was just eighteen miles. But the high-frequency waves were capable of directing anti-aircraft fire toward enemy aircraft with unrivaled precision.

Behind such a promising start, Germany had drawn first blood in the radar war. As the Third Reich marched into battle in the late summer of 1939, eight *Freya* stations guarded the country's North Sea coastline between Holland and Denmark. On December 18 stations on the resort island of Wangerooge and the naval outpost of Heligoland picked up a British raiding party of twenty-four Wellington bombers some seventy miles distant. At this time the Royal Air Force operated under strict admonitions against bombing the German mainland and risking civilian lives and property. The aircraft were winging to the port city of Wilhelmshaven looking for enemy warships at sea.

Some twenty minutes after a *Freya* operator sounded the alert, the first of fifty Messerschmitts roared off in pursuit. The day was crisp and clear, and Luftwaffe pilots had no trouble picking up the British formation as it headed home from Wilhelmshaven, where the only warships present had lain too close to the docks to risk attack. Lacking fighter protection, the Wellingtons were sitting ducks. Only twelve returned, including two that had aborted the mission early on. The Oslo Report had correctly attributed the British disaster to German radar.

After Hitler's blitzkrieg across Western Europe in May 1940, *Freyas* were positioned all along the French, Belgian, and Dutch coasts. As the war heated up, and the British launched their own nighttime bombing attacks against the European mainland, it became apparent to Luftwaffe chief Hermann Göring that his forces needed an even better air defense system. This realization brought *Würzburg* into widespread use.

Göring turned the night defense problem over to Colonel Josef Kammhuber. The methodical and industrious officer formed a special nightfighting force with headquarters in a seventeenth-century castle at Zeist, in the occupied Netherlands. Kammhuber, soon promoted to major general, devised a unique sentry system that ultimately deployed radars in a vast belt any attacker would have to cross to reach German industrial targets. In its full evolution, the Kammhuber Line looked like a huge upside-down question mark that ran through the middle of the Jutland, arched westward toward Britain, then curved back through northern Germany, the Netherlands, Belgium, and northeastern France.

By late 1941, Kammhuber had divided his line into a series of boxes, each twenty-seven miles wide and twenty-one miles deep. Every box housed a *Freya* set for early warning and two *Würzburgs*. The *Würzburgs'* decimeter waves proved effective at distinguishing between closely grouped planes. At a given time one set was used to track an enemy plane, the other to vector a fighter to the attack — essentially as Jones had surmised.

Ground controllers worked off a more ponderous scheme. Unlike the British Ground Controlled Interception radars, the *Würzburgs* did not employ Plan Position Indicators. To convert range and bearing data into a useful plot, each command post contained a Seeburg Table, which showed aircraft positions as spots of light — red for enemy bombers, blue for friendly fighters — superimposed on an area map. The lights were projected from underneath the table's ground glass surface by two operators, each linked to one of the *Würzburgs* by telephone. Back above the table, other men traced the flight paths with blue and red crayons, making it easier for the ground controller to broadcast updated vectors to his pilots.

Toward the end of 1941, Kammhuber moved to strengthen his defenses with the giant *Würzburgs*, which offered improved range and the ability to follow enemy bombers flying low to the ground. A redesigned *Freya* set called *Mammut*, or Mammoth, extended early warning range to an impressive two hundred miles, while a beautifully engineered sister radar called *Wassermann* added height-finding capabilities reaching out to 150 miles. Finally, the German nightfighters got their first airborne interception systems — a Telefunken-built radar named *Lichtenstein*. All these systems appeared slowly. Still, by late March, as Bomber Command turned to massive air raids, the German defense scheme came into its own. More and more of Butch Harris's boys were not making it back home.

As Kammhuber wove radar into a highly effective air defense system, Wolfgang Martini, promoted from his Luftwaffe post to head of the German signals establishment, plotted the first great countermeasure stroke against British radars. Martini's early suspicions about the Chain Home towers had long since been confirmed. By early 1942, technicians had also identified the Chain Home Low net, along with various enemy airborne and shipboard radars that in one combination or another kept a constant watch on the English Channel. While not exactly disrupting military activities, the British radar web was enough to give Germany's military machine headaches, and a jamming network had been established from at least Ostend in Belgium to the Cherbourg peninsula.

One of the pressing issues confronting German military planners involved using the jammers to help free two of Hitler's most powerful warships, the battle cruisers *Scharnhorst* and *Gneisenau*, from the British radar sentries and the corresponding air and sea patrols that kept them pinned in the French port of Brest. Following a two-month sortie in which they sank or captured twenty-two vessels, the ships had huddled in Brest's battered docks since late March 1941. A substantial portion of British air and sea forces was tied up standing guard, and Churchill would spare almost nothing to destroy the pair. Hitler vowed just as strongly to protect his valiant warships and bring them safely home. Rather than waiting any longer, he ordered a breakout. The ships would leave port at night so as to pass up the English Channel in broad daylight — something the Führer felt certain the British would not expect.

The ships slipped the dock around 8:30 P.M. on February 11, accompanied by the heavy cruiser *Prinz Eugen* and a bodyguard of destroyers and minesweepers. Low-flying air escorts appeared periodically along the route. Up in the English Channel, shore jammers stayed silent until the next morning, when the flotilla passed into position: the fact that the jammers had long been in place was intended to obscure the real purpose of their activity during the breakout. Meanwhile, two countermeasures aircraft, fitted with equipment that simulated the radar echoes of twenty-five bombers, flew parallel to the British coast to further divert attention from the armada.

British observers did not spot the warships slipping out of Brest. Although both the Chain Home and the Chain Home Low networks would be somewhat affected by enemy jamming, radar did pick up German fighter escorts accompanying the armada; and a patrol plane apparently noticed unusual shipping activity. However, because of breakdowns in communication and failed radar on at least one other aircraft, the actual war party was not detected until 10 A.M. on the twelfth, about the time the German shore jamming began. Even then, the Admiralty didn't receive word of the enemy campaign until 11:25, as the *Scharnhorst* and *Gneisenau* steamed within twenty miles of Boulogne. Just after noon the Dover batteries opened fire. Successive waves of bombers and torpedo bombers descended on the enemy ships until nightfall, but in the ensuing air battles the British took losses heavier than they gave, and the warships moved up the Channel into German waters.

An indignant British public was told the *Scharnhorst* and *Gneisenau* had made it all the way unscathed. But the Secret Service soon reported accurately that both ships had been hit by air-laid mines. The *Scharnhorst* took six months to get back into action. The *Gneisenau*, bombed again a few weeks later in the Kiel docks, never returned to war. All the same, the escape marked a countermeasures victory for the Germans. And when he heard the news, Churchill reportedly flew into a rage.

•

The daring *Scharnhorst* and *Gneisenau* escapade, together with the Bruneval raid, brought the slowly escalating countermeasures battle toward a crossroads. Both sides knew a lot about the other's radar systems. However, both had resisted employing widescale electronic disruptions. For its part, the British high command preferred not to up the ante before it had time to develop counter-countermeasures — antidotes to things like jammers and Window that might be turned around and used against it. This misgiving proved especially strong in the case of Window. By spring 1942, enough aluminum had been collected for Bomber Command to start dropping the foil. But Lindemann, Watson Watt, and the Fighter Command hierarchy all feared a German version could totally disrupt the British air defense network.

Jones found such fears ludicrous. As he learned more about Kammhuber's radar-based air defense scheme, he wanted to strike early and hard to disrupt it. Besides, he reasoned, the *Scharnhorst* and *Gneisenau* episode showed the enemy would jam when it saw the need. His view did not hold sway. However, while Window was kept in abeyance, officials moved ahead on several other fronts.

In August the British tried one more raid on a German radar. This time the target was a *Freya* station riding the French coast north of Bruneval. As an auxiliary part of a much larger assault against the city of Dieppe, a special team of Canadian soldiers was ordered to storm the station, pilfer documents, and remove transmitting tubes, anti-jamming devices, and other components. The entire raid was doomed. The 237-vessel main invasion force was detected twenty miles offshore and decimated. Some nine hundred lost their lives, and another two thousand were captured. The Canadian detachment storming the radar fared little better. With it was Flight Sergeant Jack Nissenthal, a twenty-three-year-old radio expert who had started radar work as a Bawdsey Manor volunteer. He got close enough to the machine-gun–encrusted *Freya* to become convinced the station could not be taken. After briefly studying the aerial in action, he returned to the main force seeking help, but the situation deteriorated and the commandos withdrew. Bullets splashing around him, Nissenthal plunged into the sea, dived under water, and swam as far as he could. He was rescued by a landing craft and ferried by destroyer to southern England. From Newhaven, the erstwhile guerrilla took a commuter train back to London.

Even as the Dieppe failure unfolded, *Freya* stations came under a different sort of attack. Cockburn's TRE countermeasures group developed its first airborne jammers, code-named Mandrel, which produced an electronic raspberry of noise designed to overwhelm the sets. Beginning that summer, the jammers were deployed to Fighter Command. Bomber Command later put them into widespread service as well.

At virtually the same time, the tactic of spoof made its operational debut. In contrast to jamming, spoof sought to fool an enemy's radar without operators knowing they had been fooled. The first spoofing device, Moonshine, also targeted *Freyas*. An airborne repeater fitted into Defiants, Moonshine was triggered by the pulse of an enemy radar and responded with a wide-ranging pulsed reply on the same wavelength that simulated a large concentration of aircraft in tight formation. The initial Moonshine raid took place on August 17, the day before Dieppe, as cover to a B-17 Flying Fortress attack by the Eighth Air Force. Three bombers, accompanied by the Moonshine Defiants and various escorts, drew off an estimated 150 German fighters, while the main force reached its target at the French rail yards at Rouen without loss. Moonshine would prove extremely successful at taking the heat off American daylight raids for several months. But it worked only against limited numbers of radar stations, and as the Germans added more *Freyas* its effectiveness waned and the technique was pulled from service.

Spoof appealed to R. V. Jones's nature. In particular, he wanted to target the *Würzburg*, the most important of the enemy radars because it directed nightfighters and gun batteries. An exhaustive evaluation of the set stolen from Bruneval turned up no design weakness that could be exploited by jamming. The radar's narrow beamwidth and small side lobes made it impossible for a few Mandrel-type aircraft to provide cover for a whole bomber force. For such blanket protection, Cockburn's team estimated each bomber would have to carry its own jammer, wildly impractical at the time.

All these factors pointed to Window. So, by early 1943, did developments in America. A small countermeasures arm established at the Radiation Laboratory in the wake of Pearl Harbor had grown so large it had moved over to Harvard University and become a separate NDRC division. Frederick Terman, the respected head of Stanford University's electrical engineering department, directed this Radio Research Laboratory, which took over a wing of the Biological Laboratories on Divinity Avenue: he had been appointed the day the *Scharnhorst* and *Gneisenau* escaped the British quarantine around Brest and sailed through the English Channel back to Germany. Throughout the second half of 1942, RRL engineers had produced their own series of jammers aimed at *Freya* and *Würzburg*. But developing Window, known to the Americans as chaff, formed the lab's chief occupation. One of Terman's recruits, Win Salisbury, showed that if cut to a length equivalent to half the radar's wavelength — in *Würzburg*'s case about twenty-five centimeters — the foil would resonate to the incoming radio waves and reradiate them to look as if a much greater object had been detected. Bound up in one-pound packets of several hundred metal strips, chaff could be dropped from an aircraft every minute or two, opening to fill radar screens with false echoes.

With both the British and American countermeasures groups improving the science behind Window, R. V. Jones kept pressing for the foil's immediate use. In making his case the physicist drew ammunition from the fact that other options were already losing their effectiveness. One tactic was to simply overload a given German defense zone—a move that prompted Kammhuber to place more stations behind the originals. Another tactical countermeasure involved flying north around the line, as done with great effect in raids against Lübeck and Rostock in spring 1942. Kammhuber responded to this tactic by lengthening the line with twenty-nine additional stations in north Denmark. On the electronic front, Mandrel, first used to support massive bomber raids that December, had indeed disrupted *Freyas*. But the Germans had quickly changed the set's wavelengths.

The stalemate over whether to use Window lasted throughout the spring of 1943. Finally, on June 23, the issue came before the British Chiefs of Staff. Churchill chaired the meeting. Jones had seen the prime minister only once since the beam meeting some three years earlier. But his leader, clad so commonly in his blue boiler suit that the intelligence man initially took him for a maintenance worker, had greeted him brightly: "Mr. Jones, very glad to have you here."

After this rousing welcome Jones laid out the reasons for using Window. Watson-Watt once again stressed the case for holding off for fear the Germans would develop their own version. But by this time Leigh Mallory, the current Fighter Command chief, had come around, agreeing that the savings in Bomber Command planes and crews outweighed the dangers to his own service arm. Churchill considered briefly, then issued his pronouncement: "Very well, let us open the Window."

Following their debut at Hamburg in mid-July 1943, the tiny metallic strips accompanied every British night-bombing sortie. Window not only disrupted *Würzburgs* but the *Lichtenstein* airborne interception radars. At first the Germans had no counter for the chaff, which they called *Düppel*. Amidst the furious Allied bombing campaign that ensued, Kammhuber was replaced as head of nightfighter command by General Josef "Beppo" Schmid. That fall, with Bomber Command losses under 4 percent, Göring proclaimed in frustration: "In the field of radar they must have the world's greatest genius. They have the geniuses and we have the nincompoops. . . . The British would never have dared use the metal foil here if they had not worked out one hundred per cent what the antidote is. I hate the rogues like

TANGLED WEB • 209

the plague, but in one respect I am obliged to doff my cap to them. After this war's over I'm going to buy myself a British radio set, as a token of my regard for their high frequency work."

The Eighth Air Force first dropped chaff late that December, using a novel design a twentieth of an inch wide and less than a hundredth of an inch thick, with a V-shaped crease down the middle to add rigidity. Each bundle weighed a mere three ounces. The United States rushed over new chaff-making machines designed at the Radio Research Laboratory and produced by International Paper Box Company in Nashua, New Hampshire. In February the Eighth dropped forty tons. The monthly figure hit 355 tons by the end of spring. Window took over a good deal of the British aluminum foil production, while in America cigarette packages and Mars bars candies reportedly gave up their foil wrappers.

Steadily, the Germans began to adapt. A first step was *Würzlaus*, a coherent pulse Doppler radar system that tracked an echo's movement and could therefore distinguish high-flying planes from at least a light rain of chaff. An adjunct to the *Würzburg*, it came into widespread use by November 1943. About the same time, the Luftwaffe adopted *Zahme Sau*, Tame Sow, fighter tactics. An innovation of Colonel Viktor von Lossberg, the technique could work with even sporadic radar information. By monitoring weather reports and British radio activity, and then combining them with any radar reports, German signals officers could get a good idea of the timing and direction of attacks, allowing ground controllers to issue running commentaries that guided pilots to a raiding party. Fighters soon got another *Lichtenstein* airborne radar as well. This was the SN2, which operated on longer wavelengths between 3.7 and 4.1 meters, rendering it immune to the types of Window then employed. Even as Allied aircraft flew through cloud cover via their bombing radars, German pilots found them.

The various jukes and fakes surrounding Window made up just one aspect of a frenzied outpouring of countermeasures and counter-countermeasures lasting from the closing months of 1943 into the following spring. The Radio Research Lab set up an offshoot at TRE's stately Great Malvern campus to handle electronic warfare for the Eighth Air Force. Among the more effective German creations was *Flammen*, or Flames, which located Bomber Command aircraft through the Identification, Friend or Foe signals meant to make it easier for *Allied* forces to identify them. *Flammen* drove British losses steadily upward in December and for several months beyond as many pilots ignored orders to turn off the sets. Yet when Hitler employed *Düppel* in the Baby Blitz launched against London in early January, the British reversed the tactic and began homing in on German IFF.

• •

Amidst the wizardry, H$_2$S had not been ignored. As the Allied air offensive thundered on in late 1943, Bomber Command relied increasingly on the radar to guide planes beyond the nearly three hundred–mile range of Oboe and other precision navigation devices. The magnetron's ten-centimeter waves never provided the resolution to pinpoint targets, but they were good enough for all-weather area bombing. Radar became Butcher Harris's bread and butter: H$_2$S was used on 32,000 of 53,000 sorties in 1943.

The physicist Freeman Dyson, then a civilian scientist analyzing bombing trends in Command headquarters, quaked over finding himself trapped in "this crazy game of murder." Boys in bombers killed thousands by aiming at fuzzy splotches on radar screens, while the hierarchy, he concluded, "failed utterly to distinguish between ends and means, measuring the success of squadrons by the number of sorties flown, no matter why, and by the tonnage of bombs dropped, no matter where." Dyson found a soulmate in the poet Gerard Manley Hopkins, in particular these lines:

> See, banks and brakes
> Now, leaved how thick! Laced they are again
> With Fretty chervil, look, and fresh wind shakes
> Them; birds build — but not I build; no, but strain,
> Time's eunuch, and not breed one work that wakes.
> Mine, O Thou lord of life, send my roots rain.

German engineers, though, did not seek refuge in poetry. The purpose of the H$_2$S set recovered near Rotterdam in the first days of February 1943 had been quickly divined. Barely a week after the transmitter was retrieved, a high-level meeting convened in Berlin to discuss the *Rotterdam-gerät's* implications. Engineer-Colonel Dietrich Schwenke, a Luftwaffe intelligence officer specializing in captured equipment, delivered the bad news. "I have to report that a new device has been found in a Stirling bomber. . . . It is a centimetric radar device mounted beneath the rear of the fuselage. We have not yet established what exactly it is, but the device is exceedingly costly. . . . It is suspected therefore that it is a night-fighter search- or warning-device, and simultaneously a navigation and target-finding device."

The newly discovered radar was sent to Telefunken experts in Berlin for a more detailed appraisal. Professor Leo Brandt, a company specialist, was placed in charge of a top-priority project to reconstruct the heavily damaged magnetron so German factories could copy the device for the country's own radars. On March 1 a bombing attack on the Telefunken works destroyed the transmitter. However, that same night another magnetron was salvaged from a Halifax shot down in Holland. Brandt's next reconstruction effort took place in a concrete flak tower. As engineers coaxed the unit back to life, British airmen taken prisoner began to confirm that H$_2$S was used by Pathfinder units for marking bombing targets. In May, after reading a report on the *Rotterdam-gerät*, Hermann Göring issued another solemn proclamation: "We must frankly admit that in this sphere the British and Americans are far ahead of us. I expected them to be advanced, but frankly I never thought that they would get so far ahead. I did hope that even if we were behind we could at least be in the same race!"

Göring's overall assessment was correct. Although German radars based on the transmitter would be developed for bombing, anti-aircraft, and airborne purposes, none were ready in time to see combat. The Reich's engineers had more success, however, designing countermeasures to Allied microwave radar. One was a basic detector code-named *Naxos*, the second a more elaborate device called *Korfu* capable of homing in on H$_2$S transmissions. Both were initially designed to spot radars operating around ten centimeters in wavelength and could not detect three-centimeter signals.

These sets, the first designed to thwart the cavity magnetron, trickled into operation early in the fall of 1943. One version of the *Naxos* detector was delivered to U-boats to warn of ten-centimeter air-to-surface-vessel transmissions. Another model was fitted into fighter aircraft to help them locate bombers. A *Naxos* modification incorporated into the *Würzburg* provided short-range ground tracking. At the same time *Korfu* receivers began picking up the enemy from long range, with monitoring posts set up across western Germany. Bomber Command's practice of switching on H$_2$S sets upon takeoff made it possible for German operators to follow the British bomber stream from takeoff to touchdown. As Brandt later boasted: "One may well claim that during this period the employment of H$_2$S was extremely useful to the German defence. . . ."

The turnabout was just as I. I. Rabi and Ed Purcell had predicted in the summer of 1942 when TRE officials showed off an early version of H$_2$S. As 1943 drew to a close, however, the Americans were gearing up for the debut of the Rad Lab's three-centimeter H$_2$X set developed under George Valley. A dozen B-17s equipped with crash production models joined the Eighth Air Force that September: two Rad Lab staffers met the planes at the 482nd Pathfinder Group headquarters in Alconbury to handle maintenance. Almost simultaneously, Dave Griggs left Ed Bowles's office in the Pentagon to provide advice about the radar's capabilities to the command staff—he would later be awarded a Purple Heart after being wounded during a bombing mission—and the Rad Lab opened a British Branch alongside the American countermeasures group at TRE. The staff swelled to around thirty and carried a broad mission that included helping install, maintain, and modify any Rad Lab sets used in combat. Servicing H_2X dominated its early days.

After equipment woes and lack of training forced two stand-downs in four days, the inaugural H_2X combat mission took place on November 3, when nine Pathfinders guided sixty bombers against the Wilhelmshaven docks. The same target had been missed entirely by eight previous visual bombing missions. After the H_2X raid, holes in the cloud cover revealed widespread damage. From that point on H_2X formed the centerpiece of the American bombing campaign, despite a paucity of sets and the repeated maintenance problems of early models. Nicknamed Mickey by the Air Forces for some long-lost reason, H_2X provided both greater target discrimination and immunity from the growing German ten-centimeter detection network. On December 10, Ira Eaker, by then a lieutenant general about to move from the Eighth Air Force to take command of the Mediterranean Allied Air Forces, cabled his assessment to Hap Arnold back in the States: ". . . Tests prove conclusively that H_2X equipped airplanes be expedited on highest priority. Our present plan is standardized on H_2X equipment. In view of situations this theatre, have ultimate goal and present operation requirements for six H_2X equipped planes per heavy bomb group. . . ."

Mickey was adopted so vigorously that by year's end the original dozen B-17s equipped with H_2X were leading 90 percent of the U.S. bombing missions. Total bomb tonnage dropped via H_2X in the last two months of 1943 surpassed the amount dropped by visual sighting over the entire year. The set became the primary Allied bombing radar, completely stealing the thunder from the ten-centimeter H_2S. *"The availability of this equipment has changed the status of the 8th AAF planes from that of being grounded for all but one or two days per month during the winter to that of being able to bomb at will the most distant enemy objectives through heavy overcast,"* an early assessment read. The Telecommunications Research Establishment soon rushed a half dozen of its own three-centimeter sets into service, but no more than a few hundred were ever produced and they do not appear to have seen service outside the Pathfinder force.

Despite its enhanced resolution, as George Valley had known from the outset, H_2X offered precious little real improvement in bombing accuracy. Its circular probable error — the radius of a circle with the target at the center, in which half the bombs would fall — was a whopping two miles, not precision bombing by any stretch of the imagination. On January 17, 1944, Valley was in England to help assess the status of his creation, joining a group that included two major generals — Earle E. Partridge, commander of the Eighth Air Force's Third Bombardment Division, and Jimmy Doolittle, a top Eighth official most famous for his April 1942 morale-building B-25 raid over Tokyo and three other Japanese cities. Dale Corson, out of Washington on a long fact-finding mission that had already carried him through Italy and North Africa with Dave Griggs and other advisors, took notes that showed the Air Forces leaders were committed to H_2X, regardless of its limitations: "Doolittle is prepared to support the effort required to knock out pin point targets using H_2X. He said he was willing to send 100 planes to do a 10 plane job. He would rather have a few H_2X's now than hundreds later or more accurate equipment later."

In the end, the radar, manufactured by Philco for B-17s and B-24s and designated the AN/APS-15, never approached the performance of visual bombing on clear days: early models encountered problems with low pressure at high altitudes, training was not yet up to speed, and supplies lagged far behind demand. The first production models began to arrive in the Eighth Air Force in February 1944, but by April only four bombardment groups were equipped, usually with twelve aircraft in each group getting a set. Still, like H_2S, the radar had an immense impact on the war by keeping planes flying on days they normally would have been grounded. All told, the three-centimeter radar enabled the United States to launch roughly double the number of missions that might otherwise have taken place.

Yet, while buckling in the face of this unprecedented bombing campaign, Hitler's defenses did not break. German inroads against H_2S, as well as other innovative techniques dreamed up to locate and shoot down planes, at least evened the slate in the radar and countermeasures duel during the spring of 1944. One Allied attack would prove devastating: on March 1 only four of 557 Bomber Command planes were lost in a raid against Stuttgart. On another, the tables would be completely turned. A 795-plane assault targeting Nürnberg on the unusually bright night of March 30 saw an appalling total of ninety-six bombers fail to return.

All told, Bomber Command lost more than a thousand planes between mid-November 1943 and the end of March 1944. That represented a 5.1 percent loss rate far in excess of Arthur Harris's plans. Counterbalancing these figures, though, Germany was being crushed between two giant vises. Russians pressed forward on the Eastern Front, British and Americans on the West. In battles over Berlin one night early in 1944, Germany lost two legendary aces: Major Prince zu Sayn Wittgenstein, with eighty-three confirmed kills the country's highest scoring pilot; and Captain Manfred Meurer, who ranked third with sixty-five tallies. By that spring, despite the fact that the German aircraft industry kept churning out Messerschmitts, the supply of skilled pilots was being steadily, bloodily, drained away.

Allied raids deep into Germany eased greatly after the Nürnberg disaster. From mid-April 1944 onward, Doolittle and Harris were forced to take orders from General Dwight D. Eisenhower, who curtailed the area bombing campaign in favor of strikes chiefly along the six hundred–mile coast from Cherbourg to Antwerp. Instead of bombing cities, Ike wanted to soften up the enemy in advance of an invasion — and sow confusion as to the intended Allied landing site.

•

The D-day assault on Normandy catapulted the Wizard War to its pinnacle. On an unheard-of scale, radars and electronic countermeasures converged in a tempestuous fury for the June 6 invasion. German jamming and early warning stations, along with radar-guided guns and nightfighters, lined the Dutch, Belgian, and French coasts. Allied forces installed the latest radars on bombers, fighter escorts, ships, and landing barges. British and American paratroopers were guided to drop zones by so-called Rebecca-Eureka sets: Pathfinders parachuted into place carrying the Eurekas, which, when triggered by Rebecca pulses aboard the main troop transports, broadcast a series of tones that revealed their positions. Finally, countermeasures experts from both main Allied camps banded together to create new standards in electronic disruption and spoof.

Never were the stakes higher. With Allied war plans hinging on Operation Overlord, depriving the enemy of early warning and the ability to pinpoint the landing sites ranked paramount. As invasion plans crystallized in spring 1944, the montage of *Freya* and *Würzburg* stations painstakingly patched together by R. V. Jones and Claude Wavell loomed increasingly important. The British routinely dropped carrier pigeons in special containers equipped with questionnaires about enemy forces and installations: Helpful locals who found the cages would answer what they could and release the birds. Jones inserted questions about rotating aerials, unearthing several more German radars. Enigma traffickers provided the intelligence specialist with still other clues to radar locations by foolishly assigning stations animal code names starting with the same letter as the nearest towns. Combining such tricks with photographic reconnaissance and radar detector data, he and Wavell plotted and catalogued virtually every enemy station — some two hundred sites, containing six hundred radars — running from the northern tip of Denmark to the French-Spanish frontier. Signals Intelligence and photo recon added further information on German jamming posts.

On May 10, three weeks before D-day, bombers began striking the stations. To convince Hitler that the brunt of the invasion would come along the Strait of Dover, two outposts were attacked outside the Normandy landing area for every one inside. By June 3, every jamming site had been destroyed, and all but one of the ninety-two targeted radar stations drenched with bombs. The exception was a *Seetakt* site at Fécamp, on the coast some twenty-five miles north of Le Havre. A great part of the invasion plan involved deception, and at least one station had to be around to see what Allied commanders wanted it to see.

The spoof was largely Robert Cockburn's baby. He had been briefed on Overlord details in the waning days of 1943. Not long afterwards, the American countermeasures lab at TRE was brought into the fold. The two groups began amassing communications and radar-jamming equipment bound for airplanes, warships, and landing craft. Special detectors designed to home in on any German jammers still in operation were installed in a handful of additional planes.

As a crowning touch to the witchery, plans called for two phony seaborne invasion forces code-named Taxable and Glimmer. These ghost fleets, the most intricate electronic skulduggery yet conceived, aimed primarily at the Fécamp radar, but also any station not destroyed. Cockburn tested his ideas with captured German equipment, as well as against an unsuspecting British radar outpost. On May 26 word reached TRE that the Germans might have radars operating at two thousand megahertz, a much higher frequency than anything they had yet deployed. If true, it could ruin everything, since the spoofs Cockburn had planned were aimed only at known enemy frequencies. Within three days, a detector, Coal Scuttle, was flying up and down the Channel to check the rumor. Nothing was found. On June 5, the invasion eve, Cockburn's mythic armadas set forth as planned.

Taxable headed for Cap d'Antifer, well northeast of the true landing zones. Eight Lancaster bombers disbursed bundles of Window: the planes flew slow rectangular orbits and traced out a front about fourteen miles long by eight miles wide so that the center point of their pattern inched forward at seven knots, as if providing a screen for an advancing fleet. Long-ignored Moonshine simulators were redesigned to cope with a much broader range of German radar transmissions and installed on motor launches. These picked up and amplified German search pulses, so signals returned looking like echoes from large objects. A flotilla of other boats towed Filberts, large naval barrage balloons with special radar reflectors designed to make them look like large warships and troop transports.

The smaller Glimmer force, similar in its essentials, was guided by six Stirlings heading still farther north, toward the beaches of Dunkirk and Boulogne. To add realism to the mock attacks, jamming aircraft hovered nearby, putting up a thin curtain of noise designed to let German technicians just make out the ghost fleets. When the two phantom armadas reached designated points ten miles off the French shore, the Filbert launches put out smoke screens, and broadcast the taped clanks, rumblings, and splashes of ships dropping anchor over giant loudspeakers.

Rounding off the spoof, a force of twenty-nine Royal Air Force bombers dropped massive loads of Window farther west near Caen, still several miles east of the real assault areas. The planes went on to the simulated landing areas, where they dumped hundreds of dummy parachutists rigged with fireworks to deceive ground observers into thinking the battle had been joined; the presence of a few live Special Air Service paratroopers added realism. A separate group of bombers dispersed more chaff over the eastern force to tie up fighters in that region.

Although German stations briefly monitored Taxable, Glimmer was a genuine success. Observed on radar, it prompted a full alert in the Calais-Dunkirk area, freezing enemy actions while reconnaissance craft investigated. At the same time German nightfighters duly attacked chaff clouds in eastern France. None of the thousand aircraft carrying real airborne teams that night and the next day would be lost to enemy fighters.

Meanwhile, commensurate with the feigned Taxable and Glimmer attacks of June 5, the first of 6500 actual warships, merchantmen, and troop carriers began crossing the English Channel toward the Normandy beaches behind a blizzard of electronic static. As it turned out, sixteen of the ninety-two targeted German radar stations had managed to stay on the air. Only one observed the main invasion force, and its report was ignored in the gathering chaos. Not until around two the morning of June 6, some four hours before the initial landings, did German commanders accurately gauge the landing site, and then it was because lookouts heard the rumble of approaching ships. "No conceivable jamming effort could have achieved more than had been done, for the invaders' approach had passed undetected until then," writes the military historian Alfred Price.

It was not yet over. In advance of the assault, Eisenhower planned to saturate the enemy beachheads with bombs. Late the night before D-day, considering the likelihood of bad weather, the Allied commander decided to rely exclusively on Eighth Air Force bombers led by H_2X-equipped Pathfinders: radar crews had been sharpening their skills on French beaches outside the targeted area. Anticipating the call, the Rad Lab's British Branch staff had passed out radar scope photos of the landing beaches to aid recognition, as well as of the invasion force at its moorings, so H_2X operators could be mentally prepared for the confusing clutter they would see on their screens.

On invasion day, nearly 450 bombers set out under complete cloud cover, groups of between eighteen and thirty-six planes flying abreast, with the Mickey Pathfinders in the middle. An exacting timetable called for hitting initial objectives between 0555 and 0614. The first landing troops waited from a mere thousand yards offshore and began the assault just five minutes after the last bomb fell, while German defenders still reeled from the concussions. After that, other missions would target road junctions and choke points.

Initially, British and American high commands were enthusiastic about the result. However, the air bombardment was effective only at Utah Beach, where B-26 Marauders destroyed fixed fortifications and shattered German morale. Otherwise, the H$_2$X error range, the fact that many bombers never saw the radar planes' marks and either dropped blind or not at all, and the fear of releasing payloads on Allied troops just offshore combined to ensure that the campaign did little to disrupt the defenders. Things went especially badly at Omaha Beach, where hundreds of tons of bombs fell onto fields behind the German front lines. Assailed by troops of the First U.S. Infantry Division, the five-mile strip between Bayeux and Insigny became the site of the invasion's most intense fighting. The misplaced payloads, coupled with stronger-than-expected defenses and choppy seas, left thousands of American soldiers pinned on the beaches for hours. Many others were cut down before leaving their landing craft. Tanks wallowed in the churning waves like strange, forgotten animals. All told, U.S. forces suffered some two thousand casualties that morning, and only a series of freelance charges drove the Germans off the bloody beachhead.

As it was, though, the fabled German defenses — the Atlantic Wall, with its miles of trenches, tons of reinforced concrete, pillboxes, mines, and barbed wire — turned out to be colossal military puffery. Only at Omaha Beach were landing troops held up for more than an hour and a half. Even there, German resistance lasted less than a day. Hitler still had some surprises in store, especially for Britain. But the Allies and their wizardry were on the continent, driving for the Reich.

10 · Victory

"So while the bomb itself was new and extraordinary, no new or extraordinary use was made of it. Nuclear fission replaced incendiaries and high explosives in an ongoing, escalating air offensive. To many partisans of strategic bombing, the mushroom cloud was like the fire storm, a milestone in the realization of a doctrine a half-century old."

LEE KENNETT

IN THE PREDAWN HOURS of June 13, 1944, a loud buzzing, like the sound of a powerful outboard motor turned full throttle, rattled the English south coast region. Suddenly, the noise stopped, creating an eerie silence that was shattered a few moments later by a thunderous explosion.

Arriving a week after the Normandy invasion, this was the first of the engineering masterpieces Hitler called *Vergeltungswaffen*, or revenge weapons. Built by Volkswagen and soon more widely known as buzz bombs, or doodlebugs, the pilotless V-1s looked like twenty-five-foot-long cigars with wings and could carry a one-ton warhead at least 130 miles. After flying a predetermined distance, based on the number of nose propeller revolutions and usually aimed to bring the bombs over London, a guillotine blade severed the pipe to the elevators, causing the odd-looking invaders to plummet to earth.

The attacks broke the nation's growing sense of complacency and impending victory. Over the first years of war, spurred by the Battle of Britain and the Blitz, the country had fused its radar early warning, airborne interception, and fighter control systems into a highly efficient network. Between the air defenses and the mounting Allied offensive that had just culminated in the D-day landings, nearly three years had passed with German aircraft rarely mounting a serious challenge to the Royal Air Force over its own turf. But the drones promised to change the odds. They cruised only a few thousand feet above the ground at close to 350 miles per hour. Fighter aircraft, the first line of defense, could get some. But the V-1s, with their jet-like reaction engines, flew too low and too fast for the never terribly deadly British anti-aircraft guns to aim accurately.

Three nights after the first sporadic attacks, the V-1 campaign escalated dramatically. One hundred forty-four buzz bombs reached Britain, nearly half striking the London area. Between June 15 and the end of August, some 8600 would be launched from German positions in occupied Western Europe. Although V-1 payloads ran under those carried in a conventional bomber, the automatons awoke a special terror. Many Londoners fled the city in panic or returned to sleeping in bomb shelters.

British intelligence already knew many of the weapon's particulars when the first V-1s arrived. Putting the Oslo Report's details together with other tidbits, R. V. Jones had located the research facility where the flying bombs were developed: Peenemünde, on the Baltic island of Usedom. He had initially been led to the site while tracking down rumors of German rocket experiments — what turned out to be the V-2 — and his information had convinced Churchill to order a raid on the night of August 17–18, 1943, that had killed 130 German scientific staff and six hundred foreign workers. As part of his V-2 investigations, figuring the Germans might track test launches by radar, Jones had ordered a special watch put on Martini's top signals outfits. It was a stroke of inspired deduction that turned out to be as good for flying bombs as it was for rockets. Not long after the August raid, the 14th Company of the German Air Signals Experimental Regiment took up positions along the Baltic coast not far from Peenemünde. By listening to radio reports of the German plots, Jones got a ringside seat on V-1 trials late that year, and had divined the approximate accuracy, speed, and range of the flying bombs. Meanwhile, photo reconnaissance had located the launch pads, gigantic sloped tracks shaped like skis along the French coast facing Britain. In late December 1943, American bombers pounded the sites into ruins. This had bought a few months while the enemy constructed much smaller mobile launchers difficult to spot and bombard. These were precious months.

Soon after the first attacks Rad Lab steering committee member Louis Ridenour, still overseas, pressed to deploy one of the first operational Microwave Early Warning stations for tracking doodlebugs. The radar, a training center for other MEWs trickling into service, stood watch in Devonshire, facing the Cherbourg peninsula from the island's southwestern slope. Up and running as a routine observation post the night of June 5, 1944, it had captured an unprecedented radar view of the Normandy invasion: operators, including the Radiation Lab's Ernie Pollard, had rigged a special camera that photographed one of the several Plan Position Indicators every thirty seconds, providing a unique time-lapse record of the historic event.

After Ridenour's plea, the MEW was resituated near Hastings at the entrance to the Strait of Dover. The sprawling installation, which required several support huts and a staff of around 150, had to be disassembled and moved on more than a dozen trucks. But it was back in operation for American Independence Day on July Fourth. From its new vantage point, the MEW spotted buzz bombs as far as 130 miles away, beating every other radar by some twenty miles and affording two or three minutes' grace to get interceptors into position. Within about a week, round-the-clock fighter patrols were placed under MEW control. A typical patrol — P-51s and Spitfires during daylight, P-61s and Mosquitoes at night — cruised at six thousand feet. Controllers waited until the invaders closed to about a half mile behind the planes, then ordered the interception. At low altitudes the V-1s were faster than almost any plane. But by angling down toward the drones, pilots could build speed and gain at least a sporting chance of shooting one down.

Over the first month of attacks, facing some three thousand buzz bombs, fighters brought down more than nine hundred, compared to roughly three hundred for the anti-aircraft guns and barrage balloons that formed a second ring of defense near London. Similar success against bombers would soon break any attacker's back. However, with the mass-produced and pilotless V-1s, Germany could go on indefinitely. Meanwhile, thousands of people were being killed or seriously injured by the nearly daily attacks.

On July 4, the same day the MEW took up its new position, a desperate Britain shifted hundreds of anti-aircraft guns and rockets from London to coastal areas: one advantage of shooting over water was that shrapnel and shards of broken V-1s would not fall on civilian areas, a significant cause of casualties. The frantic move occurred during a two-day melee of sweat and strain involving more than twenty thousand workers. Britain, said one major general, was like an imperiled maiden in dire need of assistance.

Into the turmoil, ready to earn its stripes as one of the war's most impressive technological innovations, rode the SCR-584 anti-aircraft radar. The pioneering automatic tracking and fire-control set had evolved with remarkable speed since its conception in early 1941. The first factory prototype arrived in May 1943, barely a year after the Army's initial order, with production of what would be nearly 1700 sets under way two months later. At the time of the buzz bomb attacks, several hundred 584s were in England, most awaiting transfer to the continent to help Allied forces under General Dwight D. Eisenhower. As a result of an urgent plea from Churchill to Roosevelt, Ike lent some two hundred sets to the British to direct 3.7-inch anti-aircraft guns. At least two dozen others controlled American 90mm heavy artillery deployed against the V-1s.

The SCR-584 represented almost everything dreamed of in an anti-aircraft radar. Its conical scan and ranging system could track objects in any direction with near pinpoint accuracy and only minimal interference from

German chaff. Married to the Bell Labs predictor, it fed information to batteries of four guns that fired based on where the target would be when the shells arrived. And, against the buzz bombs, it came with one final innovation, the holy grail of automatic gunlaying — the proximity fuze.

The key to the proximity fuze was a miniature radio transmitter incorporated into the nose of a projectile. It sensed when a target was nearby, then caused the shell to detonate. No longer did gunners need to hit enemy airplanes and buzz bombs directly, or rely on alarm clock fuses that exploded at a certain time after firing. As with the cavity magnetron at the heart of the SCR-584, the basic design had arrived in the United States as part of the Tizard Mission in September 1940, with the Americans picking up development at a laboratory paid for by Vannevar Bush's National Defense Research Committee.

The guiding light behind the proximity fuze, which also went by the names Pozit and VT fuze, was Carnegie Institution physicist Merle Tuve, whose early ionospheric experiments with Gregory Breit had led to pulse radar. Growing up in South Dakota, Tuve and boyhood pal Ernest Lawrence had linked their houses by telegraph line. During the war Tuve ran NDRC Section T — named after himself — which had been set up in the summer of 1940 to develop some sort of influence fuse that could sense nearby aircraft. He was considering a variety of triggering mechanisms: acoustic, photoelectric, and radio. But when the Tizard Mission extracted a radio circuit design by the brilliant engineer W. S. Butement from its famous black box, he had adopted it at once.

The fuze, in essence a tiny continuous wave radar, would screw into the front of an artillery shell to form its nose, emitting a steady, oscillating 180 to 220 megahertz signal. The shell body served as one electrode of a dipole and the fuze cap as the other. The nose cone functioned as receiving antenna. If any target came within a few wavelengths of the oscillator, it changed the signal in a way that diverted current to a detonator. The neat electrical trick was accomplished with just four vacuum tubes.

Tuve's people altered very little of the basic design. But before the fuze could be mass-produced, tubes needed to be shrunk from about the size of light bulbs to something more like thumbnails, and made rugged enough to withstand firing from a big anti-aircraft gun. Developing long-life batteries posed another problem, since the fuzes needed to be shipped worldwide and stored for months on end. The problem was solved eventually by developing a chemical mix that remained inert until firing, at which point a barrier broke and zinc, carbon, and chromic acid came together to make a short-lived, fly-as-you-go battery. In just a few months Tuve's staff surpassed a hundred, overwhelming Section T headquarters at the Carnegie's Department of Terrestrial Magnetism. In April 1942 the enterprise shifted to a heavily guarded garage building in Silver Spring, Maryland, taking the name Applied Physics Laboratory. Operating under an OSRD contract with Johns Hopkins University, the project was kept top secret: plastic noses for the shells were shipped to Johns Hopkins Medical School under the label "rectal spreaders." By September 1942, American manufacturers led by Sylvania churned out four hundred of the deadly devices a day. The figure would climb to 70,000 daily as the war approached its end.

The proximity fuze, the SCR-584, and the Bell Labs M-9 gun predictor invented by David Parkinson had already seen combat when they joined the anti-aircraft cavalcade rushing to the English south coast in July 1944. But all three had never been used together. The fuze was considered so vital that the Combined Chiefs of Staff banned its use over land, where the enemy might recover enough pieces to design jammers or build a rival version; the decision restricted deployment almost exclusively to the Pacific. Late the previous February, a handful of gunlaying sets and M-9 predictors had helped bail out the ill-planned Allied landing at Anzio, an Italian coastal town thirty-seven miles south of Rome. Enemy jamming and chaff had severely affected longwave fire-control radars brought ashore, allowing the Luftwaffe to attack with much greater effect. Rushed to the scene, the 584s, joined by some SCR-545s that also operated on ten centimeters, evened the odds. William N. Papian, a Signal Corps radio intelligence officer, followed the action via the illumination of tracer bullets and had been amazed by the instant change in the defenses. "One night I noticed that the aim was very well coordinated. It had never been that well coordinated. It started knocking planes out of the air," he recalled. By May 6, guns directed by the SCR-584s had claimed thirty-seven of the forty-six German planes shot down at Anzio.

The pilotless V-1 attacks, however, presented the perfect opportunity for combining the three technologies. At the time of the buzz bomb's debut, a consignment of proximity fuzes had already arrived in Britain in anticipation of their being approved for use against Hitler's latest weapon, since any dud fuzes would fall at sea or over friendly territory. Moreover, the buzz bombs flew in a straight line that made fire-control prediction relatively easy. Churchill's plea shook loose any remaining qualms, and the fuzes were deployed along with the SCR-584s and their Bell Labs predictors.

The SCR-584s were housed in a special trailer, the radar dish mounted on top. A gasoline-fueled generator parked nearby supplied power. Cables connected the radar to the predictor, which rode on its own small trailer. After passing through the M-9, data were relayed to a battery of four guns positioned about fifty feet away. Everything seemed ready to go almost immediately, but in the confusion of war the payoff took an agonizingly long time. The installations had been set up with little time to familiarize gun crews with their operations. Lee Davenport, in charge of 584 development under Getting, visited an American battery and found troops trying to work the machinery while reading instruction manuals. "Seven or eight buzz bombs came over within range while I was there," he remembered. "And the crew never got a single shot off at any one of them." He and Rad Lab colleague Leo Sullivan toured the coast conducting on-the-spot tutorials from a borrowed ambulance with its red cross painted over.

As gunners became more competent, and the proximity fuze was deployed to nearly every heavy gun on the Channel coast, the kill rate shot upward. On August 12, General Sir Frederick A. Pile, commander of the British Anti-Aircraft Command, passed on congratulations to General George Marshall, the U.S. Army Chief of Staff: "the curve is going up at a nice pace, and already we are far away ahead of the fighters. As the troops get more expert with the equipment I have no doubt that very few bombs will reach London."

Pile's prediction proved as accurate as the guns. By the fourth week of August, Allied forces were shooting down close to two-thirds of all buzz bombs fired. On the twenty-eighth, the last day V-1s were aimed against Britain in force before Allied troops overran launching sites near the Strait of Dover, the Germans fired 104 across the Channel. Artillery got sixty-eight. Fighters shot down fourteen. Barrage balloons accounted for two more, and sixteen crashed before reaching their targets. Only four hit London. All told, over the nearly three-month campaign, anti-aircraft batteries brought down 1629 buzz bombs, almost equaling the fighter total. After the move to the coast, the guns outshot aircraft by 50 percent, registering some 1300 kills, virtually all with 584-controlled guns and the proximity fuze. Through the end of March 1945, nearly 1200 additional buzz bombs would be launched against Britain, mainly in sporadic bursts from airplanes. Anti-aircraft fire got only about a third. But the major onslaught was already broken. The maiden's prayer had been answered.

The Rad Lab was on a roll. After receiving General Pile's laudatory message, George Marshall passed it on to Vannevar Bush. Bush wrote Alfred Loomis, and the SCR-584, which had started out as the Rad Lab's biggest gamble, officially became one of its greatest success stories.

As a result of such successes, the watershed year of 1944 marked what one observer called "the almost total conversion of the American radio and electronic industry to the manufacture of a secret war weapon." The weapon was radar, which accounted for $1.2 billion of the annual total of $2.7

billion in industry sales. Throughout the year's second half, a plethora of systems joined the Allied push — watching for enemy aircraft, guiding bombers through overcast, directing gun batteries, and continuing to ferret out the U-boat. In contrast to the early war years, when a few hand-built systems were rushed into service, the highly accurate radars arrived in combat by the thousands, replacing increasingly outdated meter wave systems, which were often rendered useless by enemy jamming. Military planners waged campaigns with radar in mind, and time after time, against a highly trained and resourceful enemy, the technology provided the critical margin for victory.

While both Bell Labs and TRE could claim a large share of the credit for the emerging era of microwave radar, the Radiation Laboratory was the real force behind the renaissance. Its developments could be traced to $640 million worth of radar produced by the United States in the peak year of 1944, better than half the total output and accounting for the vast majority of microwave systems. Virtually all the lab's major remaining projects — Eagle, three-centimeter submarine hunters, and Ground Controlled Approach — reached the front lines in late 1944 and early 1945. Although Luie Alvarez, Hans Bethe, Robert Bacher, and other top members had left for the atomic bomb project in New Mexico, the lab counted nearly four thousand employees and was constantly in demand by industry and the military. In July 1944 an astonished Lee DuBridge wrote Vannevar Bush that 23,355 visitor's badges had been issued in the just-ended fiscal year, averaging out to 74.4 guests each day.

Capping the year, I. I. Rabi won the Nobel Prize in physics for his prewar work using atomic beams to measure nuclear magnetic resonance. He was the first, but not the last, lab figure to win the award. Even though Rabi spent part of his time at Los Alamos, the laurel served as a point of honor for the Rad Lab, whose members were already starting to feel overshadowed by the bomb project. Less than a month after Rabi's award was announced, Ivan Getting griped that the lab was short on top talent compared to the A-bomb effort. "Los Alamos has all those Nobel Prize winners out there," he noted.

"We have one Nobel Prize winner," DuBridge corrected.

"Yeah," Getting quipped, "but we've only had him for three weeks."

Firmly established as a world-class industrial research and development organization, the Rad Lab ran field stations at military bases and other sites from Maine to Florida to help test, install, and maintain equipment. Overseas, the British Branch mushroomed in the days after D-day to about a hundred workers. Headquartered in six steam-heated brick huts with corrugated iron roofs, the men routinely ventured to air bases and even combat arenas to fine-tune their systems and introduce new practices. In September 1944, soon after Allied forces entered Paris, the lab opened an Advanced Service Base in the liberated city.

Access to top military personnel became almost commonplace. British Branch member L. J. Henderson took over as radar advisor to Ira Eaker in the general's new role as commander of the Mediterranean Allied Air Forces, setting up shop in the eighteenth-century palace of the king of Naples. In the Pacific, a small Australian branch opened to provide microwave expertise to General Douglas MacArthur's forces. Rad Lab members traveled with special priorities that allowed them to bump high brass. They hailed colonels and generals by their first names and wielded a card that afforded them a lofty assimilated rank. One staffer, finding himself impeded by an Army colonel, reputedly countered with: "Where do I go to get you fired?"

On the European continent, microwave radar was in increasing demand during the Big Push that followed D-day. An entire MEW installation was crated, waterproofed, and shipped across the Channel from Britain to serve as a mobile fighter direction post; it landed on Omaha Beach on D-day plus 6. The set, borne into battle by its armada of trucks, proved especially useful in spotting low-flying German aircraft trying to slip behind front lines and strafe advancing support columns. Dave Griggs, the former Rad Lab pilot turned Air Forces radar advisor, claimed it also caught enemy planes trying to surprise Allied fighters as they circled airfields low on fuel and out of ammunition after a day's run. Placed in control of a squadron of P-51s assigned as bodyguards for the returning planes, the radar picked up about seventy German aircraft, enabling ground staff to vector the Mustangs into a straight-on attack from below, silhouetting the enemy against the sky. Griggs wrote home to Ed Bowles with the blow-by-blow account. "The Huns were caught flat-footed and in the dogfights that ensued they were dispersed all over the sky. They all broke and ran for home and never again did they try that tactic."

Griggs may have embellished his account, since German aircraft were in short supply in western France at the time. In any case, the MEW proved effective enough that General George Patton screamed profanely in protest when the set was diverted from fighter control to tracking buzz bombs; his Third Army was driving along the Loire's north bank with its right flank exposed and needed dependable air cover. The irascible commander later got his wish, apparently just in time. On a particularly overcast day not long after the V-1s were defeated, Griggs again reported to Bowles, the Third wilted in the face of a fierce tank-led counterattack. Patton raised Opie Weyland, commander of the Seventeenth Tactical Air Force, which ran the MEW, begging for air support even though planes normally would have stayed grounded. The highly accurate radar enabled controllers to direct fighters over the battlefield. Dropping down under the low cloud ceiling, pilots saw German tanks out in the open, not expecting Allied aircraft. The planes picked off many and scattered the rest.

By late summer, a second MEW arrived in France. The sets were supposed to advance with the front. But it took three days to move a station, and often Allied forces pressed ahead so quickly the pair could not keep up. The SCR-584, more mobile and available in far greater numbers—thirty-nine had come across on D-day to protect American beaches—did a better job of staying with the troops. Modified versions took the place of MEWs, guiding friendly aircraft on low-level bombing runs, and standard sets could even watch for clusters of German tanks and other vehicles. The radars were positioned on high ground behind Allied artillery lines, the scan turned off, and the antenna pointed at enemy crossroads whose coordinates were known. Moving targets showed up as oddly fluctuating signals recognizable through ground clutter, and operators passed on the news to already zeroed-in batteries. In one case, a few 155mm howitzer shells were lobbed toward an enemy convoy spotted by a 584. On the radar scope, the vehicles could be seen detouring to another road. A second salvo prompted a second detour, and German prisoners later told interrogators that SS troops had searched house to house for the person radioing artillery data.

Other SCR-584s were hurried into Belgium when a renewed buzz bomb assault began that fall. The principal target was Antwerp, which has a vast harbor desperately needed by the Allies to solve a crunching supply problem for the push across the Rhine. The port city had been occupied in early September, but because German forces held the key approaches would be opened only in late November—and at great cost. It first came under V-1 fire on October 24. With fighter aircraft busy supporting the front lines, radar-controlled batteries formed almost the only defense. The proximity fuze had not yet been cleared for general use over land. But the conventional defenses managed to bring down 40 percent of the targets engaged and, by concentrating on only those drones likely to hit their targets, spared the hub from major damage. Still, the guns were not nearly as effective as they had been in Britain two months earlier, and anti-aircraft crews, according to one account, "literally cried and begged for VT fuzes."

It rankled Vannevar Bush that the proximity fuze card was being played so close to the vest. The OSRD director expected fierce enemy opposition as the Allies advanced on Germany, and he wanted the United States to pull out all the stops. He saw the fuze as vital both in anti-aircraft fire and as an anti-personnel weapon fired from howitzers. In the latter case, the miniature radar could ensure that shells exploded at just the right height over enemy troops — a feat hard to manage with timed fuzes — sending out a devastating spray of shrapnel. Tests against targets dispersed in foxholes or trenches showed the fuze could hike a howitzer's long-range killing power by a factor of ten.

In the fight to free the fuze, Bush had an old nemesis: Ernest J. King. The Pacific fleet almost lived by the proximity fuze, and the admiral didn't want that advantage risked. Since the Navy had pushed the device's development and paid the bulk of the bills, King's view counted. Bush thought the old salt was being overly cautious and sought him out for another man-to-man talk.

The session began stormily. King told Bush civilian opinions didn't interest him. The engineer countered that the issue was technical, and on that front the admiral was a babe in the woods, not entitled to any opinion at all. After that, things gradually settled down. Bush swore it would take the Germans or Japanese at least two years to reproduce the shell, and by then it presumably wouldn't matter. Finally, about the time the buzz bomb attacks began against Antwerp, King acquiesced. Within forty-eight hours of formal Combined Chiefs approval on October 25, Bush jumped on a plane for France. He personally visited the front lines, reviewing the fuze's capabilities and limitations for chief ordnance officers. Some 210,000 rounds fitted with Pozit were on the Continent by mid-December, soon helping the Antwerp guns shoot down buzz bombs approaching the city with better than double the previous success rate.

More critically, the proximity fuze arrived just in time for the Battle of the Bulge. As Allied forces pressed toward the Rhineland in December 1944, Hitler had managed to assemble — much of it in secret — ten panzer divisions and fourteen infantry divisions in the Ardennes forest in southeastern Belgium. Early on the sixteenth a massive German counterattack surprised Allied commanders and forced a general retreat. The attack precipitated the release of the proximity fuze more than a week ahead of schedule. Initial deployment was for anti-aircraft fire. But by December 21, once the American Army had caught its breath, artillery unleashed the proximity fuze against enemy ground troops as well, helping stem the advance and turn the tide of the overall campaign. The first action came around Malmédy and the Amblève River, where the famous German 150th Panzer Brigade ran up against elements from the 120th Infantry. Proximity-fuzed shell bursts killed a hundred and threw Hitler's forces into momentary panic; some troops reportedly ran to the American lines screaming "Kamerad." Similar stories peppered battle accounts through the end of the year, as thousands of rounds rained down on road junctions, bridges, and highways. Later, German POWs called the fuze-laden cannonades the most demoralizing action they had experienced. Patton reported, "The funny fuze won the Battle of the Bulge for us."

In the fierce close-quarters Ardennes fighting, the SCR-584 backed up the proximity fuze by controlling Allied fighter-bombers. A blanket of snow covered the region, making it hard to distinguish Allied and Axis vehicles from the air. American forces used the radars to track friendly and enemy ground transports, keeping fighter-bombers from attacking their own troops. "The number of American lives saved by our ability to stop attacks on our own columns cannot be measured," reported Major General E. R. Quesada, leading the IX Tactical Air Command. At the same time, the Eighth Air Force used its airborne radars to run rare tactical missions in heavy overcast and drizzle. In twelve days of flying between December 16 and year's end, it launched some five hundred radar-guided missions against marshaling yards, air fields, and communications centers supporting the German counteroffensive.

In the face of German air activity, a like tale unfolded. Bad weather held the Luftwaffe in check until December 23, when it unleashed husbanded resources in a last great fury. Allied tactical air forces fended off an estimated thousand sorties. But on the morning of January 1, Hitler struck again, targeting airfields and forward troops with some eight hundred planes hugging the hills and treetops to avoid radar contact. Although carefully planned, the German air raids did not accurately assess the strength of Allied anti-aircraft artillery, including the SCR-584 reunited with the proximity fuze. Radar apparently detected many low-flying enemy planes, while the M-9 computer directed a massive concentration of heavy ack-ack fire. When the smoke cleared three hours later, Allied guns claimed 394 enemy planes, with another 112 probably destroyed. The German counterattack was broken.

The final act in the European radar drama was played out almost exclusively in the air. In the months following D-day, the Eighth Air Force and Bomber Command resumed their punishing offensives with fresh vigor. American bombers focused more attacks on German synthetic oil plants, and by September the monthly aviation fuel production — 180,000 tons not long before the Allied invasion — had dwindled to a mere 10,000 tons. Output crept back upward late in the year, but never again approached the previous levels. Combined with a dearth of experienced pilots, the fuel shortage dramatically restricted Luftwaffe air operations. Neither Bomber Command nor the U.S. forces, though, forsook the area bombing campaign.

Microwave radar continued leading the way. Bomber Command, active at night, relied on ten-centimeter H_2S whenever weather or distances dictated. To avoid enemy detection devices, air crews were ordered to keep sets switched off until within forty miles of German territory. The Eighth Air Force, flying in daylight through cloudy fall and winter skies, depended on three-centimeter H_2X for about two-thirds of its raids over the last quarter of 1944. By year's end, thirty-nine of its groups had been fitted, along with close to two hundred Pathfinders in the recently activated Fifteenth Air Force operating out of Italy.

A number of technical innovations, such as improved antenna patterns and Plan Position Indicator presentation, increased accuracy. Aeronautical charts were stripped of clutter to look more like what the radar operator would see on his screen. Training was vastly improved, with centers in both England and Italy, as well as back in the United States. Special radar-intelligence officers flew mock high-altitude bombing sorties over the flat Midwest and the Colorado Rockies to learn to read targets in different terrain. Offset bombing also became a mainstay. With this technique, when an objective didn't present a good radar return, bombing runs were based on easy-to-find sites with a known position relative to the target. *Radar*, a confidential war publication, likened it to golf: "From the fairway a golfer can't see the cup, but can place his shot accurately because he knows where it is in relation to the flag."

None of these steps offered a panacea. Getting within a mile of a target remained fairly rare. Still, H_2X kept bombers bombing. Of the 290,000 tons of bombs dropped on Germany in 1944 by the Eighth Air Force, 147,500 were attributed to H_2X. Louis Ridenour, returning to the Rad Lab in early November so much thinner his clothes had to be altered, reported: "Every mission is planned with radar in mind."

The two air forces, thousands of bombers strong, overwhelmed Hitler's early warning network. The elaborate jamming and spoofing contests also went the Allied way. German engineers introduced several effective electronic twists. However, R. V. Jones and other operatives amassed a detailed dossier on enemy radars and countermeasures that enabled the gurus at TRE and the Radio Research Lab to strike back harder with a series of powerful and versatile jammers. Dense concentrations of chaff could still saturate the German system, so air crews also stepped up their use of foil. By early 1945, the arrays of electronic noisemakers and chaff dispensers routinely carried in Allied bombers provided a formidable security blanket in the once-deadly Reich skies, reducing the effectiveness of Hitler's radar defenses as much as 75 percent.

A glaring pinnacle of the mismatch came on February 13, the first of two successive nights of Bomber Command attacks that leveled downtown Dresden and killed up to 100,000. The various special effects that night rendered the German ground control system a shambles. Only twenty-seven fighters rose to oppose the nearly 1200 bombers pummeling the city. Just three British planes were lost to opposing fire. A Luftwaffe pilot recorded the scene in his diary: "My saddest day as a night fighter . . . Take-off too late. Huge firework display over the city. . . . No communication with divisional headquarters. Apparently Division was in the dark . . . Result: major attack on Dresden, in which the city was smashed to smithereens — and we were standing by and looking on. How can such a thing be possible? . . . Feeling that things are approaching an end with giant strides. What then? Wretched Germany!"

As the spring of 1945 blossomed, the devastation of Germany's re-sources — men and critical supplies like coal and oil — approached cata-strophic proportions. The Axis defeat was inevitable. Still, the area bombing campaign extended well into April before the Allied Chiefs of Staff called a halt, vast palls of fire and smoke spiraling up over the landscape in mute testament to needless destruction. In the war at sea, U-boats managed a brief resurgence, almost like the spurt of energy and clarity of thought that sometimes comes just before death. Fitted with *Schnorchel*, a "breathing" device that allowed batteries to be recharged while subs stayed underwater, Dönitz's fleet could often remain safe even from the Rad Lab's latest three-centimeter radar — the ASD, or AN/APS-3 — of which close to five thousand had been built by Philco by the end of 1944. But like the continuing area bombing campaign, this last-ditch effort was senseless. Thousands of Allied ships plied the oceans unscathed.

On all sides of the Reich, the walls were caving in. The afternoon of April 30, with Russian troops in the Berlin streets, Hitler shot himself. A week later Karl Dönitz presided over the unconditional surrender of the German forces; and the once seemingly invincible nation was carved into American, British, French, and Russian zones of occupation.

The radar war in Europe was not quite ended. Allied experts wanted an autopsy, a complete assessment of enemy accomplishments and the effectiveness of their own inventions and countermeasures. The dissection, done with the cooperation of German experts, proved an almost chilling vindication of past work. An inventory of German radar showed that Allied intelligence had known every piece of equipment in the enemy nightfighter system, and identified forty-eight of fifty-one ground radars. Beginning that summer, fighting still hot in the Pacific, a series of exercises code-named POST MORTEM re-created the Wizard War. Prisoners manned their old radar outposts, while the RAF flew mock raids, complete with jamming and Window. The exercises showed that the jamming screens accompanying each attack, while not completely effective, had successfully disrupted the enemy fighter-control system. On no occasion during the trials were German radar operators able to reliably tell real attacks from spoof raids.

A few months before Germany's surrender, Lee DuBridge toured Europe behind the advancing Allied lines. Flying into the Reich on military transports, he passed over once-thriving cities the Rad Lab's radars had helped flatten. Overriding that sobering reality was the lingering image from a visit to the liberated Buchenwald concentration camp.

DuBridge laid plans for closing the Advanced Service Base in Paris and the British Branch of the Radiation Laboratory. He gave no thought, though, to terminating the main operation in Cambridge or the recently opened Australian offshoot. The director could see no immediate end to the war against Japan. And although the lab had been slow getting up to speed in the Pacific theater, things were finally moving into high gear.

•

During the war's first years the Rad Lab's marginal role in the Pacific campaign was largely a matter of focus. The Army Air Forces constituted its best customer, and the fight against Japan was initially a Navy affair. Moreover, the Joint Chiefs had agreed early on to defeat Germany before turning to the Japanese threat. So while a few of the enterprise's microwave radars found their way westward — Bell Labs had finally produced the SCR-720, a much-improved version of its first ill-fated airborne interception set, while PT boats also carried Rad Lab–spawned systems — Lee DuBridge assigned the highest priorities to beating the U-boat, developing H$_2$X, and getting the SCR-584 gunlaying system ready for the Luftwaffe.

That was not to say radar itself had not been vitally important. In spring 1942 the Navy's meter wave CXAM sets had provided early warning of Japanese air attacks at both the Coral Sea and Midway, and thousands of ASV units, direct copies of Taffy Bowen's Mark II or improved versions using a single antenna for transmitting and receiving, came off the assembly lines at Philco and Canada's Research Enterprises Ltd., many to be installed on Pacific-bound aircraft. Microwave radar outside the Rad Lab's purview had also joined the Western Front in force. Bell Labs had produced the first handful of several hundred highly effective ten-centimeter SJ models to help submarines find Japanese merchant ships and evade enemy patrols. Bell also led the way with standout naval fire-control radars adapted to employ the cavity magnetron almost as soon as Bowen had demonstrated the device at its Whippany lab back in October 1940. The ten-centimeter Mark III was used by main batteries against surface vessels, while the forty-centimeter Mark IV directed secondary batteries tracking air and sea targets. Both sets, ultimately integrating range, elevation, and azimuth information with gyroscopic data that measured a ship's roll and pitch, had proven critical in neutralizing Japanese superiority in naval nightfighting around Guadalcanal and Savo Island in late 1942, and were so advanced that three hundred remained in service on V-J Day.

At the Radiation Laboratory, even in the flurry to fight the U-boat and develop H_2X, Navy officers and other dignitaries coming through town had held staff members riveted by stories of the crucial roles played by other people's radars in the Pacific, especially at Midway and Guadalcanal. As the Allied drive in the West picked up steam, there had lingered a nagging sense of more to accomplish in the war against Japan.

That misgiving had begun to evaporate toward the end of 1943. With its primary goals well in hand, a larger share of the lab's attention had gravitated toward the Pacific. Taffy Bowen left Cambridge just after Christmas to take charge of research in the emerging Australian radar program, hunkering down in the bomb bay of a B-24 for "a wild ride" on which "I got the distinct impression they used cowboys as pilots." As the Welshman explained his decision to move: "By mid-1943 most of the radar devices likely to be used, either in the defensive or the offensive mode, had already been thought of or were being worked on; it was apparent that putting these into practical use was much more important than the development of entirely new devices."

When Bowen arrived Down Under, some two hundred staff members were working out of the government-run Radiophysics Laboratory on the grounds of the University of Sydney. The lab specialized in developing rugged, lightweight early warning sets needed for island hopping in tropical climates. Not long after he took up residence, the Rad Lab sent a few staffers to help bring the Australians up to speed on microwave radar. By May 1944, the loose collaboration had evolved into an official seven-man body, the Australian Group of the Radiation Laboratory.

Almost in concert with the Australian branch opening, the Rad Lab's first significant contributions to the Pacific war arrived in combat. As in the European theater, the Army was the main customer. The Air Forces had recently begun operating its first few Snooper aircraft — Liberators patrolling Japanese shipping lanes with the three-centimeter Low Altitude Bombing set. Although the production version, from Bell Labs, was greatly modified in form, LAB was the end product of the work Dale Corson and Byron Havens had started back in mid-1942 while with the First Sea Search Attack Group to develop a more accurate method of dropping bombs once a microwave ASV set picked up an enemy ship or submarine. At its core was a Corson-patented way to slow down the rapidly moving oscilloscope sweep and link it to the bombsight. In the final product, the bombardier's station carried a special radar scope. When closing on a target, the bombardier guided the plane through the autopilot, turning knobs to keep the radar signal centered by causing the sweep rate to match the plane's ground speed exactly. Meanwhile, using this speed information, a horizontal line moved up from the bottom of the cathode ray tube. This showed the bomb release point. When the line reached the target signal in the middle of the screen, bombs released automatically.

This was precision equipment, deadly accurate from two thousand feet or less under conditions of zero visibility. Corson, Havens, and Colonel Stuart P. Wright had tested the concept dropping test bombs — blue devils — on shipwrecks off Florida that summer and fall. Havens improved the device after Corson went to the Pentagon later in the year, and the set had been sent to Western Electric for production. Wright ushered the first models into the Pacific theater in August 1943, staging out of Guadalcanal. Snoopers were carried in Liberators at the expense of firepower and armament. To compensate, the planes flew at night, reducing exposure to enemy fighters.

Low-flying radar-guided bombers formed the perfect complement to the American submarines preying on *marus*, Japanese oil tankers and cargo vessels bringing vital supplies from conquered Malaya, Indonesia, and other countries to the resource-starved home islands. Unlike U.S. and British forces combating the U-boat in the Atlantic, the offensive-minded Imperial Navy never developed a strong convoy system or an integrated hunter-killer force to defend against submarine attack. The LAB sets terrorized *marus* from another dimension and forced the deployment of valuable nightfighters for tanker defense.

By spring 1944, Snoopers were elevated to squadron status, with units in the Thirteenth Air Force around the Admiralties and the Carolines, the Fifth Air Force in New Guinea, and the China-based Fourteenth Air Force. As the year progressed, the Snoopers' effectiveness mounted. In the Chinese theater, a patrol typically covered waters near Formosa, Hong Kong, and the Gulf of Tonkin. In the five months ending in October 1944, the planes sank 250,000 tons of enemy shipping, three times the figures scored by daylight squadrons in the seven previous months, and in barely half the sorties. Only five aircraft were lost. That September, in support of the upcoming Philippines invasion, Snoopers bivouacked in the Leizhou peninsula sank thirty-two merchant vessels and a destroyer — one ton of enemy shipping for every gallon of gasoline expended. Remembers Kenneth "Kit" Carson, a Snooper pilot who hailed back to the First Sea Search Attack Group's earliest days and helped sink an estimated five merchants around the Philippines: "We flew in the dark most of the time and we'd attack at about eight hundred or a thousand feet, and you couldn't miss at that altitude, you know."

A second major piece of the Rad Lab's growing contribution to the Pacific conflict was the SCR-584. Early in 1944, as MacArthur's forces leapfrogged up northern New Guinea in preparation for his promised return to the Philippines, the first 150-odd sets joined Army anti-aircraft units. Lab engineer Hank Abajian toured the area and found the same disappointing situation being encountered in Britain. Training facilities were virtually nonexistent, and anti-aircraft detachments had set up the guns as if they still depended on optical sighting, virtually ignoring the automatic tracking features.

Abajian established a training school at Guadalcanal, and wrote manuals that were dispatched to Australia for printing and sent back. He and Albert Paul, a General Electric engineer, then personally inspected every set deployed in the theater. Abajian visited anti-aircraft battalions in Fiji, Munda in the New Georgia Islands, the Green Islands, and New Guinea. Finally, he hopped a convoy to the Philippines, arriving on the central island of Leyte amidst heavy fighting shortly after the Allied landing that October.

The Philippine invasion included what was probably the world's largest radar army — and Leyte provided a prime example of the latest techniques for melding the technology into mobile operations. The main assault wave included twenty-three radar units, with the first air warning and fighter control teams going ashore just behind initial infantry landings. "That was the first real action that I actually saw, where we were shooting at aircraft and they were shooting back at us," Abajian recalled. It was also the first great Pacific success for the SCR-584. Over the course of a few weeks artillery brought down close to three hundred enemy planes, with guns directed by the microwave set taking a large share of the credit. MacArthur sent his congratulations to the crews: "This is superior shooting."

From Leyte, Abajian hopped an LST — landing ship, tank — around Bohol Island and up the Sulu Sea to Mindoro, an island on the western side of the Philippines, due south of Luzon and the capital city of Manila. Positioned on hills overlooking a small valley, four 584s, each still typically controlling a foursome of big guns, were again distinguishing themselves. After assessing the situation, Abajian convinced one commander to let him hook up a spare unit to direct a searchlight to verify hits and aid lighter gun batteries that were not radar controlled. "Anytime we turn on that light I guarantee you that there'll be an aircraft in that beam," he promised.

Abajian knew he was taking a chance. Searchlights could tip off an enemy pilot that he was being tracked, lowering the odds for the 90mm batteries already controlled by the radar. The engineer worked out a deal with the battalion commander to keep the searchlights off until the first radar-directed volleys were fired. At that point enemy pilots knew the score, and the lights could be illuminated.

One night a high-flying Japanese aircraft passed over Abajian's position. The Rad Lab man listened to gun revetments radio in the situation.

"A-battery locked on target. Data smooth."

"Fire when ready."

"Commencing fire."

An instant after the initial salvos, the radar-guided searchlights snapped on. "Sure enough right there in the middle of that beam there was this Jap bogey," Abajian remembered. The second gun burst arrived right on the mark, and the spotlight stayed on the target as it tumbled toward earth. Cheers rang out across the valley.

Back in Cambridge, as the drive on the Japanese homeland picked up steam, and with stories of the first dramatic Low Altitude Bombing and 584 successes pouring in from the West, Lee DuBridge turned his attention to doing for the Pacific war what already dominated the lab's agenda for the European conflict: helping bombers bomb. When it came to pouring destruction on the enemy, the two theaters faced similar problems. As in overcast Europe, the Japanese home islands were often cloaked in clouds. Moreover, the wide H_2X error range left much to be desired, and the U.S. Army Air Forces screamed for something better in the way of blind bombing.

The Rad Lab's top priority project in the spring of 1944 was simply labeled H_2K. It involved taking bombing radar to an even higher frequency — the K-band, around one centimeter in wavelength — where resolution should be at least a factor of two greater than the three-centimeter H_2X. With such a radar, the Air Forces' dream, even railroad tracks would be magically visible through thick overcast.

The K-band bombing project had opened as a low-priority item the previous summer, when DuBridge returned from England impressed by the need for greater resolution. At first, developing the system seemed to pose no great problem. The magnetron scaled linearly with wavelength, and the Rad Lab had already opened a splinter facility at Columbia University expressly to develop K-band transmitters. To link up with their New York colleagues, Cambridge staffers would leave Boston on the Merchant's Limited, arrive at 125th Street on the East Side, then cut across Morningside Park to the Columbia campus. The work had progressed so quickly that by

April 1944 test systems were picking up Boston landmarks beautifully from nearly sixty miles away. DuBridge promptly bumped H_2K to the top-priority slot.

Things soon turned sour, however. Despite the arrival of improved components, including a vastly more powerful magnetron from the Columbia lab, performance dropped precipitously. In May the test set couldn't pick up Boston from more than twenty miles out. At first, staffers suspected equipment failure. But at the same time, similar problems cropped up in the Rad Lab's Group 41, Fundamental Developments, where generic studies on one-centimeter systems were led by Ed Purcell, the Harvard instructor who had joined the lab early in 1941. Particularly interested in developing a countermortar radar that tracked incoming shells to their point of origin, Purcell had set up a crude K-band radar on the top floor of Building 6. To calibrate the system, he would throw open the plywood windows and train the set on Boston landmarks across the Charles River. By early 1944, the radar was picking up a water tower six miles south of the city. As spring bloomed, however, the tower faded into the background.

The simultaneous regression of two separately designed systems virtually ruled out equipment failure. From this point it didn't take long to identify the real culprit: Mother Nature. The theory had already been advanced that water vapor in the air might absorb very high frequency radar signals. As soon as their attention turned away from hardware snafus, it was easy for the Rad Lab physicists to see what had transpired. The bonds making up the water molecule give it a symmetry that determines its energy levels, or quantum states. For K-band, hundreds of possible states exist, but only two where the energy difference between them — the energy gap — corresponds to the energy of microwave signals. Absorption occurs when the radar wave is of a frequency where its quantum energy closely matches, or is in "tune" with, the energy gap between the two quantum states. This condition is called resonance. It's much like what takes place when a singer hits just the right note, causing a piano string tuned to that note to vibrate as some of the energy from the singer's voice is translated, or absorbed, into the string.

For any given molecule, absorption occurs in a band of frequencies. But the effect is greatest at a specific point. When the K-band work got under way, no one knew the exact frequency where resonance would peak. Nor had researchers deduced the breadth of the peak, which could determine how much of the signal would be absorbed. However, a special report by the well-known Harvard theorist J. H. Van Vleck concluded that wherever the effect peaked, it should be fairly weak for wavelengths greater than one centimeter, simply put, because only a relatively few molecules would be involved. True to this prediction, things had gone smoothly in the dry winter air. But since absorption is directly proportional to the atmosphere's water vapor content, the increasing humidity of spring had changed the playing field. It appeared that Van Vleck, who would later win a Nobel Prize, had been off the mark.

The Rad Lab was thrown out of kilter by the water-vapor absorption problem. No one knew for certain how bad things would get. But it was easy to see that the effect could prove particularly devastating to one-centimeter radar systems bound for the humid Pacific. DuBridge placed his pet K-band bombing project on indefinite hold until some basic questions could be answered. Theory aside, could it be proven scientifically that moisture in the air was the signal robber? If so, where did the resonance peak? What were the breadth and magnitude of the absorption?

Special teams were dispatched to Brazil, Puerto Rico, and Florida to study the effect under conditions of higher humidity. Separate measurement projects were launched at both the Rad Lab and the Columbia branch, as some of the lab's brightest minds — among them Purcell and Robert Dicke, the young recruit from the University of Rochester — strove to establish the effect's exact boundaries. By the time the conditions under which a K-band bombing radar could be used were well established, it was so late in the war that none of the Rad Lab's systems saw action. As far as combat went, one-centimeter bombing radars were effectively finished. Few events during World War II, though, would have farther-reaching scientific consequences than the water-vapor absorption mystery, a classic problem of quantum physics.

That story would play out long after the fighting ceased. In the meantime, DuBridge and research director Rabi had a few aces up their sleeves. A quick-fix project strove to improve H_2X's accuracy through a better system of correcting for wind drift. The sophisticated, high-resolution Eagle system designed by Luie Alvarez was also nearing production at a Western Electric operation in the Lane Bryant Building on Forty-second Street in New York. With or without K-band, bombing would dominate the Rad Lab's agenda until the end.

•

Even without atomic weapons, the Pacific war escalated into a fiery cataclysm exceeding the European area-bombing campaign. By mid-1944, MacArthur had reached the north tip of New Guinea and naval forces were closing in on the Mariana Islands some 1500 miles from Japan — close enough to conduct regular B-29 Superfortress bombing runs against Tokyo and other large cities, and set the stage for the destruction to come.

The road to the Marianas, and by extension the razing of Japan, had been paved at the Battle of the Philippine Sea that June—and this was another great victory for Allied radar. Early on June 19, four days after the Allied landings on Saipan, two vast carrier fleets, several times the size of the Midway forces, met ninety miles northwest of Guam. Vice Admiral Jisaburo Ozawa trusted in superior aircraft range and the element of surprise as he launched his attack. He was denied his hopes.

The Philippine Sea engagement illustrated just how far American naval strategy, tactics, and technology had come in a few short years. The battles at the Coral Sea and Midway in spring 1942 had been tremendously encouraging, but while CXAM search sets had helped ward off disaster, much had gone wrong on the radar front that could be improved. In those early days, to complement its warning units the Navy had established radar and fighter-control rooms equipped with consoles, plotting tables, radios, and an intercom link to the bridge. Naval radars did not provide aircraft heights. Instead, operators consulted special mathematical charts, designed around the CXAM's tendency to show blind spots at different altitudes depending on range, that enabled a crude guess of an approaching plane's elevation. This height-finding deficiency, coupled with the lack of Identification, Friend or Foe signals from American aircraft, had caused problems in both campaigns, especially at Coral Sea, where Lexington's plotting room proved too small to handle a complex battle situation, and its sister carrier, Yorktown, had her own air warning radar knocked out of action. Lexington was overwhelmed and had to be scuttled.

By spring 1944, many of these shortcomings had been addressed. Raytheon had delivered some 955 models of its highly touted SG, the ten-centimeter surface search radar descended from early 1941 Rad Lab trials aboard the USS Semmes. All told, a fleet carrier hummed with a half-dozen types of radar—for early warning, surface search, fire control, and fighter control—and more sophisticated systems resolved the critical height-finding problem. The concept of the Combat Information Center had also blossomed. Housed inside the main armored box of each ship, from cruiser up the food chain, and manned by two dozen officers and crew, the centers contained a whole battery of radar, Identification, Friend or Foe, and communications equipment complete with plotting areas for navigation and tactical deployment. In a radical shift of shipboard talent, the executive officer's battle station had moved to the CICs from a conning tower. Meanwhile, fighter direction officers, early in the war based solely on carriers, were present on every combatant ship, enabling primary fighter control to continue virtually uninterrupted even with the loss of the flagship.

Combined with the proximity fuze, deployed in the Pacific in early

1943, the moves provided ships with a formidable defense against air attack, and everything came to a head in the Philippine Sea. The American shipboard radar grid detected the Japanese planes from 150 miles off. Hellcats shot down forty of the sixty-nine attackers before the enemy even reached its objective. As more raids materialized, the slaughter worsened, turning into the famous Great Marianas Turkey Shoot. Guided from the shipside Combat Information Centers, U.S. planes routinely intercepted their foes with a precious altitude advantage. Guns fitted with the VT fuze picked up the slack. The carnage continued the next day, until Ozawa finally retreated. When the Japanese fleet anchored at Okinawa on June 22, it possessed only forty-seven of the 473 planes that had started the battle. Another fifty Guam-based aircraft had also been destroyed. The Imperial Navy's three largest carriers rested on the Pacific bottom, while the American Fifth Fleet lost just 130 planes and no ships.

Never before had a major Japanese air attack been so totally thwarted. In two short days, the Imperial Navy's air losses surpassed those of the Luftwaffe in the entire months-long Battle of Britain. In testimony after the war, one Japanese vice admiral identified the single greatest U.S. strength as its ability to control fighters by radar.

And so the gateway to Japan had been flung open. By November, the United States had established air bases for its Superforts on the Mariana islands of Saipan, Tinian, and Guam. Initially, the B-29 attacks were aimed at specific military and industrial targets: Bomber crews drilled on the Continental Can Company's Houston plant to practice hitting the vast Musashi aircraft engine works in a crowded northwest Tokyo suburb. But after three months, with only 10 percent of payloads falling near their objectives, all top-priority targets remained intact. In January 1945, Brigadier General Heywood S. Hansell Jr., head of the Twenty-first Bomber Command and an advocate of precision bombing, or as the French put it, striking *points sensibles*, was relieved in favor of Major General Curtis E. LeMay.

LeMay — the joke was that even his friends called him "Sir" — brushed aside old concerns about the civilian population. By early March, barely two weeks after the Iwo Jima assault and just in advance of the Okinawa landings conceived to provide fighter-control bases and additional safe havens for bombers, he had revamped the entire American bombing program. War tacticians had seized on the fact that Japanese structures were highly inflammable, and LeMay adopted a straighforward doctrine: since many key targets rested inside large cities, set the cities afire. To accomplish this he shifted to night operations, stripped planes of guns to accommodate more bombs, and made good use of microwave radar, primarily the Bell Labs AN/APQ-13, a high-altitude bombing set for B-29s that combined a modified

Rad Lab H$_2$X unit and a Bell airborne search and intercept system. Nearly eight thousand were shipped by war's end.

On the night of March 9–10, LeMay sent 334 aircraft loaded mainly with incendiaries to Tokyo. Pathfinders marked a vast rectangle three miles by four, inside which resided more than a million people. Radar bombing still left much to be desired, with targets often unresolvable or blurred in with other ground features on the scope. Japanese targets, however, tended to be near starkly "scopogenic" coastlines, made to order for offset bombing. This was the war's most destructive air raid — counting atomic bombs. The official count was frozen at 83,793 killed, but some maintained more than 100,000 died. Sixteen square miles of city vanished in the inferno. A quarter-million structures. There was no firestorm. Instead, the *Akakaze*, or Red Wind, billowed across Tokyo, creating a tidal wave of flame.

Many similar raids followed. The B-29s hit Nagoya, Osaka, and Kobe. LeMay ran low on incendiaries; he favored the M69, containing thirty-eight small cluster explosives in a single casing. The bomb went off at 2500 feet, releasing its cargo randomly, burning out mile after mile of city. The attacks continued throughout the battle for Okinawa during April and May, where radar-directed aircraft, fire-control sets, and the proximity fuze helped Allied ships supporting the invasion defend against *kikusui* raids, massed suicide missions usually by more than a hundred kamikazes.

By mid-June, LeMay's bombers had burned all the biggest Japanese cities to rubble, and the Air Forces was addressing secondary targets. Children were evacuated from the large remaining metropolises, and blocks inside them were sometimes razed to serve as firebreaks. It did little good. Overhead, the skies belonged to the Americans. The B-29s encountered nothing like the initially deadly German anti-aircraft artillery, and nothing like the Luftwaffe, to prevent the completion of their missions. The Japanese, as soon became evident, had forsaken the Wizard War.

•

Japan entered World War II only a hair's breadth behind the United States and Britain in the potential for radar development. Although the country's overall scientific standards paled against those in Europe and America, several of its physicists and engineers ranked as world class. Kinjiro Okabe, a disciple of Hidetsugu Yagi, dean of science at Tohoku University and inventor of the common television receiving antenna, had developed a continuous wave Doppler radar in 1936. On the general technical front, Yagi, Okabe, and Shintaro Uda experimented with various types of magnetrons as power sources for microwave communications systems — and the

Japanese seem to have invented a low-powered cavity magnetron before Henry Boot and John Randall did it in England in February 1940.

Yet for all the technical underpinnings in place before the war, radar attracted little attention in the offensive-minded Japanese military. Both the Army and Navy operated electronics laboratories. University scientists, along with researchers at a few industrial labs, were eventually co-opted into mag-netron development. But a bitter rivalry between the services, coupled with the lack of a centralized civilian facility like the Radiation Lab or TRE, handcuffed radar production from the onset. A U.S. survey of the Japanese electronics industry just after the war turned up "a few things of major interest. . . . One was that the contrast between the technical abilities of the Japanese scientists and engineers and their lack of effectiveness in applying these powers to the war efforts was very striking. The other was that the production radars of the Imperial Army and Navy were four to five years behind ours in their design and engineering."

The Army and Navy did not even turn to pulsed systems until May 1941, when an observation group visiting Germany heard reports of British radar. Especially startled, the Imperial Navy quickly launched development of both meter wave and centimeter models. In November 1941, just before Pearl Harbor, a three-meter early warning set took up duty on the Japanese coast at Katsuura, some sixty miles southeast of Tokyo. The following April, the first shipboard radars were installed on the warships *Ise* and *Hyuga.* One set, operating at 1.5 meters in wavelength, could detect aircraft some thirty-three miles away. The second radar, a ten-centimeter Nihon Musen —Japan Radio—system, could spot ships up to fifteen miles off. But with no evidence that the United States or Britain possessed microwave radar, naval officials placed a greater emphasis on developing longwave sets, which posed less of a challenge to Japan's already-strapped engineering force. Moreover, none of the radars were carried on ships central to the Coral Sea and Midway clashes.

The devastating Midway defeat, with four carriers lost to American planes toward the end of a seemingly certain victory, might have been averted with better early warning. The Imperial Navy at once laid plans to upgrade its radar program, run mainly by the Second Naval Technical Institute about thirty-five miles south of Tokyo and supplemented with university and industrial work. Production was principally vested in telephone and switchboard giant Sumitomo Tsushin Kohgyoh; the Nihon Musen Company, an affiliate of Germany's Telefunken; and Tokyo Shibaura Denki, which had prewar ties to General Electric. But the effort fell short from the start. While the institute boasted a few highly capable physicists and engineers, military labs tended to be staffed by underpaid, second-rate technicians. Already facing long odds, employees tackled development of virtually all air, land, and seaborne naval radar and communications gear through an electronics division that employed just 350 technical people, among them only eighty engineers and scientists.

As the Japanese Army geared up its own radar effort, a bad situation grew worse. Nihon Musen's main plant ran two production lines — one for Army radars, the other for Navy sets. Often the two military branches developed similar models, but discussions between employees on different sides of the fence were strictly forbidden. Exacerbating matters, a staggering maze of six ministries, counting the Army and Navy, oversaw the research and production of communications and electronics equipment.

The work done by the various research arms could be exemplary. At Nihon Musen's research center in Tokyo's Mataka suburb, a vacuum tube division led by the engineer Shigeru Nakajima produced advanced one-centimeter magnetrons. But the cumulative effect of the stifling bureaucracies and ultra-secretive military policies proved devastating. After 1942 a variety of Japanese Army and Navy meter wave radars, along with several electronics countermeasures, began to roll into service. By then, though, the United States had largely gone microwave, and the technological gap with the Allies, a matter of months at war's outset, had widened to a few years. That was the situation encountered by Curtis LeMay's bombers in the early months of 1945, as they winged largely unchallenged to unload a maelstrom of fire and death.

In defending against enemy air raids, the Imperial Navy's prime task was to protect its bases and large seaports, while the Army guarded metropolitan areas, with little information shared between them. The Army controlled most of the country's anti-aircraft guns and fighter aircraft, and it bore the brunt of the U.S. attacks.

Long-range radars lined the coasts, nearby islands, and occupied China. A loosely knit mesh of airborne patrols, ground observers, boats, and radio monitoring stations supplemented the radar chain. Area control centers, at Tokyo and Osaka on Honshu, and another on the southernmost island of Kyushu, coordinated the data stream. Centers exchanged information by telephone and radio, with Tokyo serving as the central hub that broadcast orders to ground forces and phoned reports to fighter airfields.

The complex Tokyo Information Center, perched just inside the outer moat of the Imperial Palace, showcased a curious mixture of high technology and backwardness. It looked and functioned much like a British operations room. Filtered radar reports glared from electric plotting boards that were remotely operated by switchboard. A gigantic, lightly gridded map on the north wall, probably superior to anything in the Allied repertoire, presented a comprehensive picture of the unfolding situation that controllers used to scramble fighters. Within this umbrella, the longwave radar net provided adequate warning of B-29 raids. However, the system quickly broke down after that. Ground observer reports were displayed on a different board from the radar data in their own unique presentation form, and without indicating the approaching bombers' bearing. Reports from anti-aircraft radars blared over a loudspeaker and appeared on a third display.

Adding to the confusion, both the Japanese Army and Navy lacked a good ground-controlled interception system to feed pilots situational updates. Not until spring did nightfighters carry airborne radar, and even then it was crude by Allied standards: Japanese copies of the ten-centimeter cavity magnetron, based on details provided by German sources, never made it out of pre-production. Allied jamming of Japanese fire-control radars became so pervasive that anti-aircraft guns were often useless unless a searchlight first picked up the bombers, and American losses rarely surpassed 3 percent of the bomber fleet. In June, when LeMay introduced Porcupine aircraft, modified bombers stocked with jammers, losses dwindled to less than half of one percent.

Eagle made its combat debut in late June. Although a few systems had been shipped to Europe, the sets had arrived too late to see action. Early in 1945, Colonel William "Bid" Dolan of First Sea Search Attack Group fame was lost in a storm over Newfoundland while attempting to deliver the second Eagle model produced to the Eighth Air Force. In the slow drive toward Japan, however, fierce enemy resistance had convinced Allied war planners that only an invasion would end the conflict — prompting LeMay to step up the B-29 assaults. Luie Alvarez's high-resolution radar fit perfectly into the scheme. With a peak power of fifty kilowatts and a scant .4-degree beamwidth, the radar could pick up cities up to 160 miles off. The irregular Japanese shorelines showed up beautifully.

The first Eagle-led raid, against the Utsube River Oil Refinery southwest of Tokyo near Ise Bay, took place on June 26. Results were difficult to gauge, since the refinery had already been hit several times. Two other missions were also judged questionable successes. But on the night of July 6, a few Eagle aircraft almost completely razed the Maruzen oil refinery, earning LeMay's commendation as "the most successful radar bombing of this Command to date." An Air Forces observer rated the set as 98 percent as good as visual bombing, although postwar analysis of fifteen Eagle raids found this far from the case. Eagle *was* the most accurate bombing radar, but it proved hard to use and would never displace optical bombsights.

With Eagle and without, American planes delivered 42,000 tons worth of payload on Japan during the month. The scale of the attacks was increasing so vigorously that the 100,000-a-month mark would have been eclipsed by summer's end. Then, early in August, the Americans tried something horribly different.

Shortly before three the morning of the sixth, the first nuclear device used in combat rose from the island of Tinian, some 1500 miles southeast of Japan, aboard a B-29 piloted by Lieutenant Colonel Paul Tibbets. Among the *Enola Gay*'s eleven other crew members was a small, stringy technician named Jacob Beser, whose job was to man an array of electronic countermeasures and make sure the aircraft eluded the Japanese radar net. Lieutenant Beser had selected three jammers covering the range of known Japanese frequencies.

Little Boy, as its designers called the bomb, stretched ten and a half feet in length. It carried four radar fuses, dubbed Archies, mounted at intervals around the upper part of its twenty-nine-inch-wide girth. The radars' tiny clock wire antennas pointed outward, constantly bouncing signals off the ground. The bomb was designed to explode 1900 feet above the Earth. When any two of the radar units agreed that altitude had been reached, a firing signal set the fusing mechanism in motion. Archies, adapted from an aircraft tail-warning system, operated at a frequency between 410 and 420 megahertz. American intelligence had no hint of enemy transmissions in that range which might accidentally set off the fuses. But anything was possible, including a weird harmonic signal from a radar at 205 megahertz. To guard against such eventualities, Beser also operated a receiver designed to detect transmissions in the critical range. If something appeared, he could switch off the radar fuses, leaving detonation to a barometric sensing device or, if that failed, a mechanical fuse designed to go off on impact.

The first bomb fell on Hiroshima through light cloud cover around 8:15 A.M. local time. A second nuclear weapon, Fat Man, exploded over Nagasaki three days later. Beser was the only man to fly on both strike planes. Luie Alvarez and Larry Johnston, former Rad Lab partners and co-inventors of the Ground Controlled Approach landing system that would save so many lives, rode in an observation aircraft for the Hiroshima attack. Johnston was also aloft for the Nagasaki mission, making him the only person to witness the world's first three nuclear explosions — the two over Japan and the Trinity test in the New Mexican desert a few weeks earlier.

Although enemy flak and fighter interceptors challenged the second run, Beser never detected any untoward signals that might detonate the bombs, and the radar fuses worked perfectly in both cases. The second mission's primary target was Kokura Arsenal on the northwest rim of Kyushu. But when the strike plane arrived, the city was shrouded in haze and smoke. The formation turned toward the secondary target, Nagasaki, a hundred-odd miles southwest, but found it also ensconced in thick clouds. Rather than jettison the bomb over the sea, Navy Commander Frederick L. Ashworth, the weaponeer, authorized a radar approach; this was via the Bell Labs version of H_2X, the AN/APQ-13.

Nagasaki, a busy port on the East China Sea, is split almost in two by the Urakami River. The water boundaries should have provided a distinctive mark on the radar screen. Low on fuel, pilot Charles Sweeney decided to approach over land rather than swinging around to come in from the sea. When a hole opened briefly in the cloud cover at the last minute, bombardier Kermit Behan could see he was several miles upriver from the main target area. At twenty-two kilotons, Fat Man almost doubled its predecessor's destructive powers. Yet the initial mortality figure is estimated at around 70,000, roughly half the Hiroshima number. Owing to the navigational error the bomb fell about two miles off target, and the blast was confined by the steep slopes overlooking the scenic city.

Luie Alvarez never quite believed Behan's story, though Beser later confirmed it. "I've always taken this hole in the clouds with a grain of salt," Alvarez writes, "since Behan, one of the best bombardiers in the Air Force, missed his target by two miles, a reasonable radar error in those days." He left it at that, intimating that Behan concocted the story to "fulfill" orders not to bomb blind. Larry Johnston saw no reason to dispute his mentor's conclusion: from the scientific observation craft, he could not see much, but figured Alvarez, who played poker with the pilots, knew the score.

In any case, pinpoint bombing was not a requirement. After the Nagasaki blast, regular B-29 attacks were dramatically reduced in scale while the Americans waited for the Japanese response to an offer of total surrender. On August 13, with nothing forthcoming, the incendiary area bombing resumed. A day later, Japan agreed to Allied terms.

11 • Emergence

"There was nothing, after our Radiation Lab experience, that we weren't willing, within reason, to tackle."

LEE DAVENPORT

"They can do all because they think they can."

VIRGIL, *Aeneid*

THE ATOMIC BOMB only ended the war. Radar won it. That judgment already echoed through the Radiation Laboratory ranks as members filed across the Massachusetts Institute of Technology campus toward the sprawling lawn of the Great Court. The date was August 14, 1945, V-J Day, a typical drippy, white-shirt hothouse of a day with temperatures creeping toward ninety degrees. Virtually the entire Rad Lab staff gathered on the sun-baked space, backs to the Charles River and Boston, to await the words of their director.

The Japanese surrender, in effect, marked the end of the unprecedented U.S. microwave radar effort. In contrast to the Telecommunications Research Establishment, which would remain the chief British radar house, the Rad Lab would soon close its doors permanently. Vannevar Bush had months earlier sent word not to take on any new long-range commitments and to begin wrapping up existing projects. Three days prior to the convocation, the lab's steering committee had instituted its standing termination order. Those assembled in the courtyard were told the last day of operations would be December 31.

Lee DuBridge, dwarfed by the huge Ionic columns guarding the court, stepped up to a lectern poised atop the short, wide stairway that led from the lawn to one of the school's entrances. The stately dome of the MIT rotunda rose behind him. His prepared remarks ran just five paragraphs, but in them was embodied, not so much in words, but in spirit, the sentiment coursing through the crowd: radar, not nuclear weapons, had put the Allies in position to defeat the Axis forces.

"On this occasion, the first and probably the last all–Radiation Laboratory convocation, I cannot resist adding a few personal words of appreciation to each and every member of the Laboratory who has served so loyally in putting this great undertaking across," DuBridge began. "This Laboratory has not been run by any individual or group of individuals; it has been a cooperative enterprise, in which every staff member and employee has been an essential part. The spirit of cooperation you have all shown, and the enterprise and initiative which has been characteristic of the entire group have made the Director's job a relatively easy one. I think that none of us can help experiencing, even during our exuberance at the end of the war, a slight feeling of regret that this marvelous group of people is now to be disbanded. Few, if any of us, will ever again have the experience of working in such an exciting, fast-moving undertaking. We will all remember Radiation Laboratory days as a momentous experience in our lives."

The director singled out his right-hand man, Wheeler Loomis, for a stalwart job running day-to-day operations. Then, with characteristic humility, he thanked everyone again — from division leaders on down — for making his own role so easy. "It has been a great privilege for me to be the nominal head of such a group, but you know, and I will always know, the success of the Laboratory depended upon you and not upon me." Finally, DuBridge concluded by noting the grave responsibility lying before them all in the atomic era. "This war, and especially the developments of the past week, have brought home to the entire nation that one of the nation's greatest assets is its great body of scientists and the scientific facilities which they have created. In peace and in war science and technology will play an ever-increasing role in national affairs. . . . We must see that in coming generations science serves to better the condition of mankind and not lead to its destruction. To this end we should here and now give our pledge."

As the assembly broke up, people milled about chatting. Even though the work of wrapping up operations would still take months, emotions ran high. Mostly, staff members were happy, basking in the glow of victory. Hank Abajian, who had traveled the globe to service the SCR-584 before returning to the States for a much-needed Vermont vacation, was ecstatic because the sudden Japanese capitulation meant he didn't have to go back to the Pacific.

Still, the mood permeating the crowd had a bittersweet quality. Edythe Baker, who would remain DuBridge's personal assistant for another thirty-three years, and a friend far beyond that, remembered herself and many others as sorry to lose their comrades in arms. Often, these feelings would be magnified as reality hit. When the day came to actually say goodbye, Abajian recalled, "I got around about three people — and I couldn't do it anymore, couldn't do it." Ragnar Rollefson, a University of Wisconsin physicist who among other things had worked on countermeasures in conjunction with the Radio Research Lab at Harvard, searched out Wheeler Loomis.

"Is now the time you say what you think to the management?" he asked. As Loomis took up a mock defensive stance, Rollefson hurriedly explained. "Oh no! I don't mean that. I mean to say that this is the best place I ever worked and I think it is largely on account of the management."

Such tributes were common. Decades later, even after a lifetime of rewarding labors, many still claimed the period spent in the tight-knit organization, conducting vital war work, as the best they had known. Nostalgia, perhaps, accounted for some of the pronouncements. In truth, though, the lab *had* soared to tremendous heights. Beginning in November 1940 with a core of around twenty members, it had grown to a peak of 3897 workers on August 1, 1945. Fully 30 percent were scientists and engineers, and perhaps five hundred of these held Ph.D.s or other advanced degrees. That likely made the Rad Lab bigger, in sheer numbers, than any other wartime endeavor — including the Manhattan Project. The physicists who formed the soul of the enterprise had known next to nothing about microwaves when DuBridge and Ernest Lawrence mysteriously called them to Cambridge: long before V-J Day, they ranked as world experts. During the war years nearly one million radar sets, encompassing more than 150 models, were churned out by American industry. Microwave units, nonexistent going into 1941, accounted for just over half the $3 billion in radar equipment produced and nearly two-thirds the country's total output when the hostilities finally ceased. The Radiation Laboratory had a hand, in one form or another, in developing more than 90 percent of those systems.

One of the top priorities in the last days of war, and in the months afterwards, was to secure for the Rad Lab the recognition it deserved. For years, employees had labored in secret, keeping even their spouses in the dark about the nature of the work. It was high time that their efforts — not to mention the roles played by MIT and the National Defense Research Committee — were brought to light. Some things remained classified: specific wavelengths, power output, and various circuit techniques. But the generalities about microwave radar, as well as the size and scope of the Rad Lab and the names of its members, could all be released for public consumption and appreciation. The lab's publications group, under former New York City adman Charles Newton, began painstakingly piecing together the story for the press, which had long given up trying to breach the security restrictions. Staff members were encouraged to produce capsulized accounts of their work to aid reporters.

In the final weeks of the Japanese conflict, the PR machine churned into high gear and began to attract the media's attention. *Time* readied a cover story about what seemed to be the determining technological factor in the impending Allied victory: radar. But as the deadline approached for the issue dated August 20, 1945, the two atomic bombs fell in quick succession. The radar story was quickly bumped from the cover and replaced with one celebrating the sudden end of the conflict and the saga of J. Robert Oppenheimer and Los Alamos. In the three pages on radar that survived deep inside the issue, only a handful of names were mentioned: Robert Page, Albert Taylor, and Leo Young received credit as radar's American pioneers. Robert Watson-Watt was mentioned as their British counterpart. No one from the Rad Lab got a speck of ink. The facility itself was reduced to "a great anonymous army of scientists. . . ."

With its explosive fury, the Bomb had spoiled the Rad Lab's public debut. DuBridge once reflected on the turn of events with disarming matter-of-factness. "We expected that the announcement of the radar and the work of the Radiation Lab would be big, a big story. But it was submerged by a bigger one." There were a few bright spots. In November, *Fortune* ran a ten-page spread with nearly a dozen photos. The director and Louis Ridenour teamed up to write a story about the lab for *Technology Review*. Numerous smaller articles appeared, and staff members were occasionally invited to talk about their experiences in public forums. Still, it was nothing like what might have been. In the words of Rabi's biographer, the physicist John S. Rigden: "Oppenheimer became a celebrity. DuBridge remained virtually unknown. Scores of books would be written about the bomb and the people who built it; the radar story would remain in the archives at MIT."

The lack of public attention did nothing to diminish the experience of working at the Rad Lab. Still, it hurt. No one could sort out the relative contributions of code breaking, massive air attacks, Hitler's stupidities, the Russia card, and American shipbuilding and industrial output. But one fact seemed irrefutable: *Time* had been right in deciding on a cover story about radar. No single technology was more versatile, or had a more pervasive reach or a more devastating effect on the enemy. Longwave radar had largely won the Battle of Britain in 1940. Two years later it played a critical role in crushing the Japanese at Midway and in overcoming Imperial Navy superiority in nightfighting around the Solomon Islands. Microwave radar dogged the U-boats and helped send the buzz bomb down in flames. It enabled the Allied strategic bombing campaign to continue round the clock, in any weather. Radar countermeasures and navigation systems contributed mightily to the success of the Normandy invasion. Centimeter fire-control systems protected ship after ship from air and sea attack. Radar-guided submarines and low-level bombers decimated the Japanese merchant shipping fleet.

The enemy could boast of many unsurpassed technological feats at war's end — jet fighters, rockets, and the *Schnorchel* device that let U-boats remain submerged far longer than in the past. But neither the Japanese nor

the Germans had matched the Allied developments in microwave radar —
the cavity magnetron, the Radiation Laboratory, TRE, or their many prog-
eny. "Surely radar was the most important thing for winning the war. . . . It
is a pity that at the end of the war, it was overshadowed by the bomb," noted
the physicist Hans Bethe, who spent two years at the Rad Lab before moving
on to Los Alamos.

So even the bombmakers helped the Rad Lab veterans keep up their
chant: "The atomic bomb only ended the war. Radar won it."

Lee DuBridge did not let the Rad Lab linger in disappointment over
the lack of public kudos. In the wake of V-J Day, he threw himself into the
crucial task of extracting the maximum value out of the wartime effort, even
as the facility was broken apart. The task was not unlike that confronting a
1980s corporate takeover artist who had successfully brought down his target.
The director didn't want money. But his job was to ensure the best possible
future for the component pieces that had made the facility so special: the
people, projects, equipment, and vast storehouse of knowledge accumulated
in five full years of research and development.

On that last front, matters were well in hand. The monumental chore
of preserving the fruits of the Rad Lab's efforts had actually begun almost a
year before the war ended, in fall 1944, after I. I. Rabi had seen that the
tide had turned in Europe and wondered about the enterprise's "legacy to
posterity." The gist of his concern was that although basic science had largely
marked time during the war, the five years of intense radar activity had given
birth to something like twenty years of normal progression in crystal theory,
antenna development, radio signal propagation work, and general microwave circuitry. ". . . I realized," Rabi later explained, "that we had amassed
so much knowledge that unless we put it down in the form of books, then,
after the war, there would only be one group who would know all this
technology — the Bell Telephone Laboratories."

Rabi's plan was to produce a series of highly technical volumes that
would lay out, subject to Army and Navy clearance, everything from Loran
to magnetrons, crystal rectifiers, mixers, vacuum tube amplifiers, cathode
ray tube displays, servomechanisms, computers, triodes, and electronic time
measurement. Work had started almost immediately under the general edi-
torship of Louis Ridenour, son of the correspondence school doyen and
possessor of a meticulous command of English. At first, it was not difficult
to integrate the writing tasks with normal lab activities. But when the Battle
of the Bulge broke out that December, some outraged staff members berated
Rabi. "People came to me and were saying, 'You son of a bitch. You've got
us working writing books and over there the boys are dying,' " the physicist

recalled. He ducked the heat by taking a brief leave to join Oppenheimer at Los Alamos.

Once the war ended, any lingering resistance vanished. Ridenour's charge, on which he ultimately spent $495,024.07, was to produce what became known as the Radiation Laboratory Series—twenty-seven books plus an index volume published by McGraw-Hill. An army of stenographers and proofreaders materialized. Some 250 Rad Lab staffers agreed to stay on and work as editors and authors. By the time the Rad Lab officially closed on December 31, 1945, much of the work was complete. The series, beginning with the volume *Radar System Engineering*, would debut in 1947 and go on to serve as the occupational bible for at least a generation of physicists and engineers studying microwave electronics. Rabi termed the effort "the biggest thing since the Septuagint."

The series alone was enough to cement the lab's legacy. But everything else about the facility—scientists, equipment, holdover research, even its very style—emerged from the war to help transform the shape of U.S. science and technology. At the highest reaches, the critical successes of the Rad Lab and Los Alamos, in particular, had made a deep impression. The military, for one, dreaded the impending departure of civilians from defense research. Navy Coordinator of Research and Development Admiral Julius A. Furer called the scientists "ambassadors of national preparedness." The Army's Dwight Eisenhower had this to say in a widely circulated memo to the War Department and commanding generals: "The lessons of the last war are clear. The armed forces could not have won the war alone. Scientists and business men contributed techniques and weapons which enabled us to outwit and overwhelm the enemy. . . . This pattern of integration must be translated into a peacetime counterpart. . . ."

Both services moved quickly to keep the scientists at least on tap. Led by the freshly minted Office of Naval Research, and free to spend millions of dollars in just-canceled procurement contracts, the armed services became the chief patrons of academic research, supplanting industry and the private foundations that had led the way before the war. The pace of the peacetime transformation was so rapid that by August 1946, when the federal bill creating ONR became law, the organization had already doled out 177 contracts worth $24 million to eighty-one universities and nonacademic laboratories.

A step down from this level, the wartime labs deeply influenced the style of research—interdisciplinary, cooperative, hard-driven—that would be conducted at academic, industrial, and government facilities. By January 1946, I. I. Rabi, assisted by Norman Ramsey and others, was already planning a nuclear research facility to be shared by East Coast universities. This became Brookhaven National Laboratory on Long Island. "We were both basing it," Ramsey asserted, "on our knowledge of Radiation Lab and Los Alamos." The trend held true across the Atlantic. During the war, writes TRE veteran Bernard Lovell, "We had to learn to work in the disciplined world of the armed services and the industrialists . . . and in teams of scientists and technicians. When the war ended, we may well have tried to shelve these associations and attitudes, but they had become ingrained in our outlook and were to be fundamental to the emergence of the projects of big science."

Far more directly, the industry of radar hit the mainstream. Companies like Raytheon, AT&T, Westinghouse, RCA, General Electric, and Sperry kept on producing military sets for planes, ships, early warning, and even guided missiles. But the technology quickly spread to the civilian sector. Raytheon, GE, and Westinghouse scurried to build radars for ships and fishing boats. Raytheon was the wartime leader in surface search technology, and its Pathfinder® marine radar was made to standards as rigid as those required by the Navy. Later, the Massachusetts-based company also led the way in producing air traffic control systems. Luie Alvarez's Ground Controlled Approach system was manufactured by Gilfillan Company in Los Angeles. Until replaced several years later by the far simpler Instrument Landing System, it served as the military and civilian standard. During the 1949 Berlin airlift it proved critical in landing planes amidst heavy fogs. The Loran navigational network — one of the Rad Lab's three original projects — found even wider appeal. At war's end the grid covered a third of the globe, including the most trafficked Atlantic waterways and nearly the entire Pacific. The system won praise from pilots and sea captains, and was taken up at once by commercial marine fleets.

Radar was also the driving force behind a planned microwave communications and video revolution: just pushing the radio spectrum to shorter wavelengths opened more than two hundred times as many channels for radio communication as existed before the conflict — then came huge improvements in transmitters, receivers, and everything in between. Months before the war ended, wartime radar manufacturer Philco had filed plans with the Federal Communications Commission to establish a microwave television network linking Washington, Baltimore, and Philadelphia — and the technology had been demonstrated before a packed audience at the capital's Statler Hotel on April 18, 1945. From the building's roof, television signals were beamed across the Potomac to an Arlington hilltop, then sent northeast through a series of relay stations roughly 150 miles to Philadelphia. It was the first TV broadcast that didn't depend on wires or cables. Meanwhile, RCA announced plans for a TV network of its own, GE and IBM teamed up to plan a multifaceted microwave system that included high-definition color television and facsimile channels, and Raytheon sought approval to build a nationwide microwave communications web to compete with AT&T.

A variety of less anticipated markets cropped up to widen radar's impact on society. The technology could be used to survey inaccessible areas, track migrating flocks of birds, or spot highway speeders. One of the biggest areas of interest was in meteorology. In 1942 the Rad Lab's Propagation Group had noticed that radar could spot storms: their fuzzy appearance contrasted sharply with the stark outlines from stationary objects like mountains. It turned out water droplets in clouds scattered more radar beam energy at shorter wavelengths, rendering storms most visible in the microwave region. Late in the war, especially in the Pacific, the military used the AN/APQ-13, the Western Electric production version of the Rad Lab's H₂X bombing radar, to help route aircraft around bad weather. Until a specialized weather-watching radar could be developed, the practice was continued in the civilian sector during peacetime. Sets were mounted near control towers and turned upside down so they could look skyward. At the same time, the U.S. Weather Bureau funded the University of Chicago's Thunderstorm Project, which studied storm patterns from radar-fitted P-61 Black Widows.

Beyond any specific applications, the war-honed techniques and left-over equipment of microwave radar provided a multitude of pathways into unchartered scientific and technological waters. Fire-control radars like the SCR-584 showed the way to precise measurement of time intervals as small as one thirty millionth of a second. Meanwhile, radar's subminiature baseless tubes saved volume, power drain, and weight without sacrificing performance, while crystal diode rectifiers and cathode ray oscilloscopes of greater flexibility and frequency range made it possible to conduct, monitor, and reproduce sensitive measurements at microwave frequencies. Through a Navy-organized giveaway, MIT dispensed equipment to a score of fellow academic institutions. Other university arms that hosted wartime radar work — Columbia, Harvard, Oxford — held onto supplies of their own, ensuring students and staff the state of the art. The transformation even reached the man on the street: Manhattan's peerless secondhand electronics stores teemed with magnetrons, klystrons, and other once-secret treasures.

Lording over the whole bountiful crop, making certain it was properly nurtured, came the scientists and engineers who had lived, eaten, and breathed radar throughout the war. Here, again, the Rad Lab led the way. Toward the end of the struggle, Lee DuBridge was so worried about his charges' futures that he opened a placement program and set up seminars to help bring staff up to speed on the hottest scientific topics. Robert Kyhl, a former University of Chicago graduate student who had been recruited to the lab after I. I. Rabi stumbled across him building a conventional magnetron, later reflected on the lengths to which the director had gone. Just before the war ended, the noted physicist Wolfgang Pauli was brought in to lecture on modern quantum mechanics. "A couple of weeks later, they got Julian Schwinger to come and tell us what Pauli had said. A few weeks after that, they had Ed Purcell come and tell us what Schwinger had said."

In truth, not many of the scientists and engineers needed a lot of help finding jobs. At the least, for those who had established themselves before the war, their old positions were usually reserved. Ernie Pollard returned to Yale to find this carried to the extreme: his lab coat was hanging just where he had left it five years earlier, DuBridge's telegram calling him to the lab still in one pocket.

For the throngs of other staff members who had not yet chosen a career path when national need intervened, the future was still rosy. All had benefited immensely from the cool wisdom of DuBridge and Loomis, the weighty responsibilities laid on their young shoulders, the hobnobbing with generals, and the opportunity to drink beer and brainstorm with giants like Rabi, Bethe, Alvarez, and Schwinger. Often, what emerged from the radar houses — not just the Rad Lab, but Bell Labs, TRE, Oxford's Clarendon Lab, and others — was a hybrid scientist capable not only of probing nature's mysteries, but of building the equipment for the job. Radar vets even *thought* about things differently. Rather than viewing circuits in terms of resistance, inductance, and capacitance, the physicists, mainly, approached their tasks with concepts derived from electrodynamic laws. Moreover, they fairly crackled with confidence. Curry Street and Ken Bainbridge asked Lee Davenport, a former X-ray spectroscopist, to build a cyclotron for Harvard after the war. "I didn't know enough about a cyclotron to build anything of the sort," Davenport related. But he built it.

Industry and academe recognized the potential. Long after the war Denis Robinson, the Englishman who had come to America in 1941 to convince Lee DuBridge to design sub-hunting radars, found that just mentioning his Radiation Laboratory past was like an "open sesame" to leading physicists in the United States and Britain. Such a view paid off over and over again. The Rad Lab would eventually count ten Nobel laureates among its alumni and consultants, nearly as many university presidents, three presidential science advisors, and a plethora of other key figures. TRE hosted a similar body of future notables, including at least three Nobelists; and many others, though less acclaimed, made significant contributions. For years, generally, they had put aside their true love — research and experimentation. Like racehorses, they burst out of the gates when the war ended, free to do anything—anything they could dream up. "There was an enormous pent-up set of ideas," remembered Jerome B. Wiesner, the former Library of Congress sound engineer who was on his way to becoming scientific advisor to John F. Kennedy and president of MIT.

Nowhere was the dawning era better embodied than at MIT, which got a big chunk of everything the Rad Lab had to offer. Immediately after DuBridge closed up shop, Vannevar Bush's NDRC, itself soon slated for extinction, established a Basic Research Division that continued the lab's work in such fundamental areas as investigating the electromagnetic properties of matter at microwave frequencies. Six months later, on July 1, 1946, the division became part of MIT, forming the core of an innovative interdisciplinary program that incorporated aspects of both the electrical engineering and physics departments. Modeled after the Rad Lab and housed inside the dusty plywood walls of Building 20, the enterprise was called the Research Laboratory of Electronics.

Julius Stratton, an MIT professor who had worked during the war at the Rad Lab on Loran and in the Pentagon under Ed Bowles, turned down a Bell Labs management offer to direct the fledgling facility. A separate freshly hatched body, the Laboratory for Nuclear Science and Engineering, was soon brought into the fold, widening his domain. RLE got going with a million dollars' worth of surplus Rad Lab equipment, $50,000 from MIT, and a $600,000 annual budget from the Joint Services Electronics Program, part of the military's postwar strategy of fostering campus research. All told, twenty-six former Rad Lab staffers signed on, many benefiting from a special rule that allowed people with war experience to become graduate students while serving as research associates. Among the charter members: Jerrold Zacharias, Ivan Getting, and Albert G. Hill, who had taken over Division 5, Transmitter Components, when Zacharias left for Los Alamos. George Valley, leader of the H_2X effort, joined the nuclear lab to take up cosmic ray research.

RLE's establishment catapulted MIT overnight to the forefront of electronics and microwave physics, soon giving rise to a host of important innovations in radar, secure communications, electronic aids to computation, and atomic clocks. By the early 1960s alone, it had awarded some three hundred Ph.D.s and six hundred master's degrees, and helped secure for decades the institute's place as the nation's leading nonindustrial defense contractor. The lab, though, was just one venue where the radar veterans were helping transform science, technology, and society.

A handful of researchers, including the Rad Lab's Malcom Strandberg and Bell Labs radar bombing expert Charles Townes, were quick to recognize the potential radar offered for amazingly precise microwave spectroscopy. An extension of Robert Dicke's work analyzing the water-vapor absorption curve, such experiments involved bombarding a gas with microwaves instead of visible or infrared light; the resulting spectral lines revealed essential properties of molecules. Among those joining Strandberg and Townes in opening the new avenue of pursuit were Westinghouse researcher William E. Good and Brebis Bleaney, who had spent World War II developing reflex klystrons and microwave receivers at Oxford's Clarendon Laboratory.

At Columbia University, Willis Lamb, co-director of the Columbia Radiation Lab during the war, armed himself with leftover microwave radar equipment and set out to attack basic questions about the meanderings of the electron in the fundamental hydrogen atom. The result: the so-called Lamb shift that resolved subtle discrepancies between theory and experiment and garnered him the 1955 Nobel Prize.

TRE's Lovell, after viewing a classified British war report that noted radar from two V-2 tracking sites had captured the stream of radio emissions set loose by meteors, began thinking about using the technology for astronomical observations. Denis Robinson and John Trump, onetime head of the Rad Lab's British Branch, co-opted space in Building 24 after the war, but before the lab closed, and began churning out van de Graaff electrostatic generators for commercial sale; among other things they borrowed from radar, magnetrons provided the high energies to accelerate particles to new levels. The partnership formed the basis for the High Voltage Engineering Corporation, originally operating out of a rented storefront on Cambridge's Mount Auburn Street. Robinson became president, Trump the board chairman, with the pioneering physicist Robert J. van de Graaff the third co-founder.

Late in 1945, Raytheon's irascible genius Percy Spencer, who never finished grammar school, reputedly stood in front of a magnetron under testing and felt his hand growing warm, then noticed that a candy bar in his pocket had begun to melt. Fascinated, Spencer sent out for a bag of popcorn and watched the kernels pop madly when placed near the device. Company president Laurence K. Marshall proclaimed a revolution in furnishing hot food to the rubber chicken circuit and launched an initiative to produce a magnetron-based oven for industrial kitchens. A contest was held to name the insurgent appliance. The winning entry: Radar Range, with the two words later merged as Radarange®. Raytheon later bungled the marketing, but the technology became a household word as the microwave oven: fifty years later, magnetrons still lie at their core.

Everywhere, it seemed, somebody had a hot project that could make money or boost science, or both. So frenetic was the activity that the impact of radar was immediate, immense, and practically immeasurable. And this merely scratched the surface. Besides the more or less obvious pathways to explore, who knew where things might lead? Scientific curiosity had been largely penned up, even suppressed, in order to beat back a horrifying tyranny. Now the gates were open—and what the radar men had learned about electronics engineering could be adapted to their chosen professions in order to help reveal surprises in nature, create powerful new materials and technologies, and tap into the limitless cosmos to unravel fundamental mysteries of creation.

Among those gleefully riding the wave into the unknown was Ed Purcell. The rangy six-footer, whose hallmark was an unbridled enthusiasm for exploring any scientific problem, was on the verge of getting hold of a big one. As early as V-J Day, that bittersweet time when Lee DuBridge stood before the convocation to announce the Rad Lab's closing, and the world was at once peaceful and horrified by images of Hiroshima and Nagasaki, the nucleus of a grand idea swirled around in his mind, under the stark gaze of the MIT rotunda and the glare of a hot August sun.

12 • Snow on the Doorstep

"His soul swooned slowly as he heard the snow falling faintly through the universe and faintly falling, like the descent of their last end, upon all the living and the dead."

JAMES JOYCE,
"The Dead," in *Dubliners*

"If winter comes, can spring be far behind?"

PERCY BYSSHE SHELLEY,
"Ode to the West Wind"

FOR NEARLY FIVE YEARS, ever since joining the Radiation Laboratory in January 1941, Ed Purcell invariably had walked the same route to work each day. From his house on Wright Street, the young physicist traversed Massachusetts Avenue, cut across the Harvard campus, and headed east the two miles to MIT. The path took him within ten yards of his prewar office in Harvard's physics building, but it rarely occurred to the single-minded scientist to stop in and say hello. All that changed after August 14, 1945.

With the war ended, Purcell could hardly suppress thoughts of returning to academia. The problem was, he had agreed to stay on and help grind out the Rad Lab Series. While never doubting the task's importance, he nevertheless found it hard to concentrate on writing when the long-suppressed world of research beckoned. It would have been bad enough just writing. But the stultifying bureaucracy associated with the series heightened his angst. Drafts had to be fed to a cadre of nitpicky English teachers who hammered away at the physicists' syntax and structure. They never let up. As the writing slipped further and further behind schedule, a flood of memos arrived about editorial principles.

One day Purcell received a two-page, single-spaced bulletin on the proper use of the semicolon. On another occasion, the first draft of a chapter he had written on general radar problems came back marked up by the copy editors. Purcell had penned some advice for operators trying to determine whether a cluttered receiver display was just random noise or contained a hidden signal. In cases where the power of the return exceeded a certain level, he wrote, "one is justified in *betting* that the peak was due to a combination of signal and noise, and not to noise alone." In the edited copy, the word "betting" was crossed out, apparently because it was too undignified or jargonish. He put it back in, with the italics. The next time around, the word was stricken again. That was the last straw. Purcell stormed down the Building 22 hall to the editor in chief's office and shouted at Louis Ridenour: "Louis, 'betting' stays in or I quit."

Purcell won the skirmish, but the victory hardly seemed satisfying. Like so many other Rad Lab alumni, he represented a strange mix of war-experienced veteran combined with academic novice itching to start his career. He had grown up in the central Illinois farming and coal town of Taylorville, the first son of a Vassar-educated woman trained in the classics and a businessman who managed the local telephone exchange. As a teenager in nearby Mattoon, doing odd jobs for Illinois Southeastern Telephone Company, where his father had taken over as general manager, Purcell nurtured his scientific bent reading the *Bell System Technical Journal.* His career as an experimentalist was anchored in that reading and in a high school physics problem presented by the only physics teacher, Miss Edwards, who didn't know much about the subject but understood her limitations.

Purcell and another boy, Dunlap McNair, were grappling with the classic textbook exercise of the man raising himself up a flagpole by sitting on a seat connected to a pulley that itself ran over the top of the pole and back down. The central question is, how much weight does the man have to pull?

The two friends thought they had it figured out: half his own body weight. But Miss Edwards insisted that the text said a fixed pulley held no mechanical advantage, so the man must pull his whole weight. This news didn't sit well with her students. As Purcell remembered, "Well, we couldn't accept this, and as it happened, we had a barn and Dunlap had a scale for weighing ice which would weigh up to 100 or 150 pounds. . . . So we went into the barn after school and rigged this thing up with a seat and hooked the spring scales to the upgoing rope and then pulled on the downcoming rope."

Purcell weighed 120 pounds. He sat down and started pulling, and when the scales read sixty pounds, up he went. "And then Dunlap got in and tried it and we ran all the way to high school to announce our triumph, and when we got there Miss Edwards was still up in her room grading papers, where you were likely to find a conscientious teacher in those days. And so we told her all about it." Her response was remembered for a lifetime.

"Well, I must have been wrong and you must be right because you did an experiment and proved it."

Purcell considered that a gallant reaction. "I've always felt that she did me a real service at that moment."

Since that experience science had taken the young man out into the world, first to get a bachelor's degree in electrical engineering at Purdue University during the Depression, then to study physics for a year in Hitler's Germany, at Karlsruhe, near the southwestern border with France. Purcell returned to the United States in 1934 to work on a doctorate in physics at Harvard and help build the magnet for the school's first cyclotron. After receiving his degree, he had stayed on as an instructor and was still teaching at the equivalent of the modern-day assistant professor when the chance late-night meeting with Ken Bainbridge and I. I. Rabi had precipitated his migration down the Charles River to MIT.

At the Rad Lab, Purcell solidified his reputation in the physics community. Though not even thirty when the war started, he rose quickly to head Group 41, Fundamental Developments, directly under I. I. Rabi. His hazel eyes, framed by wire-rimmed glasses, glimmered with intelligence — and his voice carried weight with all the lab's senior statesmen: Rabi, Zacharias, Wheeler Loomis, and DuBridge. Yet he had only five research papers to his name.

For the Radiation Lab Series, Purcell had signed on to co-author, with Robert Dicke and Carol Montgomery, a volume on the principles of microwave circuitry. He was also writing sections of two other books — the radar systems piece that had prompted the explosion over "betting," and a discussion of the water-vapor absorption problem that had disrupted one-centimeter radar signals in spring 1944.

Of all these things, and the multitude of other possible avenues of exploration open to him, it was the water vapor mystery that fired Purcell's imagination and had him yearning to get back to physics.

When the absorption mystery surfaced, still in the days before D-day and with the war's outcome uncertain, the effect on the Rad Lab had been tremendous. DuBridge had been forced to place his top-priority program, development of a one-centimeter bombing radar, on indefinite hold while the extent of the effect was investigated. As head of Fundamental Developments, where much of the underlying one-centimeter work was conducted, Ed Purcell had taken it upon his shoulders to direct the inquiry.

As a first step, he organized a series of tests that summer, including some in the superwet tropics of Brazil and Puerto Rico, that removed whatever doubts remained that water vapor was absorbing the radar signal. But the tests couldn't answer the critical questions of where resonance peaked, and whether changing the wavelength would make any difference. For that, someone needed to measure several points around 1.25 centimeters so that a detailed water-vapor absorption curve could be plotted.

Almost by default, the job had fallen to Robert Dicke. Short, thin, and shy, with a thick wave of hair riding an impish face, Dicke was just twenty-eight years old. He had been born in St. Louis, but grew up in Rochester, New York. He had breezed through Princeton and moved on to study for a doctorate in physics at the University of Rochester under Lee DuBridge, completing his dissertation by December 1941. He was preparing to stay on at the school as an instructor when DuBridge telephoned from Cambridge to say that it might be better if the young man came to MIT.

Even in the midst of stars like Rabi, Ramsey, and Purcell, Bob Dicke was already something of a Rad Lab legend. His talents ranged from theory to hard-wired engineering, and he dazzled his co-workers with a series of ingenious inventions — everything from an improved amplifier to the Magic T, a highly versatile symmetrical junction.

Owing to his versatility, Dicke split time between Purcell's Fundamental Developments outfit and the receiver group, and he liked to keep both somewhat in the dark about his comings and goings in order to leave time to follow whatever ideas struck his fancy. In spring 1944, as the water vapor problem surfaced, Dicke thought a lot about the nature of microwave noise that interfered with radar signals. A main source of such noise, he knew, was thermal radiation — escaping from buildings, chimneys, the sun, and even the Earth. In fact, thermal sources over much of their spectrum emit a radio intensity or power which is directly proportional to their temperature. In other words, if a body is twice as hot as something else, it will emit twice as much microwave energy in a given frequency band.

As he pondered these well-known facts, probably late that summer, it had suddenly occurred to Dicke how to make an extremely sensitive type of receiver — in effect a thermal radiation thermometer — that could record the amount of microwave noise pouring out from various sources. Such a device carried no direct radar application. But its intrinsic appeal proved irresistible.

Dicke called his invention a radiometer. He soon built a crude bread-board model, consisting of a plywood plank with the components screwed or nailed into it. The dominant feature was a horn-shaped antenna that looked much like a gramophone speaker, and it was just the sort of thing that charmed a physicist. People dropped by his office all the time to see the radiometer in action. Sometime that fall Dicke demonstrated the device before three hundred or so people gathered in the Rad Lab's main conference hall. He set up two antennas across the room from each other, connecting his radiometer to one, then lighting a cigarette and holding it in front of the other. The microwave radiation emitted by the burning cigarette was transmitted from this antenna to the one connected to the radiometer, and the increased noise dutifully registered.

Dicke had no particular use in mind for the radiometer. "It just seemed like a cute trick," he recounted. However, his friend Purcell, still trying to unravel the water vapor mystery, immediately saw that the odd-looking creation might be able to measure the absorption effect. He had suggested as much to Dicke, who was looking for an excuse to push the radiometer beyond the breadboard stage. Between the two, and because of the straightforward relationship between absorption and emission of energy, it hadn't taken long to work out a strategy. First, the radiometer would be pointed straight up, where the atmosphere was thinnest and little absorption would take place. Then other measurements could be made at an angle, and by comparing the intensity of emissions at different angles, a fairly accurate water-vapor absorption curve could be pieced together.

As a first step, Dicke had lugged the breadboard device up to the Building 6 rooftop and demonstrated its feasibility for the job with a short series of sample measurements. With the concept proved, an encouraged Purcell ordered the young inventor to Florida to conduct more comprehensive studies. By this time, late in 1944, winter was approaching around the Rad Lab. Besides having higher humidity, Florida played host to extensive Army meteorological testing that charted water vapor content at various altitudes and provided the fine points needed for accurate results. Dicke enlisted the aid of staff member Robert Beringer and constructed several more radiometers. Each was attuned to a different wavelength in the region of the water vapor resonance — one at the Rad Lab's new 1.25 centimeter wavelength, another at 1.00 centimeter, and the third at 1.50 centimeters.

The necessity for a better engineered set than his breadboard model forced Dicke to redesign some components. Work on the technological arsenal was not completed until at least the following February. At that point Dicke, Beringer, and Robert Kyhl — known to Rad Lab staffers as the three Bobs — rode with Arthur Vane by train to the central Florida town of Leesburg, where tests would be conducted at an abandoned Army Air Corps weather station airfield. The men set up shop in a rickety shack equipped with two woesome cots. Since it was wartime the equipment was classified. Before leaving Cambridge the men were handed a pistol and warned that the radiometers must be guarded round the clock. The foursome took turns pulling guard duty, one man staying alone with the pistol, while the other three slept in a nearby motel.

Away from the harsh Cambridge winter, reveling in sunshine and fresh orange juice, the men had taken just a few weeks of extensive experiments to crack the water vapor mystery. Each of the three radiometers had been deployed simultaneously, measuring temperatures at the zenith and at various angles reaching down toward the horizon. With the data, Dicke produced the first accurate absorption curve, pinpointing the effect's peak at 1.33 centimeters. The finding was soon further refined by more accurate laboratory measurements at Columbia.

The upshot of the tests, in Purcell's view, had been that the attenuation effect, although bothersome, was not a great enough calamity to warrant all the fuss, especially at the short ranges employed by the anti-mortar radar he was developing. He noted long after the war, "Given the things that we might have been able to do there in radar with the shorter wavelength, we still would have gone ahead because the fruitful, or the attractive, applications of K-band radar really were only for short range anyway." The one-centimeter bombing system was an iffier issue, and he wished dearly they had chosen a slightly longer wavelength farther removed from the effect. As it was, the set had been built but had not seen significant service during the war.

In the meantime, though, the water vapor mystery had put Purcell onto an entirely different, far deeper, line of thinking. In one crystallizing moment, the problem had brought back to his awareness the basics of quantum theory. In quantum terms, absorption, or resonance, had taken place because the trespassing wave of electromagnetic energy — the radio signal at just the right frequency to match the energy gap between two distinct quantum states — had caused some water molecules to absorb energy. More precisely, these molecules had usurped part of the signal to move from what was called a lower-energy spin state to a higher-energy state.

Visualizing this alone involved no special insight. Purcell had learned about such quantum effects in graduate school. But the young physicist saw deeper. What he called "the more subtle part" was the idea that a quantum switch had occurred. Just as the radio wave caused molecules in the lower state to move upward, those in the higher level had released energy and moved down — a process called stimulated emission. Moreover, if the water molecule's atoms had been equally distributed between the two states to begin with, the energy emitted would have been the same as that absorbed, and no change would be noticed in the radar signal. The net absorption of energy had only come about because slightly more atoms existed in the lower quantum state than the upper level. "This difference in the populations of the two levels is a prerequisite for the observed absorption because of the relation between absorption and stimulated emission," Purcell explained.

The conclusion rested on the Boltzmann distribution, a well-known scientific principle that held that a slightly greater population of atoms should exist in the lower-energy level. In fact, the relative populations of the states could be calculated accurately from the related equation. Again, any good physicist knew Boltzmann. What made Purcell's insight special was that in a rare burst of serendipity and intuition, he saw a small piece of the universe unfolding and *understood* in his bones how it fit into the cutting edge of physics.

A central feature of the nucleus is its intrinsic angular momentum, or spin, which gives rise to a property called the magnetic moment. As quantum theory came into its own during the century's first decades, it was realized that when a proton is placed in a magnetic field, this magnetic component aligns itself in predictable ways in relation to the field—and that the orientation it chooses determines the quantum states, or energy levels, in which the proton resides. Purcell realized that he might be able to study this effect in the laboratory with protons—which form the nuclei of the fundamental hydrogen atom—in much the same way Mother Nature had tinkered with radar beams using water molecules. He could apply a magnetic field to split the energy of the proton's minute magnetic component in two levels. Then, once the two energy states were set up, introducing a radio signal of the resonant frequency between the two states should cause some protons to absorb energy. The physics were very similar to what happened when the radar beam induced a jump between the energy levels of atmospheric water molecules, causing a disruption in the signal.

Charting this effect—called nuclear magnetic resonance because it examined resonance in the atomic nucleus—was of extreme value to physicists. Less than a decade before the war, in 1932, the British scientist James Chadwick had discovered the neutron—and understanding such strange goings-on inside the atom was central to figuring out how the new particle fit in with already known protons and electrons in determining nuclear structure.

Although it arrived with gestalt-like suddenness, this line of reasoning was not a blind jump for Purcell. Before the war he had co-authored an analysis of magnetic cooling experiments and so was generally familiar with the terrain. He had also benefited immensely from his Rad Lab association with some of the world's greatest experts on magnetic resonance, among them Zacharias, Ramsey, and especially Rabi, architect of the first successful nuclear magnetic resonance experiment. In fact, only a year earlier Rabi's landmark prewar work on the subject had garnered the Nobel Prize.

The problem with Rabi's experiments, as great as they were, was their reliance on atomic and molecular beams, which required complex vacuum chambers and could not easily distinguish more mundane molecular changes—such as the absorption taking place when a spinning water molecule absorbs part of a radar signal—from nuclear events. What physics awaited was a means to take the revolution out of vacuum chambers to any laboratory setting, and beyond the domain of beams to any element in solid matter — or even a glass of water. Many scientists recognized this need. The Dutch physicist C. J. Gorter had set up very well designed experiments in 1936, and again in 1942. But so far nothing had materialized.

Purcell had caught a glimmer of a method that could be extended easily to any atom in bulk matter. The big question in his mind was whether detectors could be made with the sensitivity to gauge the energy absorption. In a typical laboratory setting, the Boltzmann distribution held that for every million hydrogen atoms, just seven more protons would exist in the lower magnetic state than in the upper level. The difference was so small that an ordinary radio receiver would never pick up the effect. Yet Purcell knew that if he was correct, and the effect could be duplicated and measured with protons in the laboratory, he would be opening an important scientific frontier. Knowing the exact amount of energy absorption — or magnetic resonance — taking place deep in the atomic lattice that held the hydrogen would mark a crucial step toward deducing masses, charges, angular momentum, and a host of other vital data for the blossoming field of nuclear physics.

For science, then, this almost absurdly small effect could serve as an enormously powerful probe of the atomic structure of matter.

In the days following the Japanese surrender, still wrestling with such seminal questions, Purcell tried to fight back the boredom of writing for the Rad Lab Series. It was no use. His mind kept returning to the resonance problem. One hot September day, when two colleagues stopped by to ask him to lunch, he gave in to the temptation to talk over his ideas.

Robert V. Pound and Henry Torrey shared an office down the hall and around a corner from Purcell. Although both labored on the Rad Lab Series, during the war they had served in Group 53, Radio Frequency. Pound, a thin man with a dapper mustache, was a world authority in microwave techniques and toward the end of the war had guided the lab's program in mixers, the crucial initial stage in receivers. He had been born in Ontario, Canada, but moved to the United States as a toddler and was naturalized as an American citizen while still in his youth. He studied physics at the University of Buffalo and was set to attend graduate school when the war called. He joined the Rad Lab in March 1942, after first working on an early radar program at Submarine Signal Company in Boston, an aboveground extension of the company's specialty in developing underwater sound detectors. For his part, Torrey had already distinguished himself as a theorist. He had grown up in Yonkers, earned a chemistry degree at the University of

Vermont, then switched to physics and received his doctorate at Columbia under Rabi. He had been teaching at Penn State before being recruited to the Rad Lab to investigate crystal detectors.

All three were friends. Pound and Purcell, especially, shared a mutual love of talking technology. On this day the trio headed for Pound's favorite lunch spot, a sandwich and confectionary shop in nearby Central Square called Hennessy's. The place offered a daily forty-cent lunch special — a designated sandwich on its hallmark molasses-baked dark bread, topped off with Boston cream pie and ice cream. The sandwich changed every day, but Pound knew the waitress so well he could always substitute his favorite ham and cheese. "What's your special today?" Pound would ask. She would answer, and the physicist would pretend to mull it over. "I'll take the ham and cheese."

The men probably walked through a parking lot bordering Vassar Street and then straight up Massachusetts Avenue, which runs in front of MIT into the heart of Central Square. During lunch Purcell indulged himself. Torrey had completed his doctorate under Rabi and was familiar with the famous resonance experiments. Why, Purcell queried the Columbia grad, couldn't nuclear magnetic resonance be observed in solids? After all, he reminded Torrey, the physics were similar to the water vapor problem. Both involved two quantum levels. You could apply a uniform magnetic field to produce the energy gap between two states, then zap the atomic lattice with a radio frequency beam of the same magnitude as the gap, bringing about the switch in the same basic way the radar signal had turned water molecules topsy-turvy.

"My first reaction to it was no," recalled Torrey. The key was the Boltzmann distribution. He just didn't see how the net energy absorbed by such a small population of protons — seven for every million hydrogen atoms — could be detected in this manner. He later mentioned the idea to Rabi, who was astonished to think anyone would even try. But Torrey's had been a gut reaction. Meanwhile, the idea took root in his head. "That night I went home and started making some calculations," he related. The next morning he went straight in to see Purcell at the Rad Lab, apologizing for making a hasty judgment.

"I believe it would be possible to detect the effect," he said.

Purcell digested this news. The attitude at the Rad Lab during the war had been to ask three basic questions before diving into a project headfirst. Was it worth doing? Had anyone done it? Would it even work? It took Purcell only a moment to process Torrey's statement in the same terms: Yes-No-Yes. "Okay," he decided. "Let's do it."

Despite the moment's alacrity several weeks passed before the project really got going. Torrey handled the theoretical end. The way it should work was that when randomly oriented protons, in water or some solid medium, were subjected to a large and nonoscillating magnetic field, their magnetic moments would immediately begin to precess, a technical term referring to the fact that the particles would travel around much like a spinning top. This precession has its own frequency, the Larmor frequency, generally much lower than the original spin frequency — just as a top's precession is slower than its initial rate of spin. Here, though, the analogy of the top breaks down. As a spinning top slows and eventually comes to a stop, its precession, or wobble, encompasses many levels, from barely noticeable to barely functional. The laws of quantum mechanics, however, allow the proton fewer options. The act of spinning releases energy to the crystal lattice that makes up the hydrogen source so the protons settle into thermal equilibrium in only one of two quantum states, or Zeeman levels — either with their spin and intrinsic magnetism oriented in the same direction as the magnetic field, or opposite it. Again, it was critical to Purcell that slightly more protons settle into the lower energy level, with their spin parallel to the field. At that point the men could bring about the quantum switch — and the resonance they hoped to record and measure — by introducing what is called a radio-frequency magnetic field whose radiation ran perpendicular to the magnet's own energy field. It was kind of like jump-starting nature.

The men weren't thinking of precession at the time. But the term was known and the result they envisioned the same. A second vital consideration was relaxation time — how long it would take the protons to come to equilibrium once the steady magnetic field had been applied. If the experimenters turned on the radio signal too early, resonance would not occur. Torrey dug up the only theoretical paper that directly touched on the subject — written in 1932 by the Swedish scientist Ivar Waller. The Swede had calculated the relaxation value for electrons, so Torrey recast the problem for protons. It was dicey business, taking into account a whole range of different interactions. But it looked as though equilibrium would come in an hour or so. On the other hand, it might not come for a hundred hours. He didn't really know.

Pound, well versed in signal-to-noise considerations from his war work, designed the radio electronics that would detect the faint, hoped-for signal. He scrounged almost everything the trio needed from the Rad Lab's rich stocks. Purcell, meanwhile, rooted around for a large electromagnet: Torrey had calculated that a field of 7000 gauss would provide the force to split the protons into two energy levels. A search of MIT turned up nothing. But

Harvard did have a large magnet resting in a wooden shed appended to the Research Laboratory of Physics. The magnet even had a historic past; it had been used by J. Curry Street and E. C. Stevenson to make the first determination of the muon's mass.

Purcell contrived to stop by his old haunts and look at the aging device. The yoke had begun life as part of a generator for the Boston Elevated Railway, the predecessor to the modern Massachusetts Bay Transportation Authority that ran the subway and bus system. The magnet itself had lain dormant for six years. Still, the engineering was sound. The small group spent several evenings shaking mothballs out of the thing. Current came courtesy of a motor generator housed next door in the Cruft Lab, with a rheostat remote control mounted on the shed wall.

Street and Stevenson had tracked and photographed their muon in a cloud chamber inserted in the twelve-inch gap between the magnet's poles. Purcell designed pole pieces to cut the gap down to about four inches, the size of the cavity, or container, that would hold his proton source. He recruited the help of the resourceful Carl Johnson, head of Harvard's physics department shop, who turned the fittings out from high-permeability iron stock.

After several modifications to the pole pieces, necessary to improve magnetic field homogeneity and make any signal sharper, the three partners pronounced themselves satisfied. Their calibrations showed that the magnet should produce the 7000-gauss target when the current reached about 73 amperes. The generator could deliver 100 amps at 500 volts, so their goal rested well within its range.

The trio agreed to assemble in the shed the evening of Thursday, December 13. Pound had already ferried the bulk of the electronic equipment over from the Rad Lab in his faithful '39 Buick coupe. The only Harvard contribution besides the magnet was a signal generator borrowed from the Psychoacoustic Lab.

That night Pound dropped his wife, Betty, at the Torrey house on Camden Place so she and Helen Torrey could spend the evening together. Then the two men drove the few short blocks to the lab. A steady snow fell as they cautiously maneuvered into the north yard then shared by the physics department, the law school, and the graduate school of engineering. Pound's mind turned to a recent New Yorker "Talk of the Town" item that retrospectively described how Columbia physicist John R. Dunning, after listening to the startling news early in 1939 that it might be possible to release atomic energy, had gone back to his lab and reproduced the sobering evidence on his own oscilloscope. Just before midnight Dunning had walked home from his lab, bent contemplatively into the cold January wind. Pound thought that as a backdrop for a momentous discovery, snow would do even better than cold and wind.

Purcell still had a key to the physics building. He met Pound and Torrey at Harvard around 7:30 P.M. Earlier, he had stopped by a small grocery on Massachusetts Avenue and bought several boxes of paraffin wax — Gulfwax brand, from the Gulf Oil Company. Such a substance was commonly used to seal jam and jelly jars — and was favored by physicists as a rich source of hydrogen and protons. About two pounds' worth was melted and poured into the magnet cavity.

The famous cosmic ray magnet was turned on around 10 P.M. After allowing what they hoped was time for the sample to reach thermal equilibrium, the team threw on the radio bridge circuit. Pound's radio frequency coil was designed to pick up the unbalanced bridge circuit signal induced by the rotation of the protons' magnetic moments as resonance occurred and energy was absorbed, causing a jump in an output meter. Slowly, they maneuvered the rheostat's knob to adjust the magnetic current up and down around their predicted resonance value of 73 amperes — when the frequency of the radio signal should match the frequency of the proton's precession, or wobble — hoping to see an upward deflection of their detector-gauge needle as energy was absorbed. They tried for five or six hours, almost immune to fatigue. But the needle barely flickered.

Partly to keep their spirits up, the men theorized that it might take much longer than they thought for the sample to reach thermal equilibrium. They resolved to try again Saturday afternoon, first allowing the paraffin to sit in the magnetic field for several hours. It was a faint hope. The time just spent tinkering with the current and staring at the needle amounted to basically the same thing. Still, Purcell insisted, it was worth a try. All three were supposed to work at MIT on Saturday morning, but as ringleader, Purcell proposed to play hooky. He would stop by the shed early in the morning to turn on the magnet, then work on his writing while watching the controls with one eye just in case something happened. Torrey and Pound would go to MIT as usual, then join him that afternoon.

It was close to 4 A.M. when the men called it quits. All three piled into the Buick for the short ride home. At that time the parking lot led onto Massachusetts Avenue, but the rocks that marked the exit were covered in pristine snow. Pound had to creep along, with no other tire tracks to follow. After dropping Purcell off, he inched back to Torrey's house. Their wives were awake, anxious. After only the briefest explanations, the couples said their good nights. Pound went to sleep, sad that the snowfall had not after all been the harbinger of great events.

Early on Saturday, Purcell arrived back at the shed as planned. When Torrey and Pound showed up around two that afternoon, they found him diligently working away, manuscripts spread across a desk. The magnet had been going seven hours. The men turned on the radio circuit and repeated their experiment, staring at the output meter while slowly adjusting the magnetic current up and down in a range of 10 percent from their expected value of 73 amperes — from about 65 to 80 amps. Again, the needle barely flickered.

Almost desperately, the men fiddled with the bridge circuit to try to bring about the switch from the other end. But all three felt it was set correctly: if there was anything to see, the needle would jump. By about four in the afternoon, they were ready to go back to the drawing board. Then, on a lark, someone — Pound thought it was he — suggested turning the generator to its maximum 102 amperes and adusting it slowly downward. As it reached 83 amps, barely outside Torrey's margin of error, the needle spiked.

Was this it? The men cautiously tested the results, their excitement growing as the sequence occurred over and over. To determine whether the hours of extra exposure to the magnetic field made any difference, they shut down the current for a few minutes and then took it back up to 83 amps as quickly as possible. There was the resonance. Next, the three physically removed the paraffin-filled cavity from the magnetic field for ten minutes. Again the resonance showed up.

Triumphantly, Torrey snatched up Pound's spiral notebook and scribbled: "Dec. 15 — 1945 proton resonance found at 83.0 amperes   magnet turned off — back on again still there   cavity removed from field for 10 min. reinserted — effect still present about same magnitude."

•

The letter reporting the pathbreaking success was received by the editor of *Physical Review* on December 24, 1945 — Christmas Eve. By physics convention, the date of reception becomes the date of discovery. The men had spent several days in careful quantitative measurements and made clear in nine terse paragraphs that they realized the experiment's far-reaching value. "The method," they wrote, "seems applicable to the precise measurement of magnetic moments . . . of most moderately abundant nuclei." The three co-authors went on to correctly predict that nuclear magnetic resonance could be used to investigate crystalline structures and even standardize magnets. Beyond that, to the world of chemistry and medicine that made magnetic resonance imaging a household phrase, they couldn't see.

On the heels of the discovery, word soon went around the physics community that a Stanford University team had also measured nuclear magnetic resonance of protons, but through an entirely different type of experiment using simple water at room temperature. Felix Bloch, William Hansen, and Martin Packard reported their findings in a short letter found in the next issue of *Physical Review*. Their note trailed the Purcell group's by thirty-six days.

The Stanford researchers shared a curiously overlapping history with their Cambridge counterparts. Electronics guru Hansen, a key figure in the Varians' invention of the klystron, had lectured regularly at the Rad Lab about receiver development. Indeed, R. V. Pound had learned many tricks of the trade from him. Zurich-born Bloch, the first graduate student of Werner Heisenberg in Germany, had fled the Nazi regime because of anti-Semitism. He had joined Los Alamos, but was unhappy and transferred to Harvard's Radio Research Laboratory in 1942 to work on electronic counter-measures. In the process, he learned a lot about the techniques employed in his own experiment. Like Purcell, Bloch knew Rabi and his pioneering resonance work well, and had lived within shouting distance of the great physicist's home on Avon Hill Street in Cambridge. Despite their proximity and shared interests, however, Purcell and Bloch had met only briefly during the war — at a party to celebrate Rabi's 1944 Nobel Prize.

The Harvard and Stanford groups first interacted in February 1946, when Hansen traveled to the East Coast and met with Purcell. Their experiments were so different that the men talked things over for a half hour before either fully grasped what the other had done. In late April both Purcell's team and Bloch's presented reports at the spring meeting of the American Physical Society, which was switched to Cambridge from Washington in deference to the tight concentration of physicists still left over from the war. In many ways, the meeting was a celebration of a just-born field of research.

Rabi was in the audience, listening first to Bloch and then to Pound and Torrey discuss their nuclear magnetic resonance experiments. Purcell spoke last on the side issue of radio frequency spectroscopy. Rabi, who had only a few months earlier brushed aside the whole basis of Purcell's idea, chewed him out for not sticking to the main subject: "You should have used your time for more of the NMR."

Purcell and Bloch would share the 1952 Nobel Prize. For the former Rad Lab group leader, science's top prize had come with just his sixth research paper. As it turned out, too, the snow R. V. Pound had so cherished on their first attempt did prove relevant after all. In his Nobel laureate lecture, presented on December 11, 1952, almost seven years to the day from the landmark experiment, Purcell described his mind as working through pictures, not mathematics. He recalled that soon after the initial success, he looked at the snow and was rewarded with a unique glimpse into the strangeness and wonder of nature. "There the snow lay around my doorstep," he related, "great heaps of protons quietly precessing in the earth's magnetic field."

13 · The New Astronomers

"At the beginning of the meeting on that Friday afternoon we were strangers to the astronomers, aliens infiltrating a privileged assembly. As we showed our slides, the mood changed. The fellows of the society began to grasp that this was a new astronomical technique, and by the end of the meeting we were part of the astronomical community."

SIR BERNARD LOVELL

CONTACT WITH MOON
ACHIEVED BY RADAR
IN TEST BY THE ARMY

Signal Sent From Laboratory
in Jersey is Reflected Back
2.4 Seconds Later

VAST POSSIBILITIES SEEN

Mapping of Planets, Defense
Against Bombs in Cosmic
Space are Suggested

As ED PURCELL PRIED LOOSE the secrets of the atomic nucleus, other radar men turned their attention to the limitless cosmos. The news of the moon's detection by radar jumped off *The New York Times* front page on January 25, 1946, firing imaginations around the world to a far greater degree than the hard-to-envision magnetic resonance experiment. Here was something people could understand, and a stark black-and-white photograph accompanying the article made the event even more real. A long, squiggly line like that on a modern electrocardiogram display zigzagged across the picture from right to left, hovering over a distance scale that ran out to 300,000 miles to make it easy to gauge how far the radar pulse had traveled to find its quarry. Readers saw the sharp spike of the transmitted signal settle down to a meaningless quiver as it spread across space. Then, just at the 240,000-mile mark, came the unmistakable blip of Earth's satellite.

The image fairly embodied the promise of the golden age of science and technology that arose from World War II. It demonstrated conclusively that radio waves suitable for carrying long-range communications could penetrate the Earth's atmosphere and return — detectable. More than that, for the first time researchers had pushed past the protective envelope of the Earth's ionosphere and interacted with the solar system, not just stargazing, but actively probing. As *Time* later extolled dramatically: "Man had finally reached beyond his own planet."

It was somewhat ironic that for all the attention paid to university scientists rushing back to academe after the war, the moon detection was an Army coup. The landmark radar experiment had been carried out at the Evans Signal Laboratory at Belmar, New Jersey, fifteen days before the *Times* article appeared. Outside experts, among them George Valley, fresh from the Rad Lab's closing, had checked the work upside down and sideways. Then, beginning with the annual meeting of the Institute of Radio Engineers in New York on January 24, the Signal Corps had trotted out a bevy of officers and technicians to bask in the limelight — everyone from Chief Signal Officer Harry C. Ingles to a onetime New York diamond dealer named Jacob Mofsenson, who played a supporting role in the experiment.

The star of the show was an unassuming Tennessean with a longtime fascination for the moon. In 1922, as a sixteen-year-old ham radio operator, John H. DeWitt Jr. had built Nashville's first broadcasting station, a fifteen-watt outpost with the call letters WDAA. For nearly two decades he had worked in radio while cultivating a love of astronomy. In May 1940, DeWitt had attempted to detect radio signals reflected off the lunar surface by a self-made eighty-watt transmitter. The experiment failed because of power losses and lack of receiver sensitivity. Before he could try again, the war had intervened. DeWitt joined the Army, first as a civilian consultant, then as a uniformed Signal Corps major. By the end of the war, he was a lieutenant colonel and running the Evans Lab.

The Army had given DeWitt a second excuse to find the moon. Immediately after V-J Day, worried that the Soviets had captured enough German rocketry expertise to build ballistic missiles capable of reaching the United States, the Pentagon directed the Evans group to study how such invaders could be detected and tracked. At that point, for all radar's effectiveness against German and Japanese aircraft, scientists were not convinced that radio waves could penetrate the ionosphere, bounce off a target, and be detected back on Earth. DeWitt seized the opportunity to settle the matter. With no space-going rockets in existence to test his radar systems, he recalled, "I thought, well, if we could hit the moon with radar, we could probably detect the rockets." Lieutenant Colonel DeWitt dubbed the unofficial study Project Diana, after the virgin Roman goddess of the moon and hunting.

As the Evans director, DeWitt supervised seventy officers and 1100 civilians. At war's end, most had little to do. He selected diminutive, bespectacled E. King Stodola as chief scientist of a five-man team and got down to business. The main piece of equipment was an SCR-271 early warning radar operating on a wavelength of 2.6 meters. The set had been modified by Major Edwin H. Armstrong, an Army consultant during the war famous for pioneering frequency modulation, the FM type of transmission common in modern radios. The enemy capitulation had ended the major's Army work, but not before he had cobbled together jump wires and temporary connections to turn the once ordinary radar set into a powerful transmitter and sensitive receiver. DeWitt's group further modified the receiver by employing a tunable crystal that could pick up an elusive return signal whose frequency would be Doppler-shifted by the relative motions of the Earth and moon. Two conventional antennas were married into a forty-foot-square bedspring affair and mounted on a tower overlooking the ocean. The high-gain antenna could be steered in azimuth only, not elevation, and so could be used only for half-hour periods at moonrise and moonset, when its target rode near the horizon.

On January 10, 1946, everything finally came together. Several minutes after the moon rose at 11:48 in the morning, the first radar pulses were beamed skyward. Rather than the microsecond bursts used in wartime tracking, the set belched out a quarter-second-long signal with a great clatter of the mechanical transmit-receive switch. To allow time for the moon's return echo to arrive, the transmitter waited in silence for 4.75 seconds before sending out another pulse.

DeWitt had driven into the nearby town of Belmar for lunch and cigarettes, and missed the big event. Harold Webb and Herbert Kauffman were hunched inside the large control shack when the first faint lunar echo arrived just before noon. Besides viewing the signal on a cathode ray screen, the men listened over loudspeakers to the response—a half-second-long hum like the buzz of an untuned radio set. Although they couldn't be absolutely certain it was the moon answering their call, Webb and Kauffman grew excited. The historic signal had arrived about 2.4 seconds after a pulse had gone out, and at the speed of light that would be just about the right amount of time for a radio wave to journey 240,000 miles or so into space and back. Besides, it had to be the moon they detected, DeWitt later stated, "because there was nothing else there but the moon."

The *Times* played the story on the front page again on January 26, and followed DeWitt's exploits for several more days. High-frequency radio waves waltzing back and forth through the ionosphere heralded a revolution in long-distance communications. One popular idea in the days before satellites was that the moon might reflect signals from point to point around the world. The War Department talked, too, about radio control of missiles orbiting Earth above the stratosphere and of spaceships sallying forth to unknown adventures. Major General Harold McClelland, the Air Forces' air communications officer, felt emboldened to suggest that a radar code might be developed to enable interactions with other species millions of miles across space. "If intelligent human life exists beyond the earth such signals could be answered," he theorized. "We might even find that other planets had developed techniques superior to our own." Harry Edward Burton, the distinguished sixty-nine-year-old principal astronomer of the Naval Observatory, went so far as to postulate that probing the lunar surface by radar might reveal hidden moisture — a hint that life could be present.

Such comments provoked a warning from Sir Robert Watson-Watt, who had arrived in New York the night of the announcement and was never one to shy away from publicity. While congratulating DeWitt on his achievement, the British radar pioneer admonished that the technology's primary mission was to aid life on Earth by allowing for safer travel on land, sea, and through the air.

The hullabaloo soon died down. For all its remarkability as an engineering feat, the DeWitt experiment held next to nothing of scientific interest. Nor did the stunning work of Zoltán Bay, an enterprising Hungarian wartime radar researcher who detected the moon with a crude fifty-centimeter set within two weeks of DeWitt's announcement: lacking a sensitive receiver, Bay arranged an ingenious assembly of water voltmeters that indicated the moon's reflected signal through a slowly built-up excess of hydrogen.

Still, the pundits were on the right track in predicting that the moon detection foretold a vast bounty about to descend upon astronomy. The visible portion of the electromagnetic spectrum, far and away mankind's main observation post since the days of the ancient Chinese astronomers, spans only the single octave between .4 micron and .8 micron in wavelength. The radio band, by contrast, covers some ten octaves, from ten meters to .1 centimeter. Breaking into this vast, essentially untapped region was just the sort of swords-to-plowshares trick radar could perform, whether it be to actively send out pulses and study the returns, or simply listen passively and chart the radio noise streaming in from space: radio astronomy as opposed to radar astronomy.

Even as DeWitt began his work in the summer of 1945, and while Ed Purcell still puzzled over the water vapor problem, a tiny contingent of scientists seized the potential radar offered for astronomy. American sets like the SCR-271, British early warning systems, and especially German

Würzburgs — were all salvaged and transported to academic centers to provide the foundation for a promising field of study. Somewhat surprisingly, established optical astronomers did not embrace the emerging branch of their science. Like John DeWitt Jr. and Zoltán Bay, who had blazed the trails, the new astronomers were radar men.

The scientist who most cleverly seized the opportunity to use military radar equipment for astronomy was J. S. Hey, a physicist recruited during the war to study jamming and counterjamming for the British Army. Hey, a civilian, trained a network of anti-aircraft artillery radar operators to note the characteristics of enemy jamming and report the details. In the last days of February 1942, a week after German jamming of British coastal radars had aided the escape of the *Scharnhorst* and *Gneisenau* from Brest, Hey's listening posts operated at a heightened state of alert. Suddenly, he began receiving reports of an intense daytime enemy jamming campaign that completely obliterated anti-aircraft radar signals between four and eight meters in wavelength. When no air raids materialized, Hey studied the jamming patterns and noticed that the directions of the most intense disruptions seemed to follow the sun. After telephoning the Royal Greenwich Observatory and learning of a recent large sunspot group in the sun's central meridian, he concluded the solar outburst had accounted for the jamming.

If Hey was correct, it meant that sunspot activity and other astronomical phenomena such as meteors and even cosmic rays could be studied by radio. Many notable experts questioned his conclusions about the sunspot interference, reasoning that, if real, the phenomenon would have shown up before. Still, the physicist kept at it. In the spring of 1945, just before the war ended, he used modified radar from two unneeded V-2 tracking sites, along with a third experimental set, to watch the stream of radio emissions from meteors crashing into the ionosphere.

Hey's classified reports went the rounds of the radar establishments. A few investigators had made pioneering radio wave studies before the war — most notably Karl Jansky of Bell Labs and the electronics engineer and amateur astronomer Grote Reber, who built a parabolic dish in his backyard in Wheaton, Illinois, and used it to develop the first radio maps of the sky by cataloguing emissions at about two meters in wavelength. But the British scientist's work sparked a new race to investigate the heavens through radio technology, one made feasible by the dramatic advances in precision and sensitivity brought about by wartime radar, as well as by the well-stocked pool of young people trained in radio science and engineering still out to make their professional marks.

The gun went off as soon as the war ended. Hey quickly produced his own detailed radio map of the cosmos, following Reber's footsteps on a longer wavelength. At Cambridge University, J. A. Ratcliffe returned from the Telecommunications Research Establishment, where he had worked on airborne radar, to lead the school's radio and ionospheric research. He brought with him TRE colleague Martin Ryle, a brilliant, if highly strung researcher who in a fit had once hurled an inkpot at countermeasures guru Robert Cockburn's head. As a young physicist before the war, Ryle couldn't have asked for anything better than joining Ratcliffe's ionospheric group. By late 1945, however, that shopworn idea seemed so boring that he cast about for something fresh and exciting. Ratcliffe, who had access to Hey's reports, suggested investigating the still-controversial association of intense solar radio noise with sunspot activity—as well as determining whether the sun emitted radio noise in the meter wave band even in the absence of such disturbances. With the sunspot cycle only a few years from its maximum, he advised, the field seemed ripe for major discoveries.

Ryle liked the idea. Money was tight, so he commandeered surplus wartime radio and radar equipment and enlisted the aid of a few other TRE grads, among them Graham Smith and Tony Hewish. Bettering Hey's resolution dictated building an aerial system with the infeasible diameter of five hundred feet. To get around that logjam, Ryle hit upon the idea of a radio interferometer, much like, as he later learned, an optical device erected in the early 1900s on California's Mount Wilson by A. A. Michelson and Francis Pease. By cabling two small aerials to a shared receiver, he could achieve the resolving power of a gigantic antenna with a diameter as great as the distance separating the two small arrays. With his interferometer, Ryle was able to narrow in on the solar region from which radio emissions arose and determine that it corresponded closely to the area of sunspot activity.

Almost simultaneously, a pale and exhausted Bernard Lovell left TRE's centimeter radar program and returned to cosmic ray research at Manchester University. Although the city had been bombed, the school's grime-covered Victorian structures stood unscarred by war. It was a trying time. The university salary ran at half Lovell's TRE wages. Bread, clothing, and gasoline were rationed, and much else in short supply. Still, the excitement of getting back to science soon energized the thirty-two-year-old physicist. Lovell and prewar boss Patrick Blackett, back at Manchester after wrapping up his operational research duties, had once written a paper theorizing that radio echoes might be spotted from the columns of ionized molecules produced in the atmosphere by cosmic ray showers. Such showers had been keenly studied by scientists investigating fundamental physical events since the 1930s, and the men were interested in adding the radio dimension to the body of work. Blackett reminded his co-author of the paper and told Lovell to get to it. The task, which required a high-gain antenna, sensitive receiver, and powerful longwave transmitter, appealed to Lovell. "I began to feel at home again," he writes. "This was exactly the gear I had been dealing with for six years — except the wavelength." He contacted Hey, a Manchester alumnus, and expressed the hope of using some gunlaying radars on 4.2 meters. One day that September, less than two months after the war ended, an Army convoy arrived in the quadrangle outside the school's physics department, dropping off three trailers stocked with the necessary equipment, including a diesel generator.

Lovell coaxed the radar to life, only to find that the screeching electrical noise from trams running near the university grounds obliterated any signals. He rooted around for a better site, finally discovering the Jodrell Bank Experimental Grounds twenty-five miles south of the city, where the university's botanical department operated an experimental horticultural plot. The botanist who ran the site was an amateur radio enthusiast and quickly accepted Lovell's proposal to park the trailers at Jodrell Bank. The eager physicist set up shop in a muddy field, near some ramshackle wooden huts and a manure heap.

As Bernard Lovell at Manchester and Martin Ryle at Cambridge began exploring the promise that radar and radio offered to astronomy, they kicked off a keen competition — mainly scholarly, but occasionally personal — that would last for years. Ryle's interferometer continued to listen for solar radio emissions. By spring 1946, though, Lovell had switched his main focus from unfruitful cosmic ray work to tracking the ionized trails left by meteors. The revitalized scientist sped between campus and Jodrell Bank in an open Triumph sports car, quickly supplementing his equipment stocks with more than a million pounds' worth of surplus military transmitters and receivers due to be warehoused or dumped down abandoned mine shafts. His setup recorded the first meteors charted in daylight, and the muddy outpost began to amass a definitive body of evidence on the numbers and habits of the fiery heavenly visitors. He plowed on in relative obscurity for months. But everything changed that December, when Lovell and two colleagues, C. J. Banwell and John A. Clegg, summarized some of this early work before the Royal Astronomical Society. "By the end of the meeting," Lovell later wrote, "we were part of the astronomical community."

Curiously, no one emerged in America to join in the hunt. As in Britain, the mainstream astronomical establishment wasn't interested in radar and radio technology, and none of the World War II radar veterans were willing or able to take up the gauntlet in the face of that intransigence and the rich attraction of nuclear physics.

Other centers for the new astronomy were starting to crop up, however. One was in The Netherlands, where the theorist Jan Oort, director of the Leiden Observatory, and researcher H. C. van de Hulst began musing about radio astronomy. But the main challenger to the British did not come from any of the science-rich European nations, as might be expected. Instead, it arose from Down Under in Australia, where no Ph.D. had yet been awarded in any academic field, and wouldn't be until 1948. The upstart Aussies made their mark almost by hook and by crook, rounding up comrades who had studied abroad, luring Britishers being discharged from the service, and through raw native talent. When combined with the Cambridge and Manchester groups, their efforts were about to expose the world to an unsuspected universe laced with radio stars, hidden galaxies, and other phantasmal bodies. At the center of the scientific upheaval, beginning a third career on a third continent, was Lovell's old boss from the early days of British radar, Taffy Bowen.

•

Somewhat to his surprise, the Welshman was getting to like Australia. He had closed out his days in America despondent over the lack of challenges remaining for him at the Rad Lab. And when he arrived in Sydney on the first day of 1944, after stops in exotic venues like Hawaii, Canton Island, and Nouméa, the airborne-radar pioneer hadn't expected things to be much better in the Southern Hemisphere. He planned on staying only for the duration: help the Aussies get started on the microwave front, then hurry back to real life in Britain. But the rewarding job and pleasant sea-oriented life he found had quickly begun to work their magic.

A few months after Bowen's arrival, Vesta had come over. The couple's first son, Edward, had been born the previous November back in Boston. As soon as Vesta deemed the infant able to travel, she had given up the apartment on Brattle Circle, disposed of their furniture, and taken a train to San Francisco accompanied by a representative of the Australian embassy in Washington. An unescorted Swedish freighter had ferried mother and son to Brisbane, where Taffy waited. Vesta was thrilled when he arranged for a flying boat to wing the reunited family on to Sydney.

The couple settled in a two-bedroom flat in the prestigious Double Bay suburb on the coast a few miles southeast of downtown Sydney. The fear of Japanese mini-submarines sneaking into the calm harbor and lobbing a few shells had left many spacious houses for rent, but Bowen preferred the security of an apartment building. When a second son, David, arrived at the end of 1945, the couple had felt cramped and eventually rented a house in nearby Bellevue Hill. As his sprouting family grew up — there would be one more boy, John, born in 1947 — Bowen cultivated a love of sailing and boat building, hobbies he would share with his sons late into life. Nearly every weekend when the boys were little, the family would go out: David and John sailed together, with their father following in a rowboat.

Australian radar development took place in the government-run Radiophysics Laboratory, which shared a castle-like redbrick edifice on the eastern edge of the University of Sydney with the National Standards Laboratory: radar workers entered around the back of the three-story structure, down a slight incline and a floor below the building's main entrance. The Radiophysics Laboratory had been formed in late 1939 as a secret arm of the Australian Council for Scientific and Industrial Research, after a leading physicist, D. F. Martyn, had been called to England for a radar briefing and returned with a lead-lined trunk — eerily reminiscent of the Tizard Mission's black box — crammed with a few valves and two thousand–odd reports and blueprints. The first job had been to develop an Army shore defense radar. But the lab, stocked with Australia's top engineering talent, had quickly moved on to bigger things. Hallways and labs teemed with electronics, and an unmistakable spirit of camaraderie crackled in the air. With the payroll held steady at about two hundred, it was a mini–Rad Lab, bottled up and preserved in its formative years. When work ended at 4:50 each evening, staff members dashed kitty-corner across City Road to the Lalla Rookh hotel and pub, downing beers to beat the 6 P.M. closing before heading home in cars and commuter trains.

Unlike its American counterpart, Radiophysics did not close after the war. Unlike the Telecommunications Research Establishment in Britain, it did not continue with military radar development, which was shifted to other agencies. Instead, the lab was transformed into a research and development center for the civilian technologies Australia hoped would carry it into the modern age.

Bowen inherited it all. Promoted to be the division's acting director in 1944, and later given the directorship outright, as the war ended he found himself in charge of a powerful group. Of the two hundred staffers, about 70 percent remained on board, including the bulk of the sixty-six-person scientific core. As a nation, Australia was still a technological neophyte. But those staying on formed the cream of its radio science and engineering crop. Several had earned doctorates in Britain or studied in top overseas laboratories. Others had received a special joint engineering and physics master's degree from the University of Sydney that seemed to tailor recipients for R&D. Instead of hurrying back to Britain, Bowen found himself turning down job offers in his native land to stay at the helm of this talented bunch.

Radar was the clay with which Bowen and the Radiophysics staff sought to mold Australia's scientific future. A world of equipment was left over from the war: klystrons, magnetrons, pulse counting circuits, "the whole paraphernalia of a new electronic era," Bowen writes. Large stocks of radar and communications gear were stored in Sydney. More equipment arrived on the U.S. Pacific Fleet, which assembled in the port before returning home. Ordered to destroy the surplus, the Navy began dumping tons of electronics a few miles outside the harbor. Aided by friends in both the American and Australian governments, Bowen scrambled to preserve the rest. "After a few frantic weeks loading our own trucks on the dockside, we ended up with a cornucopia of invaluable equipment, often brand new and in the original crates," he relates. "I seem to remember two huge warehouses full of these good things near Botany Bay, which we were to draw on for many years to come."

So equipped, Bowen drew up plans for research programs in vacuum tube technology, radio propagation, radar navigation, electronic surveying, meteorology — and astronomy, though not one staff member had even taken a college course in the subject. As he organized the efforts, he was influenced by the postwar goings-on in Britain. Watson-Watt, who had set up a London-based research consulting firm with Robert Hanbury Brown as one of his partners, appeared to be everywhere, reveling in his role as the perceived inventor of radar. It seemed to Bowen that the Scotsman took credit for everything, including his own field of airborne radar, and it rankled. In running the Radiophysics Lab, the Welshman vowed, "I was going to be quite certain of one thing . . . I was not going to jump in and claim credit when somebody else did the work."

Over the next few years, under his leadership, the lab made steady gains on several fronts. Radiophysics developed a method for precise distance measuring that helped pilots navigate between airports. Bowen himself sought methods of cloud seeding that could bring relief to his adopted land's vast arid regions, thereby cultivating what turned out to be a lifelong interest in rain and cloud physics. It was the astronomy effort, though, that put Australia on the world's scientific map.

Bowen had left this end of things almost entirely to his best researcher, a gangly native Australian named Joseph Lade Pawsey. Coming out of the war, Pawsey ranked as one of the world's experts in antenna development. In the early 1930s he had earned a doctorate from Cambridge University, where he had studied radio wave propagation at the Cavendish under J. A. Ratcliffe. Afterwards, he had worked at Electrical and Musical Industries on antenna systems for early television, only returning to Australia at the outbreak of war. A brilliant researcher, Pawsey possessed a knack for homing in on the right questions and recognized immediately the vast possibilities radar offered for astronomical studies.

In particular, Pawsey was fascinated by the idea of using antennas to pick up the thermal noise that revealed the temperature of distant objects and set out soon after the war to determine the sun's temperature through its thermal radiation. He designed an experiment at the standard radar frequency of 200 megahertz, or 1.5 meters in wavelength, co-opting for his eyes a series of existing Air Force radar antennas lining the Australian coast around Sydney harbor. By late October 1945, after coaxing military personnel into making some observations for him, he had enough data to announce that solar temperatures reached a million degrees kelvin. That was almost improbably higher than optical spectrum values of 6000 kelvin, and only later did researchers develop a full theory explaining the temperature gradient from optical wavelengths to 1.5 meters as largely due to partially ionized atoms of solar radiation that cause the very thin outer layer of the sun's atmosphere to heat to enormous temperatures visible only at longer wavelengths. At the same time Pawsey loosely verified J. S. Hey's wartime work and concluded that the intensity of the sun's radio emissions seemed to vary with sunspot activity.

Encouraged by Australia's initial foray into radio astronomy, published as a letter to *Nature*, Bowen and Pawsey decided to ramp up solar noise studies. In February 1946, during what was then the century's largest sunspot eruption, Pawsey set out to locate more precisely the origin of the intense solar radio bursts and determine once and for all whether the noise came from sunspots. Joining him were receiver expert Lindsay McCready and Ruby Payne-Scott, a talented researcher who for nearly two years had escaped a rule barring married women from serving on the Radiophysics staff by keeping her own marriage a secret.

The experiment would put Australian radio astronomy on a par with the group in Cambridge, where Martin Ryle was deploying his radio interferometer to resolve the same question. Like the Britisher, Pawsey saw that pinpointing the source of the radio emissions with significant accuracy depended on achieving a much greater angular resolution than possible with a single aerial. His elegantly simple solution took advantage of the military radars positioned on high cliffs overlooking the Pacific. On calm days, the sea itself could act as another antenna by reflecting the sun's emissions back to the cliffside outposts. The signals picked up from the rising sun would therefore show a fringe pattern resulting from the interference of the reflected and direct rays, making it possible for Pawsey to later get a good fix on the point of origin. In classical optics, the effect was known as Lloyd's mirror: instead of being stuck with the antenna's six-degree beamwidth,

Pawsey could in theory locate celestial objects to around ten arc minutes, about a third the size of the visible sun.

He took measurements for weeks. Processing the results took still more time and consideration. The seminal paper submitted to the *Proceedings of the Royal Society* in July 1946 accounted for the Earth's curvature, reflections from chopping seas, tides, and refraction. But the conclusion was sound: intense solar radio emissions, Pawsey asserted, followed the sunspot group as the sun climbed over the Pacific.

Bowen summarized Pawsey's work in a talk at the Cavendish Lab at Cambridge University about two months later, telling an assembly of thirty to forty scientists the details of Pawsey's discoveries, which still awaited publication. Martin Ryle sat in the audience. Although the Australians had beaten him to the draw proving that the increased radio signals originated near sunspots, Ryle's two-aerial interferometer suffered less severe refraction effects than the sea-cliff interferometer, and he recently had been able to pin down the source of the emissions more precisely. As Bowen finished, Ryle rose and took issue with several points, including Pawsey's repeated claim of a million-degree corona. Bowen boiled. It was a minor thing, a scientific spat. But from that point on, he suspected the Cambridge man of striving to undermine the reputation of Radiophysics—and the tension between the two groups grew as time went on.

•

Dover Heights is one of a series of rocky outcroppings that tower above the Pacific north and south of the entrance to Sydney harbor. Poised a few miles below the scenic gateway, halfway between South Head and the famous sandy stretches of Bondi Beach, it rises some 250 feet above the ocean's rounded expanse to offer a commanding view of the sea. During World War II the Royal Australian Air Force built a radar station on the cliff edge to provide warning of anticipated Japanese attacks.

After the fighting Dover Heights became a favorite spot for Joe Pawsey's sun watchers. Working from a three-room blockhouse on the sloped cliff edge that served as office, equipment room, and bivouac, researchers conducted a variety of sea-cliff interferometer studies. Sometimes swollen waves crashing into the cliff face kicked a cold mist of salty spray onto the bluff. In those days, long before a pipeline carried sewage farther out to sea, the city's effluence was dumped just to the south. Clusters of seagulls hung around picking orts, and when the winds blew wrong the stench permeated the radar station. All the same, Dover Heights was good duty, away from the lab and Pawsey's supervision, a spot where a daring soul could sneak time for a bootleg experiment. Settling in at Dover Heights in March 1947, that was the way John Bolton seemed to view it, anyway.

The highly talented Bolton was another one of Taffy Bowen's up-and-coming stars. A native Yorkshireman who had studied physics as a Cambridge University undergraduate, he had served during the war as a Royal Navy radar officer on the carrier HMS *Unicorn*. Demobilized in Sydney, he had hooked up with Radiophysics as its second postwar recruit the previous September. Bolton was a small man with close-cropped hair that made his large ears stand out, and he could chill people with an intense, piercing stare. He loved precision hand labor and was happiest holding a drill or welding torch. And he vastly preferred to work away from the madding crowd.

The autumn assignment was Bolton's second tour of duty at Dover Heights. As his initial Radiophysics task the previous November, the former radar officer had been sent out to the cliff site to study solar noise. But on a surprise visit, Pawsey had caught him aiming the aerials at the stars, trying to detect radio emissions from objects listed in *Norton's Star Atlas*. Bolton and technical assistant Bruce Slee were reassigned to projects back at the Sydney lab.

It had taken several months, but Pawsey had finally decided to grant his insubordinate subordinate a reprieve. Bolton returned to Dover Heights with Gordon Stanley, an electrical engineer from New Zealand who had joined the lab three years earlier after an Army stint. Obediently, the men spent several months enmeshed in solar observations. But when the sun lapsed into inactivity in June 1947, they turned to the stargazing that had landed Bolton in hot water a few months earlier.

Bolton was particularly intrigued by an observation J. S. Hey had made soon after the war. While producing his meter-wave radio map of the sky, Hey had accidentally discovered an oddity in the direction of the constellation Cygnus, a rapid fluctuation in the signal intensity that he concluded must be a discrete source like a star. Bolton and Stanley set out to measure the Cygnus source on two frequencies, 100 and 200 megahertz. On the first night, a cable snapped on the array for the longer wavelength. When the men attempted repairs their only soldering iron broke, so they dumped the aerial over the cliff and continued the search with their remaining eyes, a pair of Yagi antennas attached to a converted radar receiver.

Almost immediately the relatively smooth straight line being drawn on the blockhouse recorder changed into a regular sinusoidal pattern, indicating that direct rays from some distant source were being interfered with by those reflected off the sea — and telling the men something was out there.

For several months, they studied the phenomenon, adding observations from other cliff sites. Cygnus never rose more than fifteen degrees over the northern Sydney horizon, so it was hard to be precise. Nevertheless, the bumpy measures of fringe maxima and minima indicated that whatever was causing the signal had a dimension of something less than eight minutes of arc. That indeed made it small and discrete by galactic standards, like the star Hey had hinted might exist. Bolton called it a point source.

Nothing of the sort had ever been positively identified. Even more intriguing, the radio emissions the Australian pair recorded were of an intensity comparable to that streaming in from the sun. Coming from stellar distances, that was jolting. But Bolton and Stanley got a far greater shock when they moved to identify the star or nebula responsible for such activity. Their calculations put the mystery radio emitter in Cygnus — at 19 hours, 58 minutes, 47 seconds, plus or minus 10 seconds, +41 degrees, 47 minutes, plus or minus 7 minutes. It was not exact, but far better than Hey's five degrees of uncertainty. However, when they consulted star catalogues, the region came up virtually empty. No bright stars. No nebula. Just a bland, nondescript area of the Milky Way. The mystery deepened when they attempted to pin down its distance. From the lack of any position change in the source as the Earth orbited further away, they calculated it had to lie at least ten times the distance to Pluto. As an upper limit, the men figured that a star with a total power output similar to the sun's, but with all its energy channeled into the radio spectrum, could be as far away as three thousand light-years. But how such an object could exist was beyond them.

Once known to the general astronomical community, the findings engendered a host of questions. Did other such sources exist? If so, how many? Could the cosmic static that pervaded the universe arise from these unknown objects rather than expanding clouds of ionized gas as commonly believed? At least one eminent optical astronomer, Rudolph Minkowski in America, grew intrigued. Just as Bolton and Stanley were wrapping up their reports, for publication in *Nature* and the premier issue of the *Australian Journal of Scientific Research*, they heard from Pawsey, who was visiting Mount Wilson Observatory near Pasadena. He told them Minkowski had suggested searching for radio noise among a handful of galactic curiosities: the Crab Nebula, the white dwarf near Sirius, or perhaps the strange nebula in Orion.

Others might have taken up the distinguished Minkowski's suggestion without a thought. Returning to the hunt that November, however, Bolton and Stanley, joined by Bruce Slee, chose to launch an empirical survey of the entire sky, instead of focusing exclusively on the optical astronomer's list. The New Zealander first laboriously stabilized the receiver's power supplies to cut noise fluctuations and increase its sensitivity to faint signals. Within a few days of being put into service, the improved apparatus picked up a second strong radio source, in Taurus. By the following February, the team had painstakingly plotted half the southern sky, accumulating evidence for five additional objects. As more contacts appeared, Bolton devised the nomenclature that became for a time the standard means for identifying radio sources: the strongest emitter in a constellation was called A, the next strongest B, and so on.

Cygnus A, the original curiosity, thus turned out not to be so unique. Moreover, none of the newly charted radio emitters could be squarely lined up with known stellar objects. Convinced they were hot on the trail of something big, Bolton dashed off a short note to *Nature* announcing a novel class of astronomical oddities.

Immediately, the men moved to draw a tighter bead on the origins of their puzzling radio emissions. From Australia's east coast, Bolton, Stanley, and Slee were only able to follow rising stars. They proposed setting up additional aerials on the western rim of New Zealand so sources could be tracked coming and going. Caught up in their excitement, Taffy Bowen okayed the funding and arranged for logistics support from the New Zealand government.

At the end of May, the researchers shipped an old Army gunlaying radar trailer packed with 100 megahertz Yagi aerials, receivers, recorders, and other equipment from Sydney to Auckland. Slee stayed in Australia to man Dover Heights, while Bolton and Stanley went over to set up their installation at Pakiri Hill, a farm roughly forty miles northwest of the capital that was situated on a bluff nearly a thousand feet above the sea. From three times the elevation of Dover Heights, the site gave significantly better angular resolution on sea-cliff interferometry measurements. After two months of observations the men carted the equipment over to Phia, a surfing haven halfway between Pakiri Hill and Auckland. Finally, following three more weeks of near-nightly studies, they returned to Australia for the long process of reducing the data.

Astronomy had reached a turning point. When the data were reduced back in Sydney, good fixes came through on four sources — Taurus A, Centaurus A, Virgo A, and the original puzzler, Cygnus A. More important, all but the Cygnus source were associated with optical objects, the first time such a link had been established.

For starters, Taurus A emanated from the Crab Nebula, the expanding gaseous remnant of the supernova explosion observed by the Chinese in 1054. Since the radio emissions pouring forth from this galactic landmark would be of intense interest to optical astronomers like Minkowski and Jan

Oort, who strove to understand all processes surrounding the Crab, in this discovery alone Bolton, Stanley, Slee, and their ilk seemed justified in claiming membership in the inner circle of astronomy. Even more exciting, while the origin of Cygnus A remained elusive, the Virgo and Centaurus sources seemed to line up with the external galaxies M87 and NGC 5128. The former was a bright elliptical entity a staggering five million light-years distant. NGC 5128 ranked as one of the most curious nebulosities known, a hazy sheet of stardust so elusive in its properties astronomers couldn't even be certain it did not actually reside in the Milky Way.

The idea that radio astronomy, initially concerned mainly with the sun, could look far beyond the galaxy and root out the physical processes of some of the most distant and peculiar objects in the universe proved daunting. Before submitting a paper, Bolton summarized the findings in letters to a handful of leading astronomers, among them Minkowski, Oort, and Bengt Strømgren in Denmark.

The initial feedback shook their confidence even further. While the Crab Nebula data was viewed with great interest, no one believed radio noise could be picked up from external galaxies: the sheer power implied was mind-boggling. Bolton was cowed. Especially when it came to claiming they had recorded extragalactic signals, the three men hedged enormously in their final paper, afraid, as Bolton later said, of being denied publication because it was too unbelievable. But they believed it.

As it was, the relatively firm identification of the Crab Nebula as the source for one set of strong radio emissions solidified the Radiophysics Lab's position at the forefront of radio astronomy, and the early successes of the Pawsey and Bolton efforts begat more success. Astronomical studies flourished, dominating the Radiophysics agenda at the expense of virtually everything else: only Taffy Bowen's pet project of rainmaking and cloud physics also continued to receive funding at a strong level.

Most of the radio astronomers were young men working in small teams and doing all the dirty repair jobs themselves. "Each morning people set off in open trucks to the field stations where their equipment, mainly salvaged and modified from radar installations, had been installed in ex-army and navy huts," recalled W. N. "Chris" Christiansen, one of the group. Paul Wild, another recruit, also considered the early days of Australian radio astronomy a special time. He writes glowingly: "With the flow of new discoveries from a brand new science a special atmosphere developed—perhaps, though on a humbler scale, something like that at the Cavendish in Rutherford's time."

By the time of Bolton's announcement, Martin Ryle's pathbreaking interferometer was paying its own dividends. Besides taking advantage of his

Northern Hemisphere position to improve on the Cygnus fix, he picked up another, even stronger radio signal in the constellation Cassiopeia, and by 1950 had produced a catalogue of fifty distinct radio sources, to go along with twenty-two identified by then in Bolton's group. Still, only one of the mysterious fountains of radio waves had been firmly — though even then not conclusively — linked with an optical object: the Crab Nebula. In contrast to what Bolton, Stanley, and Slee believed, but were afraid to publish, Ryle proposed that nothing outside the Milky Way galaxy could emit such intense waves, deducing instead that the signals stemmed from a class of undiscovered dark stars. Thrilled at the prospect, the popular press gave the idea great coverage. The supposed stellar bodies became known as radio stars.

The almost palpable existence of a hidden galactic neighbor set off a heavenly fox hunt. Bernard Lovell's group at Manchester jumped into the fray, led by Robert Hanbury Brown. The radar veteran had left Watson-Watt's consulting business after the Scotsman met a Canadian woman and, still married, decided to uproot the firm's main operations to Ontario; Sir Robert would later divorce Margaret and marry Jean Drew Smith. Brown had refused to move overseas and instead accepted a research fellowship at Jodrell Bank, where Lovell had built a gigantic 218-foot-wide parabolic reflector to further his radar studies. By the time Brown arrived, however, Lovell had decided to turn the massive aerial into a radio astronomy telescope that could scan the heavens for cosmic noise. The newcomer's first major task lay in building a more sensitive receiver, improving the aerial feed, and reorienting the massive paraboloid and its spidery array of guy wires toward the zenith. When he finished nearly a year later, Jodrell Bank boasted the world's largest radio astronomy dish.

Brown and research student Cyril Hazard immediately launched a time-consuming, detailed survey of everything in the telescope's field of view. Beginning in August 1950, hoping to determine whether other galaxies also hosted radio sources, they trained the scope toward M31, the spiral nebula in Andromeda, an extragalactic cousin to the Milky Way two million light-years off. After an exhausting ninety-night survey, the pair produced contour maps that proved M31 did emit radio waves of much the same intensity as those coming from inside the Milky Way. With the information, Brown was able to say what Bolton dared not: radio astronomy could see far beyond the local galaxy.

There was one catch. The Andromeda signals were faint and hard to pick up, even for the huge 218-foot telescope. Therefore Brown and Hazard concluded that Ryle must have been right in arguing that the far stronger sources discovered by less sensitive arrays in Cambridge and Australia emanated from radio stars hiding somewhere in the Milky Way. There arose around the issue a furious debate — widening the Radiophysics feud with Ryle — that would not be settled for many more years, when the long-elusive Cygnus source was dramatically identified as two distant galaxies in collision. On the heels of that finding, more precise studies from Australia and Britain established that what were once thought of as radio stars were really radio nebulae. So while the astronomical world lost the radio stars that had kindled so many imaginations, it gained a window on supernovas and what looked to be colliding galaxies, some of the great catastrophes of the universe.

Late in 1950, even without this critical piece of the puzzle, radio astronomy was making its presence felt. Receptive optical astronomers like Minkowski took serious note of the upstarts and treated them as equals. For the diehards who continued to look with disdain on the emerging discipline, an event was about to take place that would shatter their complacency and elevate the field into an even more important tool for probing the cosmos than its optical counterpart.

Wartime radar had bequeathed the people and equipment that formed the foundations of radio astronomy. The coming breakthrough was a second great gift.

14 • Twenty-one

"Here was the K-band radiometer just on his desktop. . . . All of us who saw it then knew that this would be an important instrument."
EDWARD PURCELL of Robert Dicke's invention

ED PURCELL, his mind attuned to the atomic nucleus and not the heavens, could hardly recognize a constellation. He knew the bare bones of the revolution in radio astronomy that other wartime radar men like Taffy Bowen, Bernard Lovell, and Martin Ryle had kicked off, but that was about it. For nearly four years since his pathfinding magnetic resonance experiment, often in collaboration with his old Rad Lab friend Bob Pound, the Harvard researcher had devoted his energies to investigating the phenomenon in the various states of matter—gaseous, liquid, and solid—and extending his resonance technique to elements beyond hydrogen. So while he could be called a physicist, engineer, even chemist, since unraveling atomic structures formed an extension of the nearly century-old technique of spectroscopy, Purcell was not an astronomer.

Then, in the fall of 1949, came word of a remarkable prediction. The gist was that interstellar hydrogen, the invisible building block of the universe, might be detectable through an almost infinitesimal radio signal as it changed quantum states. This idea of a quantum flip-flop was not at all unlike what Purcell routinely studied in his laboratory, only now the lab had been extended into deep space. And, suddenly, Purcell was an astronomer.

Just how Purcell learned of the prediction is uncertain. He recalled the news reaching him via Harold Ewen, a graduate student who had heard about it at a scientific meeting held at nearby Yale University. Ewen maintained it was Purcell who told *him*. In any event, both men were intrigued. Ewen undertook a quick literature search that turned up two papers discussing the concept and suggesting a hunt be mounted. But it remained unclear whether anything was actually afoot. By Ewen's account, he and Purcell then talked things over and decided the doctoral candidate should attend the Yale get-together to try to resolve the question.

It would have been hard to find a more perfect choice for the assignment than Harold Ewen, a handsome, ruggedly built former Navy officer with an unusually wide variety of interests and talents. The native of nearby Springfield, Massachusetts, had graduated with honors from Amherst College in 1942, then become the youngest faculty member in the institution's history. For eighteen months Ewen had taught math, physics, astronomy, and an occasional course in celestial mechanics to West Point and Naval Air cadets attending Amherst during World War II; one of his pupils had been Boston Red Sox star Ted Williams, who once brought Harvard physics to a standstill by stopping in to visit.

Midway through the war, Ewen had joined the Naval Air Corps and served three years as an officer on a patrol bombing squadron, including a stint hunting U-boats in the English channel. He planned a career in law and business, and before leaving the service had been accepted at Harvard Law School. However, a prestigious National Research Council fellowship for the study of nuclear physics had caused him to shift gears temporarily and enter the physics program in the fall of 1946. Three years later, when the hydrogen detection problem came up, the twenty-seven-year-old Ewen was senior to virtually every other physics graduate student. Even professors naturally called him Doc, his nickname since grade school. The trouble was, Doc still didn't have a dissertation topic. His undergraduate honors thesis at Amherst College had involved calculating orbits of comets and asteroids. So a hunt for interstellar hydrogen offered the prospect of a celestial subject that would play to his interests in quantum physics, mathematics, thermodynamics, spectroscopy, astronomy, and radar. He jumped at the chance.

At Yale, Ewen found considerable interest in the hydrogen line, both within the astronomical community and among radio scientists attending the meeting. However, the line's very existence in the galactic radiation was extremely doubtful, and no one knew of any plans under way to try to detect it. As Ewen recalled, "There was so little known about this that you might even get drummed out of town if you were found to be working on it." But the graduate student didn't worry much about that. Purcell, too, agreed it was worth a shot. Failure, the advisor warned, would make it necessary to determine the absolute sensitivity of the equipment in order to establish the precise level at which the signal was not detected — a process that might take two years. But what's two years? Purcell quickly added: "If you do detect it, you'll be in *Life* magazine." Doc Ewen finally had his thesis subject.

The notion that scientists might be able to reach into interstellar space and study the most abundant element in the cosmos had been developed some four years earlier by the Dutch theoretician Hendrik C. van de Hulst. Although Holland choked under German occupation, smuggled copies of the U.S.-based *Astrophysical Journal* had reached Jan Oort at Leiden Observatory with news of Grote Reber's pioneering radio maps, and through Oort the information had been passed on to "Henk" van de Hulst.

For decades, optical astronomers had been photographing the stars through telescopes and studying the telltale spectral lines produced by different elements. The work followed Niels Bohr's 1913 demonstration that through photons of light emitted on characteristic wavelengths as electrons jumped down the quantum ladder toward the ground state, each element produced a unique spectrum of energy. And through this spectroscopy, as well as Doppler shifts in the spectral lines, astronomers had gleaned much about both the makeup of stars and their motions.

Probing the dark, cold void of interstellar space was a far different matter, however. Light absorption studies showed it consisted of a yawning vacuum laced with extremely sparse populations of gas molecules and dust. Where the intense light of very hot stars ionized the gas, the freed electrons eventually recombined with atoms and emitted energy, so that the gas glowed with its own light. Analysis of this outpouring had shown that neutral hydrogen made up the plethora of gas particles. But beyond this limited view, the extent and reaches of interstellar matter remained impenetrable to even the best optical telescopes.

In war-scarred Holland, at a 1944 colloquium held to discuss Reber's observations, a fascinated Oort pointed out the important scientific horizons that would open if the same monochromatic lines relied on for optical spectroscopy were found to thrive in the radio spectrum. Not only could radio waves often be detected farther away than visible light, they also might help illuminate the vast cold and dark tracts of the universe devoid of starlight, an unexplored region thought to be punctuated with gaseous clouds that eventually congealed into stars and planets. Van de Hulst, a research staff member, had studied the matter at Oort's request and in his own talk concluded that a tiny radio signal—a photon hiccup, really— might be observable from the neutral atomic hydrogen making up the bulk of this invisible star stuff. His musings were published immediately after the war in the Dutch journal of physics, *Nederlandse Tijdschrift voor Natuurkunde.*

When Purcell considered the idea several years later back in Cambridge, it almost immediately struck a nerve. The whole thing hinged on the hyperfine transition, essentially the same spin-flip maneuver between two quantum states that he had induced in the laboratory with a magnet and a bridge circuit in his landmark nuclear magnetic resonance experiment. The rule-of-thumb difference in this case is that it isn't the proton that emits and absorbs energy, but the electron. Moreover, the transition is very subtle, involving quantum states arising from the so-called hyperfine splitting in the ground state of the hydrogen atom, a phenomenon caused by the interaction between the spinning electron and the proton's magnetic moment. As Purcell once explained for a television documentary, "Ring of Truth": "The amount of hydrogen out there and its temperature was such that the radiation at this frequency that we're concerned with, this very special frequency, amounted to only one watt landing on the entire Earth."

The two papers Ewen had dug up about the hydrogen line were of different minds about the difficulty of the challenge. One, in Russian, was translated for him by Ed Purcell's wife, Beth, who had a proficiency in languages. In it the Soviet theorist Iosef S. Shklovsky concluded finding the line would be easy. As Ewen read it, it was almost as if you could "lick your finger, hold it up, and detect hydrogen." At first, Doc worried that any Russian engineer reading the paper would easily beat him to the punch.

But the other account painted a different picture. Translated from the Dutch at Cornell University, this was van de Hulst's original 1945 work, with various notations added three years later. These calculations showed that the emission should occur at a wavelength of 21.2 centimeters. Although such frequencies were easily within the sphere of the current micro-wave art, the depressingly small energies involved still made detecting the spectral line a long-shot proposition. In fact, picking up such low-level energy with current electronic techniques seemed almost impossible. It wasn't even clear whether the signal would be detectable in emission — a peak in the amount of energy recorded — or absorption, which would show up as an energy dip. As Doc Ewen recalled, the unknowns seemed almost too daunting: "Would the signal appear in emission? Would it appear in absorption? What would the bandwidth be? Would the signal be coupled to the cosmic noise in a manner that would make it indistinguishable from the background, and hence, undetectable?"

Balanced against those long odds were the immense scientific ramifica-tions of being able to track the density distributions of nature's most basic element into the vast dark radio region where optical astronomy could not peer. As it turned out, van de Hulst had tried to follow up his theoretical musings. After the war, in the summer of 1946, he had accepted a postdoc-toral post at the University of Chicago's Yerkes Observatory in Wisconsin. Soon after his arrival in America, van de Hulst traveled to Illinois to enlist Grote Reber's help in finding the twenty-one-centimeter line. But the pion-eering radio astronomer had pursued the idea sluggishly and abandoned the project altogether in 1947, when he took a job at the National Bureau of Standards. A year or so later Oort's team at Leiden had formally taken up the charge and begun retooling an old *Würzburg* radar abandoned by Hitler's troops, funding the work with Holland's first grant for studies in radio astronomy. But progress was exceedingly slow.

Not so at Harvard. For one thing, Ewen began to feel that while the idea of the line's existence in space had been around for years, he and Purcell might still have the jump on the few competitors who might be looking for it. Van de Hulst had concluded the line should exist at 21.16 centimeters — which Ewen calculated corresponded to 1417.777 megahertz — with a fairly wide margin of error. However, the Harvard student was aware of more recent atomic hydrogen measurements at Columbia University, which showed that in the laboratory, at least, the line existed at precisely 1420.405 megahertz. Although seemingly small, the difference between the theoretical and the real was significant because it greatly simplified the job of building a receiver to detect the line. "So I knew where the line was," related Ewen. Anyone relying solely on van de Hulst's value faced the difficult proposition of scanning a wide frequency range to make sure nothing was missed.

Ewen had another edge — his advisor. Inside a few weeks of their first discussing the idea, Purcell had conceived a way to possibly detect the signal. His mind hearkened back to the old Rad Lab days and the water-vapor absorption mystery — in particular, the remarkable microwave radiometer that he had brought to bear on the problem. Only Bob Dicke's device, he realized, could provide the needed receiver sensitivity. Purcell had always been a huge fan of his friend's invention. He had known instantly that it would be an important instrument; now he had glimpsed just how important.

Purcell had a trick or two in mind about how to adapt the radiometer for the twenty-one-centimeter line hunt. But since Ewen would be building and operating the instrument, he urged the graduate student to make the pilgrimage to Princeton and consult Dicke directly about the ins and outs of the device. "I don't know anything about that — better talk to Bob," he advised. Ewen went.

•

Robert Dicke had returned to Princeton University in the spring of 1946, armed with a radiometer and, as it turned out, dreams of doing some astronomical trailblazing of his own. In addition to his coup solving the water vapor mystery, Dicke had found time at the Rad Lab to use the device for a variety of extracurricular dabblings. He spent a day at Boston harbor detecting the presence of ships against the water by the thermal radiation emanating from their hulls. Another time he scanned the Cambridge sky-line, developing a building-by-building chart of thermal radiations accurate enough to reveal, from the extra-hot chimney stacks, which boilers were being stoked. Finally, a month before the war ended, he and one of the other Bobs — Beringer — had dragged the 1.25-centimeter radiometer to an MIT rooftop and turned it toward outer space.

The pair observed the sun during the partial eclipse of July 9, putting the temperature of the solar disk at 10,000 degrees kelvin. Three days later Dicke and Beringer charted the microwave radiation from the unobscured sun and found the orb had grown a thousand degrees hotter. After that expedition they had waited until mid-October to observe a nearly full moon, busying themselves in the meantime writing parts of the Rad Lab Series. Still months before Lieutenant Colonel John DeWitt Jr. made headlines by bouncing a radar signal off the moon, the two Bobs put the lunar tempera-ture at a cool 292 degrees kelvin.

These successes had fired Dicke's imagination. Beyond the relatively uninteresting moon and the increasingly well-studied sun, if he could mea-sure the microwave "noise" emitted from any number of atmospheric or celestial objects, it would be theoretically possible to deduce their tempera-tures, unlocking a scientific treasure trove. Knowing the temperature of Venus, for instance, one could begin to answer questions about the planet's makeup and climate. Dicke had already, somewhat benignly, uncovered an important cosmological clue. While studying the water-vapor absorption mystery in Florida several months earlier, he had measured the omnipresent cosmic noise down to the limits of the radiometer's sensitivity, fixing an upper threshold of 20 degrees kelvin.

Upon taking a post as a junior faculty member in the Princeton physics department in March 1946, Dicke approached a colleague in astronomy and offered up the radiometer for a joint study. The idea got nowhere, as Henry Norris Russell, the department's chairman, was about to retire and showed little interest in accommodating novel techniques. Disappointed, Dicke concentrated on getting old radiometer results published. Within a few weeks he and Beringer, installed at Yale, dashed off a two-page note on their sun and moon observations for *The Astrophysical Journal*. The two men, joined by Kyhl and Vane, sent their Florida water vapor results off to *Physical Review*. And Dicke alone submitted a third paper, really a rewrite of an old Rad Lab report, giving the full details of the radiometer to *The Review of Scientific Instruments*. Then the young physicist decided to forget about the radiometer and astronomy, turning instead to atomic spectroscopy. "I thought it would have been rash for me to go at this myself, in the physics department, and start doing astronomy, so I didn't do it," he explained decades later. "I should have done it, though, in retrospect. It would have made good sense to open up this new field of research."

It turned out to be a fateful decision, almost folklore among a select group of physicists and astronomers, because Dicke's failure to enter radio astronomy might well have cost him the Nobel Prize. He didn't use a radiometer again until 1964. Concerned by then with gravitation and cosmology, and completely forgetting his own crude measurements of nearly two decades past, Dicke suggested to two young colleagues, Peter G. Roll and David T. Wilkinson, that they employ a vastly improved device to measure the cosmic microwave background radiation. Such radiation, Dicke had come to realize, would be left over from an extremely hot and dense phase of the early universe, a time when matter could not yet have congealed into galaxies and stars, and even electrons and nuclei had not formed their atomic partnership. Accordingly, a radiometer was mounted on the roof of Princeton's Palmer Physical Laboratory. Just as the work got started, two Bell Labs scientists, Arno Penzias and Robert Wilson, found this blackbody radiation first, a largely accidental discovery that ranked among the most important of the twentieth century. Dicke and P. J. E. Peebles, also of Princeton, provided the landmark theoretical interpretations, but Penzias and Wilson got the Nobel.

As things stood, however, even with Dicke out of the picture the publication of the radiometer's schematics should have put the astronomical world on notice. The Rad Lab veteran had found an ingenious way around the formidable problem of receiver gain instabilities to enable measurements with unheard-of radio sensitivity. Typical microwave receivers available at war's end generated internal noise equivalent to a thermal temperature of about 7500 degrees kelvin, with fluctuations in gain of about a tenth that figure, or 750 degrees K. This meant that objects with thermal temperatures lower than the gain instability could not be detected. The moon, for example, has a radio temperature of less than 300 degrees kelvin and would be extremely difficult to study in the microwave region.

Dicke had managed to exorcise this considerable demon by dramatically improving on a technique developed by Bell Labs researcher G. C. Southworth. During the summers of 1942 and 1943, as Dicke was well aware, Southworth had measured the sun's microwave radiation by simply pointing his radio antenna alternatively at the solar disk and a black region of interstellar space. Even if an individual comparison was not sensitive enough to detect any difference between the two, by repeating the back-and-forth measurements and looking for an average difference, Southworth could reduce the level of discrepancies through the process called integration. The Bell engineer manually performed hundreds of side-by-side measurements at three wavelengths between one and ten centimeters to come up with a crude guess that ultimately fixed the sun's microwave temperature at around 20,000 kelvin.

Dicke's vastly more accurate technique performed the same switching between an object of interest and a thermal reference source, but did it so quickly that he avoided aberrations in signal amplification due to receiver gain variations — a phenomenon sometimes referred to as flicker noise that had plagued Southworth's measurements. To that end, instead of physically moving the antenna back and forth between the target and the reference, a slow process that opened the door for gain variation problems, he designed a system in which the comparison was made without moving the antenna at all.

The waveguide running from the radiometer's horn-shaped antenna to the heart of the system contained a narrow slot, into which fitted a rotating wheel. The wheel itself, made of radio-transparent material, was half coated with resistive carbon so that it would provide a thermal reference called a black body, which emits a precisely known amount of microwave energy for a given temperature. The disk was mounted on a shaft, with part of it always resting inside the wave guide and the rest sticking up outside. The wheel then was driven around at a rate of thirty times a second. Each time the carbon-coated portion journeyed inside the guide, the wheel effectively disconnected the antenna and connected itself, with the net effect that the signal traveling down the waveguide to the receiver alternated between being energy reaching the antenna from space and thermal radiation from the wheel's resistive coating.

Instead of employing just one mixer to convert the radio input to a lower intermediate frequency, Dicke split the signal into a pair of balanced mixers, a nice trick that enabled him to cancel out local oscillator noise. From there, the signal was rectified and the 30-hertz modulation extracted, amplified again, and passed to what is often called a synchronous detector, or lock-in amplifier, that measured the difference in amplitude between the antenna signal and that coming from the reference. Ideally, any differences translated into the temperature spread between the target and the black body. Dicke came close to this ideal because the radiometer alternated so rapidly between the antenna signal and the reference source that both would be affected equally by gain fluctuations, effectively eliminating the gain variation problem. In the days that preceded the computer, the radiometer kept track of the vast amount of data it processed by converting the 30-hertz differences it detected into a direct-current signal that was integrated by a resistor-capacitor and read on an output meter.

As a whole, the Dicke radiometer amounted to nothing less than a revolution in receiver sensitivity, and Dicke was so proud of it that he would keep one of the original Rad Lab models in his house for over fifty years. Despite its potential, however, radio astronomers only slowly awakened to the device's true worth. Joe Pawsey stopped by, possibly in 1947 on the same trip that took him to see Rudolph Minkowski in California, to learn more about the instrument: Taffy Bowen had known about the radiometer since the war, and probably tipped off his star researcher. The following year, two other Bowen men, J. H. Piddington and Harry Minnett, built their own microwave radiometer to try to refine Dicke's earlier measurements of the lunar temperature.

Such studies, though, proved few and far between. As the 1940s played out, virtually all the action in radio astronomy, from sun-watching to probing distant radio galaxies, took place at meter wavelengths below 300 megahertz. At these low frequencies, because the radio sky shone surprisingly bright, anyone able to sling together a few wires could summon up plenty of signal. Receiver noise wasn't as big a problem, either. Experimenters could get enough relief by stabilizing power supplies. Further reductions à la Dicke were pointless, since at low frequencies omnipresent background noise set the limits on what could be detected. Up in the microwave region, where the radiometer could really help, there simply didn't seem to be much to find. After all, even the sun put out only modest amounts of microwave radiation, and there weren't many objects up there as bright as the sun.

Such had been the view of most astronomers, anyway, until Henk van de Hulst theorized about interstellar hydrogen, and Ed Purcell and Doc Ewen dug the radiometer out of mothballs.

•

By the time Ewen caught up with Dicke late in 1949, the Princeton man had already put the radiometer well out of his mind. The graduate student found him far more interested in discussing the Big Bang and other cosmological questions than an old invention. At a dead end, Ewen returned to Cambridge and looked up the original Rad Lab report spelling out the device. Only then, seeing Purcell's name on the cover as Dicke's group leader, did he realize his advisor had known all along how to build the radiometer. It was typical Purcell, he decided, making students do things the hard way.

Purcell's brainstorm had been that instead of simply building a twenty-one-centimeter version of the original radiometer, Ewen should modify the fundamental Dicke idea of rapid switching between a target and a blackbody reference so that the device switched instead between the expected hyperfine signal at 1420 megahertz and an adjacent frequency far enough removed that none of the emissions spilled over onto it. Dicke's "switched-load" system worked beautifully to analyze broadband energy, like that streaming in from the sun or the cosmic background. But to have a fighting chance of quickly detecting the hyperfine transition's relatively narrow signal, Purcell had told Ewen, the receiver should be frequency modulated. That way they could eliminate cosmic background noise; the difference, if it showed up at all, could only be the faint whisper of neutral hydrogen.

Ewen set about designing the system and collecting materials and components. Together with the Rad Lab report and his conversation with Dicke, Ewen got the foundation he needed from several volumes of the Radiation Laboratory Series. In particular, for help with the balanced crystal mixer, he turned to R. V. Pound, who handed over *Microwave Mixers*, volume 16 of the technical series. The book had been authored by Pound himself, and the Harvard professor deftly underlined a section about the S-band, or ten-centimeter mixer. He told Ewen it scaled linearly: virtually every dimension could be doubled to work at the hydrogen line's lower twenty-one-centimeter frequency. The one crucial exception, he warned, was the crystal diode mixer that sat on a tiny crossbar waiting to pick up any signals. Ewen would have to taper the waveguide to funnel the energy efficiently from a copper-lined, horn-shaped antenna into the detector.

All this was relatively straightforward. But as soon as Ewen started to calibrate the receiver, he ran into problems. Frequency switching of the first local oscillator, as required by Purcell's original concept, proved too difficult. After several days of thought, Ewen came up with a hybrid design that involved a second local oscillator in what was called a double-heterodyne configuration. It was this second oscillator that performed the frequency switching.

Doc got the parts together with some creative scavenging. From Harvard's Nuclear Lab he borrowed a telecommunications receiver made by the National Radio Company to serve as the second local oscillator and second intermediate frequency amplifier. To provide the rapid-fire frequency switching, he rigged a metal propeller to spin at 30 hertz in the vicinity of the capacitor components that determined the second local oscillator frequency. Even then, the frequency shift was only about 10 kilohertz, much less than desired to guarantee an adequate reference point.

Finding a good crystal diode to capture the faint signal Ewen sought took a bit more hunting. Purcell had told him to try MIT's Research Laboratory of Electronics, which had a stockpile of standard 1N21-B crystals — or, if he liked, go straight to Bell Labs. Ewen did both. After first gathering a few of RLE's crystals, he visited Bell's facility in Holmdel, New Jersey. As a first order of business, he consulted well-known radio noise authority William W. Mumford about his newly developed technique for calibrating receiver noise levels. Following Mumford's advice, Ewen would mount standard household fluorescent tubes — General Electric Cool Whites — inside his waveguide and use their near-constant noise output to accurately establish his receiver's sensitivity. Then, he stopped by the office of director Harald Friis to discuss the need for something better than the garden variety crystals he'd already collected. Talking with "Papa" Friis — the equally legendary Mumford was known as "Moms" — it soon became evident Purcell must have greased the wheels. Friis reached down into a desk drawer and held up two crystals, announcing: "Right here are two of the best diodes ever made by Bell Laboratory." He told Ewen to tune up the radiometer with MIT crystals and save the Bell products for the real thing. Doc took the advice.

By June 1950, Ewen was ready to put his listening post on the air. He worked out of a fourth-floor room in Lyman Laboratory that in later years became a women's lounge area, then was split into two physics offices. The plywood antenna, lined with copper sheeting and painted red, was big enough for a man to crawl inside, and ignobly poked out of a window onto a parapet-like ledge just around a corner from the cosmic ray shed central to Purcell's nuclear magnetic resonance experiment and Curry Street's discovery of the muon. The big horn-shaped opening pointed into the air at a fixed forty-three-degree angle above the horizon, a position dictated by the bend connecting it to the waveguide. To allow some room for scanning, the bulky object was supported by a makeshift wooden platform that itself could be raised or lowered a few inches. All in all, the whole thing looked more like the wild product of a mad scientist than a highly sensitive instrument that could listen across the cosmos.

It *was* a seat-of-the-pants effort. Much of the equipment was begged and borrowed from physics department stocks. To pay for the antenna and main receiver parts, Purcell finagled a $500 grant from the American Academy of Arts and Sciences, tapping into the Rumford fund reserved for highly sensitive heat measurements by couching the search for the twenty-one-centimeter line as a cosmic temperature experiment. Signal generators and other necessary electronic test equipment were liberated each Friday afternoon from the Nuclear Physics Lab: Ewen rolled the booty across Harvard in a wheelbarrow, then rode the Lyman elevator up to the fourth floor. Since he had to have everything back by Monday morning for physics duty, this experiment was only operational on weekends.

Sometime in mid-June, Ewen switched his apparatus on for the first time. A variety of gauges and instruments spilled over into the office, but all roads led to the Esterline-Angus recorder that would trace out whatever was picked up. The experiment was set up in an ingeniously simple way. To ensure that he didn't miss anything, Ewen decided to tune his receiver over a band about 200 kilohertz wide on either side of the hoped-for signal. If the hydrogen line appeared when the receiver was tuned to the first frequency, the synchronous detector was set up to provide a positive voltage output. A negative voltage would appear if the signal appeared when the receiver was switched to the second frequency. That way, as he slowly tuned the entire reception band through the signal, the recorder would trace out an S-curve.

Ewen got nothing like the expected wave and trough. Instead, he saw slight bumps on both frequencies, with a flat line in between, so that the printout looked like the top half of a pair of spectacles. He scribbled in his logbook that the bumps might indicate a spectral line. But before hazarding a surer guess about what was happening, he would have to check all his equipment and redo the tests several times. That would take at least a few weeks.

In the end, it took months. At the time, in addition to the hydrogen line experiment, Ewen worked with Norman Ramsey on an effort to get an external particle beam from the Harvard cyclotron. Although Ramsey strongly backed the Purcell-Ewen effort, Doc pressed to wrap up the beam work before Enrico Fermi finished a similar project at the University of Chicago, putting his thesis on the back burner. The decision was made easier when Purcell left that summer for a military consultancy that would keep him largely off campus for the next seven or eight months.

It wasn't until the fall that Ewen convinced himself the equipment was fine and realized the June data might suggest a different kind of mousetrap. Specifically, he concluded that the radiometer's two observing frequency bands — just 10 kilohertz apart — must be too close together. In other words, the hydrogen line signal was broad enough to cover them both at the same time. This would explain the spectacles-like recorder image.

Ewen tried a slightly wider spread of 25 kilohertz, but failed again to detect a signal. He explained his frustration to Purcell during one of his advisor's brief returns to Harvard, concluding that he needed $300 to buy another receiver core capable of separating the two observing frequencies by at least 75 kilohertz. That way he could minimize the possibility of the hydrogen signal appearing simultaneously in both bands. If he could not get the money for a better receiver, Ewen asserted, he would write up a "negative results" thesis that would spell out the precise level of sensitivity at which the signal could not be detected. While a far cry from finding the hydrogen line, establishing such parameters was worthwhile scientifically by delineating a new starting point for the next person to take up the hydrogen hunt.

Purcell told Ewen he wanted to think things over and advised the student to come back the next day. When Doc returned, Purcell said: "Tell me again about this." Once more Ewen explained the situation. When he had finished, Purcell pulled out his wallet, and without saying where it came from peeled off $300. "Well, go for it."

On the heels of this exchange, early in 1951, van de Hulst arrived in Cambridge as a visiting professor at Harvard College Observatory. The Dutch effort to detect the hydrogen line had been stymied by a lack of technical know-how and a serious fire that destroyed vital pieces of equipment. However, a talented new engineer, C. Alexander Muller, had been hired, and the project was finally picking up speed. Ewen, though, had not met the Dutchman and remained ignorant of the rival effort in Holland. Still, the race was on.

Ewen had finally succeeded in coaxing an external proton beam from the Harvard cyclotron sometime in late February 1951. As Doc shifted his attention back to the radiometer, R. V. Pound, who with his own unrelated experiment in the same frequency band had proven a steady source of good advice, left for an Oxford sabbatical. Meanwhile, Purcell returned to Harvard from his military consulting job: Ewen remembered his just-arrived mentor throwing snowballs down the horn antenna until he put a flap over the thing. By this time, few in the physics department paid the twenty-one-centimeter line project much attention. Even Ewen considered it a ho-hum affair with little likelihood of success. To add to the air of ambivalency, the graduate student decided to finally take up his planned career in law and business and enrolled in the Harvard Business School for the coming fall term. At the same time, the Navy called him to reserve duty as part of the Korean War buildup. He was due to report to Los Alamos in April 1951 to join the ongoing nuclear bomb project. If Doc was to detect the hydrogen line and earn his doctorate, it would have to be soon.

He returned to the fourth floor of Lyman the night of March 23, a Friday night on Easter weekend. The big horn was trained south, so that it peered in the direction of Aquila and Ophiucus when the galactic plane crossed in front. Ewen turned on the system around midnight. To stabilize his radiometer's front end — including the first local oscillator, mixer, and battery-powered intermediate frequency amplifer, which was shock-mounted on wads of cotton to minimize vibration-induced noise — he had conspired with the janitor to keep the heat turned up to around eighty-five degrees. He took a seat on a comfortable old couch dragged up to the office, clad only in his Skivvies — no shirt or pants — listening for the heterodyne beat of the oscillator as it tuned itself: he could feel things running properly. On the first try the tracing from the Esterline-Angus recorder showed an abrupt rise near the end of his scan. Over the next two hours Ewen saw some of the S-shaped curve he sought, but again couldn't be absolutely sure he was seeing the hydrogen resonance because the signal was coming at a lower-than-expected frequency near the end of the tuning range. Ewen worked through most of the night to try and retune the receiver to center on the signal, around 1420.6 megahertz, but before he could complete the adjustments the Earth had rotated enough that his antenna no longer pointed at the Milky Way core. The effect did not reappear.

The next night, approximately between midnight and two in the morning on Easter Sunday, as the plane of the Milky Way rotated overhead and with his receiver focused on the lower frequency, Ewen finally saw the full S-curve. At first he was puzzled about why the value rested so far off the laboratory figure. Then he grew excited, concluding the effect must be real, with the frequency variance due to the Doppler shift as the Earth moved away from the hydrogen cloud. "It's just the way you designed it, it's just the way you thought about it, the chill goes up your back and you say, 'I got it,'" Ewen once remembered. To see this special line, on the surface like a million other spectrographic squiggles produced on Earth, was like being allowed into God's inner circle. It had come to Lyman Lab from the nether space between the stars, riding some invisible crest so far across the light-years, a million millionth of a single watt, to reveal a glimpse of the handiwork of creation.

Ewen was still reluctant to approach Purcell until he had more definite data. He came back to the lab for another night or two of checking. At last he was ready to bring in his mentor. As Purcell remembered events, he got an early call. It was Doc: "I think I have a thesis." When Purcell got over to Lyman, Ewen was waiting outside his advisor's second floor office. The excited graduate student unfurled his recording paper down the hall. The roll hit Purcell in the foot. The paper showed a slightly wiggly line that might or might not contain the faint, telltale jumps they sought. "And we rolled out about twenty feet of it and got down inside it, along it . . . and then we could see this bump like that," Purcell related in a videotaped lecture years later.

Purcell felt little doubt now that the recorder had captured the hydrogen line. But he grew instantly cautious. "One of the things that goes through your head is what an ass I'll be if this isn't right," he remembered. He led Ewen into his office and telephoned Jerry Wiesner at MIT's Research Lab of Electronics. "What are you guys doing over there on 1420?" Purcell asked. The answer came back: nothing. Purcell shot off a few more questions. Finally, satisfied that no easily explainable outside effect had interfered with Doc's experiment, he called in van de Hulst and Frank Kerr, visiting Harvard from Australia's Radiophysics Lab on a Fulbright grant. Purcell asked the pair to wire home and get people working to verify the results. Meanwhile, he and Ewen planned a further few weeks of observations to clinch their case.

Using very similar equipment, both groups soon flew into high gear. From Holland, because they had been working on the problem already, Muller and Oort reported their first confirming signals just six weeks later; the gracious Purcell insisted that publication of his own results in *Nature* be held up until Leiden was ready. In Australia, no such project had previously existed. About two years earlier Taffy Bowen had been urged by I. I. Rabi and Zacharias to look for spectral lines at radio frequencies, but he and Joe Pawsey had not known about van de Hulst's prediction and never looked in the right place. Upon receiving word from Kerr of the Harvard detection, two separate teams, led by Chris Christiansen and James V. Hindman, launched all-out efforts to verify the results. The investigators soon combined their projects, and their July 12 cable reporting a successful confirmation arrived in Britain just in time to make it into *Nature* alongside the two more complete reports.

The papers immediately and dramatically enhanced the picture of the universe. Since its first detection the Harvard group had scouted around the sky for more signals as the astral geometry changed, putting together a hydrogen-based picture of the galaxy that showed the fundamental element was concentrated near the galactic plane in the constellations of Aquila, Scutum, and Ophiuchus. But Purcell and Ewen saw only the tip of the iceberg. Their radiometer's frequency switching interval of 75 kilohertz had caught some of the emissions, but was too small to see the edges of hyperfine emissions. Besides, the rigid horn antenna could scan only the portion of the sky exposed to it by the Earth's rotation. Better data from Leiden, which employed a parabolic antenna that could be positioned in any direction above the horizon, pinpointed the bulk of the hydrogen in a thin plane within 1200 light-years of Earth.

Purcell always considered the detection of the hyperfine transition more important to astronomy than his nuclear magnetic resonance experiment was to physics. A whole new universe awaited, for the moment surpassing anything optical astronomy had to offer. The makeup and kinematics of the local galaxy were now within reach. Dust clouds that played havoc with optical signals in the galactic plane no longer posed a problem. Observers could calculate the neutral hydrogen's mass by changes in intensity of the twenty-one-centimeter line. More important, Doppler shifts in the line provided astronomers with a ringside seat on the gas's wanderings: clouds moving toward Earth showed a higher frequency signal, while those speeding away shifted to a longer wavelength.

Life did run a story, in its February 1952 issue. But the scientific euphoria remained long after the public excitement died down. In Manchester, Bernard Lovell planned to construct a giant steerable dish 250 feet in diameter to replace his older fixed antenna—the first radio equivalent to the 200-inch Hale telescope on Mount Palomar in California. The parabola had been intended for studies at wavelengths of one meter and longer. But when news of the Purcell-Ewen discovery reached Jodrell Bank, Lovell redesigned the reflector to reach into the microwave region. Meanwhile, joint teams from Leiden and Sydney began mapping the pattern of hydrogen emissions in the Northern and Southern Hemispheres, revealing for the first time the galaxy's long, spiral shape.

Over the next two decades, as radio astronomy flourished in Britain, Australia, and the United States, something like forty other molecular lines were detected. Besides helium, the second most-abundant element, scientists reported the stellar presence of oxygen-hydrogen molecules, carbon monoxide, and ammonia. Then came polyatomic molecules, among them formaldehyde, H_2CO; and methyl alcohol, CH_3OH. The existence of such complex organic structures in space ran contrary to prevailing theories, sparking fresh ideas about the fabric of the stars.

Almost immediately after the hydrogen line detection, scientists found better ways of detecting spectral lines using banks of filters instead of radiometers. Still, lurking quietly behind the scenes, Robert Dicke's invention played a large role in the radio astronomy revolution. After the twenty-one-centimeter line's detection, astronomers dived into the once-ignored higher frequencies, and nothing could beat the radiometer for detecting broadband noise sources. The instrument's sensitivity was so acute, wrote J. S. Hey in a classic book about the origins of radio astronomy, that "the method afterwards came to be almost universally applied in one form or another in making radio astronomical measurements."

The device was there to help open the world to quasars, pulsars, and a host of revelations about the cosmos. And, as the twenty-first century approached, the COBE satellite that performed extremely accurate measurements of the cosmic background radiation relied on the latest generation Dicke radiometer, one improved to the point that it could detect temperatures within tens of microkelvins. In the modern incarnation, rather than physically inserting a blackbody into the waveguide, a "true" switch alternatingly sends the antenna signal and a reference output to the mixers, which have themselves been miniaturized and built into a single device rather than two as Dicke originally had it. The bandwidth is far greater, thanks to modern intermediate frequency amplifiers. And the amplifier noise has been reduced over the years. All these changes, though, are evolutionary, as opposed to Dicke's revolutionary concept.

Edward Mills Purcell, whose own scientific career traced back to the seat of a pulley in an Illinois barn, retained a humble respect for the strange twist of fate that had begun in World War II with the water vapor mystery, and eventually played a dramatic role in nuclear magnetic resonance, the evolution of the radiometer, and modern-day radio astronomy—all milestones in which he was intimately involved. In a 1986 Harvard University lecture commemorating the influence of radar on physics, he looked back on the thwarted Rad Lab effort to develop one-centimeter radars and marveled over "that measly little line" that had upset the applecart and at the time seemed only to be ruining the lab's reputation.

"So hurray," Purcell said, "for the water vapor line . . ."

15 • The Magic Month

"Most important, and most difficult to create, is 'the will to think'—the theme that runs through 'the magic month. . . .'"

WILLIAM SHOCKLEY

"The initial impact of the public announcement of the transistor was disappointing to those of us who were most intimately involved."

WILLIAM SHOCKLEY

THE NEW YORK MEDIA piled into the small auditorium at 463 West Street on the western edge of Greenwich Village, scores of uncomfortable-looking scribes perched elbow to elbow in their theater seats. The press had arrived at the yellow brick headquarters building of Bell Telephone Laboratories expecting a dramatic announcement. Fittingly, recently installed Bell Labs research director Ralph Bown took the stage. At his feet rested a rectangular placard, perhaps four feet long by a foot high, that would mean more to history than to the reporters in the crowd: *"The* TRANSISTOR."

Clad in a bow tie and light-colored suit, Bown looked like an advertising pitchman. In a way, that's what he was. The lab's public relations staff was pulling out all the stops. Behind the research boss stood a blown-up schematic drawing. Two other, even more gigantic posters towered in the background and depicted the device's possible uses in telephones, radio, and television. The PR folks had rolled out a 100-to-1 scale model of the transistor on wheels, a futuristic-looking apparatus seven or eight feet tall, since the real thing stood less than an inch in height. The company bragged in its literature that more than a hundred could fit into the palm of a hand.

The date was June 30, 1948. Sandwiched in the midst of the radio revolution sweeping astronomy, Bell had organized a series of live demonstrations to showcase its own coup: astronomers looked to glory, AT&T to the bottom line—but science gained from both approaches. The transistor, with no glass envelope, grid, or cathode, was primarily a rival of the vacuum tube amplifiers that permeated the electronics of the day. To highlight the tiny workhorse's advantages over the well-established triode, engineers had arranged for Bown to talk through a telephone handset into individual receivers in the audience, rigging the apparatus so that the transistor could be inserted and removed to elucidate its amplifying prowess. Reporters were also invited to listen to a radio broadcast through a receiver whose vacuum tubes had been replaced entirely by transistors. Marvelously, the thing ran on batteries that consumed less than a tenth the energy needed to light a flashlight bulb. For visual edification, a cathode ray oscilloscope allowed attendees to see the changing wave forms, with the creation and without. A television set was turned on, and an attenuator adjusted to render the picture all but invisible. When the transistor was plugged in — voilà! The image glowed back to its original brightness.

It should have been a great day for Bell. But not all the media were terribly impressed. The next day *The New York Times* devoted only a tiny blurb to the demonstration in its radio column, noting in what amounted to a rewrite of the press release: "A device called a transistor, which has several applications in radio where a vacuum tube ordinarily is employed, was demonstrated for the first time yesterday at Bell Telephone Laboratories. . . ." The *New York Herald Tribune* ran an article that duly mentioned the minor revolution predicted by Bell. "However," it added, "aside from the fact that a transistor radio works instantly without waiting to warm up, company experts agreed that the spectacular aspects of the device are more technical than popular." The trade press paid more attention. *Electronics* magazine made the transistor the cover story of its September issue, predicting the device was "destined to have far-reaching effects on the technology of electrons and will undoubtedly replace conventional electron tubes in a wide range of applications." This, though, was the exception.

As it turned out, both views of the transistor — as consumer-unfriendly oddity and eventual revolution — were correct at that time. For several years, demonstrations of quick-starting transistor radios drew gasps from audiences accustomed to waiting for tubes to warm up. Yet early versions were limited in frequency and ability to handle power. Six years after the dramatic public announcement, when the first mass-produced commercial transistor radios appeared, their cost was a staggering $49.95 — nearly $300 in 1995 money — and the product hit the market with a resounding thud. It would take roughly a decade before transistors could compete with the triode on a cost and performance basis, and not until 1959 did such solid-state devices overtake the sales of receiving tubes. Still, unlike vacuum tubes, which routinely burned out because of thermionic emissions from their hot cathodes, the solid-state transistor dissipated no heat and contained no inherent mechanism that would wear out. Coupled with the incredible miniaturization and power savings the transistor enjoyed over its competitor, these features made clear to the engineering intelligentsia that Bell Labs had achieved something remarkable.

The transistor, the foundation for the coming solid-state era of portable radios, television, and computers, proved the ultimate technological payoff of the wartime microwave radar development. Its invention hinged on the marriage of several strands of research tracing back to before the war. But at its soul were the old-fashioned cat's whisker crystals transported to America in July 1941 by Norman Ramsey, after the showdown between Rad Lab and Telecommunications Research Establishment airborne interception radars. The fragile devices had largely sparked the U.S. conversion to crystal detectors, unleashing a fervid thrust of research into solid-state materials like silicon and germanium that would form the basis for the later Bell Labs innovation.

During the war Bell had played a key role in the country's microwave detector investigations. Even before the conflict ended, wagering heavily that something vitally important would emerge from these studies, then research boss Mervin J. Kelly, Bown's predecessor, had reorganized Bell's already eminent research division to concentrate on solid-state devices. In so doing, Kelly assembled perhaps the greatest corporate physics and engineering team in history. Drawing purposefully on the interdisciplinary success of war research, he put these scientific stars in touch with his metallurgists and chemists, weaving the loose threads that made up the transistor into a cohesive fabric with just the blend of talent and radar knowledge I. I. Rabi had feared Bell alone would possess when he conceived the Radiation Laboratory Series.

The Rad Lab's technical books had done their job spreading the microwave gospel. But Bell Labs still held many cards. In two short years the long-sought dream of developing a high-frequency amplifier was transported to the verge of reality, then catapulted over the edge in a stunning month of inspiration and invention that changed the face of the modern world.

•

On the eve of America's entry into World War II, the Bell Telephone Laboratories was well established among the global leaders in corporate science and technology. No organization could rival it in telephone engineering, and few in communications of any sort. The lab, formally established on New Year's Day 1925, employed one Nobel laureate, Clinton Davisson, who had discovered the wave nature of the electron. And its widely read *Bell System Technical Journal* had helped educate a generation of scientists and engineers at home and abroad. On October 3, 1940, after dropping off the cavity magnetron at West Street in anticipation of its first test on U.S. soil three days later, Taffy Bowen lunched in a Greenwich Village restaurant with the physicist and Bell journal writer K. K. Darrow. "He had been a revered figure from my youth for the wonderful summaries of various branches of physics which he wrote about so elegantly. . . ," Bowen related. "I cannot remember in detail what we talked about, but it was a very pleasant lunch and I felt a better man from just talking to such a person."

As Bowen and Darrow dined, two strands in the transistor tapestry were coming together at Bell. The first revolved around the lab's ongoing quest for an economical way to amplify long-distance telephone signals. The current choice was the grid triode conveyed to AT&T in 1912 — some six years after its invention — by Lee De Forest. But despite three decades of improvements and study, triodes continued to suffer from burnout and a glaring lack of energy efficiency. The second thread hinged on what had brought the Welshman to the company in the first place: microwave radar. Since early 1938, Bell had maintained a fledgling effort to build a naval fire-control radar that operated at a wavelength of about forty centimeters.

These widely separated areas of investigation found common ground in solid-state semiconductors like the silicon and galena cat's whisker crystals once common in radio sets. Semiconductors are a class of solid-state materials that lie between conductors, which offer little resistance to electric currents, and insulators such as glass that restrict flow. This middle road promised a high probability of controlling their ability to conduct electrical current without moving parts or the vacuums associated with electronic tubes, and over the century leading up to World War II scientists had learned enough about semiconductors to begin exploiting their uniqueness. In 1874, Ferdinand Braun demonstrated semiconductor rectification, the ability to conduct electricity far more readily in one direction than another. In this nonlinearity lay the key to their utility as radio detectors — by allowing the alternating current of a radio transmission to be transformed into an audible direct current — and, as it turned out, the electronics renaissance to come. By the early 1900s, investigators had conjured up rectification by placing a tungsten steel needle lightly against a semiconductor crystal. The needle, or cat's whisker, functioned as an antenna to pick up radio signals. A chunk of crystal was typically soldered into a cup or other receptacle, with the whisker adjusted to find the sweet spot — often called the hot spot — offering the best rectification. As current flowed magically from needle through crystal, many a budding scientist huddled in bed listening to radio transmissions through an earplug.

As far as Bell's hunt for what became the transistor went, the ball had been set rolling by Mervin Kelly in 1936, when the worst of the Great Depression appeared over and the company was free to resume hiring. The fundamental shortcoming of semiconductors was that, unlike the triode, they had no amplifying powers. Kelly didn't know whether that was a necessarily permanent condition, but he felt the materials needed more exploration, and like most Bell managers, he knew his science. Valedictorian of his Missouri high school class at age sixteen, he had studied under the renowned physicist Robert Millikan and earned a doctorate in physics at the University of Chicago. A tireless but often abrasive personality, Kelly was not one to shy away from a challenge, either. One of his nicknames was Iron Mike. His son-in-law dubbed him Hotspur, after the keenly ambitious nobleman from Shakespeare's *Henry IV*.

Convinced that a much deeper, fundamental understanding of the basic physics of solids was critical to Bell's long-term success, Kelly moved first to strengthen the company's theoretical foundation by bringing in the first of an emerging generation of scientists well versed in quantum theory. A top recruit was the theoretical physicist William Shockley, who had trained under John C. Slater at MIT, one of the country's prime breeding grounds for solid-state specialists. Although assigned to Davisson's auspices, Shockley was almost immediately put on loan to the vacuum tube department to learn the core of Bell's business. After a short time there, he was allowed to switch his emphasis to solid-state matters and joined a group set up under the benign auspices of Harvey Fletcher, Bell's director of Physical Research. There, he came in contact with experienced veterans like Bell physicist Foster Nix, as well as other oustanding fresh recruits, among them Dean Wooldridge from the California Institute of Technology, later the "W" in the military and electronics conglomerate TRW.

It was an exciting time at Bell. The group, loosely charged with investigating basic areas of "electronic conduction in solids," was handed an unprecedented amount of freedom. As Nix recalled, "When Kelly created this little group of independent people—there were Shockley and I and Wooldridge under Fletcher—we were told, 'You do whatever you please; anything you want to do is all right with me.'" The three men helped kick off a small, informal study group that included several other Bell legends-in-the-making: Walter Brattain, Charles Townes, James Fisk, and Addison White. Over company-supplied tea and cookies, the high-powered body met one afternoon a week for more than four years to discuss developments in solid-state physics.

Buoyed by the stimulating sessions, Fletcher's solid-state group labored to puzzle out the still-abundant mysteries of semiconductor behavior in quantum terms as a prerequisite for coaxing the materials beyond rectification to amplification. Shockley took particular interest in Brattain's ongoing work with copper oxide semiconductors. Early in the recruit's Bell tenure, Mervin Kelly had mentioned looking forward to the day when the metal contacts used to make connections in the telephone exchange would be replaced by electronic devices. In the last days of 1939, inspired by that conversation and drawing largely on the latest copper oxide studies, Shockley took his first stab at designing a semiconductor amplifier that might fulfill Kelly's wishes with solid-state devices. Heading into the New Year, he carried out a few unsuccessful experiments. But the pace of the work proved so slow, the urgency around it so minimal, that it took two months before the disclosure was even witnessed. By that time, late in February 1940, the next major piece of the puzzle arrived from the radar side.

In contrast to the work of Shockley and the rest of the Fletcher group, Bell's radar team eschewed theoretical studies and dealt only with hard-wired reality. In the long road to the transistor, though, its investigations would have the dual effect of adding to the theoretical body of knowledge while also guiding researchers away from copper oxides toward ultimately more feasible materials like silicon.

The small-scale radar project had started in late 1937, when reseachers at Bell's radio laboratory in Holmdel, New Jersey, had been asked by the Naval Research Laboratory to help design a shipborne fire-control radar on what was then the extremely short wavelength of forty centimeters. As part of its long-standing interest in communications, Bell had vigorously investigated higher-frequency radio systems. As a consequence, the company possessed a wide range of knowledge in such areas as centimetric waveguides, antennas, transmitters, and converters that the Navy lacked.

In early 1938, agreeing to the Navy request but deciding to pay its own costs in order to safeguard the rights to any resulting technology, Bell broke off a small engineering team and assigned it to the Whippany Radio Laboratory, another New Jersey facility about thirty miles west of Manhattan. In July 1939 the group would demonstrate its first crude radar from an eighty-foot bluff overlooking New York harbor. However, while work was under way, members appeared to have been in regular touch with their colleagues at Holmdel, where it was realized that the triodes then used as radio detectors probably would not be nearly as good at picking up high-frequency waves as they were at finding AM radio signals. The basis for this realization could be found in the chief limiting factor of any receiver — the noise inherent in the circuitry itself. Such interference, which increases with the thermal temperature of various components, is present everywhere a current flows. Because crystals are solid-state devices, and relatively cold compared with a vacuum tube's cathode, they offered a way to reduce noise. Moreover, since crystals depended on a steel whisker only a few thousandths of an inch in diameter, electrons passing through the contact junction traveled a much shorter distance than in tubes, reducing capacitance and increasing the frequency of signals the device could detect.

By 1939, both the Whippany radar team and the Holmdel group were engaged in crystal investigations. George C. Southworth, the radio astronomy pioneer, seems to have been a member of the Holmdel camp. He rummaged around the dusty offerings of the famous secondhand radio market on Cortlandt Alley, just off Canal Street and the confluence of Tribeca and Soho, for the supplies needed to construct high-frequency detectors. Initial tests confirmed that common galena crystals did seem to make better microwave detectors than vacuum tubes. However, their nonuniformity, expressed in the need to hunt for a hot spot, prompted Southworth to goad Bell electrochemist Russell Ohl into studying other semiconductor materials. It was Ohl's investigations that bridged the gap between the radar team and the solid-state theorists.

Ohl was a native of Macungie, Pennsylvania, who had whiled away his youth pestering the local watchmaker, blacksmith, furnace caretaker, and electrician with technical questions. From boyhood he had the feeling that he was destined to accomplish something technically outstanding, and he claimed to experience visions of the future and sudden insights that helped in his endeavors. Whole circuits would flash periodically into his mind. And Ohl felt guided by an unknown force into radio work, where it was revealed that his mission in life was to push the shortwave radio spectrum to higher frequencies.

Whether such powers were real or imaginary, Ohl was a talented investigator with extensive experience in semiconductor crystals. Working on high-frequency communications in the early 1930s, long before Southworth approached him, he had concluded that elements from the fourth column of the periodic table made the best shortwave detectors. Nevertheless, to help answer the Holmdel man's queries, he laboriously studied some hundred prospects, including molybdenite, zincite, galena, and iron pyrites, before settling on silicon as the best bet, although, like galena, it behaved erratically. Sometimes his samples rectified signals in one direction. Sometimes the silicon worked its magic in the opposite direction. Occasionally, nothing happened at all.

Commercially available silicon at the time was not better than 99 percent pure, leaving plenty of room for contamination. Trying to increase consistency, Ohl asked two Bell metallurgists, Jack Scaff and Henry Theuerer, to further purify samples, a process that involved melting ingots at high vacuum in special furnaces, then slowly freezing them as they were removed from the heat.

Lacking a diamond saw, Ohl sent the samples to a jewelry outfit, where they were cut and ground into thin cylinders about an eighth of an inch wide and an inch long. At first, the purified product showed the same basic traits as lower-quality silicon. When a cat's whisker wire was placed in contact with the crystal, occasionally the ingots rectified in one direction, occasionally in the other. Then, in September 1939, Scaff and Theuerer accidentally produced one sample in which the top portion, the first to solidify, rectified in the opposite direction as the center area.

Ohl apparently didn't get around to examining the bizarre sample, a chunk of silicon as black as coal, until early in 1940. After a series of tests proved inconclusive, he finally held it up to an ordinary incandescent light and noticed that an unusually large voltage developed across the junction between the two regions. Intrigued, Ohl next corralled a neon laboratory lamp whose light passed through a chopper and shone that on the silicon. Voltage changes followed the chop in the light, proving that the sample was extraordinarily photosensitive.

Ohl noticed the effect, which later proved pivotal not only in the transistor but in producing solar cells, on a Friday afternoon. The following Monday, he began showing his sample up the chain of command. Each step of the way, no one knew exactly what to make of it, so the enigmatic silicon was passed along until it reached Kelly. Ohl marched into the research boss's office and showed him the photovoltaic effect by shining a flashlight on the sample. An impressed Kelly immediately called in a handful of other researchers to discuss the meaning of the phenomenon. Among those joining the huddle were semiconductor expert Walter Brattain and his supervisor, Joseph Becker, the witness to William Shockley's solid-state amplifier disclosure. Recalled Brattain, "we were completely flabbergasted at Ohl's demonstration. The effect was apparently at least two orders of magnitude greater in room light than anything we'd ever seen . . . I even thought my leg maybe was being pulled — but later on Ohl gave me that piece or another piece cut out of the same chunk, so I was able to investigate it in my own laboratory."

Silicon's increasingly curious properties, rectifying this way and that, or across the newly discovered junction, hinted at an important role for the material if only its behavior could be understood and controlled. Down the road, Ohl and Scaff would coin the terminology for what they had discovered, calling one type of rectification "n," for negative, the other "p," for positive, based on the polarities that developed. In the meantime, while preparing samples one day, Theuerer detected the odor of acetylene, a smell he associated with old carbide automobile headlights contaminated with phosphorus. This caused him to suspect minute traces of that element might be present in some ingots. Further experiments indeed showed the so-called p-n junction arose from the segregation of impurities during freezing, although the exact concentrations of the contaminants remained too low to be picked up by spectroscopic analysis. Based on these findings, Theuerer and Scaff theorized that p-type conductivity resulted from traces of elements like boron from the third column of the periodic table. N-type rectification, the men argued, came from elements such as phosphorus that occupied the fifth column. Such reasoning did not fit people's intuitions. When Scaff put the idea to Walter Brattain, the physicist told him he was crazy. "Well, I may be crazy," Scaff huffed back, "but that's what I think it is."

Despite such disagreements the work with silicon was drawing the theorists and experimentalists, semiconductor experts, metallurgists, and chemists closer together. On February 29, 1940, in the midst of all these deliberations, Brattain joined William Shockley in a revised disclosure of the latter's copper-oxide amplifier idea and the two began lab tests. The pace of the research remained leisurely. Moreover, Shockley writes, "The early results that Brattain and I obtained in experiments related to these disclosures were not encouraging." Either in the course of these investigations or their normal interactions, the men would have discussed developments on the p-n junction front. From there it seemed only a short step to actually doing something about it. However, no action was taken. Instead, Taffy Bowen arrived with the magnetron, the Rad Lab formed, and Bell took on more pressing activities. As Shockley continues, "Shortly after, we quite willingly responded to calls to apply ourselves to different areas of research and development related to America's entry into World War II."

The approaching war broke up any union about to take place at Bell Labs in the hunt for the solid-state amplifier. Ohl, Scaff, and probably Theuerer were assigned to a greatly expanded radar detector program. Dean Wooldridge and Charlie Townes got shunted over to Bell's efforts to develop a radar bombing set. Walter Brattain and William Shockley soon left the Labs altogether, the former to work for the National Defense Research Committee on magnetic airborne detectors, his partner for the Navy's operations research group.

At the same time, though, the rise of microwave radar focused unprecedented attention on semiconductor crystals for use in receivers. Since these would be going into combat, samples suddenly had to be produced by the millions in far higher quality than previously available. Rectification needed to be improved, vulnerability to shock and burnout reduced, and variations between batches of crystals minimized. Achieving these aims dictated an intense examination of semiconductor properties that extended far beyond anything one lab, even Bell's, could hope to achieve. As talented researchers from university, industrial, and government labs in both the United States and Britain got into the act, the unforeseen payoff was to provide the next great step in the forthcoming solid-state revolution.

One of Bell's researchers, the multifaceted J. R. Pierce, who wrote science fiction under the pseudonym J. J. Coupling and later carried to reality Arthur C. Clarke's 1945 musings of a communications satellite, put it succinctly: "Nature abhors a vacuum tube." So, increasingly, did engineers and physicists.

•

At the Air Ministry Research Establishment early in 1940, about the time Russell Ohl showed off his mystifying silicon crystal junction in Mervin Kelly's office, Denis Robinson browsed through the small technical library searching for a shortcut in his quest to find a suitable microwave radar detector. It was there his eyes fell on what was likely Hans Hollmann's *Physik und Technik der Ultrakurzwellen* (Physics and Technology of Ultra Short Waves) and its assertion that cat's whisker crystals constituted the best high-frequency detectors. Thus the British radar effort started fortuitously down the semiconductor path without the laborious trial-and-error studies under way at Bell Labs.

One of the main stumbling blocks to overcome in building a microwave radar receiver, as opposed to a simple detector, lay in the fact that faint signals had to be amplified for ultimate presentation on a cathode ray tube. While the triode could suitably enhance longwave radio broadcasts, no satisfactory microwave amplifier existed. The favored way to pick up centimeter-wave radar echoes was through a trick called the superheterodyne receiver, similar to those used in everything from Robert Dicke's radiometer to 1990s television sets. The incoming microwave signal arrived at the detector, where it had to be mixed, or combined, with a preselected transmission produced by a local oscillator, usually a klystron. The result was an intermediate frequency signal that could be passed through standard amplifiers. Also called the lower beat frequency, this was what, after being converted to a video signal, actually appeared on radar screens. Although detectors and mixers operated as two separate components of a receiver system, in radar the same entity — vacuum tube or crystal — performed both functions.

Robinson at once set out to build a centimeter radar detector and mixer system using crystals. To help with the project, his boss, W. B. Lewis, brought in Herbert W. B. Skinner of Bristol University, a physicist trained at the Cavendish Laboratory in Cambridge. Skinner, scion of a wealthy shoe company family, could have been a beatnik long before the beat era of the fifties. Starting with his self-cut ragged hair, he seemed to cultivate a disheveled appearance from top to bottom. "Everything about him was untidy," Robinson remembered. "His suit was unpressed and he had a filthy pipe, and he dropped tobacco all over himself and everything he was doing."

But Skinner knew the barrier layer, the layer of impedance that formed when the cat's whisker touched the crystal and in which lay another key to the later solid-state revolution. He demonstrated a hands-on, intimate knowledge — one of the byproducts of Rutherford's insistence that his boys do their own dirty work. After the British radar effort moved from Dundee to Worth Matravers that May and became the Telecommunications Research Establishment, Skinner would sit in his laboratory hour after hour making microwave detectors from silicon or galena. These he would seal, whiskers and all, in quartz glass tubes that he would blow himself, and then tap with a knife until the sensitive crystal settled into what was hopefully its optimum mechanical and electrical state. Skinner measured the resistance with a standard laboratory meter. When the value was five times greater in one direction than the other, he carefully put it aside and said: "That one's all right, nobody touch it." Similarly, Skinner wouldn't let anyone clean off his knife. "No, no, no, don't touch that knife, that's perfect — its momentum is just right." Meanwhile, the room filled with his nicotine.

By the time Taffy Bowen, Dale Corson, Norman Ramsey, and other Rad Lab emissaries arrived at TRE in mid-1941 to compare progress in early microwave radars, their hosts had made significant strides on a variety of interconnected fronts. Taking up a Skinner-inspired design from Marcus Oliphant's lab at the University of Birmingham, British Thomson-Houston engineered a more reliable version with a ceramic casing fitted with brass endcaps to secure the cat's whisker and crystal in place. Workers still tapped the assembly à la Skinner until the tungsten found a sweet spot and the desired resistance was reached. Then a wax was poured into the casing to hold things in place and reduce vulnerability to shock. Around the same time, an effort at Oxford's Clarendon Lab led to the improved transmit-receive box, based on the reflex klystron, while a quieter local oscillator, again employing the klystron, was produced by Admiralty researchers at Bristol University.

Carrying word of such advances to America, Ramsey at first had been greeted with skepticism, prompting his tour of key manufacturers and industrial labs. Only at Bell did he find much of an existing crystal mixer program, but this lingered more in the research stage, far behind the British practical end result. The young physicist had then gone back to the Rad Lab to find the samples he had so urgently shepherded across the Atlantic locked in Lee DuBridge's safe. About this time Dale Corson returned from England. Deciding their colleagues needed a push, the pair set up a rooftop lab demonstration to illustrate the benefits of the British crystals and TR box.

After that, men like Ramsey, a future Nobel laureate, and R. V. Pound, a world-renowned expert in microwave techniques, watched the spark provided by the British crystals and TR box turn into a flame. Stated Ramsey bluntly, "That started this country on condensed matter work — on solid-state electronics." By that fall, various types of crystal investigations were ongoing in the Rad Lab's rectifier group, a mixer development effort under Pound, Ed Purcell's Fundamental Developments, and in the Theory group, where researchers included Hans Bethe, recently arrived from Cornell. Bethe, a German immigrant who had received his security clearance the day after Pearl Harbor, worked by himself. He showed up a few hours after the 8 A.M. rush and stayed longer in the evenings to talk fundamentals with Julian Schwinger, a fellow future Nobelist famous for pioneering quantum electro-dynamics theory. Like a fish swimming upstream, Schwinger checked in as others left for the day, working through the night and arriving back at his rented house around five in the morning, just as his roommate, a farmboy named Carlson, woke up. Each day Carlson fixed breakfast for himself, dinner for Schwinger.

The Rad Lab soon became the focal point of the nationwide investigation, most of it centering on silicon and germanium, nature's two most-abundant semiconductors. To tap other talent around the country, the OSRD contracted for companion studies at the Purdue Research Foundation and the University of Pennsylvania. Purdue took germanium, while Penn concentrated mainly on silicon. Meanwhile, industrial work got started at E. I. du Pont de Nemours and Company, Westinghouse Research Laboratory, and other facilities. Through shared reports and personal visits, these groups stayed in constant touch with one another.

Bell Labs was a major part of the attack. Crystal work played just a small role in AT&T's overall war effort — perhaps twenty or so people at its height — but it covered a broad front that ran the gamut from engineering to applied science. In the latter half of 1941, possibly through Ramsey, Mervin Kelly's operation learned of British Thomson-Houston's improved rectifier cartridges and soon launched its own project. Researchers retained the idea of housing crystals in ceramic cartridges that could be plugged into receivers, but rearranged the positions of the silicon wafer and whisker to boost performance. By December, around the time of Pearl Harbor, Western Electric churned out the first commercial versions. The assembly was still hand-tapped, although research showed that rather than moving the whisker to a hot spot, tapping improved rectification by increasing the forward current and decreasing the reverse current.

Other Bell investigations, as well as studies at the Rad Lab and elsewhere, centered on purifying semiconductor samples, then "doping" them with minute traces of impurities to try to enhance conductivity. The Ohl-Scaff theory — that third column elements gave rise to p-type conductivity, and fifth column elements to n-type — fast gained acceptance and the mechanism behind it was largely understood in quantum terms. Third column substances such as boron carry one fewer electron in their uncompleted shell than silicon. Once injected into a silicon crystal, the boron atom can accept another electron from higher in the atomic lattice. This frees up the possibility of developing a current in the occupied bond, as atoms in successively higher positions in the lattice dispatch emissaries to fill the void in the one below. The end result is a vacancy or hole that gives the structure a positive charge and engenders a stronger current flow across the crystal. Conversely, a fifth column element such as phosphorus holds one more electron than silicon. Rather than taking up an electron, it donates one to the crystal lattice. This sets up a negative charge that increases conductivity in the opposite direction.

For the war work, remembered the Rad Lab solid-state physicist Henry Torrey, it didn't matter which way the current flowed; one could just flip the machinery around so that the signal moved in the desired direction. But techniques for purifying and treating silicon and germanium needed to be honed, and the most effective doping agents identified, both in type and quantity.

In the face of this broad-based assault, the body of semiconductor knowledge grew rapidly. Where germanium was concerned, the Purdue effort, led by the autocratic Viennese Karl Lark-Horovitz, put together an especially strong theoretical foundation, showing how to produce germanium in relatively pure form and demonstrating that electrical properties could be accurately predicted by impurity content. The small crystal team at the University of Pennsylvania, no more than eight scientists under the theoretical physicist Frederick Seitz, made similar contributions on the silicon front. Seitz, who would go on to a distinguished career as president of Rockefeller University and of the National Academy of Sciences, had taught for a few years at the University of Rochester under Lee DuBridge. Early in 1941, probably, before the Rad Lab temporarily abandoned crystal research for triodes, DuBridge had asked him to look for ways of improving silicon rectifiers. This had led to an important collaboration with du Pont, which Seitz had already helped find a better way to produce lead-free house paints. The result was a "five nine" radar detector — meaning the crystal's spectroscopic purity reached 99.999 percent — that became the U.S. standard.

Seitz left Penn at the end of 1942 for the Carnegie Institute of Technology. However, his old group had already demonstrated that crystals of such purity were required for optimum doping, and that the addition of small traces of boron dramatically aided conductivity. This fit well with a late 1943 report from Bell Labs' Theuerer, who found that adding boron in quantities on the order of a thousandth of a percent reduced burnout and resulted in dramatic advances in conductivity. Boron thus became a standard doping agent.

On the practical front, the reward of the combined American and British efforts could be seen in compact, rugged, and reliable crystal rectifiers that outclassed vacuum tubes in many applications, most often as microwave radar detectors and mixers. On a deeper level, the roughly four years of intensive wartime study rendered scientists and engineers intimately familiar with solid-state materials and the melting, etching, and doping techniques that formed the basis for the sophisticated clean rooms of modern semiconductor manufacturers. Detailed theoretical pictures arose to explain the interplay between metal contact and semiconductor, as well as the frantic dash of electrons across invisible crystalline pathways spiked with boron, aluminum, and beryllium. Like artists at one with oil and brush and thought and canvas, solid-state scientists began to develop a feel for their materials, and with that came the ability to bring control and order where neither had been readily apparent.

As their understanding grew, reseachers began wondering, too, what could be done beyond making better radar receivers. Crystal detectors formed solid-state counterparts to the vacuum tube diode invented in 1904. Many scientists drew the connection and took the next step. The diode had been turned into the triode amplifier with the implantation of a third electrode between the cathode and anode: a secondary current dispatched through the addition produced a strong electric field perpendicular to the main current, allowing the original signal to be modulated and amplified. It was only natural to wonder if a third electrode could be introduced between the cat's whisker and the semiconductor surface.

The trouble was the whole area, from tungsten point to crystal surface, ran barely ten thousandths of a centimeter thick. It was simply impossible to insert an electrode into such tight quarters. At the Rad Lab, Henry Torrey had spent much of his time studying the nature of crystal burnout, earning himself the nickname Crystal Crackin' Papa, an epithet inspired by the popular song "Pistol Packin' Mama." He good-humoredly remembered discussing the idea of creating a semiconductor amplifier with several colleagues, but then dismissing it out of hand: "We sort of laughed and said, 'Yeah, that would be great.' "

In other wartime semiconductor labs, many investigators reached the same conclusion. Bell Labs was the most notable exception. There, Mervin Kelly gazed into his crystal ball and saw a different future.

•

Iron Mike Kelly had been extremely busy throughout the war. Bell Laboratories had entered the conflict with roughly 2300 members in its technical staff, of which a tenth had been involved in earlier defense projects. By late 1943, the height of military demand for its services, the overall number of workers was about the same, but almost 90 percent of its scientists and engineers could be found on the defense side. Bell's war work ultimately spanned some 1500 projects, including atomic bomb research, secure communications, sonar, the bazooka, analog computers, homing torpedoes, and microwave radar. Only the Rad Lab did more on the radar development front; and no radar manufacturer surpassed Western Electric, which turned out nearly half of all sets used by the United States.

As research director, promoted in 1944 to executive vice president, and, effectively, the right hand to Bell president Oliver Buckley, Kelly had overseen much of the lab's sweeping activities. It was also his responsibility to position the labs to help AT&T benefit commercially in peacetime from the fruits of its wartime labors. As the fighting drew to a close, Kelly began to take firm action.

Some things must have come fairly easily. Bell almost alone retained intact the highly skilled engineers and scientists, the experience, and the right business to exploit the military work—especially when it came to microwave radar and high-frequency communications. Fittingly, it kept a radar corps that continued with postwar military development. At the same time, improved waveguides, aerials, crystal oscillators, and transmitters all impacted telephone technology, and they were applied with vigor. Not long after V-J Day, odd-looking towers began popping up at roughly twenty-five-mile intervals on hills around the country, heralding the nation's first extensive microwave telephone and television network: towers packed twelve channels, each capable of carrying 480 voice circuits or one TV program. Within a few years, as the Federal Communications Commission extended Ma Bell's monopoly on telephone service and ignored the cost benefits of competing plans from Philco, GE-IBM, Raytheon, and others, the system would become the dominant carrier of long-distance calls.

Few subjects, however, lay as near and dear to Kelly's heart as the solid state. Throughout much of the war he had visited Bell's crystal program once a week, continuing to emphasize fundamental research whenever possible. As the conflict wound down, he became even more convinced that Bell's vast experience in solid-state physics, with all the improvements in theoretical understanding and in the ability to manipulate and control semiconductors, held the key to getting a jump on postwar rivals. He couldn't see the final payoff clearly. But he sensed something was there, almost as if someone had laboriously laid out the basic ingredients of a mouth-watering new dish but omitted the recipe.

In July 1945, about a month before the Japanese surrender, Kelly acted on his feelings by creating a wholly revised framework for postwar research. With Buckley's blessing, he reorganized the Physical Research Department, clearly setting out solid-state research as the chief aim in the peacetime environment. Physical Research was divided into nine new departments. Three of these — Solid State Physics, Physical Electronics, and Electron Dynamics — were given the go-ahead to conduct basic solid-state materials studies. The authorization for work, signed by Kelly, laid out his specific hopes: It read in part:

> *Subject:* Solid State Physics — the fundamental investigation of conductors, semiconductors, dielectrics, insulators, piezoelectric and magnetic materials.
>
> *Statement:* Communications apparatus is dependent upon these materials for most of its functional properties. The research carried out under this case has as its purpose the obtaining of new knowledge that can be used in the development of completely new and improved components and apparatus elements of communications systems.
>
> We have carried on research in all of these areas in the past Large improvements in existing types of apparatus and completely new types have resulted. . . . The quantum physics approach to structure of matter has brought about greatly increased understanding of solid-state phenomena. The modern conception of the constitution of solids that has resulted indicates that there are great possibilities of producing new and useful properties by finding physical and chemical methods of controlling the arrangement and behavior of the atoms and electrons which compose solids.

The stage was set for the final, dedicated run to the transistor. Sometime near the spring of 1941, with Bell's West Street headquarters spilling out into neighboring buildings along Bethune, Bank, and Washington Streets, its staff besieged by the mounting New York City grime and noise, the labs had begun moving many research operations to a beautifully landscaped parcel of land in Murray Hill, New Jersey. The site, already home to a small scientific outpost, had long been intended as the setting for a campus-like research facility, and it was there, after V-J Day, that Mervin Kelly pulled together his vision.

Murray Hill, later to expand into a megacomplex, at the time consisted of just two significant structures — a small auditorium and acoustics facility, and the central research operation. The latter was a four-story structure shaped somewhat like an H, but with the sides mismatched in length, and the crossbar, which held the main entrance, high up toward the east end of the shorter leg. Physical Research, labs stocked with the latest equipment, covered several floors on the long leg of the H.

The core of the transistor work took place under the Solid State Physics group, which occupied a corner of the Physical Research space, just above an expansive employee lounge, and below that, the cafeteria. One co-supervisor was Stanley O. Morgan, a chemist of long-standing repute. His partner was William Shockley. After leaving the company in 1942, the keen-minded theoretical physicist had become research director of the Navy's operations research effort. Two years later, disgusted with the service's distrust of civilians, he had jumped ship and signed on as an Army Air Forces radar advisor under Ed Bowles. Late in the war Mervin Kelly wooed him back to Bell for a series of visits; and, well aware of Bell's plans for the future, Shockley had rejoined the labs' staffs immediately after the atomic bomb fell on Nagasaki.

The two co-supervisors seemed hand-picked to complement each other. Morgan was open and affable. Shockley, while brilliant, proved more mercurial. He was a talented amateur magician and inveterate practical joker, rumored to have rewired MIT elevators during his student days so that pushing a button for one floor delivered riders to another. His secretary, Bette MacEvoy, once opened a desk drawer at his request and shrieked at finding realistic-looking plastic dog feces. Shockley could also be exceedingly friendly, for a boss, inviting staff members rock climbing and to barbecues. Against that, however, loomed a dictatorial and abrasive side: in later years he tape-recorded all telephone conversations, screened workers for psychological problems, and amassed evidence he claimed showed that blacks were not as adept as whites at certain mental tasks. At Bell, anyway, it took Morgan to round off his rough edges.

The organizational chart released by Kelly in July 1945 divided Physical Research into four subgroups covering subjects as diverse as magnetism, physical chemistry, crystal oscillators, and the propagation of sound in solids. Each of these groups was sketched out as a multidisciplinary enterprise, able to draw on a wide range of expertise and thereby attack problems from a variety of viewpoints. Kelly had learned the strength of such teams by following the successes at Los Alamos and the Rad Lab, and he made no bones about it.

A fifth subgroup took shape later that year and appeared on organizational charts in January 1946. Small, packed with talent, and devoted exclusively to semiconductors, the upstart endeavor was entirely Shockley's to run. One of his top recruits was Walter Brattain, his old partner from the early days of trying to find a semiconductor amplifier. Other staffers included Gerald Pearson, a Bell experimentalist with years of semiconductor experience; physical chemist Robert Gibney; electronics expert Hilbert Moore, and two technical assistants, Thomas Griffith and Philip Foy. There was one fresh face — John Bardeen, a theoretical physicist from Princeton. Shockley had known the new man since his graduate student days at MIT, when Bardeen had been at Harvard on a postdoctoral fellowship.

For the work they were about to undertake, Bardeen and Brattain would share with Shockley the 1956 Nobel Prize. However, the two had much more in common with each other than with their boss. Both were products of Big Ten universities, and both had learned the basics of quantum theory from John H. Van Vleck, who had already emerged as one of the leading American-bred theorists of the times.

Bardeen was a native of Madison, Wisconsin, a former child prodigy who had skipped from third to seventh grade. He had come across Van Vleck's path in the late 1920s at his hometown university, where he swam on the varsity team and paid school fees out of poker winnings while pursuing undergraduate and master's degrees in electrical engineering. In 1930 he landed a job as a geophysicist with Gulf Research Laboratories but soon decided to try for his doctorate. He applied only to Princeton, attracted by Albert Einstein and the newly created Institute for Advanced Study, and once noted that "if they had not let me in, I probably would not have left Gulf." As it was, he found himself in the thick of perhaps the country's most invigorating scientific community: Princeton teemed with brilliant European émigrés escaping anti-Semitism. Einstein, John von Neumann, and Eugene Wigner were all on campus; and teachers and students congregated on given afternoons in the common room of Fine Hall for tea and shoptalk.

Young Bardeen, bespectacled and with a high forehead that accentuated his sharp intellect, became Wigner's second student — after Frederick Seitz — and began studying mathematical physics. In 1935 he moved to Harvard on a prestigious junior fellowship, where he was reunited with Van Vleck and crossed trails with Shockley. Afterwards, he taught physics at the University of Minnesota, then spent the war years with the Naval Ordnance Laboratory in Washington working on magnetic mines and torpedoes. No matter how far he advanced in his career, though, Bardeen remained a man of quiet and legendarily modest personality. In later years, after he had won a Nobel Prize for his theory of superconductivity and become the only person to win the physics prize twice, his longtime golf partner had buttonholed him: "Say, John, I've been meaning to ask you. Just what is it you do for a living?"

Walter Brattain was more outgoing and gruff than Bardeen but shared his basic modesty. Born in China to American missionaries, he grew up in Washington State, where his father ran a flour mill. In high school, he took apart a Dodge engine and put it back together, then matriculated to tiny Whitman College in Walla Walla, Washington, where his parents had studied. Brattain successfully completed his degree in physics and math, but did not go on to doctoral studies at the University of Minnesota until his professor assured him he had the makings of a better-than-average scientist.

In Minneapolis, Brattain learned quantum mechanics from Van Vleck, just before the theoretician moved to Wisconsin. In 1931, after a brief stint at the National Bureau of Standards, he joined Bell Labs, spending years studying copper oxide rectifiers and thinking fuzzily about making solid-state amplifiers. It was his recognized expertise in semiconductor materials that caused Mervin Kelly to call Brattain into his office to view the famous p-n junction early in 1940. Once the war started, Brattain took leave for a few years to develop magnetic detectors for hunting submarines. Itching to get back to research, he returned to Bell in 1943 after the U-boat's defeat and worked on infrared detection devices.

With the buttressing presence of Bardeen and Brattain, the group coming together under Shockley amounted to a solid-state hit squad. Close quarters and the boss's intensity built familiarity and kept things focused. Bardeen, the newest recruit, had to share an office with Pearson and Brattain, joined his suite-mates at golf, and was soon caught up in trying to explain their experimental data. The rest of the team would race down the hall between offices, or yell back and forth, "Come in here, I want you to see this." All met regularly for bull sessions in Shockley's office, where he kept a big blackboard: Bette MacEvoy could hear them in heated but friendly discussion, emphasizing points with chalk scribbles, and she never erased the board without permission. Walter Brattain confirmed the picture of conviviality: "I cannot overemphasize the rapport of this group. We would meet together to discuss important steps almost on the spur of the moment of an afternoon. We would discuss things freely. I think many of us had ideas in these discussion groups, one person's remarks suggesting an idea to another. We went to the heart of many things during the existence of this group, and always when we got to the place where something needed to be done, experimental or theoretical, there was never any question as to who was the appropriate man in the group to do it."

Later, Shockley's manner would drive a wedge between himself and other team members. But from those days, at least, colleagues remembered a keen sense of humor that could defuse tense situations. Bill Shockley was famous for whipping out strings of esoteric equations during the planning sessions. Brattain, always pacing and moving, was just as well known for an inability to sit still. One day Shockley turned and snapped at his colleague.

"Walter, I wish you'd quit jiggling those coins in your pocket. I can't think when you make money jingle."

"Look, I can't think when I don't have money jingling," Brattain replied.

"OK," said Shockley. "Will you please only jingle bills, after this?"

Shockley's first instinct was to tackle the solid-state amplifier problem head-on. Months before World War II ended, during one of his visits back to Bell Labs to discuss research with Mervin Kelly, the physicist had excitedly studied such a device built by Russell Ohl, who demonstrated that amplified radio broadcasts could be heard over a loudspeaker. Ohl's invention, depending on a so-called negative resistance effect, was far too unreliable for practical use. But it seems to have inspired Shockley. In April 1945 he sketched out what looked like a far better answer in his laboratory notebook. In Shockley's circuit, a thin layer of silicon and a metal sheet formed a parallel plate capacitor. His speculations suggested that charging the metal would produce a strong electric field perpendicular to the silicon slab, increasing the number of electrons on the semiconductor surface and modulating the current flow substantially enough to produce amplification. Completed in April 1945, this was, in fact, the first workable transistor design. However, a series of experiments, conducted by Brattain, Ohl, and others, failed to produce any observable modulation.

When John Bardeen came on board that October, he verified Shockley's formula. Tests, though, continued to flop. It wasn't until the following March that Bardeen finally broke the logjam by unveiling a theory that explained the mysterious failures. His basic concept was that various states on the semiconductor surface served as dead-end canyons that trapped electrons so their flow could not be modulated, essentially canceling out the benefits of adding the external field. It took until the 1950s for scientists to overcome this obstacle and make Shockley's original design work. Then called a field-effect transistor, since it depended upon the external electric field, it ultimately became the most common type of high-frequency amplifier.

In the meantime, Bardeen's theory of surface states led the group toward a different line of thinking. Both Bill Shockley and Igor Tamm in Russia had shown theoretically that different semiconductor surface states should exist, but no significant observable implications had been proposed. Now, thanks to the Princeton theorist, the real-world implications became

quite clear. Confidence swelling, the Bell team decided to temporarily abandon practical attempts to build a solid-state amplifier and try to shore up
their understanding of the basics. In so doing, they employed a research
strategy Bardeen had learned firsthand from Eugene Wigner back in graduate school. "Reduce a problem to its bare essentials, so that it contains only
as much physics as necessary." The bare essentials came from World War II
radar work.

The first step was a strategic assessment of all wartime semiconductor
progress. Shockley recalled that others at Bell chided them for forsaking
practical goals. However, he writes, "Our group was of one mind and we
followed the wise course of working, not upon such practical but messy
semiconductors as selenium, copper oxide and nickel oxide, but instead on
the best understood semiconductors of all — silicon and germanium." These
were the two most studied crystal rectifiers for microwave radar, where
extensive progress in purification and doping techniques had provided a
perfect foundation for the marriage of theory and experimentation. As
Shockley continues: "For these semiconductors, not all of the theoretical
concepts, developed largely during World War II, had been experimentally
verified; accordingly, we elected to concentrate upon the resulting gaps in
this branch of science, among them the recently proposed surface states. We
felt that it was better to understand these two simplest, elemental semiconductors in depth rather than to attempt to add piecemeal contributions to a
variety of other materials."

For the next twenty-odd months, until November 1947, the core team
of Bardeen, Brattain, and Shockley, with their able assistants, carried out
one experiment after another to test the surface-state theory. Although both
Brattain and Pearson had extensive prewar experience with rectifiers, none
of the team members had worked on semiconductors during the war, a fact
Bardeen saw as a tremendous strength. As he later noted in his Nobel
lecture, "We were able to take advantage of the important advances made in
that period in connection with the development of silicon and germanium
detectors, and at the same time have a fresh look at the problems."

•

The Magic Month, actually a five-week span that saw the birth of the
transistor and the genesis of two Nobel Prizes, opened on November 17,
1947. Walter Brattain had been pursuing the team's goals of building the
base of fundamental knowledge and testing the surface-state theory. To that
end, he probed what happened to Shockley's basic field-effect system when
it was doused in a bath of electrolytes and dielectric liquids: alcohol, toluene,
distilled water, and acetone.

To his surprise, when light was shone on the silicon surface, the photo-voltaic effect long associated with semiconductors actually increased. It quickly became evident that ions in the various liquids created an electric field strong enough to overcome any surface-state traps. Brattain showed his results to team physical chemist Gibney, who made the key suggestion that Brattain vary the voltage bias applied to the system and see what happened to the photovoltaic effect. Here came the first bit of magic: the phenomenon increased or was cut nearly to nil as the direct current potential was turned up and down to alter the number of electrons attracted to the surface. Brattain recorded the chain of events in a notebook entry dated the seventeenth: "In showing these effects to Gibney he suggested that I vary the D.C. bias on circuit while observing the light effect. It was found that with distilled H_2O alcohol and acetone that a plus potential as shown increased the effect and the opposite potential reduced the effect to almost zero."

Here at last was the way to overcome the surface-state traps that had blocked transistor development. With blinding speed, this critical observation transformed the solid-state group's emphasis back to practical goals, setting off a chain of events that created Shockley's "will to think," followed by "the will to think and to act." Within a few madhouse days, teeming with observations and insights, the men began imposing their will on the electron flow. On November 22, the Saturday before Thanksgiving, John Bardeen summarized much of the work while filling seven pages in his notebook. He concluded, ". . . these tests show definitely that it is possible to introduce an electrode or grid to control the flow of current in a semiconductor."

By early December, full success was staring them in the face. Bardeen had suggested switching from silicon to germanium, which was available in purer form, harbored better rectifying contacts than silicon, and was a great deal easier to do research on because of its relatively low melting point and related characteristics. Gibney doped a sample so that it was n-type — meaning that negatively charged electrons constituted the excess, or majority, carriers of electric current. On to this he evaporated a circular speck of gold with a tiny hole in its center. The gold served as the first electrode in their circuit. The basic idea was that a control voltage would cause it to emit electrons, which would be modulated through contact with the semiconductor surface and then conducted to an output by a tungsten wire threaded through the hole in the gold.

On December 11, while preparing his experiment, Walter Brattain accidentally ruined the hole by setting off a spark between the gold and the tungsten. Miffed by his own carelessness, he placed the wire's point alongside the gold instead of through it, leaving a tiny space so that the two did not touch. When he connected the system to a power supply, Brattain still found the substantial current modulation he hoped to create. However, both he and Bardeen were astonished to see that the current flowed the wrong way. That is, rather than decreasing surface conductivity as electrons were injected into the germanium via the gold contact, the experiment actually increased the current flow. More magic.

Both Brattain and Bardeen, the theorist, realized immediately something tremendously important, even revolutionary, had occurred. It took only a few days to figure out what. Modulation took place only when the gold was positively biased, in opposition to the n-type germanium. Therefore, rather than increasing the number of majority carriers by adding electrons, the gold contact effectively drew electrons out by injecting holes into the semiconductor matrix. The holes, or minority carriers, were then propagating from what the men labeled the gold emitter over to the tungsten plate collector.

The men now had in their sights a different form of transistor, a point-contact amplifier as opposed to a field-effect device. Until the Bardeen-Brattain experiment it was widely assumed that when holes were injected into n-type germanium, the surplus electrons already present would simply fill them and stop the current dead in its tracks. However, the Bell Labs trio saw plainly this was not the case. Instead, the resistance of the tungsten contact, so long as it was extremely close to the emitter, could be so lowered that it sucked up holes and carried the modulated current out.

As it stood, the initial configuration did not produce the power amplification needed in a workable amplifier. As a next step, Bardeen suggested constructing another version, with the emitter and collector extremely close together — on the order of less than two mils, or two thousandths of an inch, apart. With less distance for the holes to cover, he hoped amplification would be far greater.

Even this was a tall order. As Brattain once noted, the wires typically used for such point contacts spanned five mils in diameter: ". . . and how you get two points five mils in diameter sharpened symmetrically closer together than two mils without touching the points was a mental block." After some thought, he had an aide cut a piece of polystyrene into a triangle. Then he attached a strip of gold foil around the edges of one tip and carefully made a thin slit at the apex, turning what had been a solitary strand of gold into two barely separated slivers that could be placed against the germanium: "I found if I wiggled it just right so that I had contact with both points that I could make one point an emitter and the other point a collector . . ."

Brattain and Bardeen tried out the device on December 16. Almost upon plugging in their circuits they saw some significant amplification, recording an impressive voltage gain of fifteen times and power gain of 1.3 times. By the afternoon of Tuesday, December 23, the amplifier was working so efficiently they held a demonstration for Bell Labs research boss Ralph Bown and Harvey Fletcher, director of Physical Research. Mervin Kelly, who had set the whole chain of events in motion, was noticeably absent. As Brattain explained, Fletcher and Bown didn't want it going any higher. "It was so damned important that they were scared, if they told Kelly about it, that it might be a flop. And they didn't want to advertise it until they were convinced themselves that it wasn't a flop."

Electronics specialist Hilbert Moore helped configure the circuit for the demonstration, rigging things so that the signal could be followed audibly and on an oscilloscope. Exactly what followed has been lost, much to Shockley's regret: he viewed the event as in the tradition of Alexander Graham Bell's famous, "Mr. Watson, come here, I want you." As it was, on Christmas Eve, Brattain and Bardeen both recorded in their notebooks some vague details of the pioneering demonstration.

Writes Brattain, "The circuit was actually spoken over and by switching the device in and out a distinct gain in speech level could be heard and seen on the scope presentation with no noticeable change in quality."

Bardeen's account put the voltage gains at about one hundred, with power gains around a factor of forty. He conjectured that when the gold electrode was positive, holes were emitted into the sample, spreading out into the thin layer of p-type germanium. Those holes that came near the second contact, by then made of tungsten, were drawn into it.

The Bell hierarchy immediately labeled the discovery "laboratory confidential" while work proceeded on its perfection. Meanwhile, Brattain and Bardeen strove to concoct a fitting name for their device. Both were aware Lee De Forest's "audion" had gone the way of the dinosaur in favor of the more easily identifiable triode. So they wanted to stick with something simple. Various suggestions poured in, several ending in "itron," which they didn't like. Noted Brattain, "Bardeen and I were about at the end of our rope when one day J. R. Pierce walked by my office and I said to him, 'Pierce, come and sit down. You are just the man I want to see.' "

Brattain explained the dilemma, noting he preferred something that fit well with existing semiconductor devices such as the varistor and thermistor. Pierce considered, focusing on the most important property of vacuum tubes, the freshly hatched device's circuit soul mates. That was transconductance. Associating freely as he strove to find the solid-state equivalent, he threw out, "transresistance," and then uttered, "transistor." Brattain pounced. "Pierce, that is it!"

•   •

On New Year's Eve, William Shockley sat alone in Chicago, in limbo between two Windy City meetings so close together he decided not to return to New Jersey after the first. The discovery of the transistor had conjured up feelings of angst. "Frankly," he writes, "Bardeen and Brattain's point-contact transistor provoked conflicting emotions in me. My elation with the group's success was balanced by not being one of the inventors." Things must have grown tense between Shockley and his two colleagues because Walter Brattain later admitted the three had had a private chat sometime after the landmark invention. He kept details of the discussion tantalizingly vague, but Shockley's misgivings must have surfaced: "I told him, 'Oh hell, Shockley, there's enough glory in this for everybody.' "

But there wasn't. Spurred by his failure to be in on the discovery, Shockley developed an almost manic obsession with the transistor that dominated the next five years of his life and would alienate him from his colleagues. Alone in Chicago that night, he told himself that despite Bell's enthusiasm for the device, something better was needed before the transistor would play a major role in electronics. Eschewing any holiday revelry, Shockley sat down and tried to find it. Between that day and January 1, he filled nineteen pages of a notepad with scribblings and calculations. He mailed the pages back to Stanley Morgan, who witnessed them and asked Bardeen to do the same. Shockley would later have these pages rubber-cemented into his notebook.

Nothing palpable came out of those New Year's calculations. But, Shockley claimed, they built on his own previous attempts to create a transistor and prepared his mind for the ultimate breakthrough. As a result, on January 23, 1948, the hard-edged theorist in effect won his Nobel Prize. He had temporarily abandoned efforts to create a different type of transistor, and was musing instead about more basic experiments, when a hunch sprang into his consciousness. The essential idea was to sandwich an extremely thin region of n-type germanium, on the order of a thousandth of an inch thick, between two areas of p-type semiconductor.

This configuration created a pair of p-n junctions. On the p-side of each junction — as Russell Ohl had guessed before the war and radar-related crystal work subsequently confirmed — holes outnumber free electrons. On the n-side, the situation is reversed. By placing his junctions back to back, Shockley set up the condition whereby a voltage could be applied to one end, the emitter, and cause a torrent of holes to course across the n-type base area into the other p-region, or collector. This current can be modulated in the minority carrier region, just as the grid voltage modulates the current in a vacuum tube triode.

The junction transistor did not play a role in the extravaganza Bell Labs staged at 463 West Street on June 30, 1948. However, it was destined to surpass its progenitor in importance. The original point-contact transistor invented by Bardeen and Brattain proved a nightmare to manufacture. When Western Electric turned out the first ones in 1951, they were plagued by the same problems of unreliability and wide variance in performance that haunted early crystal detectors.

Meanwhile, Bell physical chemists Morgan Sparks and Gordon Teal led the efforts to produce a feasible junction transistor. By 1954, Teal had moved on to Texas Instruments and orchestrated production of the first commercial transistor out of silicon, which was ultimately cheaper and far more stable than germanium. In this form, the junction transistor, produced by Texas Instruments, Raytheon, and others, became synonymous with portable radios and the symbol of the solid-state revolution.

Long before the transistor gained a firm hold in commercial markets, the importance of the Bell Labs inventions, missed by the popular media in 1948, had become starkly evident. In November 1952, the prestigious *Proceedings of the Institute of Radio Engineers* devoted a special forty-eight-paper issue to the transistor. Various contributors depicted a future world where the devices controlled everything from the nation's telephone dialing system to the latest military systems. Four years later Brattain, Bardeen, and Shockley were jointly awarded the Nobel Prize in physics.

Russell Ohl did not take part in the celebrations. Some seventeen years earlier, around the time he discovered the p-n junction so vital to Shockley's version of the transistor, Ohl had seen in one of his visions that a solid-state amplifier was there to be found. Framed in the same burst of insight, he relates, "I got a flash that I was never going to invent the transistor, the amplifier. . . . It wasn't to be me."

Early on that December day in 1956 when the Nobel Prizes were granted to his old Bell Labs colleagues, another vision came to Ohl. "That morning before I woke up, the whole history of that flashed before my mind," he notes. In his dream-state, Ohl saw the Nobel committee deliberating, considering his own contributions, but deciding only three people could share the prize — and they were Bardeen, Brattain, and Shockley. Not long after he awoke, Bell co-worker Homer Dudley telephoned and asked if Ohl had heard the news about the prize. "Yes," Ohl replied, "I'd known about that."

16 • A Sense of God

"And so while one can wonder whether such a figure as God can in fact exist, we may sense him strongly—both at the moment and in reflecting on the events of a lifetime."

CHARLES TOWNES

CHARLES TOWNES AWOKE to a shimmering Washington spring morning. It was still early in the nation's capital, probably around 6:30. Surveying the small hotel room, he could see another figure slumbering contentedly in the twin bed next to his own. This was Arthur Schawlow, a fellow physicist and friend who also happened to be engaged to his sister. Not wishing to disturb his future in-law, but feeling unable to go back to sleep, Townes rose, dressed, and headed out for a walk.

As he left the hotel, the city had not yet come alive. Amidst only a few passing strangers and the pleasant good mornings of cavorting birds, Townes proceeded to nearby Franklin Park. He had always loved nature, and the tiny federal reserve three blocks northeast of the White House at 14th and I Streets acted as a magnet. The azaleas were in bloom, and he inhaled the fresh-scented air. He never forgot the beauty of that day. It was the last week of April in 1951, only a few weeks after Doc Ewen and Ed Purcell discovered the twenty-one-centimeter line of interstellar hydrogen and about the time the transistor made it to mass production. So, perhaps, it was a good time for science and technology.

Townes at this point was a few months shy of his thirty-sixth birthday and a full professor of physics at Columbia University. Several years earlier, in the wake of the Magic Month and well aware of the transistor, he had left Bell Labs and the tea klatsches with old colleagues Shockley, Brattain, and the rest to join I. I. Rabi on New York's Upper West Side. His specialty was microwave spectroscopy, the probing of gaseous molecules with radio waves to divine the basic properties of atoms and molecules; and he ranked as a world authority. He had come to Washington as chairman of a special committee that advised the Navy on ways to produce high-frequency energy that might eventually lead to secure communications systems or highly accurate short-range fire-control radars.

The committee, due to meet in a few hours, had spent the last eighteen months exploring ways to push the state of the art in millimeter waves, which were perhaps as much as ten times shorter than current technology conveniently allowed. But a witches' brew of physicists' tricks — things with mysterious names like magnetron harmonics and coherent Cerenkov radiation — had met with a brick wall. Shorter wavelengths, so far as science understood, started with smaller transmitters. And those that were small enough to churn out millimeter-wave signals simply could not produce the desired energies. It was a bit like the problem haunting radar developers seeking to delve into the microwave region at the onset of World War II, before the arrival of the cavity magnetron, except that unlike John Randall and Henry Boot, Townes and his group had no specific goal and could not even see any obvious payoffs for their efforts.

As Townes, a strapping young man a shade over six feet, took a seat on a bench perhaps fifty feet into the park, he mused over the reasons for failure. He saw nothing on the technological horizon, no breakthrough in miniaturization, that would provide a solution. He considered, not for the first time, that the best sources of the high, sharp frequencies the committee sought weren't man-made machines like the klystrons and magnetrons permeating his own lab, but molecules themselves — specifically, the telltale signals of the molecules' own oscillating motions along the quantum ladder. Looking at the problem from this fundamental viewpoint, he wished he could become some atomic ringmaster, snap a whip, and train molecules to spew the energies he needed.

Like all good physicists, Townes was fully aware of quantum theory. In one sense, to fulfill his aim he needed to take control in the laboratory of the flip side of the same basic phenomenon involved in Ed Purcell's pioneering magnetic resonance experiment. Purcell had measured the absorption of energy as protons spinning in a magnetic field moved from one energy level to a higher one. Townes was interested in the simultaneous release of energy — in the form of just the kind of sharp radio signal he needed — that took place as atoms or molecules hopscotched to a lower state. Hydrogen wouldn't do: the proton resonant frequency was too low. But many molecules emitted a much higher signal in the elusive millimeter range.

The overriding fundamental problem was embodied in the very same Boltzmann distribution that had made Purcell's experiment possible. It held, from the second law of thermodynamics, that a slightly greater number of molecules would settle in the lowest quantum state at thermal equilibrium; any disruption such as a radio signal that brought about the switch meant more leapt up than down. So while he would get emissions, they would be of less magnitude than what he put in initially. Net energy would be absorbed, not emitted. Everywhere Townes turned, there was the second law of thermodynamics staring him in the face.

And then the light came, bursting into his awareness, and forever linked to the sweet smell of azaleas. The young scientist grabbed an envelope and pen from his jacket pocket, and began furiously calculating energies and sketching the raw bones of the vision. It *was* possible, perhaps even simple: he didn't have to obey the second law of thermodynamics.

With this epiphany, nearly six years removed from World War II and still several years from reality, Charles Townes entered on the path to the maser and a share of the Nobel Prize. Albeit the product of a longer and more convoluted route, it marked the third great payoff — after nuclear magnetic resonance and the development of the radiometer — of the water vapor mystery that had haunted microwave radar developers during World War II.

•

Charles Hard Townes was not an ordinary physicist. A devout Christian born in the Baptist heartland of Greenville, South Carolina, and a descendant of a *Mayflower* pilgrim, he brought an unusual amount of religious faith to his work — a quality not exactly common in scientists. He would never link his Franklin Park brainstorm directly to divine inspiration: to say God gave him a vision seemed arrogant. But if one believed in God, He was everywhere, so everything that happened was in some sense through His inspiration. Who could say what was a vision and what wasn't?

Townes's unshakable belief in God had made science all the more appealing from the beginning. He saw in nature the workings of a creator and had grown up reverently collecting leaves, insects, rocks, and fish. Cars, by contrast, which fascinated other boys, held no appeal. "Some of my friends seemed to know every year model of every car," he once confided in his soft southern drawl. "That seemed to me so temporary and uninteresting. Nature is such a permanent aspect of our universe, and so obviously God-made." To him, science was an attempt to understand the universe and revel in its wonder, and theology had basically the same goal. No conflict existed between the two.

Bored with high school, Townes asked his parents for permission to skip a grade. As a result, in 1931, at age sixteen, the young scientist matriculated to tiny Furman University, an all-male Baptist school in his home city. Adept at languages, he studied Latin, Greek, Anglo-Saxon, German, and French, earning a bachelor of arts degree in modern languages after just three years. But it was a different tongue — physics — that captured his heart: Townes knew it with the first elementary course during his sophomore year. The teacher stressed the logic of the physical laws, and this appealed to the student's ordered mind. "The fact that what one needed to do was think things through carefully, and then you could really prove something," he marveled. "You could prove new things."

Furman lacked a strong physics program. But by the end of his fourth year, Townes had completed a bachelor of science degree in the subject by studying a basic text on his own, attacking every problem, and presenting the solutions to an advisor. He learned about electromagnetic theory from the *Encyclopædia Britannica* in the school library and picked up the basics of the emerging field of nuclear physics from a series in the *Bell System Technical Journal*. He found *that* in the public library.

Although the homespun coursework had not rendered him particularly adept in physics, in the 1930s even the best graduate schools accepted nearly every reasonably qualified applicant. The main problem Townes faced was money. Although his attorney father could probably have supported him, it never occurred to the youth to ask his parents for aid. When Cornell, Princeton, and other top eastern schools all rejected him for fellowships or work-study programs, he fell back on a teaching assistantship offered by Duke University.

At Duke, Townes continued to study languages — this time it was Russian — while completing a master's degree in physics in one year. Eager to earn a doctorate, he applied for the sole physics fellowship the department offered. At the time, the university's physics program was adequate, but hardly first tier. When the fellowship went to another student, he bolted. Again, the top East Coast schools declined to offer fellowships. So Townes decided to use the money he had saved and go to the best school he knew. That was the California Institute of Technology, home to an extraordinary community of scientific pioneers: Robert Millikan, Carl Anderson, J. Robert Oppenheimer, and Linus Pauling, who had deftly adapted quantum mechanics to probe the mysteries of the chemical bond. Townes worked all summer, increasing his savings to $500, packed a bag or two, and rode buses cross-country, stopping off to visit the Grand Canyon and other national parks. He arrived in Pasadena in the fall of 1936, just in time for the new term.

Sharing a sleeping porch with another student for $6 a month before moving into a graduate student dorm, Townes slipped easily into the relaxed but intensely intellectual life on the pleasant campus, with its spattering of cool stucco buildings capped with Spanish-styled roofs. Like a kid in a candy store, he studied chemistry under Pauling and took a modern physics course from Millikan. Midway through the first year, he received a teaching fellowship that solved his basic financial worries. But money remained tight, so Townes pressed ahead at what had become his typical breakneck pace. He chose a thesis, under William R. Smythe, that involved spectrographic analysis of stable isotopes of oxygen, nitrogen, and carbon. The work was done in three years, and Townes received his doctorate with the class of 1939.

By the time Townes graduated, the country had turned the corner on the Great Depression, but physics jobs remained scarce. Many of his colleagues found themselves forced to sign on as high school teachers or become oil field seismologists: the physicists joked that Ph.D. stood for Post Hole Digger. Townes received a rare offer to join Bell Labs as a researcher, but could not imagine working for an industrial concern when what he really wanted to do was study fundamental physics and teach at a university. He was about to turn down Bell when one of his professors, the optical spectroscopist Ira S. Bowen, jumped in and urged him to reconsider. After days of deliberation, Townes took the job.

The position paid $3016 a year, half again what many of Townes's friends were getting. Bell sent along $100 for a first-class train ticket to New York. But the young recruit considered that a ridiculous waste of money. Instead, he took a bus to Arizona and rode a third-class train into Mexico, visiting a friend and brushing up on his Spanish while vacationing in Guadalajara, Mexico City, and Acapulco. Finally, he bused back to his home in Greenville and on up to New York, all for around the same $100.

Arriving in Manhattan determined to make the best of things, Townes took rooms rented by the week in Greenwich Village, not far from the Bell Labs headquarters on West Street. The flexibility of the lease enabled him to initiate a pattern of moving every three months to a different part of the city so he could get to know the world's most vital hub. He would simply pack his few belongings and grab a cab to the new apartment. During his tenure on the Upper West Side, near Columbia University, he studied voice at the Juilliard School of Music. On a weekend ski trip, he met Frances Brown, who was on the activities staff of International House of New York and had arranged the trip. The couple would marry in spring 1941.

On the professional side, Bell had promised Townes a research assignment in fundamental physics, and everything had started off fine as well. Under a just-launched company program, he could choose his field after a series of rotations among various technical departments. He spent time studying microwave generation, vacuum tubes, and magnetics. Then, in April 1940, the hopeful scientist moved on to a more permanent stint investigating electron emissions under Dean Wooldridge, a Caltech grad who had preceded him in Smythe's lab.

Up to this point Townes could hardly complain about his lot. He still planned on returning to a university when the job climate improved. But in the meantime, Bell possessed a wealth of equipment and engineering know-how that even Caltech couldn't rival. Another plum was being asked to join the weekly seminar group where William Shockley, Walter Brattain, James Fisk, and others brainstormed about general physics issues while sipping tea and munching cookies. In short, Townes realized, his days at Bell were spent delving into matters not far removed from the fundamental problems he would have studied at a university anyway.

Suddenly, though, the comfortable existence came to a crashing halt. One afternoon in late February 1941, after less than a year under Wooldridge, Townes was summoned with his boss to the office of Mervin Kelly. "On Monday, you start working on the design of radar bombing systems," the research czar informed his charges. "The war is coming on, and that's your assignment."

Townes was almost bowled over. In the space of a weekend he was being transformed from a physicist into an engineer. What's more, the reassignment seemed totally arbitrary, and he thought about quitting. On the other hand, he told himself, war was coming, and everyone should pitch in. On the twenty-eighth, therefore, just a few months after the Radiation Laboratory formed, Townes begrudgingly became the corporate equivalent of the university physicists rushing to join Lee DuBridge and I. I. Rabi in Cambridge. He closed out his lab notebook with the notation: "This work discontinued for national defense job." Then he joined a microwave radar program up and running at Bell's site in Murray Hill, New Jersey. His first task, still under Wooldridge, who had also been reassigned, hinged on incorporating David "Parky" Parkinson's revolutionary predictor into a ten-centimeter bombing radar. Like his Rad Lab counterparts, Townes began to learn about analog computers, waveguides, antennas, and microwaves—all key building blocks for the vision he would have a decade later on the Franklin Park bench.

Overall, the bombing radar effort did not go satisfactorily. During the war Bell and its well-trained engineers typically developed proven sets for mass production, leaving it to the Rad Lab to carry novel microwave concepts through the early stages of their incarnation. The project Townes worked on was different. Wooldridge's group sought to design its own airborne radar from the beginning. Because the unit was to house the Parkinson predictor, it depended on a complex array of tubes, mechanical relays, and potentiometers that rendered it a far more ambitious undertaking than the more straightforward systems Lee DuBridge's outfit pursued. Consequently, though the work went fairly quickly, the Bell set was not what the military bought. A prototype was ready shortly after Pearl Harbor. However, by then competing systems were also available, so fairly early in 1942, Bell was asked to extend its efforts to three centimeters, the same as the Rad

Lab's H$_2$X set. The move necessitated a new start and slowed things down a bit. But the following year Townes and a few other staff members spent months in Florida, testing both their ten- and three-centimeter systems by dropping sand bombs on shipwrecks and deserted Caribbean islands. By spring 1944, the bugs had been worked out to the point the three-centimeter set seemed ready for production. Then, suddenly, the project was terminated, and Townes and his colleagues were ordered to move to the K-band, on 1.25 centimeters.

It was annoying enough to have worked so long and hard on a system that was scrapped abruptly. To make things worse, Bell management told the team to forget about the Parkinson predictor-based system and build a one-centimeter set that the Rad Lab had already brought to the prototype stage: the unit was designated the AN/APQ-34 high-altitude bombing radar. Somewhat frustrated, Townes considered signing on as a military advisor in China but was persuaded to stay by his boss, Walter A. MacNair, who assured him the bombing radar effort was useful. At this point, the entire project was uprooted to nearby Whippany. Townes, promoted to systems planner, plunged into the work, only to come head-to-head with the water-vapor absorption problem.

Bell Labs staff members working on microwave radar had access to many Rad Lab documents throughout the course of the war. Somehow, as Townes started on the K-band radar he read the report written some two years earlier by J. H. Van Vleck in which the theoretician examined a possible water-vapor absorption problem in the one-centimeter region. The paper did nothing to lift his black spirits. Townes was probably not aware that almost at that moment the performance of the Rad Lab's own experimental one-centimeter systems was degrading in the moist spring air. But he reasoned that the tide of the European war was already turning in the Allied favor, and that the humid Pacific would become the main battleground. After more careful study he concluded that water vapor absorption might render a K-band radar ineffective on the Japanese front, or anywhere much range was required. "I couldn't tell at the time that it would really ruin the radar," he recalled. "But it seemed to me very possible." He voiced his concerns to Bell Labs superiors, Pentagon officials, and I. I. Rabi overseeing the Rad Lab's one-centimeter program, arguing for a less drastic shift to something closer to two centimeters. No one heeded the junior scientist's warnings.

Townes gritted his teeth and kept at it. After undergoing several months of substantial redesign, a few APQ-34 test systems made it into Pacific service in the summer of 1945, suffering from water vapor absorption pretty much as Townes had feared. On Monday, August 6, the Bell scientist was working late alone in a small hut at Whippany when he heard over the radio that an extremely powerful new bomb had been dropped on Japan. He knew several people at Los Alamos and deduced instantly the type of bomb. "And I said, 'Well, I don't have to keep working today.' So I shut down and went home." From then on Townes worked leisurely on the bombing radar. But he didn't forget the water vapor mystery.

Even before the war ended Townes had spent a lot of time thinking about the water-vapor absorption problem, not in terms of what it would do *to* his radar, but what it meant *for* his future career. He saw in the phenomenon a way to transport himself to the frontiers of a novel form of spectroscopy, using high-frequency radiowaves to delicately examine the basic makeup of molecules.

Spectroscopy itself was a well-established science dating back at least to the eighteenth century, when the Scottish physicist Thomas Melvill set fire to a mixture of chemical salts and alcohol and examined the blazing light through a prism. Melvill noted that each chemical gave rise to its own telltale pattern of color. Since then, scientists had developed spectroscopes that projected slits of light through a prism onto ruled lengths of film, where it was easy to measure spacing and calculate wavelengths. Each element carried its own electromagnetic fingerprint through its line spectrum.

Although spectroscopy had started off using visible light for chemical analysis and other applications, by World War II it reached beyond the optical spectrum into the infrared and radio. With his pioneering molecular beam experiments of the 1930s, I. I. Rabi had driven the technique to the atomic nucleus through studies of the interaction of radio radiation with beams of atoms and molecules. It was magnetic resonance spectroscopy, measuring the spin rates of nuclei in magnetic fields, that Ed Purcell and Felix Bloch would extend to nuclei in bulk matter and higher frequencies a few months after the war. In the meantime, scientists working with visible light and in the infrared had been busy measuring the vibration and rotation of molecules caused by such factors as their angular momentum. These motions, though, showed up as extremely small energy differences at such frequencies and could not be gauged with high accuracy. As it turned out, a far better picture was available in the microwave and far infrared regions, which had not yet been explored.

Two central factors had worked to preclude such an extension of spectroscopy. First, few scientists even appreciated the payoffs of moving to higher frequencies. A second hurdle was the technology. Prior to World War II, most off-the-shelf electronics components were made for frequencies below 700 megahertz, well away from the microwave region—the next logical front to open. A few low-power exceptions could be found, and enterprising scientists could jury-rig other alternatives. But microwave spectroscopy had only been attempted at the University of Michigan in Ann Arbor. As early as 1933, graduate student Claud Cleeton's hand-built transmitter tubes had generated waves of under two centimeters into a cloth bag filled with ammonia gas. Working under Professor Neil Williams, Cleeton had measured the molecule's absorption patterns, providing proof that ammonia absorbed microwaves in a way predicted by another Michigan physicist, David Dennison. Still, even with this pioneering advance, the lack of available electronics helped hold the technique at bay until around the time the water-vapor absorption mystery reared its head.

As Townes considered the absorption problem in New Jersey, at first unaware that Robert Dicke and others were rushing to understand and characterize it but later following those developments with interest, he saw that the ill-fated K-band radars had provided a bounty of high-powered magnetrons, stable and precisely tunable klystron oscillators, sensitive receivers, waveguides, and other components for "a spectacular kind of spectroscopy." Since radars must cope with real-world conditions, Dicke's work, as well as more detailed studies at Columbia University, had been done at atmospheric pressures, where resonance lines were relatively broad and generally unsuitable for precise spectroscopy. But Townes remembered that Van Vleck and fellow physicist Victor Weisskopf had worked out a theory of the effect of a gas's pressure on spectral line widths, building on the idea to show that as pressure was lowered the line would narrow without changing in intensity. That meant that in a controlled vacuum, precise high-frequency beams injected into a gaseous molecule would give rise to extremely narrow and sharp spectral lines, making possible measurements far more accurate than anything in the past.

Sometime in the spring of 1945, Townes wrote a memo arguing that he should begin pursuing this line of study immediately after the war. "Microwave radio has now been extended to such short wavelengths that it overlaps a region rich in molecular resonances, where quantum mechanical theory and spectroscopic techniques can provide aids to radio engineering," he stressed. Among the possible payoffs Townes foresaw for AT&T were building sensitive radio detectors and helping set frequency standards.

Policy makers had ignored Townes's warnings about how water vapor absorption might disrupt radar. Had Bell turned down his spectroscopy proposal, the young scientist might have gone into radio astronomy: a notebook entry from this period showed that he had independently considered looking for the hyperfine line of hydrogen, found six years later by Ed Purcell and Doc Ewen. As it was, events conspired against this second course. First, his old Caltech professor Ira Bowen, about to become director of the Mount Wilson Observatory in California, discouraged him from pursuing radio astronomy, insisting it would never amount to much as a field of science. More importantly, Bell did not disregard the spectroscopy memo. Both the company and the Air Forces wanted Townes to wrap up loose ends on the K-band bombing radar and turn it over to someone else: most subsequent K-band radars were built on what was called K_u band around two centimeters in wavelength. But in January 1946, as soon as Townes finally freed himself of the whole dismal venture, he turned his attention to microwave spectroscopy. Bell blessed the endeavor by assigning him two technical assistants.

The result was that Townes happily resumed his research life at the forefront of a rising field of chemical and physical study. The vast spectroscopic potential provided by radar technology gave rise to competing endeavors at Oxford, Harvard, MIT, Westinghouse, and other university and industrial labs. For the most part, the work went in two directions: Chemists sought the details of molecular structures. Physicists, fascinated by the properties of nuclei and the principles involved in this new type of spectroscopy, observed small changes in molecular spectra produced by spinning nuclei. Townes himself spent the next eighteen months straddling the line between the two approaches. Exciting ammonia gas with precise, klystron-induced microwave signals, he embarked on a series of imaginative manipulations of the molecule, which can be induced to turn itself inside out like an umbrella caught in the wind as the troika of hydrogen atoms pivot past the solitary nitrogen atom. Townes soon moved up the food chain to other interesting molecules, working out theoretical problems with John Bardeen. In particular, by measuring the absorption frequencies of OCS, BrCN, CICN, and other molecules, he and various competitors derived crucial basic information about things near and dear to the hearts of physicists and chemists: dipole and bond distances, chemical bonding patterns, and nuclear properties such as masses and quadrupole moments. In this way, the new spectroscopists competed successfully with the molecular-beam magnetic resonance technique pioneered by Rabi, which also borrowed from microwave radar to reach into the higher frequencies. Summed up Townes, "It was a very, very fruitful, exciting period."

By 1947, Townes had established himself as one of the field's elite players. But in discussing some recent observations in a summer seminar at Brookhaven National Laboratory, he found some of his conclusions about molecular structure under attack from no less an eminence than Rabi, who had returned to Columbia University as physics department chairman and was overseeing an impressive body of radio spectroscopy studies that within a decade would lead to Nobel Prizes for Polykarp Kusch and Willis Lamb. The Bell Labs man felt Rabi did not really understand molecules well enough and had defended himself by coolly pressing the senior physicist for specific objections; none had come. Shortly afterwards, Townes had been caught off guard again when Rabi accompanied him to the beach and offered up an unsolicited job on Columbia's powerhouse faculty.

Rabi's timing couldn't have been better. Not only had Townes always wanted to return to university life, he had also failed to convince Bell management his spectroscopy work was important enough to warrant additional researchers. The head of the physics department had even admonished him for persisting with requests for a bigger budget: "You've made a lot of people annoyed because you are talking about what you would like to do. You ought to be talking about what is good for the company."

A few days later, after talking with his wife and thinking the matter over in more detail, Townes accepted Rabi's offer. On January 1, 1948, he moved to Pupin Hall, on 120th Street between Broadway and Amsterdam, which had housed the old Columbia Radiation Laboratory and still played host to the school's physics department. Immediately, the new associate professor encountered a congenial atmosphere where radio frequency spectroscopy flourished, supported by a joint Navy–Army–Air Force contract and nurtured by some of the brightest minds in science.

For several months Townes continued to conduct his research at wartime frequencies around one-centimeter in wavelength. But within a year he sought to push into the millimeter region, where the interactions with molecules should be even stronger. In 1950, shortly after his promotion to full professor, he was asked to chair a Navy committee trying to determine how to distribute funds for millimeter wave research that one day might be useful for high-frequency communications and short-range shipboard fire-control radars. It was this work that had led Charles Townes to Franklin Park on the memorable morning of April 26, 1951.

After his epiphany an excited Townes hurried back to his hotel and caught Schawlow just getting up. His sleepy-eyed future brother-in-law was in Washington to attend an American Physical Society meeting, not the Navy committee, but nevertheless possessed the expertise to provide a good sounding board. So Townes laid it all out.

He was, in essence, seeking a way to coax nature into providing a source of high-frequency transmissions of energy. For this he needed to cause excited atoms to jump down to the ground state, an act that provided the release of the energy he sought. Einstein had introduced the concept of such stimulated emission in 1917. But the first and most formidable barrier to making it happen in a useful and striking fashion was the second law of thermodynamics, which held that at thermal equilibrium the energy of atoms would be determined by the Boltzmann distribution. Since a slightly higher number existed in the ground state than the excited state, any signal that caused excited atoms to jump down a level and emit energy would cause others to jump up and absorb it, with the net effect that more energy would be absorbed than emitted. Meaning Townes was out of luck.

Or, rather, he had been out of luck. There were several layers to the revelation. But the key, Townes told Schawlow, was that he had found a way around the second law of thermodynamics. He didn't need to violate the law because he didn't need to conduct the experiment at thermal equilibrium. In other words, he could isolate already excited molecules or atoms and keep them excited until radiation at their resonant frequency caused them to jump to a lower level and emit energy.

All this was clever thinking, but still basic science. In fact, though in circumstances that would not provide the effects Townes envisaged, several researchers, including Ed Purcell, Robert Pound, and Norman Ramsey at Harvard, had already achieved such population inversion in the orientation of the spins of nuclei, with more in the high-energy state than in lower ones. Purcell and Pound had even noted the appearance of stimulated emission. But the general feeling was that the intensities of such emissions would always be low, good for little except demonstrating the already accepted concept or providing a minuscule, hardly detectable amplification of the signal. Townes had certainly taken this view. In his spring 1945 memo to Bell management, in which he advocated the benefits of microwave spectroscopy, the young scientist had concluded that no very useful amount of radiation could be produced by molecules.

On the park bench, though, Townes had strung together two crucial additional pieces of the puzzle. A month or so earlier he had attended a colloquium at Columbia. There, the German physicist Wolfgang Paul, not to be confused with Austrian-born Nobelist Wolfgang Pauli, had described an innovative way of creating molecular beams, one that provided intense concentrations of excited molecules or atoms by focusing the beams with four surrounding electrodes instead of employing essentially unfocused configurations like those pioneered by I. I. Rabi. To this idea Townes added the central concept of coherence — continuing the amplification process by building up waves of other emissions exactly in phase with their predecessors. This was to be done with the addition of a resonant cavity, or chamber, that would contain the radiation and enable the maximum amount of energy to be extracted. To pull off such a feat, a threshold condition needed to be met before the inevitable losses inherent in the cavity could be overcome.

From his hasty back-of-the-envelope calculations there in the park, it seemed to Townes marginally possible that with the technique Paul had outlined, combined with the feedback resonator, he could breach the threshold level and produce a sharp and dramatic amplification of the original signal.

Nothing central to this line of thinking hadn't been around for twenty-five years. As he pondered that fact decades later, Townes came to believe that the reason his idea had not been conceived before stemmed from the fact that it straddled the gap between physics, electrical engineering, and spectroscopy. Engineers typically did not appreciate quantum mechanics, and many physicists didn't know electrical engineering. Microwave spectroscopy, where all roads met, needed to be invented and matured before the consummation could take place. Lending credence to his view, within a few months at least two other groups independently considered stimulated emission as a means for amplification. In the Soviet Union, twice-wounded World War II veteran Aleksandr Prochorov explored the issue with doctoral student Nikolai Gennadievich Basov at the Lebedev Institute of Physics. Closer to home, the challenge was taken up by Joseph Weber, a young electrical engineering professor at the University of Maryland. All those involved in the still-undeclared race for the maser were microwave spectroscopists.

As it was, back in the Washington hotel room, Schawlow pronounced the concept intriguing and worth a shot. At the naval committee meeting an hour or so later, however, Townes held his tongue. The committee had already considered a host of crackpot ideas, so he decided to give himself time to think things over before speaking up.

Like many professors, Townes typically brought in graduate students to build equipment and provide other help with research projects so that the hopeful scientists could apply the work toward a thesis. Upon his return to Columbia, Townes began looking for someone to undertake the new project, but it proved difficult to find a doctoral candidate willing to stake his degree on such an iffy venture. It wasn't until late 1951 that he came upon James P. Gordon, a twenty-three-year-old Scarsdale product who had graduated from MIT two years earlier, then moved to Columbia for his Ph.D. "I'm not sure it will work," Townes told a skeptical Gordon. But, he added, even if it failed, the experiment should lead to spectroscopic studies with much better resolution than in the past, and that alone would be enough for a degree.

A second co-worker joined the team a little later. He was Herbert J. Zeiger, a molecular beam expert who had just completed his Ph.D. requirements under I. I. Rabi but was still putting the finishing touches on his dissertation when the chance arose for a postdoctoral fellowship under Townes. Originally, Townes planned for his group to use the ammonia molecule at a high resonant frequency. The men would then try to produce amplified radiation at .5 millimeter in wavelength, solely as a tool to help push spectroscopy to higher frequencies. But within a few months Townes had realized that if the system were to work it would be best to stay in the microwave region, where equipment was widely available and general techniques well known. He therefore changed his goals and decided to build the first device at 1.25 centimeters, the frequency used by all the wartime K-band radars, and where a vast supply of components already existed. He also began to see a host of direct payoffs beyond spectroscopy. For example, stimulated emissions could be used to amplify faint radio signals — from some far-distant transmitter, or even space — with almost none of the internal noise plaguing conventional receivers. At the same time, precisely tuned emissions oscillations could form the basis of an extremely accurate atomic clock.

As the central player in his investigations, Townes still favored the well-studied ammonia molecule. One reason for working with the K-band frequency in the first place was that by a happy coincidence, the primary resonance value, or spectral line, of ammonia gas lies at 1.25 centimeters, making it a perfect match for his equipment. As a gas, ammonia takes on the shape of a three-dimensional pyramid, with a solitary nitrogen atom riding the apex and a hydrogen atom at each of the three base corners. As Townes knew from firsthand spectroscopic studies, when a chamber full of ammonia gas was irradiated at its resonance frequency of 23,870 megahertz, molecules in the ground state, or lowest quantum level, absorbed energy and flew into an excited state as the hydrogens tunneled past the nitrogen to form in essence their own mirror image. At the same time, the quantum switch took place, with originally excited molecules emitting energy on the same frequency and leaping down to the ground state.

At thermal equilibrium, the Boltzmann distribution held true, of course, so net energy was absorbed. To pick out excited atoms for his experiment, Townes set up the quadrupole focusing system described by Paul, in which a series of four electrodes created an electrostatic field with its axis pointing in the same direction as the beam. Molecules in the lower states were attracted to high-field areas created on the beam's fringes, while excited molecules wandered to the slightly charged center. In this way the electrodes acted as a lens that focused the excited molecules exactly where Townes wanted them to go.

The weeded-out beam next fed into a small cylindrical chamber, or cavity, about the size of a 12-gauge shotgun shell, that was itself attuned to the resonant frequency. If this system could not reach the desired critical threshold, then a klystron signal would cause Townes's captive army of excited atoms to briefly emit energy as it leapt down to the ground state. However, if the threshold was met, the system would produce its own continuing oscillations at exactly 23,870 megahertz. In fact, with the beam continuing to spray excited molecules into the cavity's tiny opening, Townes could spit out coherent radiation almost indefinitely, transforming the original signal into a gorgeous, self-sustaining crescendo. At least, that was the concept.

The main physics offices ringed Pupin Hall's hallowed eighth floor. The department itself splayed out between the sixth and twelfth levels, the top three of which had housed parts of the Columbia Radiation Laboratory during the war. The eleventh floor in the early 1950s teemed with spectroscopic and molecular beam apparatuses. Gordon set up shop there, occupying one corner of a large open room shared by several other experimenters. The beam operated in a vacuum, so ammonia molecules could fly two or three feet without colliding with usurpers; the ammonia itself was stored in standard laboratory bottles under pressure at room temperature. A series of pumps evacuated the chamber, the spent ammonia gas solidifying on liquid nitrogen-filled baffles while the beam traveled through the vacuum chamber into the resonant cavity.

It was an elegant experiment, but hard to execute. Townes's initial calculations showed that if all went well, the amplified energy would be small, with the cascade of oscillations barely sustained in the cavity. Full-time work began in the first few months of 1952. Gordon and Zeiger spent about a year building, tearing down, and reconstructing chambers precisely attuned to 1.25 centimeters, or 23,870 megahertz. With waves of emissions bouncing around the small container, one of the main problems lay in fabricating a cavity that would not lose energy faster than the molecules produced it. The tubular chamber was therefore molded out of copper, whose low resistance minimized losses; and the men finally left a hole in each end, so that the beam of molecules could cruise through it, shedding energy into the chamber in the process, the way a train drops off passengers at a station, then moves on down the tracks.

After about a year, by February 1953, Zeiger left for a job outside academe. Gordon and Townes struggled on, joined eventually by postdoc T. C. Wang. By year's end, Townes had plowed through about $30,000 of the department's Joint Services grant with little to show for the money.

By this time, Townes had refined the system several times and continued to believe success lay around the corner. Others seemed unanimous in believing the endeavor would fail, though the stated reasons for failure varied. One school of thought, exemplified by Michael Danos, considered the basic concept fine but reasoned that the oscillations produced would not come in the nice narrow frequency signal Townes sought. He bet Townes a bottle of scotch the project would fail. British-born theoretician L. H. Thomas didn't believe the oscillations would be coherent. Rabi and Polykarp Kusch, who had taken over as department chairman, thought Townes just would not be able to make it work. One day in late 1953, the pair strode into Townes's office. As Townes recalled, "They sort of banged on the table and exclaimed, 'Look, you know that is not going to work. We know it's not going to work. You're wasting money and time. You ought to stop it!' "

Miffed at the outright lack of confidence, Townes replied coolly, "No, I still think there is a reasonable chance that it will work." The men discussed the matter for several more minutes without settling anything. When Rabi and Kusch left, Townes went on with his work and thanked his stars for tenure.

It took just a few months for Townes to win the scotch and silence the critics. The afternoon of April 8, 1954, as his group was holding its regular weekly seminar meeting on the eighth floor, Gordon walked in to say the oscillations were being sustained.

Up to this moment anyone monitoring the experiment could see a slight bump in the power output — as measured with a crystal detector — to indicate the klystron's 23,870 megahertz input signal had been amplified by the stimulated emissions it sparked inside the cavity. Gordon himself had seen such a blip many times. But on this occasion, so long as the molecular beam itself remained in operation, power continued to pour out of the system at a very specific, narrow frequency *after the klystron had been turned off.*

Gordon was not the type to wave his arms and yell in excitement, but his news that the threshold level had been reached stopped the seminar in its tracks. Both he and Townes remembered the entire entourage trudging upstairs to witness the event. Afterwards, a smaller bunch went to a restaurant to discuss a suitable name for the device. Townes hoped for something in Greek or Latin and went to the trouble of recruiting the help of a Greek student. But he laid down the law that whatever was chosen could not end in "tron." Nothing fitting turned up at that session, but a few days later Townes, Gordon, and a few others shuffled across 120th Street to a Teachers College cafeteria. It was there that Townes suggested the device be christened the MASER, for Microwave Amplification by Stimulated Emission of Radiation. Not long afterwards, Danos presented Townes with a bottle of scotch. Gordon, too, collected a bottle of booze from a doubting postdoctoral researcher. Not being professors, neither man had a lot of money: they had bet bourbon.

The physics community reacted quickly to the maser. The first public mention came in a hastily arranged talk Townes gave that spring at the annual meeting of the American Physical Society in Washington. That June the discovery received wider readership, appearing as a letter to the editor of *The Physical Review*. Townes, Gordon, and Zeiger, who was pleased to be included despite his absence from the critical experiment, described the work. They estimated the maser produced a signal power of 10^{-8} watts, not much more than a butterfly flapping its wings. The trio predicted its implications, both for atomic clocks and low-noise amplifiers, although the maser could not yet provide useful amplification. The main stumbling block on that front was that ammonia gas emitted energy at only its precise resonance frequency; any truly functional amplifier would have to be tunable, so that it could cover a wide range of frequencies. Shortly after the announcement in *The Physical Review*, Townes took a sabbatical year at the Ecole Normale Supérieure in Paris. There, from Arnold Honig, a former student also spending time in Paris, he learned more about paramagnetic materials — solids containing atoms that can be partially magnetized. Such materials are much denser than ammonia, or any gas, and their resonance frequency varies with the magnetic field to which they are subjected — so they interact with electromagnetic radiation over a broader frequency band. In particular, though, Honig and French scientist Jean Combrisson had discovered a paramagnetic semiconductor whose electron spins could stay oriented in one direction for an exceptionally long time. Townes reasoned that if he could set up a maser-like inversion action with such paramagnetic materials, he could tune the device over a wide range of frequencies simply by varying the magnetic field strength.

Townes pursued the idea with only moderate success using germanium. Meanwhile, a similar train of thought had been set off independently. At an MIT seminar held on May 17, 1956, microwave spectroscopist Malcom "Woody" Strandberg, a former Rad Lab staff member, also suggested trying to extend stimulated emission to paramagnetic materials. In attendance was Nicolaas Bloembergen, a Dutch-born physicist who had become an associate professor at Harvard and joined Purcell and R. V. Pound in several magnetic resonance experiments. Set to thinking, Bloembergen made the next critical leap. He proposed a tunable maser based on a different type of paramagnetic material than Townes had tried — something with at least one additional free electron operating outside the crystal lattice structure. The number of free electrons dictated the number of quantum levels atoms could occupy. Germanium had one free electron, so it could occupy one level in addition to the ground state. Bloembergen wanted something like gadolinium ethyl sulfate, which played host to several free electrons. That way a magnetic field could split atoms into several states, with the preponderance occupying the lowest orbit as usual. A radio signal set at the frequency corresponding to the energy difference between the lowest and highest states could bring about the quantum switch between those two levels — in effect, cocking the quantum gun. A second, lower frequency microwave signal — the one to be amplified — fed on the now heavily populated third level and made transitions to the sparse intermediate orbit. The rain of electrons leaping downward brought about stimulated emission on the desired frequency. The trigger had been squeezed.

The beauty of this system over a two-tiered maser lay in the fact that the microwave "pump" that kicked off the first quantum switch worked on an entirely different frequency from the stimulated emissions and the signal to be amplified, and could keep up a never-ending fresh supply of molecules in the uppermost state. After he proposed the idea, Bloembergen set up an experiment. At the same time, perhaps a dozen other laboratories joined the race to build a multilevel maser, as the vast possibilities of quantum electronics became apparent. Here things came back full circle to Bell Labs. Even while in France, Townes had been in touch with his old company — which had since hired Jim Gordon — suggesting it pursue the germanium maser idea. Still in 1956, a team led by H.E.D. Scovil, and including George Feher and H. Seidel, followed the trail independently of Bloembergen and built the first solid-state maser using $Gd3+$ ions in lanthanum ethyl sulfate. As long as a signal was steadily fed into it, the maser could do service as an amplifier with nearly zero inherent noise. From this point on, the nation's applied labs rushed vigorously to pursue the technology.

The maser ran counter to the common view of invention, in which an idea in basic science spurs technological innovation. In this case, the novel instrumentation came first and generated better ways to conduct science on a variety of fronts. By spitting out almost pure beams of excited molecules, the maser enabled better spectrometers, since the exclusion of molecules in the lower states enhanced signal intensity by two orders of magnitude. At the same time, the stability of the maser's sharp beams laid the groundwork for an almost ideal source of frequency standards, while the constancy of atomic properties and its lack of inherent noise fluctuation rendered the device the basis of the world's most precise clock. Such a timepiece based on hydrogen, the best of which was built originally by Norman Ramsey and

Daniel Kleppner at Harvard, would lose a second every 300,000 years. Masers were adopted in communications satellites as the most efficient detector of microwaves carrying transoceanic signals.

For Charles Townes, the use for the maser that provided perhaps the most personal satisfaction came in radio astronomy, which he had decided not to pursue after Ira Bowen told him the field was a dead end, but which other World War II radar veterans had pioneered. Following the debut of the Bell Labs three-tiered maser, radio astronomers adopted the device as a receiver-amplifier that could detect signals with a hundred times the sensitivity of the previous state of the art. But even this was just the beginning. The maser proved so popular that by 1960 *The Physical Review Letters* banned further correspondence about its uses.

Beyond its direct applications, the maser also marked the beginning of quantum electronics — machines and instrumentation based on the quantum behaviors of molecules, atoms, and nuclei. Following this trail, in September 1957, Townes turned back to the quest that had guided him to the maser in the first place, but had been pushed to the back burner in the flurry of excitement over the discovery — the desire to reach shorter and shorter wavelengths in the far infrared. This time, rather than seeking revelation in a park, he sat at his desk and attacked the problem head-on. As he related, "I played with equations until suddenly I realized from them that it would be just as easy to go right on to still shorter waves — light waves."

By this time, while still at Columbia, Townes had reconnected with Bell Labs as a consultant. That fall he shared his notes with brother-in-law Arthur Schawlow, who worked at Bell in solid-state physics and had been considering the same problem on his own. The pair published a definitive theoretical description of an optical maser, which amplified visible light by using mirrors to bounce beams back and forth through a gaseous molecular soup, stimulating the emission of photons. The publication appeared in *The Physical Review* in December 1958, setting the stage for the first laser — Light Amplification by Stimulated Emission of Radiation — which was produced at the Hughes Research Laboratories in Malibu, California, a year and a half later.

For his invention of the maser and for providing the laser's theoretical backbone, Townes was awarded the 1964 Nobel Prize. He shared the award with Prochorov and Basov, who had separately published a paper in the Soviet Union describing the maser in October 1954, six months after Townes spoke before the American Physical Society. Subsequently, both Basov and onetime mentor Prochorov made important contributions to laser development that solidified their rights to the Nobel.

By the time he traveled to Stockholm to receive the award, Townes had left the laser far behind. In the late 1950s, during the Cold War, he served for two years as vice president and research director for the Institute for Defense Analyses, a nonprofit Washington think tank that advised the government on weapons and national security. He left after two years to serve as provost of the Massachusetts Institute of Technology, but was denied a hoped-for position as president of the school by a selection committee headed by Vannevar Bush, with whom he had had some disagreements over the course of MIT research. In the face of this disappointment, he joined the University of California at Berkeley and finally was able to take up astronomical research. In the 1960s, Townes helped kick off the cavalcade of radio astronomers discovering molecular lines in space: his student at MIT, Alan H. Barrett, detected the first line—from the OH molecule. Townes himself led the group that found the next two—ammonia and water—stable molecules astronomers had not expected to find in space. Meanwhile, a Berkeley colleague, Harold Weaver, got the first indications that Barrett's OH line did not behave quite as expected—an anomaly shown to be a naturally occurring maser bursting forth from vaporous clouds of star stuff.

Late in his career, after some fifty years of pursuing scientific conundrums and still professionally active in astronomy, Townes reflected on his life before a convocation gathered at Berkeley's Center for Theology and the Natural Sciences. He noted how dramatically things had changed that day in 1941 when he was transferred out of physics to the Bell Labs radar bombing program, and in his words could be found a truth about a good many World War II physicists schooled in microwave radar. "Now, that turned out to be exceedingly important to me; much of my subsequent work has grown out of it. How? First, I learned electronics. Most physicists didn't know electronics very well at that time. Also I learned about microwaves, and the early stages of computing. Perhaps most important of all, in the last years of the war we were learning how to produce and work with shorter and shorter wavelengths in the microwave region, in order to have better and better directivity of the radars, and from this I glimpsed my future research."

Townes spoke, too, of his continuing awe for nature and creation, and about God's role in his physics and his life. "You may well ask," he told the audience, " 'Where does God come into this?' To me, that's almost a pointless question. If you believe in God at all, he's always here—everywhere. He's in all of these things. To me God is personal, yet omnipresent—a great source of strength, who has made an enormous difference to me. When my atheistic friend asks, 'What has he done for you?' What can I say? I look at what's happened to me, and think that all of those things are what he's done."

17 • Rad Lab Redux

"For the first time in its history, as a consequence of the atomic explosion in the Soviet Union, the United States is confronted with a really serious threat of a devastating attack by a foreign power."
PREFACE TO *Final Report of Project Charles*

"Our restrained views regarding any spectacular solution of the air defense problem are counterbalanced by considerable optimism about the contributions to air defense that will be made by new basic technology. We think the electronic high-speed digital computer will have an important place. . . ."
Final Report of Project Charles

"Nihil sub sole novum (there is nothing new under the sun)."
ECCLESIASTES

AT SIX O'CLOCK the morning of August 29, 1949, an almost unbearably bright light filled the steppes of eastern Kazakhstan. The unearthly glow dimmed, then seemed to renew itself rapidly as a white-hot fireball shot upward and changed hues to orange and red, rolling up feather grass, stone structures, wood, and metal in a tumultuous fury that punctuated the morning stillness and turned sandy yellow soil to glass. Slowly, a huge mushroom cloud rose skyward and wafted away to the south, its shape vanishing like an ephemeral spirit. This was the first Russian atomic bomb test, code-named *Pervaya Molniya*, or First Lightning. At twenty kilotons, the device roughly equaled in explosive power the initial American nuclear bomb detonated in the New Mexican desert four years earlier. It should have, for the Russian offering was copied from the American design, courtesy of the spy Klaus Fuchs, a German-born British scientist who had worked on the U.S. atomic program during World War II.

The Russians had taken great pains to keep their trial secret. The test site lay nearly a hundred miles southwest of Semipalatinsk, in an isolated valley flanked by two small hills, and had been chosen to minimize seismic effects and the spread of radioactive material. Yet the precautions failed. Elements of the dust cloud rose into the stratosphere, borne eastward by prevailing winds toward the Pacific, where the Americans awaited.

It became almost standard lore that the Soviet nuclear test caught U.S.

intelligence and scientific experts by surprise. This was only partially true. The American consensus held that Josef Stalin's physicists would probably not be able to construct an atomic weapon for at least another year; but those in the know were well aware the Soviets might beat that time frame, so in late 1948 the United States began constructing a sophisticated monitoring system to stand watch. The backbone of the network, which included acoustic listening posts and rainwater collectors, was dust filters that trapped telltale particulates arising from a nuclear explosion. Such filters were carried in special weather-modified B-29s and owed a debt to Luie Alvarez, who during World War II had pioneered the concept of using aircraft for radiological surveillance by designing an airborne collection system to look for radioactive gas that might betray the existence of German nuclear reactors.

In the summer of 1949, American monitoring of Soviet nuclear activity operated under the code name Bequeath. To intercept airborne dust carried beyond Russian borders, Air Weather Service planes routinely skirted the eastern edge of the Soviet Union, ostensibly conducting regular weather reconnaissance but also bearing "bug-catcher," the particulate sampling system. The flight path running closest to Russian airspace, dubbed Loon Charlie, linked Eielson Air Force Base in Alaska with U.S. facilities in Japan. Flown by the 375th Weather Reconnaissance Squadron, the route stretched across the North Pacific, then down the east coast of Kamchatka. It was a 375th plane flown by Lieutenant Robert C. Johnson that recorded the first evidence of the Soviet explosion.

On September 3, Johnson's aircraft took off from the northern Honshu city of Misawa to fly the reverse leg of the Loon Charlie route. It was a 13½-hour marathon back to Alaska, and the craft gathered air samples at three-hour intervals much of the way. When the B-29 landed at Eielson, the samples moved along established channels until they reached a special trailer in a high-security section of the base. Sergeant Eugene W. Tews was on duty, and he placed the filters one at a time in a 600-pound lead cylinder so that they could be analyzed free from unwanted background radiation from the Earth and cosmic rays. The second filter exposed contained just enough levels of radioactivity to qualify as an official alert. Tews called his superior, Captain Carroll L. Hasseltine. The officer ordered a Teletype sent to Washington.

It was Labor Day weekend, so most offices in the nation's capital were closed. But on the other end of the Teletype, the third floor of a yellow brick building on G Street, someone was always on duty. The structure housed the Data Analysis Center of AFOAT-1, the Air Force agency responsible for the covert monitoring of foreign atomic tests. Already in the previous year, the tightly guarded center had logged 111 alerts that might have indicated a Soviet nuclear explosion. All had been dismissed as due to earthquakes, volcanic activity, or a natural cause such as normal variation in background radiation. The 112th alert, though, proved different. When he heard about it, technical director Doyle Northrup wired back to Eielson suggesting Tews periodically remeasure the filters. The sergeant complied, and soon transmitted the follow-up data to the Analysis Center, where it was plotted. From the rate of radioactive decay, it seemed evident that the air sample could contain fresh fission products, and the Air Force scrambled to life.

Over the next week or so, specially assigned planes monitored the air trail. On September 6 the first filter samples arrived for detailed analysis at Tracerlab, a radiological equipment company retained under a top secret contract to analyze dust samples brought back by the B-29s. Dr. Lloyd Zumwalt, who headed the analytic effort at the firm's Berkeley laboratory, worked through the night as he found fresh fission products of barium, cerium, and molybdenum. From their probable "birthday" it seemed certain the samples had come from an atomic bomb rather than a reactor accident, while other analytical data suggested the weapon was similar to the first U.S. plutonium implosion device. Zumwalt put the time of fissioning somewhere between August 26 and 29. Then, as the radioactive cloud drifted over North America and the polar areas, the British were tipped off, and more telling rainwater samples were collected in Alaska, Washington, D.C., and north of Scotland. By September 14, little doubt remained that the Soviets had the bomb. Insiders were already calling it Joe-1.

Vannevar Bush, back at the Carnegie Institution but still a highly influential Washington force, was asked to convene a scientific advisory panel to review the situation. Joining him at AFOAT-1 headquarters on September 19 were J. Robert Oppenheimer; Robert Bacher, a veteran of the Rad Lab and Los Alamos; and William S. "Deke" Parsons, the *Enola Gay's* weaponeer when the atomic bomb was dropped on Hiroshima. It took Bush's panel only a day to conclude unanimously that a Soviet nuclear test had taken place. President Truman delayed an announcement for a few more days to make absolutely certain, and also to give the USSR time to make its own public statement. This did not occur. So late the morning of September 23, the President informed the press: "We have evidence that within recent weeks an atomic explosion occurred in the U.S.S.R."

The news riveted the nation. "ATOM BLAST IN RUSSIA DISCLOSED," *The New York Times* screamed across its front page the next day. Seven page-one stories were devoted to the event. In Painesville, Ohio, John Rigden, later to become a well-known physicist and I. I. Rabi's biographer, was a newsboy delivering newspapers. "Johnny, quit your paper route," a customer told him. "We'll all be dead in a few days."

Back in Washington, the revelation fell into the midst of a major running debate over how to control the spread of nuclear energy, and whether to construct the Super, a thermonuclear bomb with vastly greater destructive power than the atomic weapons dropped on Japan. H-bomb proponents, led by the Hungarian immigrant physicist Edward Teller, found in the Russian explosion the leverage they needed to settle the matter. Barely four months later the nation was told that Truman had ordered the Super built.

On another, much quieter front, the Soviet emergence as an atomic power called many of the World War II radar men back from academic research into military pursuits.

In the past, America had waited until Europe was embroiled in fighting before awakening its war machine. In the nuclear age, when an unprepared nation might find itself out of the fighting before a single shot had been fired, no such complacency was possible. The Soviets already possessed a very long range bomber: the propeller-driven Tu-4 "Bull," copied from B-29s that had sought refuge in Siberia during World War II. All of a sudden, with the bomb in Russian hands, America was a vulnerable island, much as Britain had been when confronting the emerging era of airpower on the eve of Hitler's rise.

Still years before spy satellites, only one technology could keep a round-the-clock watch on the sky approaches to the United States. That was radar. Yet, America's protection scheme in 1949 was far less efficient than Britain's a decade earlier. The few warning and Ground Controlled Interception stations standing guard over the country's borders were better technically, but worse operationally — since the coverage did not extend to low altitudes, the radar sets were ill maintained, and the overall system was poorly orchestrated. Exacerbating matters, interceptors possessed no nightfighting capabilities to speak of. The best hope, World War II–vintage Black Widows, lacked both the speed to close on modern bombers and the deicing equipment necessary for bad-weather operations. So while notables like Teller lobbied to build offensive weapons, others saw a more pressing need for an ambitious modernization of America's air defenses.

One of the main catalysts for the change was George Valley, who had led the Rad Lab's drive to develop the H_2X bombing radar. Valley was no fan of atomic weapons. Although he had spent much of the war honing an all-weather way to drop tons of bombs on the enemy, the thought of going nuclear broke through what he called his "clip level." At least, he reasoned, the enemy stood a good chance of shooting down planes and surviving a rain of conventional explosives. Nukes, though, raised the stakes horrifyingly high. During the war he had refused to join Los Alamos, breaking down in tears in Lee DuBridge's office when his old mentor asked him to consider moving over to the other project: "No, I think that's filthy, I won't do it," he had sobbed. After the war, as an assistant professor of physics at MIT, Valley had joined Ed Purcell, Norman Ramsey, and others in campaigning against the May-Johnson bill, which they viewed as unwisely seeking to vest control of nuclear energy almost solely in the hands of an Atomic Energy Commission dominated by the Pentagon. "I had made innumerable speeches," Valley recalled, "to lawyer's clubs, to doctor's clubs, to chambers of commerce, to Rotary Clubs, to Lions Clubs, to the League of Women Voters, to anybody who would listen."

Once the Soviet atomic test became known, however, Valley rethought his reluctance to join official discussions relating to nuclear war. Ever since his Rad Lab days, the physicist had served on the electronics panel of the Air Force Scientific Advisory Board. The panel investigated all matters of electronics issues — from batteries on up — and things had gradually become more active as the Cold War loomed. Considering the immutably changed balance of world power, Valley decided it was time to get even busier. It didn't take long to figure out how.

Early in the year Valley had been spotted in the Pentagon taxi line by the Caltech physicist H. P. "Bob" Robertson. The respected expert in relativity theory, himself a World War II military advisor, had recently been briefed on Air Force plans to revitalize the country's air defense network. The two men didn't discuss many details then, but the service, which consistently rated Soviet nuclear potential higher than either the Army or Navy, had long fretted over the bare-bones warning net protecting the continental United States — just nine radar stations for the entire nation in June 1948. Shortly after it formally branched off from the Army in 1947, the Air Force had begun hawking an ambitious Radar Fence Plan that called for establishing some four hundred outposts to provide the country with round-the-clock protection. The staggering $600 million price tag got the plan buried. But in late 1948 the idea had been scaled back to a more austere eighty-five installations — seventy-five in the continental United States, the rest in Alaska. Even this awaited final congressional approval when Valley and Robertson bumped into each other outside the Pentagon. But the Air Force was already busy erecting a string of low-budget radar sites, the Lashup network, to hold the line until the permanent system could be established.

Although he backed the Air Force's intentions, Robertson found the plans appalling and described a system of outmoded equipment, poor maintenance, and ghastly training. He had urged Valley, as a member of the electronics panel, to investigate the matter more fully. Reluctantly, Valley had agreed. However, the Soviet threat at the time had seemed so distant that he had placed the project on the back burner.

In barely half a year things had changed dramatically. Valley wasn't the type to dwell on images of a fiery cataclysm engulfing civilization. Instead, his angst was far more personal. He and his wife, Louisa, their two young sons, and infant daughter were set to move into a home being built in the affluent suburb of Lexington, some ten miles from the MIT campus. The hilltop house, of California redwood, enjoyed a clear line of sight from its living room to the Boston skyline. Valley had looked forward to the view. But suddenly, with his physicist's eye, he saw something starkly different. "So when the bomb came, here I was with this house, and it was so constructed as to have considerable blast resistance from the west, just because of the way the topography went, and it had no blast resistance from the east at all."

"Well," Valley recalled thinking, "what the hell am I gonna do?" He was going to keep his promise to Bob Robertson, that's what he was gonna do.

A few weeks later Valley found himself tromping around a Lashup radar station on Cape Cod. The post was part of CONAC, the Air Force's Continental Air Command network set up at the end of 1948 to detect and intercept intruders into U.S. airspace. The place fit Robertson's bleak description to a T. Recalled Valley, "The site resembled one of those army camps of the Indian wars that you see in the late-night movies — except that Quonset huts substituted for log cabins, jeeps took the place of horses, and the officers didn't wear slouch hats."

The chief warning radar was the General Electric AN/CPS-5, which operated in the L-band around twenty-five centimeters. The model had first been produced toward the end of World War II, once it became apparent that rain clutter was far less obtrusive at this wavelength than on the higher frequencies of Luie Alvarez's Microwave Early Warning set. More importantly, longer wavelengths allowed for wider beamwidths, solving some problems for Moving Target Indicators that sought to identify real targets from clutter as a result of the change in phase or frequency of their echoes caused by the Doppler shift.

In Valley's estimation, the CPS-5 was better than anything standard during the war. But he found the Cape Cod radar unpolished, untuned, and being watched over by "guys fit to maybe cut grass." Beyond this dreary situation, another factor drew his concern. Operational messages were passed over low-powered, high-frequency field radios — a mistake, the visitor saw, since long-distance transmissions could be hindered by ionospheric variations. He asked the commander why they didn't just use a telephone and got back more of a sermon than an answer. "He started with the customs of Pharaoh, went on past Ashurbanipal and Darius the Persian to the Battle of Marathon, and paused for breath at the fall of Rome," as Valley remembered. "Then he quoted from Napoleon, from various Civil War generals, and wound up by reciting from the official investigation of the Pearl Harbor attack." The bottom line, through all the smoke, was that the military did not depend on civilian lines of communication.

Valley felt a surge of anger just listening to the man. Still rankled when he got back to Cambridge, he phoned the Scientific Advisory Board office and vented his spleen to Major Teddy F. Walkowicz, the top aide to Chairman Theodor von Karman. Walkowicz listened attentively and asked Valley to put his air defense thoughts down in writing. It took about a week. On November 8, Valley sent off a three-page letter laying out key areas of concern. The gist was that if the existing state of the art was found inadequate — and Valley knew it would be — then the Air Force should consider creating a committee "to find the best solution to the air defense problem." Such a body, Valley suggested, should be established in the science-rich Boston–New York area and populated by experts in guided missiles, aerodynamics, physics, and electronics.

The Air Force concurred. Soon after Thanksgiving, Valley was invited to join a special meeting of the Scientific Advisory Board's executive body, where a technical committee along the lines he had suggested was proposed. By mid-December, General Muir "Santy" Fairchild, the Air Force vice chief of staff under General Hoyt S. Vandenberg, had signed a memo to Valley bearing more definitive news: "Regarding the Air Defense Committee, the Chief of Staff has directed that it be organized immediately, and it is planned that this group will be functioning within the next few weeks."

So was born the Air Defense System Engineering Committee, or ADSEC. In a separate letter Fairchild named Valley chairman and listed the experts invited to join. The roll was dominated by people Valley knew. One was fellow Rad Lab veteran George C. Comstock, an experimental physicist and vice president of Airborne Instruments Laboratory, Inc. Then came four of Valley's MIT colleagues — Charles Stark Draper, an expert in inertial guidance; meteorologist Henry Houghton; jet engine specialist William R. Hawthorne; and aeronautical engineer H. Guyford Stever, another Rad Lab alumnus. The last two committee members were Allen F. Donovan, an aerodynamicist from the Cornell Aeronautical Laboratory; and John W. Marchetti, civilian director of the Air Force Cambridge Research

Laboratories, which occupied an old factory building on Albany Street near the MIT campus. It was all in the family. Everyone except Marchetti sat on the Scientific Advisory Board, and all had strong ties to the Air Force.

Although the initial wave of fear and trepidation stemming from the Soviet test had worn off, the group moved quickly, stirred by the importance of its task. The first of what would become weekly Friday meetings took place that same month. All the members were local except Comstock, whose company was on Long Island, and Donovan, who flew over on Friday mornings in his private plane. Marchetti offered his shop as the meeting place; and he took care of security, financial affairs, secretarial help, and all other housework.

Although the Air Force research center was a short walk from his own office, Valley had never stopped by before. He was surprised at both the scope of the projects under way and their sophistication. Johnny Marchetti himself was a first-rate electrical engineer who had helped the Army Signal Corps design the pioneering SCR-268 and SCR-270 longwave radars in the late 1930s, and he had recruited perhaps two dozen former Rad Lab and Harvard Radio Research Lab members to his several-hundred-person staff. Among the items under development, Valley saw a device for digitizing analog radar signals for transmission over voice telephone lines, as well as early light pencils, used somewhat like a computer mouse for manipulating text and images on a cathode ray screen.

As ADSEC came into being, the Lashup network was rapidly taking shape. Late in 1949 the Air Force began boosting coverage around Washington State and installed temporary warning and control systems in the technology-rich San Francisco and Los Angeles areas. Another ring of defense went up around the atomic energy facility in Oak Ridge, Tennessee, where uranium-235 was extracted. Before another year was out, forty-three Lashup stations would be in operation, most in the Northwest and industrial Northeast, but with three in New Mexico, home to Los Alamos, A-bomb development and production supervisor Sandia Corporation, and Kirtland Air Force Base, the principal lair of the Special Weapons Project, which oversaw weapons handling. Each site was manned by one or two search radars — the AN/CPS-5 or something even older — a height-finding set, and communications post. Centrally located controllers usually followed the action on vast, edge-lit Plexiglas maps that depicted the local terrain. Aircraft status and position were updated by grease pencil–wielding operators who stood on scaffolding behind the board.

The Valley Committee, as ADSEC became known, set its bearings by studying CONAC operations and reading up on RAF and Luftwaffe air defense schemes. At the same time, Lashup commanders were invited to the meetings. Valley found the military men extremely frank in laying out the network's operational shortcomings, and committee members quickly reached a consensus about the grave deficiencies plaguing the country's air defenses.

The gist of their reasoning was that a single Russian bomber, on a one-way mission of destruction, could carry enough nuclear payload to erase two large cities. In order to conserve fuel, Soviet forces would have to fly at high altitudes, probably over the North Pole, for most of the long journey to the United States. If the planes stayed high, detection and interception would be relatively easy. But at some point, it had to be assumed, pilots would drop low to avoid radar warning posts.

Just as in World War II, low flyers posed a difficult problem. Anti-aircraft batteries were most effective against high-altitude targets. Airborne interception radars, still longwave or Rad Lab–spawned ten-centimeter models, lost ground huggers in clutter. Even relatively sophisticated land-based sets such as the CPS-5, formidable against massed attacks, suffered dreadfully against the envisioned foes. The Earth's curvature puts a limit on how close radars can see to the horizon. At the distances necessary for early warning, the CPS-5 could not peer much below 5000 feet, nor could its successor, the AN/CPS-6B, earmarked for the permanent radar network. A lone Russian bomber, the ADSEC members therefore concluded, would probably have no difficulty avoiding the U.S. radar net. As Valley summarized, "To attack most of the northern cities of the United States, such a bomber would have to fly low for only about 10 percent of its journey, and therefore its range penalty would be small. If, in addition, aerial refueling were to be employed in the vicinity of the arctic circle, the entire United States would be vulnerable to low flyers. . . ." Even in the rare event enemy aircraft were detected, the cumbersome plotting and interception scheme instilled little confidence that they could be brought down.

In its first report, issued early in 1950, the Valley Committee compared manually run systems, such as those in the Lashup network and the proposed permanent stations, "to an animal that was at once 'lame, purblind, and idiot-like.'" The group immediately began considering a wide range of treatments to correct the situation. Members examined the state of rockets for shooting down enemy planes, as well as a supersonic interceptor proposal from Douglas Aircraft. Sometime in the first few months Donovan pointed out that an airplane's empennage, or tail section, formed the weakest part of the craft. He suggested equipping interceptors with steel-edged wings that could slice through a bomber's tail and built elaborate models, accurate in scale and relative strength, to show the concept was feasible. ADSEC members and various military officials liked the idea, but the Air Force's civilian bosses thought it looked too much like an American kamikaze and nixed the proposal out of hand.

From the start, Valley himself thought most about improving the radar net and integrating its data into an effective command and control system. He figured that in the nuclear age, when one bomber could equal the destruction of a 1000-plane raid, the days of massed attacks were long gone. Besides, A-bombs cost a lot to make, and the Soviets were unlikely to commit more than one aircraft to each target. It was more feasible that the enemy would disperse its craft, sending several hundred bombers to different places. Valley figured hundreds, maybe even thousands, of adjunct radar stations would be required to provide an effective low-coverage blanket around the United States. To keep tabs on the massive amount of data pouring in would oblige a filtering and ground-control system far more complex than dreamed up in World War II. Signals would have to be rapidly and accurately translated into useful velocity and position information, for the stakes were the highest. Valley realized at once that human operators could not conceivably handle the job. "The individual computations were straightforward enough, and anyone could combine the data on a map if he had enough time," he acknowledged. But in real time, and with the country's future at stake, it was an impossible chore.

Tired of grading papers one evening soon after the committee formed, Valley started doodling his way into the air defense problem. To detect low-flying planes, he reasoned, a continuous wave radar was needed. Such sets constituted the crude progenitors of modern pulsed systems. But in a state-of-the-art defense scheme they could be superior to pulsed radars. Given a steady beam, omnipresent ground echoes could be effectively ignored, so that only moving objects were picked up. The big catch, or so Valley thought at the time, was that while these electronic eyes could determine a target's speed, they could not pinpoint its position.

Addressing that failing, Valley began ruminating about deploying a network of linked, continuous wave radars that could triangulate on an enemy bomber. Working out the math, however, he realized that the calculations had to be extremely precise. He tried to envision a company of GIs whipping out slide rules or scurrying through logarithm books, then gave up in exasperation and returned to his pile of school papers.

Brainstorming with Marchetti the following Saturday morning, Valley went over his scribblings. The Air Force research man studied them intently for several minutes before pronouncing, "Say, I think you've got something there." Valley protested, pointing out the GI slide-rule factor. But after a long silence, Marchetti ventured that the crews didn't have to work things out: a machine could. Raw radar signals could be passed fairly easily over phone lines. The big stumbling block, Marchetti felt, was the processing job. Valley remained unconvinced, but he threw out some crazy ideas and finally muttered something about feeding the data into one of the newfangled digital computers grabbing headlines around the country. At that, Marchetti rubbed the side of his nose with his pipe and cut in enthusiastically: "Now you're talking, George."

Within a few hours the pair had scratched out block diagrams of a revitalized early warning and air defense system. The plan hinged on Valley's off-the-cuff idea of placing cheap, continuous wave radars on telephone poles every ten miles or so throughout the country's vulnerable areas of approach. No more than a boxful of electronics, the sets would be able to measure a target's radial velocity — a series of such measurements would enable range to be deduced via the change in Doppler shift — and shouldn't cost more than $20,000 a pole. The radar signals, barely more than an audio squeak, could be sent over an ordinary phone line to a digital computer for processing.

Much of the following week was spent drawing up a list of known digital computer projects. Valley counted less than a dozen, including a John von Neumann venture at Princeton's Institute for Advanced Study, a Raytheon endeavor dubbed Hurricane, and the well-known ENIAC — for Electronic Numerical Integrator and Computer — effort at the University of Pennsylvania. Marchetti called every outfit to gauge the enthusiasm for taking on the air defense problem. He came up empty. Most computer designers wanted their creations to perform complex mathematics and simply were not interested in connecting with the real world. An exception was Hurricane, but it was already committed to a Navy project studying computers for automatic ship defense.

Frustrated, but still convinced they had found the right path, Valley and Marchetti began exploring the rather uninspiring possibility of building their own computer. Then, at MIT one day in January 1950, Valley bumped into Jerome Wiesner and spilled his problems. Wiesner had made his reputation soon after joining the Rad Lab in 1942 by solving a critical transmit-receive problem. Still in his twenties he had gone on to join the lab's steering committee and run Project Cadillac, an airborne early warning system that didn't see combat in World War II but became the forerunner to the modern-day Airborne Warning and Control System, known as AWACS. In early 1950 he served as associate director of the Research Laboratory of Electronics under fellow Rad Lab veteran Albert G. Hill, and was considered a rising MIT star.

Long after the event Wiesner and Valley differed about the circumstances of the meeting. Wiesner remembered talking in his Building 20 office, Valley in a Building 10 corridor. Both, however, agreed on the general course of events. Valley detailed the difficulties he and Marchetti were having finding a brain for their telephone pole radar system. Wiesner retorted that he knew of an available digital computer that might just fit the bill, and it was housed right there on campus.

By happenstance, Wiesner had just finished a review of the still-experimental computer. Like so many other MIT projects, it was funded principally by the Office of Naval Research. However, the estimated cost of completing development had soared to $3 million, more than triple the price tag of its nearest competitors, and Wiesner had been asked to assess the situation. He had found the contrivance plagued by troubles — most centering around developing a reliable memory — and realized that the Navy could not go on funding it at the same level. Still, he considered the endeavor highly promising and hated to see it fade away.

Wiesner also knew something of Valley's telephone pole concept. He had been briefed by MIT provost Julius Stratton and Ed Bowles, then an advisor to the Department of Defense. Wiesner thought the idea was crazy and had sent off a one-page letter arguing against it. But he was all for an improved air warning network and had even mentioned the possibility of using a digital computer to correlate radar data. So with everything coming together in their chance meeting, Wiesner suggested that Valley look up the computer operation. The key contact was an engineer named Jay Forrester, and the machine was called Whirlwind.

•

The steady drone of Whirlwind's cooling fans permeated the Barta Building, a distinctive redbrick structure capped by a lonely minaret and a giant smokestack a few blocks west along Massachusetts Avenue from the main MIT campus, almost directly across the street from the New England Confectionary Company, which churns out the popular Necco wafers. The computer itself, electronic organs silently performing their duties, took up the back of the second and uppermost floor. Offices, labs, fabrication shops, and the power system swallowed up the rest of the building, down to the basement. The top-floor concentration of equipment was staggering. Each data bit was stored in memory on a flip-flop, a breadbox-sized rack that contained two vacuum tubes, one for each on-off state of a binary digit, and a host of other tubes for various control purposes. These ungainly elements were packaged in groups of sixteen, corresponding to the computer's sixteen-bit word size. Several rows were often stacked atop each other, crowned with logic tubes and assorted other components, making for huge electronic walls that ran from floor to ceiling. Whirlwind was a computer one walked around inside.

The brains behind Whirlwind's brain were Jay Wright Forrester and Robert R. Everett. As Wiesner had noted, Forrester was the driving force, a reedy Nebraskan who grated some with his unwavering self-assurance and midwestern taciturnity. He had grown up on a homestead cattle ranch, though both his parents had college educations and encouraged their only son to follow his inclinations rather than stay with the family business. So he had decided to become an engineer. While a senior in high school Forrester built a wind-driven electric plant that powered the house and its workshop — the first electricity on the ranch. He did his undergraduate work at the University of Nebraska, then moved on to MIT. There, as a graduate student when World War II broke out, he had joined the school's newly formed Servomechanisms Laboratory to work on devices for controlling radar antennas and gun mounts.

His partner, Bob Everett, a native of Yonkers, had earned an undergraduate degree at Duke before moving on to MIT in 1942 to study for a master's. Arriving as he did in the wake of Pearl Harbor, Everett had gotten the strong hint that students were expected to pitch in with the war effort — or enlist. So he had joined the "Servo" Lab under Forrester and soon became the Nebraskan's right hand.

Forrester and Everett were somewhat infamous at MIT. Partially by dint of being off campus, partially by running the school's only digital computer, the two enjoyed an unusual degree of autonomy. They had their pick of the best graduate students, who flocked to be on the cutting edge of an emerging technology. Yet, as Everett marveled, "Neither one of us had a doctorate, neither one of us was a professor — or even a lecturer." Such factors engendered a degree of resentment with their tenured peers, and Forrester's pervading belief in himself didn't make things easier.

They got by because they possessed the backing of Gordon S. Brown, the influential Servomechanisms Lab director, and because they had talent and outside funding. Over the years leading up to 1950, the pair had cultivated a highly effective, almost good cop–bad cop routine. Forrester ran the lab from a distance, ordering office walls kept bare of cartoons and other signs of frivolity. His agile mind got at the heart of things with sometimes brutal quickness. As one lab member put it, he had a "high signal-to-noise ratio." Everett was just as fast mentally, but without his partner's edge. He assuaged bruised egos and kept the lab on an even keel.

Whirlwind had started life a few months before World War II ended, when Captain Luis de Florez, director of the Navy Bureau of Aeronautics's

Special Devices Division, asked MIT to develop an Airplane Stability and Control Analyzer. As Jay Forrester took up the project shortly thereafter, this was intended to be a complex hydraulic system, controlled by an analog computer and capable of simulating cockpit movements for pilot training and the study of aircraft aerodynamics. However, Forrester's vision and tireless salesmanship had transformed Whirlwind into a far more ambitious undertaking. Early on, he decided that digital power was needed to direct such elaborate real-time mimicry of airplane behavior. To aid in rapid processing, he drew on the high-speed circuits developed for Moving Target Indicators, actively recruiting Rad Lab veterans to his operation. Within two years the analyzer had mutated again, into an all-purpose computer. For a long while the Navy had gone along with the expanding vision of Whirlwind, sharing some costs with the Air Force, which was studying digital computers for air traffic control. But by 1950, tired of the long delays and facing funding limitations, the Navy had let it be known the party was about to end.

George Valley showed up for the Whirlwind tour the afternoon of Friday, January 27. He and Forrester lunched with Jerry Wiesner, then walked over to the Barta Building. By this time, Valley had done his homework, phoning computer experts he knew to get an overview of what Whirlwind could and could not do. John Marchetti, meanwhile, had contacted counterparts at the Office of Naval Research to get the Navy story on the machine. None of the reports were rosy, with the main disagreements coming over just how *bad* the computer had turned out. But Valley was not particularly put off. A group of complaints amounted to nothing more than petty jealousy, so he dismissed them out of hand. Other respondents centered on the sixteen-bit register, which had been intended to reduce complexity during the prototype stage but was considered short for a big processor, even for its day. Not only did this feature curtail the computer's ability to tackle big number-crunching problems, or so the line of thinking held, Forrester's insistence on inordinately high-speed operations was also thought to raise costs needlessly. As Valley pondered the matter, though, none of this bothered him. Air defense required an extremely fast computer, and sixteen bits was good enough. So maybe something completely new held the answer. In other words, for lack of an acceptable alternative, Whirlwind seemed worth a look.

Valley was not immediately taken with Forrester or Everett, who greeted him warmly, perhaps too warmly in his view, then launched into their sales pitch with what seemed annoying cockiness. Valley recalled his exasperation: "They could piss off practically about anyone—Jesus, what a pair. They'd been put in charge of this and nobody'd ever reined them in, or done anything. They were just off on their own spending money like it went out of style — it did go out of style." Whirlwind didn't exactly bowl him over, either. "Horrifying," was how he described the sprawling behemoth.

Counterbalancing the discouraging first impression was Whirlwind's pure, unadulterated speed, though at 1 megahertz it ran at a hundredth the rate of many 1990s personal computers. In addition to the short register, or word length, the computer gained operating speed through its parallel architecture. That is, rather than processing each of the 16 bits in a number one at a time in serial fashion, Whirlwind handled all bits simultaneously. What's more, when Valley arrived the Barta brain was up and running, calculating a freshman mechanics problem and displaying the solution on a cathode ray screen.

Forrester and Everett explained that the computer suffered chiefly from shortcomings in what was later termed random access memory, which dictated the amount of information Whirlwind could handle at any given time. However, the men insisted, a more capable memory lay on the horizon. Valley then went over the air defense problem, satisfying himself that the planned improvements should juice up Whirlwind enough to carry out at least one aircraft interception as a test of his concept. The meeting broke up with Forrester and Everett turning over a stack of performance reports for review and the understanding that Valley would be in touch.

The physicist spent the weekend poring over the sheaf of reports, which included an overview of the Air Force's air traffic control project. In it, he found the computer codes to sort and land airplanes. This wasn't what he was after, but it touched on air defense needs, especially in calling for a system to quickly assemble pieces of a large puzzle into an overall mosaic. The reports were also lucid and well written, bolstering his confidence in Forrester's team. Besides, Valley told himself, he could rent Whirlwind for a year, make his report to the Air Force, and get back to cosmic ray research.

The more Valley thought about the idea, the more he liked it. The Air Force had effectively given him, through ADSEC, a budget of some $6 million to test air defense ideas. He may as well get started spending the funds. On the following Monday he returned to the computer lab with Stark Draper, Guyford Stever, Marchetti, and an AFCRL staffer. When this second tour finished Valley confronted his hosts. "So, how much money do you need?" Forrester hesitated, but Everett jumped in and said they had to have $560,000 to stay afloat another year. Valley agreed on the spot, pending Air Force approval. The men shook hands all around.

Afterwards, Valley accompanied Marchetti back to the research center to talk things over.

"George, do you really want that kludge?" Marchetti queried.

"Well, have you got something else?"

"No."

It took Valley barely a month to wend his way through the military hierarchy and iron out the contractual details. By the first week of March 1950, Whirlwind officially had a new chief sponsor — the Air Force — which picked up virtually the entire tab for the machine. Whirlwind was still not really a functioning computer. It ran elementary math calculations, showing values for x, x^2, and x^3 on the oscilloscope. But it continued to lack a viable memory and remained unproven.

As Forrester and his crew scrambled to devise an effective storage system, Valley and Marchetti busied themselves getting their other ducks in a row. The days grew long and tedious as they rented terminal equipment, simulators, and phone lines, and set up the infrastructure to test the air defense scheme once Whirlwind was ready. Corporate salesmen got wind Valley could dole out contracts and began appearing at his office door.

Sometime that spring it became apparent that the telephone pole radar concept was ill conceived. Finally able to examine the plan's nuts and bolts, Valley and Marchetti saw that whenever more than a single target appeared, it would be nearly impossible to tell which aircraft a given radar was seeing. At first, the men scurried to devise more elaborate configurations of the same basic idea, on the assumption Whirlwind would bail them out. It didn't. As Valley later wrote, "ambiguity in, ambiguity out." Forced back to the drawing board altogether, Valley invited physicists, mathematicians, and engineers to a series of Thursday night seminars designed to find another way to plug the low-altitude hole. But even though the confabs included talented Rad Lab vets like R. V. Pound and Louis Smullin, the problem was never solved and it was necessary to fall back on standard radars.

While the Valley Committee deliberated over such issues, the world reeled from events that only intensified the pressure to deliver a working system. On October 1, less than a month after news of the Soviet atomic bomb reached the West, Mao Tse-tung proclaimed the communist People's Republic of China. That same month, Time lamented the condition of the country's air defenses, and influential Americans agreed. The following April, Vannevar Bush wrote to Omar Bradley, chairman of the Joint Chiefs of Staff, declaring himself "appalled" at the state of the warning network.

The Air Force used the criticism to break loose funding to start construction on the permanent radar system, which was extended from eighty-five to 109 stations. However, air games in June 1950 with the Lashup chain made it plain to all that a manually run radar and control grid was not going to help anyone sleep more easily. Strategic Air Command bombers playing the part of Russians launched sixty "strikes" against Washington State targets. When the attackers came in between 17,000 and 25,000 feet—perfect for the radars—there was ample warning to get Seattleites under cover. But when aircraft came in low, everyone fried. Either way, the city and the Boeing works were leveled. The day after the air exercises concluded on June 24, North Korean forces crossed the 38th parallel into the Republic of Korea, threatening to draw the superpowers into a confrontation in the bargain. The Lashup network, such as it was, went on temporary twenty-four-hour alert.

World events also probably helped Valley plow through some of the remaining obstacles to the proposed automated air defense system. An early ADSEC project had involved reorienting an existing Air Force contract with the Airborne Instruments Lab so that the company undertook a more general review of the manual system's operations. Among other things the devastating report, submitted early in the fall of 1950, lent credence to Valley's inclination that the Air Force would be far better served by telephone communications than radio. Valley at once set out to convince Western Electric, which managed the Sandia Corporation, to study radar station operations and detail what it would take to install and maintain a wide-ranging telephone network to handle both voice communications and radar data. To get The Western on board, he first had to sell Bell Labs on the project.

Instead of dealing with Mervin Kelly, who had moved into Bell's top spot, Valley sought out Vice President Don Quarles, an acquaintance from wartime radar days. The two met in Quarles's West Street office. Valley recalled being greeted by an efficient secretary, whom he labeled neither pretty nor homely, "probably Bell Labs spec." He was ushered into an impressive room with high ceilings and a marble fireplace. Quarles, a future Air Force secretary and deputy secretary of defense, greeted him warmly and soon fairly glowed at the idea of sending radar data over Bell's phone lines. When the meeting ended, the AT&T veep ordered up a corporate Cadillac to take Valley back to La Guardia. On the ride, the physicist reflected with wonder on his lot. He had recently been to a Pentagon meeting where he counted twenty-one generals. Now, he was riding in style. Still three years the good side of forty, he could get used to the perks of power.

With Bell on board, Valley began selling the plan to the Air Force, whose people would actually man the phones. Inside the service, a number of changes beyond what Valley was recommending were taking shape. CONAC was run by Lieutenant General Ennis C. Whitehead out of Mitchel Air Force Base on Long Island. Beginning on January 1, this body would be greatly reduced in scope. From its loins, however, would spring a new Air Defense Command, as well as a Tactical Air Command. Whitehead would take charge of the ADC, with headquarters at Ent Air Force Base in Colorado Springs. As far as air defense went, he was the man.

Whitehead was still based at Mitchel when Valley caught up with him. A balding man barely five feet seven inches tall, the commander cultivated an image of toughness and had earned the nickname "butcher of Moresby" from the Japanese during World War II; Valley soon dubbed him Ennis the Penis. The general stood in classic military bearing and glared out at the newcomer from behind imposing gold-rimmed glasses. "Darkter, my research is on the subject of blood," he intoned dramatically. Valley vaguely expected some discourse on blood banks or controlling pilot blood pressure. Instead, he heard a lunatic's ravings. "Darkter, my research tells me that when you have bled a nation white, you have it at your mercy!"

Several more minutes went by before Valley caught snickers on the other faces present and realized Whitehead was pulling his leg. He responded by dramatically slicing up his roast beef and stuffing it into his mouth, letting juice dribble out onto his chin. "General," he chomped, "that is the best piece of military research that has been done since Clausewitz." At that Whitehead merely grunted and went back to his martini and plate of fresh Louisiana shrimp.

The rest of the day passed in straightforward fashion, with Valley making his pitch for the phone system. When it was over, Whitehead asked when the network could be put into operation. As soon as the Air Force and Western Electric signed a contract, Valley told him. "Well, Dark," said the general, "you tell them down there to harray up. Pleased to have metcha."

The deal would be done quickly. Back in Cambridge with Marchetti, Valley expressed his amazement over how easily the Air Force had been sold on the radar–phone line idea. His friend scoffed. "George, what did you expect? They may be stupid, but they're not *that* stupid."

Another major piece of the envisioned computer-run network began to come together that same fall. John Marchetti's Air Force Cambridge Research Laboratories demonstrated its technology for transmitting radar data to a computer. Aircraft echoes picked up by a Microwave Early Warning radar at Laurence G. Hanscom Field, fifteen miles northwest of Cambridge, were sent over phone lines back to the Barta Building, where they were processed and displayed on Whirlwind's cathode ray screen.

In the not-too-distant future, relaying digital information over phone lines would pose no problem for millions of computer users. At the time, however, it was a formidable accomplishment. The telephone system, designed for analog voice signals, harbored interference and various distortions that rendered digital transmissions unreliable, especially when the signals were meant to protect the country from nuclear attack.

Orchestrator of this achievement was John V. "Jack" Harrington, an AFCRL electrical engineer who during World War II had taken part in the Okinawa invasion as a naval radar and electronics officer aboard the destroyer USS *Hubbard*. Along with Radiation Laboratory veterans Horst Feistel, Edward "Ted" Samson, and Ed Staples, he was part of a small team that had started immediately after the war following up on a Rad Lab project to send radar data over microwave relays. The system had worked well, but in those days the wide bandwidth required to transmit unprocessed video images, and the high cost of installing and maintaining relay stations, had led the group onto what it called the Digital Radar Relay. The basic idea was that all the useful information on a radar screen lay in the few points containing targets. Instead of passing on details of the entire screen, the range and azimuth coordinates of any objects detected were digitized for telephone transmission, and the computer at the other end was left to infer that the rest of the tube was blank. It was a relatively simple concept, but hard to implement technically. Real targets needed to be automatically detected and separated from noise and clutter. Then their coordinates had to be encoded and stored long enough for the relatively slow phone line to clear for transmission in ten-second bursts that like a Teletype system simultaneously dispatched several packets of information at parallel audio frequencies. Under Harrington's guidance, almost all the critical hardware had been invented in house.

George Valley considered the feat proof of the air defense concept—and the end of his job, since his committee was only supposed to put the military on the right track. He did not slip off the hook so easily. In the scramble surrounding the Korean War, three old friends had been brought in as high-level Pentagon advisors. Louis Ridenour, already a leading voice in organizing Air Force research, signed on as the service's first chief scientist. Ivan Getting served as a temporary assistant for evaluation in the research and development arm. Dave Griggs, who would succeed Ridenour in the Air Force science slot, was also a high-level military advisor.

All three strongly supported Valley's idea for an integrated air defense network and wanted it developed as quickly as possible. Ridenour, especially, seeking to mobilize civilian science and technology, pushed the idea that MIT should launch a major new air defense development center. Sometime late that fall, he flew to Boston to broach the idea to MIT president James R. Killian Jr. and provost Stratton. With him were Getting and Major General Gordon P. Saville, who as deputy chief of staff for development and architect of the permanent radar fence plan greatly influenced how the service used outside scientific expertise.

At the meeting, Killian explained that MIT had recently turned down a Navy anti-submarine project because the school maintained a policy against conducting classified work on campus. However, both he and Stratton felt that the Soviet atomic card represented a problem of a different order of magnitude. They agreed in principle to manage an off-campus facility devoted to air defense.

Ridenour followed the trip up with a November 20 memorandum, in which he estimated the proposed lab required a professional staff of a hundred and a $2 million annual budget. The memo read in part:

> It is now apparent that the experimental work necessary to develop, test, and evaluate the systems proposals made by ADSEC will require a substantial amount of laboratory and field effort.
>
> All concerned agree that the necessary effort might be made available by negotiating a research contract with a suitable institution in the Cambridge area. It is important to have this work centered in Cambridge, in order to provide continuing close contact with ADSEC and AFCRL.
>
> A very tentative exploration of the matter with MIT has indicated that they would consider taking such a contract as that proposed.

Ridenour's memo couldn't have been better timed. In Korea, American forces entering the conflict had driven North Korean regulars back beyond the original boundary between the two states, overwhelming the enemy capital of Pyongyang. It had seemed the war would be over by Christmas. But then, on November 26, an estimated 400,000 Chinese soldiers spearheaded a counterattack that soon recaptured about a third of South Korea. The previously distant prospect that the fighting could bring in the Soviet Union suddenly seemed all too real. On December 15, President Truman declared a national emergency, admonishing that the communists had shown themselves "willing to push the world to the brink of general war to get what they want."

That same day Valley and Ridenour lunched at the Pentagon. After putting on a dazzling meal in the secretary's mess, Ridenour coaxed his guest into drafting a letter formally requesting that MIT set up an electronics lab to bring the ADSEC ideas to fruition. By late that afternoon the document had been recast into military language, typed, signed by Chief of Staff Hoyt Vandenberg, and dispatched to Killian.

Valley felt a kind of awe at the way Ridenour had pulled everything off. Ten years earlier a few choice words between the right people had led to the timely creation of the Radiation Laboratory. The Vandenberg memo, he felt certain, signaled the Rad Lab's resurrection.

Yet it did not happen that way. Incredibly, or so it seemed to Valley and the Air Force, MIT turned down the proposal after a group of influential professors objected. Those questioning the Air Force decision-makers were also distinguished Rad Lab alumni, so it was a clash of the titans. At the head of the pack was Jerrold Zacharias, who after the war had become the first director of MIT's newly hatched Laboratory for Nuclear Science and Engineering. He was seconded by Al Hill, then director of the Research Laboratory of Electronics, which oversaw the nuclear lab. The third challenger was Jerry Wiesner. A friend of both Zacharias and Hill, he was not as militant in his objections to the proposed air defense facility. But he stood behind his confreres.

Excluding Wiesner, who got along well enough with all the parties, an unmistakable personal animosity already existed between the two camps. In general, it was the Air Force contingent of Ridenour, Getting, and Valley against Zacharias, Hill, and Wiesner, all of whom depended on a lot of Navy funding. Stratton and Ridenour didn't particularly like each other, either. But the focal point of the tension was the relationship between Valley and Zacharias. Although the two got along well when Valley was teaching under his senior colleague's auspices, they had been at odds on other matters off and on since the Rad Lab days. Valley blamed his fellow physicist, whom he considered "a nasty son of a bitch," for getting him taken off the K-band radar bombing project amidst the water-vapor absorption woes. Zacharias, in return, seems to have harbored few good feelings about Valley.

Beyond the personality clashes lay a more practical issue. The Zacharias cabal, worried that under the proposed laboratory existing Navy-backed projects would be diluted, argued that any new facility should represent all three services and not the Air Force alone. Moreover, the men couldn't believe such a mammoth undertaking would be launched with relatively little consideration. "What stuck in my craw was that . . . here was a major project to be based on a very small study that hadn't been reviewed by anybody except the people who wanted it," griped Al Hill.

Caught in a vise, Killian went back to the Air Force for further talks. He once recalled sitting with Ridenour on a park bench in Washington's Lafayette Square and asking why the school should launch the lab at all. Ridenour's reply had made a deep impression: "It will make MIT a world center in the field of electronics."

A flurry of negotiations ensued. Saville, Getting, and Air Force Secretary Thomas K. Finletter flew north and huddled with the MIT hierarchy to iron things out. In the end, it was decided that any lab would be tripartite, although the Air Force would play the dominant role. More importantly, the men agreed that if another facility was really going to rise like the phoenix from the Rad Lab's ashes, it would not do so solely on the Valley Committee's recommendations. Instead, the lab would be established only after an intensive air defense study led by a nationally recognized, nonpartisan scientist selected by MIT, not the Air Force. The study became known as Project Charles. The man tapped to run it was the Rad Lab's former associate director, even-tempered Wheeler Loomis, who for the second time in a decade left the University of Illinois for Cambridge in times of national emergency.

George Valley had not wanted to start an air defense lab in the first place. He felt roped into the idea by Louis Ridenour, and, he admitted, by his growing yen for the Washington power game. Once on board, however, he resented the fact that he had to jump through more hoops, and the study session only increased his antagonism toward its originator. "This was pure Zacharias," he complained. "Zacharias was the inventor of MIT study groups, which consisted of getting a treeful of owls together, furnishing them with a tree, and writing a report—that did or did not have something to do with reality." Still, the decision had been made, and Wheeler Loomis was as fair-minded a mediator as Valley could imagine.

Project Charles kicked off in February 1951. Meetings took place on the top floor of MIT's recently acquired Lever Brothers building, which overlooks the Charles River on the northeast side of campus and would later house the faculty club and the School of Industrial Management. All told, Charles included twenty-eight full-time scientists and engineers, joined by various technical consultants and representatives from the main service arms of Great Britain, Canada, and the United States. Eleven of the full-time scientific members hailed from MIT, an equal number were ex–Rad Lab, including Loomis, Zacharias, Valley, Hill, and Curry Street. Other notables included Marchetti, Jay Forrester, and Polaroid Corporation founder and instant photography inventor Edwin Land. A separate economics council contained other big names: James Tobin of Yale; MIT's own Paul A. Samuelson, another Rad Labber; and Carl Kaysen from Harvard, an OSS official during World War II. Tobin and Samuelson were future Nobelists.

The first month was spent in intense briefings covering military capabilities, emerging weapons systems, and countermeasures. Charles members visited field installations, questioned officers, and generally staked out their turf. Valley couldn't conceal his relative disgust at the whole proceedings. "Had Margaret Mead attended Project Charles," he later wrote, "she might have written a sequel to her well-known book: *Growing Up Among the*

Physicists." Still, he joined the fray with Machiavellian energy. He figured Zacharias wanted a broad-based military research lab, not one concentrating solely on air defense. So, feeling the odds had been stacked on Zach's side, he listened attentively to Air Force Lieutenant Colonel Peter J. Schenk, who insinuated himself into Zacharias's favor and thereby launched an informal intelligence service to keep tabs on opposition ploys and tactics. Valley also felt he had a key ally in John von Neumann, a Project Charles consultant who embraced the still-unpopular idea of a digital computer helping protect the nation's airspace. "Having Johnny von Neumann on our side, that got me perhaps half the rest of the table — because, you know, physicists aren't used to saying no when a real honest-to-god genius pops up and says yes."

A turning point came on April 20, 1951, barely two months into the study. Forrester had Whirlwind operating to the point that a second, more elaborate demonstration was set up at Hanscom, with the MEW assigned to track a slow-flying Beechcraft "bomber." Just as before, radar signals would be piped back to Whirlwind, but this time the computer was to display intercept coordinates on its screen. Controllers would then radio the proper vectors to a single-prop "fighter."

A contingent of Project Charles members shuffled over to the Barta Building for the event. Valley still couldn't stomach what he saw as Jay Forrester's arrogance. "If I hadn't had to work with him, I'd have kicked his ass — with a steel point on my shoe, too," the physicist asserted long after the war. Nevertheless, he awaited the trial in calm confidence, aware that Forrester had been up most of the night checking the machine, and trusted that if the Nebraskan said Whirlwind would be ready, it would be. His confidence was well placed. The interceptor pilot reported that from forty miles out he had been successfully vectored to within a thousand yards of the target three times running. That was close enough for airborne interception radar to take over.

Valley felt this demonstration finally overwhelmed the opposition. Even before the study session formally terminated later that summer, it was clear the Project Charles members were going to concur with ADSEC's general recommendations and back the establishment of an MIT-run laboratory aimed primarily at upgrading the nation's air defenses. The university immediately began expanding or reorienting various Research Lab of Electronics studies — among them long-range radio communications, radar, and solid-state physics — toward air defense. These would be transferred to the new facility, which became a convenient receptacle for classified work.

The final Project Charles report, classified Secret and dated August 1, 1951, ran to three substantial volumes that examined the state of the art of all aspects of air defense and suggested promising avenues of technical pursuit. The tomes covered, in varying degrees, early warning radars on land, sea, and in the air; interceptor design and weaponry; missiles and rockets that might knock down Soviet aircraft; civil protection plans; and likely enemy countermeasures. Some of the ideas seem fanciful, especially with the passing of time. Still, the report was sobering. Charles members found the nation's air warning system full of holes, especially at low altitudes and on the coasts, with the existing radar net extremely vulnerable to jamming. It was possible, the study noted darkly, that Soviet bombers might even use U.S. navigational aids to find their targets—and that an attack might not be noticed until after the bombs fell.

Contrary to popular sentiment, nuclear warfare was not about the end of the world—or even the country. The economics council that included Tobin, Kaysen, and Samuelson found no Achilles' heel in American industry. Oil refineries, steel plants, and the manufacturers of ball bearings, pumps, compressors, and vacuum tubes were spread out well enough that it would be virtually impossible for the Soviets to halt production of key commodities. As an example, the council noted, thirty-four atomic bombs, delivered on target, would destroy the seventeen biggest steel plants, but that would erase only about half the country's steelmaking capacity. The U.S. population was similarly protected by geographic distribution. The 1950 census counted 151 million Americans. Project Charles concluded that 125 bombs would be needed to render a sixth of them "homeless." Far fewer would be killed, but after that there wasn't much point in wasting bombs on people. Another 136 nukes would add only ten million to the homeless ranks.

At the same time, however, the nation could easily be dealt a crippling blow to its fighting ability. Among Charles members there existed wide divergence about Soviet priorities and overall strategy. But the consensus was that 150 well-placed bombs could wreak enough havoc with Strategic Air Command bases, centers of government and population, and key industrial plants that the war might be over before the United States entered it.

It didn't have to be that way. The study recommended, as had others before it, further diversification of population and production centers, with stockpiling of critical goods in a variety of locations. On the air defense front, the Charles report suggested upgrading radar coverage with additional ground stations, especially in the Northwest, and destroyer pickets and radar-carrying blimps off the East and West Coasts. It wanted to reconstitute the old World War II ground observer corps. Most importantly, Project Charles urged the development of a digital computer to speedily correlate information from all these sources. It named Whirlwind as the model computer, advising that the machine be linked with a network of ten to fifteen radars for operational evaluation.

Charles went out with a bang. The charter creating the new air defense laboratory was signed on July 26, 1951, by a rear admiral and two major generals representing the Air Force and Army—nearly two years after the Soviet A-bomb test. Nathaniel McLean Sage, director of MIT's Division of Industrial Cooperation, hosted a grand dinner at the prestigious Locke-Ober's restaurant in Boston to celebrate the school's contract to run the operation. Once a hasty FBI security check cleared the chef and waiters, thirty to forty scientists and officers showed up to gorge themselves on clams, lobsters, beer, and martinis. To cap the meal, Sage called for a bottle of Napoleon brandy that cost upward of $150. The men toasted one another with thimblefuls.

A site for the facility was found on MIT property in Lexington, adjacent to Hanscom Field at the city's border with Bedford and Lincoln. Marchetti's research center was already moving right next door; and the airfield, along with some thousand acres of land, was soon deeded or leased by the state to the federal government. It fell to the Air Force to name the establishment. The service had already run a Project Bedford on submarine hunting, as well as a Project Lexington that looked into nuclear propulsion of aircraft. That left Project Lincoln.

Even as the military chewed over the various Project Charles recommendations, recruitment for Project Lincoln got under way. The facility was not exactly another Rad Lab, but it followed in the same vein—with a director, associate director, two assistant directors, and a steering committee. Wheeler Loomis agreed to run the endeavor, but for no more than a year. Zacharias was named associate director, Al Hill and Valley assistant directors. All four sat on the steering committee, along with a group of other notables that included Wiesner; University of Wisconsin physics professor Ragnar Rollefson; Malcolm MacGregor Hubbard, assistant director of the MIT nuclear lab; Carl Overhage from Eastman Kodak; British physicist H. M. Mott-Smith, who worked for the Atomic Energy Commission; and Gordon Brown. Only Smith and Brown had not worked at the Rad Lab. On the more nuts-and-bolts level, former Rad Lab members ran divisions, groups, and sections, and populated the rank and file.

By September, Project Lincoln numbered three hundred employees. Until the Lexington site was ready, classified work took place in MIT's Building 22, with unclassifed projects housed next door in Building 20. Both were pseudo-temporary structures on Vassar Street, erected for the Rad Lab, that would be in use well into the 1990s. Administrative chores were handled out of a third MIT building. A motor pool, electronics shops, and publications office were crammed into nearby commercial space.

As the endeavor spurted forward late in 1951, it even carried a sense of urgency similar to that pervading the Rad Lab in its infancy. The thought of Stalin with the bomb was as disquieting as visions of Hitler running rampant in Europe; the stakes far higher. George Valley and his fellow Project Charles members had been briefed by military intelligence on the presumed Soviet capabilities, as derived from defectors, attaché reports, and ex-POWs forced to work in Russian industrial facilities, and they were worried. The enemy was believed to possess four hundred to five hundred long-range Tu-4 bombers able to travel 4000-odd miles and strike most areas of the United States. It was not known whether the Soviet Air Force had developed midair refueling capabilities, as had the United States in 1949. "If not," Project Charles reported, "the enemy will be limited for some years to one-way missions; but we do not believe that this will deter him."

Atomic bombs were still in far scarcer supply than planes. However, considering the rate it took to mine and process uranium, then build nuclear weapons, a reasonable bet was that the Soviets would stockpile enough nukes to attempt a knockout blow by 1956. Détente notwithstanding, it looked as if Project Lincoln enjoyed just five or six years' grace.

18 · The SAGE Age

"We tend to say, 'Well, if we're going to build a defense it has to be perfect—we have to get all the bombers.' But that's not the way the enemy looks at it. The way he looks at it is, 'If the country's got any reasonable capabilities, I'll lose most of my bombers, and I won't be able to carry out my mission, and there'll be too much of the United States left and they'll beat the stuffing out of us—and so maybe I better not do it.' "

ROBERT R. EVERETT

"We invented all kinds of things, not because we were so smart, but because we were the first people who had the problem."

ROBERT R. EVERETT

LATE IN THE AFTERNOON OF APRIL 16, 1952, a junior officer at Ent Air Force Base in Colorado handed Colonel Woodbury M. Burgess, the Air Defense Command intelligence chief, a report of unusual bomber activity in the Soviet Union. The report was labeled an "indication," a strong hint it came from a secret source privy to Russian military operations. That was enough for Burgess to take it seriously. Although due to go off duty, he remained in the Combat Operations Center with his entire staff. Only after hours of vain attempts to confirm the report did Burgess call it a day.

Shortly after midnight, the long-sought confirmation arrived. An Alaskan air defense post had received a report of vapor trails spotted over Nunivak Island in the Bering Sea. Four unknown aircraft were winging southeast by east, toward the U.S. mainland. The sighting had taken place about ninety minutes before Colorado Springs got the word. The information had been passed through the chain of command to McChord Air Force Base in Washington State and onto Hamilton Field in Northern California. From there it had finally been relayed to Ent, where a duty captain called Burgess back to the operations room. This time the colonel notified Major General Kenneth P. Bergquist, the ADC operations deputy, who joined him at the center. Royal Canadian Air Force officials were also briefed.

By 2:20 A.M. Colorado time on the seventeenth, air defense forces around the United States had been alerted, with sector stations and radar outposts in the Northwest told to be extra vigilant. Bergquist attempted to get back in touch with Elmendorf, but the line went dead before his call got through. Tensions mounting in the room, Bergquist phoned General Frederic H. Smith, the ADC vice commander: "We have something hot — I think you better come over."

When Smith arrived in the control room, the two generals debated calling a full-scale alert. The procedure for taking this unprecedented step had been established earlier by Ennis Whitehead, immediate predecessor to current Air Defense Command boss General Benjamin W. Chidlaw. It was a dramatic measure that meant rousing scores of Air Force personnel and ordering them to station. The issue still unsettled, a duty officer came running up with word that the Eastern Air Defense Force was reporting five "unknowns" streaking south past Presque Isle in Maine. That ended the debate. At 3:11 A.M., Smith ordered the country's first Air Defense Readiness alert. Besides calling in Chidlaw, notification went over special hot lines to other Air Force brass, including Strategic Air Command boss Curtis LeMay and officers at the Pentagon's Air Force Command Post. Members of the Tactical Air Command, as well as various radar and fighter units, were contacted by commercial lines. The Army Antiaircraft Artillery Command got the alert at 3:41. All units were ordered to man their guns. It looked like war.

No enemy bombers materialized. The original contrail report was never associated with known aircraft. In the meantime, three of the eastern seaboard sightings were identified as airliners off their flight plans. A few minutes after Chidlaw canceled the alert at 5:50 A.M., the Pentagon telephoned to admonish the ADC for panicking. Chidlaw refused the call, ordering Bergquist to handle it: "Tell 'em if the situation occurs again, I'll do the same thing."

The incident, which Chidlaw labeled *Next to the Real Thing*, took place just as Project Lincoln moved off the MIT campus to its permanent quarters in Lexington. No other real-life experience better illustrated the huge scale of the challenge ahead. The radar and reporting net was so slack the Air Defense Command had not even learned of the first firm evidence of unknown aircraft approaching the United States until ninety minutes after other Air Force personnel had it in hand. For hours after that, officials had no idea whether the sightings were real, or if other planes were also approaching the country's borders. In the end, commercial air traffic had triggered a nuclear alert, and from that stage it had taken anywhere from thirteen to thirty-nine minutes for the ADC to notify cooperating commands over commercial telephone lines, a potentially tragic delay. In his memo to Air Force Chief of Staff Hoyt Vandenberg, Chidlaw told how the incident had heightened awareness of "the very thin margin of evidence" on which the nation's survival depended. Widening that margin was what Lincoln was all about.

The exodus from Cambridge to Lexington lasted throughout the spring of 1952, among the first high-tech migrations to the later crowded Route 128 corridor. All staff members except the Whirlwind contingent were billeted to the new digs; to accommodate the flow, the facility was constructed quickly but without imagination. A quartet of wings — three- and four-story structures labeled Buildings A, B, C, and D — sprouted off opposite sides of a long central artery. A concrete utility bunker went up nearby: Building E. Much more expansive than the just-vacated MIT space, the site soon teemed like a small city. Already by the time of the April air defense incident, Wheeler Loomis counted 550 personnel under his direction. Consequently, he decreed to MIT officials, the organization no longer looked much like a "project." From that moment on, it would be known as Lincoln Laboratory.

Lincoln's early work proceeded at almost wartime pace. Here again, consistent with its hierarchical structure, the facility deliberately copied the Rad Lab. It housed seven divisions, broken into groups and then sections. The first and seventh divisions were support — administrative and engineering design. In between, each handling a different aspect of the air defense question, came five technical branches that covered everything from communications to weapons. George Valley took day-to-day charge of Division 2, Aircraft Control and Warning, which tackled the radar data and transmission end of things. As assistant director, he also oversaw Forrester's Division 6, Digital Computers, back at the Barta Building.

Early that summer, almost as soon as things were up and running, Loomis kept his promise and stepped down as director, leaving a hole at the top. Zacharias, technically the associate director, was immersed in other activities outside the lab and apparently had no interest in the job. That meant one of the assistant directors — Al Hill or George Valley — would take over. On paper, little distinguished the two. Both had earned their doctorates in physics under Lee DuBridge at the University of Rochester. Both had risen fairly high in the Rad Lab hierarchy and taught at MIT after the war. Hill *was* a bigger deal at the school, running the Research Lab of Electronics. But Valley was inexorably linked to Whirlwind, Lincoln's main reason for being.

The job went to Hill. He was considered a better manager, both because he had an easier time delegating authority and because more people liked him. Detractors pointed to a sometimes embarrassing love of drink. But he was very sharp and far easier on the nerves than Valley, whose pale eyes and impatient manner conjured images of wolf-like unpredictability. As one insider put it, Valley was a good choice to manage the development of the computerized air defense network. But a lot of MIT bigwigs "would have been seriously concerned" if he ran the entire show.

Valley was hurt by the decision, and not a little irritated. After all, his ADSEC committee had engendered Lincoln, while Hill, a Zacharias man closer to the Navy than the Air Force, had come along for the ride. Still, the physicist consoled himself, he controlled something close to $18 million in funding, more than four times the lab's remaining budget. So long as he remained the Air Force's darling, that fact guaranteed ongoing independence from Hill. Reflecting this assessment, no sooner had the new boss taken over than Lincoln effectively divided into two labs — the one run by Hill, and the one run by Valley. The men's different styles and bare personal tolerance for each other permeated the corridors. Valley considered the director a lush whose desk was littered with burn marks from when he had apparently hit the bottle, then fallen asleep while smoking. For his part, Hill saw Valley as a loose cannon, difficult to rein in and disruptive. The rest of the lab tried to steer clear of any fallout. "Hill 'n Valley," people would quip.

Valley ran his end of things from Building B. Division 2 occupied the entire second floor. His true domain, though, extended not only back to the Barta Building, but on to the Pentagon and Colorado Springs as he tried to balance the formidable job of managing a wide-ranging body of technical work with the equally challenging task of meeting Air Force needs and concerns.

The prime goal was to examine a digital computer in a real air defense setting. To that end, Valley set his charges to developing a string of test radar stations in the Cape Cod area to serve as a small-scale model of the envisioned nationwide network. The vacationer's paradise, bounded by the sea on two sides, with hilly terrain inland, and a steady stream of commercial air traffic, provided a perfect microcosm of the real world — and was close to home. Valley also planned a series of air games in the region to assess both the network and Whirlwind's ability to control it. Nothing came easily, though. His people had to acquire real estate and erect radar outposts under the constant scrutiny of anxious congressmen and town fathers who worried that the scheme might make their home turf the target of a Soviet nuclear attack.

On the technical side of things, Valley and his staff faced far greater obstacles. Division II's job lay in finding better ways to remove ground clutter, spot moving targets, and transmit radar data. All were problems dating back to Rad Lab days, but still not solved. Even Jack Harrington's Digital Radar Relay, while useful for early demonstrations, proved too complicated and unreliable to bank on. In 1952, Harrington, who had been on loan with his group to Lincoln, left the Air Force research center to head the lab's effort on this front. His team went with him.

Whirlwind, the heart of the ambitious project, alone was enough to cause ulcers. As much as possible, Valley vowed to leave it in the hands of the computer gurus, who, anyway, preferred to have him stick to military dealings, his forte, and stay out of their hair. But even as 1952 opened, some six years into development, the lack of a reliable memory continued to plague the machine. With electrostatic tubes showing little possibility of ever amounting to much, Valley only hoped the glitches would be smoothed out enough to demonstrate the feasibility of his air defense ideas. It would be up to the Air Force to figure out the next step.

Things turned out better than he had dreamed. By the time Al Hill stepped up to the director's office that summer, strong evidence of a solution to the memory problem was in hand. The turn of events hinged on an alternative method of storing information on what were called magnetic cores. Initially, each core was a piece of material about the size and shape of a wedding ring, but they would later be shrunk smaller than BBs. The idea was that to accommodate binary data bits, the alter-ego "on" and "off" conditions on which digital computers conducted their business, these materials would retain one of two magnetic states. Wires would be threaded through the ring holes so that various types of electric signals could either "read" the data held on the cores or "write" new information by bringing about a shift in the states. Unlike with vacuum tubes, there was nothing to burn out.

The use of magnetic materials for information storage had been suggested over the years by several researchers, including Harvard's An Wang. However, magnetic tape was too slow, and efforts to use magnetic cores had been stymied by the complex electronics required. Forrester, though, had launched his own investigations in 1947, pursuing a novel way of stacking memory elements on top of each other in a three-dimensional configuration designed to cut down on sheer physical volume and wiring complexity. He had started studying glow-discharge tubes, but two years later extended that earlier work to cores. That fall he had convinced MIT graduate student William N. Papian, who as a Signal Corps officer had witnessed the arrival of SCR-584 sets at the Anzio beachhead in early 1944, to pursue the magnetic core idea as a master's thesis project. Papian had not accomplished an operational system by the time Valley hooked up with Whirlwind early in 1951, but the subsequent months had seen rapid progress. Critically, at Forrester's suggestion he had abandoned traditional metallic materials. It turned out their conducting properties set up eddy currents that tremendously slowed down the switching process. In their place he took up solid-state ferrites, which were nonconductors. For a landmark test in May 1952, Papian filled a panel with 256 cores arranged in a sixteen-by-sixteen square. He found that while the assembly still gave weaker output signals than tubes, it switched twenty times faster — on the order of a microsecond or less. That kind of speed could be critical in air defense.

A heartened Jay Forrester decided that same summer to build a more realistically sized core memory that could go digit to digit against his banks of vacuum tubes. In keeping with his three-dimensional concept, he told Papian to build another array, sixteen panels high, with each panel filled by cores thirty-two square, making for a total memory size of 16,384 bits, or two kilobytes in modern terminology. Such a novel memory could not be plopped straight into Whirlwind. Yet it had to be tested in an actual computer, to make certain, for example, that continued switching from one state to another did not change the cores' crystalline structure or alter their magnetic characteristics. So, at Everett's urging, Forrester assigned one of the lab's brightest graduate students, a former Navy radio technician and second-class petty officer named Kenneth Olsen, to work under chief engineer Norman Taylor and construct another computer much like Whirlwind but more modest in size and abilities. If cores did well in this Memory Test Computer, the group would take the final plunge and put them into the main machine.

George Valley followed these developments with fascination. It would take months to fully evaluate the three-dimensional configuration. But as the leaves of New England changed hues to herald the fall of 1952, things were looking up. For the first time, he could truly envision the whole system coming together: a sophisticated radar network driven by an electronic brain that tracked planes and derived intercept courses faster than any human.

That was when a group of Air Force officials almost got the project canned.

In parallel with the Lincoln effort, the Rome Air Development Center at Griffiss Air Force Base in New York had been bankrolling its own defense research out of the University of Michigan's aeronautical research center in Willow Run. Called ADIS, for Air Defense Integrated System, the Willow Run venture constituted an electronic upgrade of a British Royal Navy surveillance system. At the heart of its strategy was the deployment and control of a still-undeveloped pilotless interceptor dubbed Bomarc — the Boeing and Michigan aeronautical research center missile. Starting in 1951 the Air Force had supported both ADIS and the Lincoln project in a dual-path strategy meant to ensure that the military got something for its money should one fail to pan out.

An analog computer formed the brains of Willow Run's system. By dint of this feature alone, the plan offered much less flexibility than Whirlwind. The number of planes and radar stations it could assimilate, the displays and subroutines it could run, the very complexity and speed of its data-processing nerve center, were severely limited, at least in theory. But the Michigan-Rome leaders had filled a need that Valley's group had overlooked. They had worked closely with midlevel officers at the Air Defense Command headquarters in Colorado Springs — the ones who would actually use the network — to design an easy-to-understand simulated version of their final product. This step, Valley soon decided, had gone a long way in convincing the ADC that the Willow Run concept was more realistic and affordable than Lincoln's complex digital computer, with its assorted bells and whistles.

The Lincoln group, by contrast, had started with the attitude that it knew what was best for the Air Force. The military men whom Valley's people dealt with tended to be high-grade officials outside the ADC. Although many held advanced scientific degrees and didn't have to be sold on the pros of digital computers, they lacked the hands-on air defense experience and clout that could make or break a project. As Valley later wrote: "If a project isn't pushed by a competent operational type, it will not necessarily fail, but it will flounder. . . ."

He didn't realize his mistake until the last days of 1952, when several astute officers sat him down and explained that the Whirlwind program was in trouble unless Lincoln sold it better. At first Valley grew angry at the Air Force's sheer idiocy. Then, thinking the matter over, he came to understand that the very sophistication of the project held it back. Whirlwind wasn't simulating anything. Its tests hinged on forming a novel link between aircraft and controllers using radars, telephone lines, and real-time computing. Such factors were towers of strength to engineers. "But," as he once elucidated, "instead of praising us for our realistic approach to the problem, some visiting officers tended to think we were just using a particularly expensive and clumsy way to simulate, and for them this was a mark against us, as compared with our chief competitors at another university."

Resolving to give the Air Force something it could sink its teeth into, Valley enlisted Forrester's help, and the two got to work. By late January, after laboring through much of the holidays, their staffs produced a 166-page technical memo, *TM-20, A Proposal for Air Defense System Evolution: The Transition Phase*, that spelled out plans for an interim air defense system to bridge the time gap between the still-pending Cape Cod trials and the permanent air defense web. Although still hypothetical, the concept sparkled with the things that bring ideas to life: clearly defined goals, charts, even diagrams of the planned operations centers. What's more, Valley and Forrester predicted their Lincoln Transition System could be up and running by 1955, a full year before the ADIS target.

The proposal wasn't by itself enough to save Whirlwind. But, Valley felt convinced, it bought precious time. On its heels came a more concerted campaign to get the monkey off Lincoln's back. Late that January, James Killian wrote Air Force Secretary Thomas Finletter demanding a technical evaluation to decide which system the Air Force wanted. Killian reminded Finletter that the school was not particularly enamored with classified projects, and that if the evaluation showed Lincoln's system was not up to snuff, "we stand ready to withdraw. . . ."

Killian's letter rocked the Air Force. Many officials distrusted the Lincoln just-let-us-do-it attitude, but no one wanted to lose MIT's expertise. Two weeks later Finletter crafted a carefully ambiguous reply, assuring Killian the Air Force cherished its relationship with the university and that no evaluation was necessary. Surely, though, he added, Killian could see that it was too early in the game for the Air Force to put all its eggs in one basket. But neither Killian nor anyone at Lincoln was mollified. MIT meant to have its showdown.

It didn't take long. At the request of the secretary of defense, Mervin J. Kelly led an investigation of the Lincoln and Willow Run systems. His committee's report, issued early that year, was favorable to Lincoln and seemed to have helped spur an official tour of the competing labs. Several generals and their aides took part, buffered by an assortment of colonels and majors representing both sides of the debate. The first stop was Lincoln. Valley spent a day going over operational details and the basic design plan. The next morning, probably, he and Forrester ushered everyone into the Barta Building to witness live interceptions. Everything went smoothly. Whirlwind's cathode ray tubes showed the action unfold, while radio communiqués to and from the pilots spewed forth over loudspeakers.

Confident that Lincoln was about to finish off the competition, Valley joined the entourage for the trip to Willow Run. The group witnessed several simulated interceptions played out on a pen-and-ink plotter. One pen traced out the tracks of a hypothetical enemy bomber, the other represented the Bomarc interceptor as it hied to destroy the attacker. Valley found the whole thing ludicrous. But then, he related, "I was astounded to see that this trivial exercise in preprogrammed curve plotting had impressed the majority of the officers as much as had the real thing shown them at Lincoln." His frustration grew that afternoon when the Air Force announced it would continue to back both Willow Run and the Lincoln Transition System. A major explained that Michigan still had the edge because it was being designed to direct Bomarc. Valley offered to direct the missiles, too, just as soon as they existed. But things only deteriorated from there. Air Force officials indicated they preferred the Willow Run plotting board to Whirlwind's cathode ray tubes, and urged Lincoln to adopt it. Valley fumed. How could a massed attack be drawn on paper without confusing everyone? he almost shouted. But he was temporarily defeated, and bit his tongue.

Lincoln had gotten its showdown and come away with something close to a draw. Valley sensed the demonstrations did win a few more friends for Whirlwind, but the outcome was nevertheless unacceptable. It took several additional factors to finally break the dam. Eisenhower had taken over the presidency the previous January bent on slashing defense funding—a move that predisposed the military to drop some dual-path developments. At the same time, on March 28, 1953, Killian and MIT pushed again—harder. Harold E. Talbott, Ike's choice to replace Finletter as Air Force secretary, arrived at Lincoln that day with his research boss, Trevor Gardner. Hinting MIT might pull the plug, Al Hill told them the lab had serious financial woes.

That was enough for the Air Force. By April 10, it was all over. At a high-level conference, the Air Force formally went with the Lincoln Transition System. As the service's official air defense history states, "Lincoln's power play proved successful when Talbott and Gardner decided it was time for the Air Force to drop its dual-development policy and invest all its resources in one agency. Thus, what had earlier been called dual development was now denounced as a duplication of effort." Colorado Springs didn't like it. But the Willow Run program was off the books. It was up to George Valley and Jay Forrester to keep the faith.

•

The Cape Cod Network was pronounced ready for operational trials on October 1, 1953. Real estate secured, civic leaders mollified, a string of radar stations stood guard along the coast south of Boston. Lording over the web, a line officer surveying a battlefield, was a lone early warning set situated at South Truro, near the tip of the Cape. The troops consisted of a few height-finding units, supported by a gaggle of gap-fillers to scan low altitudes. At least initially, two of these were old SCR-584s. The previous summer, the core of Forrester's group had transferred to Lincoln Lab's newly completed Building D. But Whirlwind—the general—still orchestrated the campaign from back at the Barta Building. The plan called for flight testing the defense two days a week, with another couple of days devoted to simulations. Despite the net's many radars Whirlwind could handle data from only a pair of gap-fillers and the South Truro long-range unit at a time. Even with this limitation, however, it possessed the ability to simultaneously track up to forty-eight planes.

Years of planning and investigation had come together in a rush over the past few months to bring the network to reality. Benchmark trials with

Ken Olsen's Memory Test Computer had proven magnetic cores and three-dimensional stacking so reliable that Everett remembered shrugging off a technician who came hurrying up to say the lab was down to a week's supply of the old storage tubes. He and Forrester just ordered the core bank placed directly into the main computer. The effect was stunning. "We took it over to Whirlwind, put it in, connected it up — ran like a charm," he said. Access time plummeted from a typical tube performance of twenty-five microseconds to just nine microseconds. Operating speed doubled, input data rate quadrupled. Even more critical, the chronic need to replace tubes evaporated, as memory maintenance time fell from four hours a day to two hours per week. Within a few weeks a second array of cores was wired in to increase capacity. Olsen found his project usurped before his eyes: "Young kid, out of memory now," he quipped.

Behind the enhanced memory came a feast of leading-edge hardware and software technology. This was the kind of market-making result defense spending advocates liked to argue always trickled out of military research. Any future computer user who received electronic mail, deftly clipped out the communiqué with a mouse and pasted it into another document, then printed a report for his boss that relied on the incoming information to calculate some future trend, owed a tremendous debt to Whirlwind and Cape Cod.

Data streamed into the Barta Building command center over dedicated phone lines at the unheard-of rate of 1300 bits a second — slightly faster than common personal computer relays into the late 1980s. Harrington's group placed electronic translators on either end of the telephone connection to convert digital radar data to analog for transmission, and then switch it back again for entry into Whirlwind. The transmitting device was called a "modulator," the receiving apparatus a "demodulator," with the beginnings of the two words later combined to form the term "modem."

An improved Digital Radar Relay packaged information from the South Truro radar, while gap-fillers employed a more economical Harrington invention called Slowed Down Video to massage and compress their tidings into their relatively narrowband essentials that could be displayed as blips on a cathode ray tube. SDV was a particularly down-and-dirty way of telling roughly where an aircraft was — accurate enough at short ranges of thirty miles, but far too crude for tracking bombers at long distance.

Whirlwind's memory was still so limited that feeding in clutter — in the form of storms and rain, bird flocks, or echoes from mountains, trees, and buildings — wasted precious capacity. Therefore, before radar data was allowed to enter the machine, as much of this interference as possible was stripped out. Moving Target Indicators helped to some degree. But the analog MTI technology then available still failed amidst dense clutter. To get around the problem, Lincoln staff members had employed aother trick, the Radar Mapper. The mapper contained a cathode ray tube set horizontally into a display console, allowing an operator to look down and watch the incoming stream of radar data. Each report appeared initially as a bright blue flash that triggered a photocell mounted like a bedside reading lamp above the screen. The cell responded by opening a gate that normally passed all data onto the computer. But if an operator saw an area of heavy clutter, he could brush over that part of the screen with a special masking paint opaque to blue light. The mapper was not continually monitored. But if the situation changed as a result of shifting weather conditions, operators simply wiped paint from one area and applied it to another. Clutter got through, but not enough to gum up the works.

At this point Whirlwind strutted its stuff. A dozen or more radar operators and controllers waited around their own consoles, dependent on the colossal calculator. During an actual enemy attack the men would have precious few moments to assimilate details and transmit vectoring information to fighter pilots. Whirlwind not only had to rapidly store, process, and display the radar data on which to base these decisions, operators also had to be able to jump into the loop and enter their own information for the computer to consider. To that end, Whirlwind functioned interactively, although the term was not yet applied to computers. Nothing of the sort had been done before.

Another critical piece of the puzzle was the development of a high-speed digital coordinate converter that translated incoming Plan Position Indicator (PPI) data — originally in polar coordinates in relation to a single station — into Cartesian plots. This enabled the construction of a sector-wide map that drew on radar data from many sites. An individual operator was responsible for a particular geographic area. So if anything in his zone inched across the screen like an aircraft track, he would aim a light gun at the latest blip. That told Whirlwind to start a track in its memory. The computer assigned each track a number, and certain information — such as bearing, altitude, and speed — could be called up on the screen through keyboard commands. Once Whirlwind had the essentials, it flashed intercept coordinates for relay to fighter pilots. Meanwhile, to avoid mucking up the PPI face with details, a smaller tote board alongside the main console provided even more specific information about each track: hostile or friendly, whether an interceptor had been sent, the weaponry of the plane if known.

For the initial Cape Cod air games, Whirlwind's main internal memory still consisted of just four kilobytes. Before long, memory was tripled, but even that brought it up only to half the size of early 1980s personal computers. To supplement this spartan capacity, Forrester and Everett built external storage drums, or buffers, that temporarily stored incoming radar data until Whirlwind was ready to process it.

Driving the whole system was the world's largest real-time software program, a monster created under the direction of Cape Cod System development leader C. Robert Wieser, and so complex it amazed Forrester. "This was launched before there was any such thing as a computer programming profession," he observed much later in life. What's more, the whole thing was compiled in machine language, since higher-order languages had not yet been developed and probably could not have been used because of limited computer memory. The Whirlwind group used specially developed aptitude tests to find likely programmers: Forrester remembered music graduates from all-woman Wellesley College as among the best. Before long the software corps alone ran upward of forty full-time workers.

For all its complexities and all its firsts, Whirlwind did a remarkable job. When the Cape Cod trial ended in June 1954, some five hundred sorties had been flown against the network. Boston had been "destroyed" many times over. Yet, the general concept was proven sound, with Whirlwind reaching its capacity and successfully tracking forty-eight aircraft. Forrester's crew at once began bolstering their system's capabilities for the next go-round that November. Control screens were rigged to display more complete aircraft information. Whirlwind got additional storage, and its processors were beefed up to enable the machine to crunch data from more than a dozen radars simultaneously. Early the next year the problem of accurate long-distance interception was eased considerably when Harrington's group finished development of the Fine Grain Data System, an elegant enhancement of the Digital Radar Relay concept. On a given antenna sweep, targets might be hit by as many as a dozen of a radar's rapid-fire pulses. This typically resulted in a blurry smudge on the cathode ray screen, making it impossible to tell exactly where the aircraft really flew. FGD relied on beam-splitting to home in more exactly. Simply put, if the number of successive "hits" in a given range increment exceeded a predetermined threshold, then an object was declared a target — and the returns were automatically counted and divided by two, so that what appeared on the tube was a sharp dot arising from the echoes' centermost point. From two hundred miles, a radar equipped with FGD had an accuracy of about half a mile in range and azimuth.

The Cape Cod radar network eventually swelled to some thirty radars running from Brunswick, Maine, down to Nantucket and Martha's Vineyard, then across Long Island Sound to Montauk in New York. Over the next few years, as the latest B-47 bombers and jet interceptors joined the fray, ultra-high-frequency ground-to-air radio systems linked defenders to the direction center. Whirlwind evolved to the point it could conduct simultaneous running battles between up to sixty-two interceptors and fifty bombers, with the attackers flying in tight formations, splitting up, turning suddenly, and doing anything possible to raise confusion. As long as Whirlwind's capacity was not exceeded, such exercises showed that the system could vector interceptors into position for successful firing passes about 70 percent of the time. That was far from perfect. At best, George Valley and the Air Force figured such a rate would translate to shooting down perhaps a third of the enemy's bombers in a real-world attack. Still, that could save many lives while providing American bombers with a fighting chance of getting off the ground and striking back at the Soviets. That was what deterrence was all about.

•

Even as Lincoln's engineers sweated over Whirlwind and the Cape Cod system, the groundwork was being laid for meeting the ultimate goal: fashioning a nationwide, computer-driven air warning and fighter control network. The lab's shopping list was formidable: everything from sophisticated software to easy-to-use consoles for displaying and entering data. In other words, all the same things patched together for Cape Cod had to be manufactured as real-life operational systems designed for airmen, not engineers packing master's degrees.

Dominating the agenda was finding an industrial partner to help mold the ungainly Whirlwind into a lean, mean fighting machine. By June 1954, as the first Cape Cod trials ended, Jay Forrester and Bob Everett had this job well in hand. IBM had recently signed a contract to build the production version of the new brain, and Lincoln was starting construction of a two-story concrete bunker to house the first prototype. The windowless slab went up adjacent to the main Division 6 operations in Building D. It became the lab's sixth major structure: Building F.

The selection of IBM was destined to play a major role, perhaps *the* critical role, in shaping the computer industry for the next several decades. The choice had come relatively easily. Forrester had launched his hunt back in the summer of 1952, before a core memory was plugged into Whirlwind, but with that technology already shaping up as the wave of the future. For want of a better name, the still-undeveloped production prototype was dubbed Whirlwind II. To choose a corporate partner as objectively as possible, he and Everett, Cape Cod design leader C. Robert Wieser, and Whirlwind II project boss Norman Taylor concocted a numerical grading scale to rank candidate companies. The quartet agreed on a weighted standard incorporating such factors as organizational quality, top management enthusiasm for the job, technical know-how, field service ability, and proximity to MIT. After a quick survey of perhaps twenty possible candidates, the men narrowed the field to four — two separate Remington-Rand divisions, Raytheon, and International Business Machines. When the numbers were crunched, IBM's score of 1816 overwhelmed the second-place Remington-Rand division's 1374. Raytheon had managed only a paltry 1067.

Big Blue's culture and drive, more than anything else, were responsible for the yawning gap between it and the pack of competitors. Daily operations were run by Thomas J. Watson Jr., who had taken over as president the previous January. However, the climate created by his father, still chief executive and board chairman, was in stark evidence. THINK signs radiated throughout the main campus in Endicott, New York. Professional employees generally forswore alcohol in public settings and always dressed in trademark blue suits, white shirts, and dark ties. They were enthusiastic, professional, and expert in electronics and high-speed data processing, which fit in well with Lincoln's aims. Upon touring a manufacturing plant in nearby Poughkeepsie, the Whirlwind contingent found even assembly line workers neatly dressed and efficient. Factory floors seemed clean enough to eat off. Taylor was moved to burst out: "Oh, Karl Marx, where are your dark satanic mills?"

A contract for IBM to study the Whirlwind II project was signed late in 1952. The collaboration advanced swiftly, with representatives of both groups meeting at a large motel in Hartford, roughly halfway between their two headquarters. The following summer, the partners launched Project Grind, a hard-driving effort to grind out details of a prototype. By the time the Cape Cod System blinked on in fall 1953, Big Blue had already started building two prototypes — the XD-1 and XD-2. The formal contract to turn out the actual production version was signed the following January. This final model would be called the AN/FSQ-7, designating Army-Navy Fixed Special Equipment.

Big Blue reared its prototype computers in an old Poughkeepsie necktie factory, initially renting space but soon purchasing the structure outright. The plant peered down on High Street, so the operation was christened Project High. Manufacturing would later take place in Kingston, New York. Almost as soon as the nitty-gritty work began, the honeymoon with Lincoln ended in a memorable clash of cultures. The Whirlwind clique, disdainful of outside expertise, oozed MIT cockiness. The straight-arrow IBM professionals tackled problems in manufacturing terms and found their partners tragically unversed in assembly line reality. A major squabble erupted over whether to make core racks of L-shaped aluminum, as MIT preferred, or the square steel tubing favored by IBM.

On a more basic level, Jay Forrester fretted that his corporate colleagues worried more about getting a computer out the door than the critical problem of making it the most reliable electronic brain ever built. The contract gave Forrester final say-so over all design changes. Yet he recalled battling Big Blue every step of the way. MIT would submit a blueprint, only to see IBM reconfigure it for manufacture. Forrester often changed it back. At one point a High Street manager exploded: "You run your university, I'll run my plant."

Eventually, a sense of mutual respect would evolve. But the two sides inched only gradually toward that middle ground. IBM sent several representatives to Lincoln to labor alongside the Whirlwind crew. A few Lincoln staffers, among them Ken Olsen, moved to Poughkeepsie to monitor progress there. Usually the men traveled between the sites by car, a four-and-a-half-hour excursion. However, it was possible to go by rail, via Manhattan. Fairly early in the venture, an MIT entourage took the train for a firsthand look-see. Returning home after a few days of IBM's alcohol-free hospitality, the visiting dignitaries were dismayed over the lack of a lounge car or diner. The school's assistant treasurer pronounced the travel situation "intolerable" and decreed that team members could charter planes to expedite the commute. IBM soon bought an Aero Commander, while Lincoln purchased a ten-seater De Havilland that was based out of Hanscom Field.

After that, things seemed to move more rapidly, and within a few months Lincoln was pouring the foundation for Building F. By this time, in the summer of 1954, the Cape Cod trials had proven so successful that the idea of building the Lincoln Transition System was dropped in favor of moving straight to the permanent network. To that end, the XD-1 would take up residence at Lincoln, directing operational trials on an expanded version of the Cape Cod Network that would form a rough subsection of the envisioned nationwide grid. The XD-2, meanwhile, was to remain in IBM hands as a software and hardware test bed.

As the whole endeavor moved closer to reality, the Air Force launched a search for an official project name. Planners wanted to distinguish the computer-run network from the existing manually operated system while at the same time making clear that man was still in the loop. A logical choice was Semiautomatic Air Defense System. But, as George Valley related, no one wanted to ask Congress to appropriate more than a billion dollars for something called the SAD network.

One day that July an Air Force major poked his head into Valley's office and announced that the choice was SAGE, for Semi-Automatic Ground Environment.

"Where did that come from?" the physicist asked.

The officer explained that during a meeting to pick the name earlier that day he had noticed another attendee doodling a cartoon of a figure being hanged. Beneath the victim ran the name "G E VALLEY." Then the artist had sketched another effigy being beheaded by a sword. Again, the unfortunate soul was labeled "G E VALLEY." At this point, the snooping major had been inspired to add the already-discussed words "Semi-Automatic" before the initials of Valley's first and middle names. Liking the ring of the acronym SAGE, he concocted two words to fit the "G" and the "E." The best he could come up with was Ground and Environment. But when he mentioned the idea aloud, everyone had gone for it. That was the story Valley told, anyway.

Building F stood nearly complete early in 1955 when a North American Van Lines truck pulled up outside bearing the first clements of the XD-1. The computer was slated to take over the entire ground floor, its power frame consigned to the basement, while an Air Force–operated direction center went in on the top floor. The Cape Cod Network was still up and running under Whirlwind's direction. But as soon as possible, the XD-1 would take over handling of the expanded network, recently dubbed the Experimental SAGE Subsector. The SAGE age was about to dawn.

So began another long stretch of down-and-dirty engineering and in-fighting. Since the computer-driven radar network had been conceived, its parameters had expanded dramatically. The Air Force had earmarked several hundred more radars, mostly gap-fillers, for its modern air defense web. These would be augmented by airborne early warning planes, Navy picket ships, and Texas Towers, radar-fitted platforms modeled after Gulf of Mexico oil rigs that were to be erected on the shallow continental shelf about a hundred miles off the northeast Atlantic seaboard.

To extend the blanket northward, three layers of radar coverage were being constructed largely outside the United States. The first was the Pinetree Line, initially some thirty radars along the U.S.-Canada border. Concurrently, with American help Canada was putting up the Mid-Canada Line, an unstaffed microwave "fence" along the 55th parallel designed to trigger an alert when something flew past. Most ambitious of all was the Distant Early Warning Line, a necklace of radar outposts strung out from the western Alaska tundra to eastern Canada, a few degrees above the Arctic Circle. The DEW Line, airborne early warning, and Texas Towers had all blossomed in the early and middle 1950s, thanks largely to Jerrold Zacharias.

Unsatisfied with the vision of Project Charles, he had headed an influential 1952 summer study that pushed the concepts as ways to provide the United States with another few hours' warning of nuclear attack. Lincoln Lab took a central role in each, with the DEW Line forming its top priority after SAGE.

All these radar sources had to be tied into SAGE and coordinated with Colorado Springs, where the country's warning and defense resources recently had been consolidated into CONAD, the Continental Air Defense Command responsible for plotting all reports of unknown air activity. In March 1955, soon after XD-1 components began arriving at Lincoln, lab officials and Air Force staffers put together the Red Book, a complete schematic for the full-fledged SAGE system. This was modified several times as demands were piled on the system. But it laid out the essentials.

Two identical FSQ-7 computers would be installed at twenty-odd area direction centers. While one machine ran the air defenses, the other would check for errors, ready to take over should its twin go awry. Besides coordinating data from its various radar eyes, each post was to maintain links to air bases, interceptor pilots, weather stations, anti-aircraft artillery, missile sites, ground observer stations, and civil defense centers. Sector centers would deal with attacks in their area. But they also needed to communicate with adjacent fellows, as well as a regional combat center that tracked the battle on a grander scale. Combat centers, in turn, reported to the CONAD bunker at Ent.

As the prototypical cerebrum for this vastly enlarged network, the XD-1 had far more on its hands than had Whirlwind, and although heavily influenced by its predecessor, the machine was a wholly different computer with distinctive parameters. Jay Forrester had initially estimated it would take 8000 vacuum tubes to process data from 70 radars and track 1000 planes. The actual prototype held around 25,000 tubes. Memory had been expanded to a thirty-two-bit register, and mechanical details were vastly changed — factors that complicated installation and greatly extended the costly programming chore. In April 1955, for help writing software, Lincoln turned to the RAND Corporation, already busy creating simulator programs for the manual air defense system. Within a few months RAND's System Development Division had five hundred staffers developing battle simulation programs and other SAGE applications; many worked out of prefabricated Butler Buildings erected on Air Force land across from Lincoln. Within a year, up to a thousand employees and bigger than the rest of RAND put together, the division would split off from the mother ship to become the System Development Corporation.

Bob Everett oversaw the excruciating workup period, trying to tie it all together. Just hooking the XD-1 up properly took months, as workmen installed a false ceiling to make room for wiring between the computer and the control center. Every step of the way, equipment needed to be calibrated and software checked out. The job never seemed to end, filling Everett's days with stress. At one point, when finally the machine was almost ready, Lincoln brass came in for a tour. Everett warned the visitors not to pull the emergency off-switches that guarded each storage rack from fire. Somebody pulled one anyway. It took three days to bring the operation back on-line.

Exacerbating the overall situation, Lincoln seemed perpetually embroiled in a sea of controversy and innuendo that at least contributed to the departure of its three principal figures — Hill, Forrester, and Valley. Hill was first to go. The director resigned in early 1955 after finding himself in frequent contention with both Valley and Forrester over how, or even whether, to proceed with SAGE. After a brief stopover at MIT, Hill left that summer to help establish the SHAPE Technical Center, NATO's air defense laboratory in The Hague. The next year he moved to Washington as director of the Defense Department's Weapons Systems Evaluation Group and vice president and director of research for the newly created Institute for Defense Analyses, a nonprofit advisory body spawned largely by MIT.

A few months after Hill's departure, Jay Forrester quit the lab. He had long felt that the pioneering days of computing were over. And while he didn't like to talk about it, he, too, had tired of the Lincoln histrionics. In the early days his group reveled in being indispensable architects of the world's only real-time digital computer for air defense. That kept things bearable, Forrester remembered, because, "We had the authority and influence that came from being indispensable." With SAGE well under way, and with George Valley seeming more unbearable and unwilling to step aside so he could run Lincoln, Forrester saw things getting only worse.

One day as the computer man walked down a Lincoln hallway with James Killian and a slew of visiting dignitaries, the MIT president suggested that the recently created School of Industrial Management, later named the Sloan School of Management, might be the perfect change of scene. The more Forrester thought about the idea, the more he liked it. Thanks to Whirlwind and SAGE, he had gained extensive management experience. Moreover, MIT was having trouble combining its technical strengths with management expertise in a way that would differentiate the program from other business schools. Forrester made the move and within a year conceived what he initially called industrial dynamics, a specialized field that involved developing computerized feedback loops to analyze such factors as how a company's own structure and policies affected corporate growth and market share. Over the years, the idea broadened into system dynamics, the study of how the structure of any economic or physical entity — from business to city slum — causes it to change with time.

George Valley held on the longest. He had expected to be named Lincoln director after Hill's resignation, but Killian bumped him up only as far as associate director and gave the top slot to Marshall G. Holloway. A Lincoln outsider, Holloway never seemed to get his bearings and clashed repeatedly with his technical people, especially Valley. He resigned in February 1957 and was replaced by longtime steering committee and Division 2 member Carl Overhage. At about the same time an increasingly stubborn and bored George Valley was fired. His military friends offered to go to bat for him, but the physicist didn't feel like fighting. Instead, that September, after returning to the main MIT campus for only a few months, he signed on for a stint as Air Force chief scientist. He was the sixth man to hold the position, the fourth Rad Lab veteran — after Louis Ridenour, Dave Griggs, and Guyford Stever. Valley watched SAGE come of age from the other end.

In the midst of these upheavals, Bob Everett was the glue that held things together. He got more and more responsibility, but it kept wearing him down. To lessen the agitation, Everett sometimes visited the Barta Building. "Both Whirlwind and SAGE were machines you could walk around inside of," he remembered. "And while we were working on SAGE, the XD-1 was right there, right outside my office, really. And if I went out and I walked around inside XD-1, I'd get a feeling of anxiety — but if I went down on the campus and I walked around inside of Whirlwind, I got this feeling of comfort, pleasure, relief. Well, I was waiting for the day that I could walk around the SAGE machine and feel good about it." He made it through the XD-1 assembly and all the subsequent trials. But that feeling never quite came.

•

The first operational Semi-Automatic Ground Environment center hummed into gear in an imposing blacked-out concrete slab on McGuire Air Force Base near Trenton, New Jersey, on July 1, 1958. According to plan, it housed a pair of IBM's revolutionary FSQ-7 digital computers. Over the next five years, twenty-two additional direction centers, most on the East and West Coasts or along the border with Canada, would be identically equipped. IBM also built three sets of scaled-down versions, FSQ-8s, to run combat operations centers: the screens showed a wider field of view, but the system did not directly handle radar and other field data. Ultimately costing several billion dollars, SAGE constituted the greatest military research and development outlay since the Manhattan Project.

When deployed, each pair of IBM's mammoth brains weighed 275 tons, housed some two million memory cores, 50,000 vacuum tubes, 600,000 resistors, 170,000 diodes, 1042 miles of internal and external wiring and cables, and drew enough power to run a town of 15,000 at the time; the immense heat generated by the Q-7 dictated an equally huge air-conditioning system. A typical direction center, like the one at McGuire, spanned four windowless stories. The operations command post, with more than a hundred display consoles and other visual aids, took up the entire top floor, with the level below devoted largely to offices and other service functions. The twin FSQ-7s, labyrinths of tubes and memory, filled the second floor. Below them sat the air-conditioning units, data-processing equipment, telephone frames, and cables — with a separate powerhouse adjoined to one wall, and a cooling tower beyond it.

Decades later radar engineers liked to mock SAGE, pointing out among other things that its Moving Target Indicators and clutter-removal features would probably have failed during an actual low-altitude attack. As time went on, some Air Force officials even came to suspect that air games showing SAGE's effectiveness had been rigged. Proponents of the multibillion-dollar network, however, viewed SAGE in another light. The FSQ-7s functioned with an average downtime, for both machines, of just 3.77 hours a year. The system therefore contributed heavily to the nuclear deterrent, they felt, by raising grave doubts in Soviet military minds about the advisability of an attack. As Robert Everett explained, no one in the United States knew how well SAGE would perform, but it almost didn't matter. So long as the Soviets didn't know either, that may have been enough. Given this reality, and because SAGE was much more than its computers — few on the inside wanted to update consoles or write voluminous new software — six centers remained in operation into 1983. The last survivor was shut down the following year.

Whatever disagreements exist over SAGE's worth in air defense, the system unequivocally reshaped America's technological landscape. The FSQ-7 was the first mass-produced digital computer for controlling complex operations in real time — as a situation unfolded. It came with a host of important innovations that were either conceived entirely or at least made practical by the Whirlwind-Lincoln crew.

The revolutionary magnetic core memory topped the list. But one couldn't stop there. The light gun, which allowed operators to write directly on a display face, is a precursor to the computer mouse. Whirlwind and SAGE pushed the creation of modems, the concept of time-sharing from a common data base, and computer-to-computer communication. Further behind the scenes percolated other important, though less sexy ideas — marginal checking to guard against failing components, and a dual arithmetic element for more efficient parallel processing. Finally, owing to the vital need to quickly and easily keep tabs on aircraft and weapons, SAGE pushed the state of the art of computer graphics. Everett liked to claim these innovations were born of necessity, not pure ingenuity. But the Lincoln-Whirlwind teams *were* smart.

The various SAGE contractors and subcontractors — among them IBM, AT&T, General Electric, Burroughs, Bendix Aviation Corporation, Convair, Fairchild Camera and Instrument, and Hughes — wasted no time bringing the fruits of their labors to market. The intricate SAGE data tentacles, connecting radar stations with direction centers and on to Colorado Springs, constituted both the nation's first major digital network and the world's largest data transmission web: AT&T used the processing links as the foundation of its Long Lines' data communications business. Similarly, System Development Corporation, later bought by Burroughs and eventually part of Unisys, trained hundreds of programmers who became instrumental in the nascent digital computer industry.

It was on the computer hardware front, though, that SAGE made its biggest contributions. At the end of August 1957, straight-jawed, broad-shouldered Ken Olsen left Lincoln Laboratory and joined Lincoln colleague Harlan Anderson in forming the Digital Equipment Corporation. The old woolen mill in Maynard, Massachusetts, where the two set up shop had once churned out blankets and uniforms for Civil War soldiers. With $70,000 borrowed from pioneering venture capitalist Georges Doriot, a native of France who had become a brigadier general in the U.S. Army during World War II, Olsen and his partner took over the second floor of one building and began developing their first product — small, transistorized Lab Modules philosophically similar to test equipment developed for Whirlwind that could help engineers check out memory cores and other computing elements.

From the start, Olsen's vision was to create what he thought of as a personal computer that, much like Whirlwind, interacted with people in real time through keyboards, light pens, even loudspeakers. But *Fortune* had reported that the computer industry represented a financial risk, so Doriot advised Olsen against using the word "computer." Fittingly, when DEC branched out just over two years later to sell its first full-service computer — a direct descendant of the prototype transistorized brains Olsen had last worked on at Lincoln — the offering was called the PDP-1 for Programmable Data Processor. Filling its corporate roster with many of the lab's workers, DEC would soon establish itself as one of America's fastest-growing corporations. Olsen still thought of his products as "personal" — though analysts and the press classified them as minicomputers. It hardly mattered. At its heydey in 1990, the company counted 124,000 employees worldwide and reported annual revenues of nearly $13 billion. That year it climbed to twenty-seventh on the *Fortune* 500, and ranked as the world's second-largest computer maker.

Number one was IBM, which built and maintained its lead largely on the strength of SAGE. At $30 million per computer pair, the FSQ-7 and FSQ-8 alone brought in over half a billion dollars in total revenues, and the total SAGE windfall was responsible for nearly 10 percent of the company's U.S. sales during the last half of the 1950s. Big Blue also drew heavily on SAGE to fashion its 700-series computers and help American Airlines develop SABRE, the Semi-Automatic Business-Research Environment airline reservations system.

Of far greater consequence to Big Blue was the magnetic core memory, which formed the backbone of the emerging digital computer industry and remained the basis of virtually all computer storage until the rise of integrated circuits in the 1970s. The hookup with MIT was like manna from heaven. Before it was introduced to Jay Forrester, IBM had been searching for years for a reliable high-speed memory to carry it into the postwar electronics era, even to the point of investigating ferrite cores. However, in those days the company afforded no competition for Forrester's stocks of talented graduate students or virtually unlimited budget, and had made little progress. When the Nebraskan showed the way, it followed gladly.

During the first nine months of Project High, records show, Big Blue personnel spent 950 man-days at MIT getting up to speed about Whirlwind and core memories. That was nearly triple the time MIT engineers spent in Poughkeepsie. IBM, although it initially expressed much skepticism about Forrester's innovation, then built steadily on the Whirlwind foundation, improving, in Ken Olsen's words, "all the process of manufacturing, all the steps and all the controls and disciplines and techniques and tooling." In January 1956, five months before the first production version of the far more elaborate FSQ-7 was packed off to McGuire Air Force Base for initial tests, Thomas Watson's charges shipped the 704 and 705, the company's landmark commercial computers based on the Whirlwind random-access core memories.

Already by this point, other firms were on to similar technology. Both International Telemeter and Sperry Rand beat Big Blue to market with computers sporting magnetic core memories. None, though, could really compete. With the construction of its own manufacturing plant, IBM soon emerged as the world's lowest-cost maker of ferrite cores. At the same time, the company had assigned something like 1500 personnel to the SAGE project on a wide variety of fronts — not just memory — with the overall objective of mass-producing a powerful digital computer. As competitors' mainframes proved one-of-a-kind designs not suitable for widescale manufacture, the Thomas Watsons, junior and senior, found themselves in a unique position to dominate the emerging computer industry, shrewdly parlaying the air defense work to climb to the top of the heap. Forty years later IBM still reigned as the world's largest computer maker.

Jay Forrester was widely acknowledged as the critical force behind the core memory revolution. He not only experienced the key stroke of creative insight, but also wielded the will to make cores a critical part of a mainframe computer. But was he the inventor? RCA researcher Jan A. Rajchman had filed for a patent on the same fundamental concept of information storage on September 30, 1950, seven months before Forrester's own submission the following May. These rival claims sparked a divisive patent feud that slogged on for more than a decade. A major casualty was MIT's long-standing relationship with IBM.

Forrester was a meticulous worker. His notebook entry from June 1949 clearly established him as the originator of the central magnetic core memory concept, a fact the RCA claim granted. However, Rajchman contended that, after a brief exploratory foray, the MIT group had abandoned its ideas until learning he was hot on the same trail, and that, in any case, his team had been first to build a working core memory. Forrester and Bill Papian disagreed vehemently. Not only had their investigations continued without interruption, it was probable as well that Rajchman had gotten wind of their project before starting his own.

A legal stew of other factors made the truth harder to find. For one thing, Rajchman's patent did not involve three-dimensional core stacking, critical in Whirlwind and SAGE because it vastly reduced complexity and made the whole thing manageable. However, the Board of Patent Interferences sided largely with RCA. Examiners ruled that since Forrester's patent was filed after Rajchman's, the burden of proof rested with the MIT researcher, and that he had failed to prove his group had met the due diligence test in patent law, necessary to show it had not temporarily abandoned cores as RCA had claimed. On October 18, 1960, the board awarded Rajchman what seemed to be the ten broadest claims of Forrester's patent. The Nebraskan, though, retained the rights to his three-dimensional configuration.

A messy situation turned nasty. Forrester's original patent had been assigned to Research Corporation, a philanthropic foundation that routinely handled such matters for many universities. The group launched civil action to regain the lost claims. On the presumption it would win, the foundation also filed separate infringement suits against RCA and IBM, the main users of core memories. RCA countered with its own lawsuit, seeking everything the patent board had left with Forrester. In a sense, IBM was caught in the middle. To cover one flank, the company earlier had negotiated a cross-licensing deal with RCA that gave it rights to Rajchman's patents. However, similar overtures had drawn a blank on the MIT end: Big Blue's best offer of a $3.5 million settlement was rejected as far too little. In May 1962, as negotiations descended into acridity, Thomas Watson Jr. resigned as a member of the MIT Corporation. Correspondingly, James Killian left the IBM board. A few months later MIT terminated its agreement with Research Corporation.

It took until February 1964 for the university and IBM to come to terms. By that time, Big Blue estimated it would install more than twelve billion cores before Forrester's patent expired in 1973. In lieu of royalties, MIT agreed to a one-time payment of $13 million. Nearly $3 million would come up front, the remainder within thirty days of the school's settling its outstanding case with RCA — so long as at least one of Forrester's claims was upheld.

MIT settled with RCA the next month. As part of the deal, Jan Rajchman's company acknowledged the validity of the Forrester patent. IBM handed over the outstanding balance exactly a month later. The payout surpassed by some $4 million the biggest licensing accord on record. Meanwhile, IBM wasn't the only company using core memories. Before the Forrester patent expired, it had taken in about $22 million and was by far the most lucrative in MIT history.

•

Lincoln Laboratory, though immune to all the patent wranglings, had long looked forward to life without SAGE. Once the McGuire center debuted, officials did what many worn-down parents do when their child comes of age: they kicked their offspring out.

No one in the lab's hierarchy wanted to run the operational air defense network: Lincoln's business was "R," with a bit of "D." So after the Air Force vainly tried coaxing IBM and AT&T into taking the job, Lincoln set up a nonprofit body to manage SAGE. On the first of January 1959, nearly five hundred staff members, including a third of the Division 6 workforce, transferred to the neophyte MITRE Corporation, which soon became a powerful Defense Department contractor. The first board chairman was H. Rowan Gaither Jr., who had run the Rad Lab's business administration office and in the late fifties chaired both the RAND Corporation and the Ford Foundation. Luie Alvarez signed on as an early trustee as well. But it was

Bob Everett, as technical director, who provided in many regards the real driving force. The origin of the name MITRE remained cloaked in mystery. It was generally assumed to stand for something akin to MIT Research and Engineering. Wags joked, though, about Must I Trust Robert Everett.

With the departure of SAGE, Lincoln marked the end of its first stage of evolution: bomber defense. Other arrows from the same quiver, chief among them Airborne Early Warning and the Distant Early Warning Line, had also matured beyond the laboratory's purview. The AEW concept, a tactician's dream that promised to bring radar and fighter control over the ocean or even enemy territory, had posed an especially knotty technical challenge. Back in the Rad Lab days, Project Cadillac had been designed on ten centimeters but suffered from ground and sea clutter that limited range and made it hard to track ships and low-flying aircraft. Lincoln succeeded, in part, by retreating toward the meter wavelengths used in early airborne radars. The move reduced clutter, especially over the ocean, since choppy seas produce less backscatter at longer wavelengths. It also extended range, because water becomes mirror-like in the ultra high frequency, enabling radar waves to dance ahead of the beam: forward scatter. Testing of the system was carried out at MITRE as the Airborne Long-Range Input program. There, the effort evolved into the famous AWACS.

The DEW Line, by contrast, was more of a construction feat. The string of Far North radar outposts flickered into operation in 1957 after thirty months of icy hard labor by American and Canadian crews. Initially, the line stretched from Cape Lisburne, at the tip of Alaska, to Cape Dyer in eastern Canada. Over the next few years it was extended eastward, first to Greenland, then Iceland. The main network consisted of manned stations every hundred miles or so. Each held quarters for a dozen or more staff in a central structure, capped by a Raytheon-built, radome-encrusted early warning radar. A separate receiving antenna picked up signals from unmanned, continuous wave transmitters placed between the main posts. This arrangement, with transmitter and receiver widely spaced, is known as bistatic radar; Lincoln's version was called Fluttar. The sets all employed Lincoln-designed automatic alarms. To link stations, both with one another and the continental United States, the lab pushed scatter communications. A prime feature of modern military long-distance command and control, scatter techniques involve bouncing voice and data signals off the troposphere or ionosphere; much of Lincoln's work took place at the old Round Hill estate where Ed Bowles and other MIT researchers had conducted prewar microwave studies.

Riding the wave of SAGE, Airborne Early Warning, and the DEW Line, Lincoln Lab in just a few years had catapulted to the top tier of civilian-run military research centers. But even as its three big bomber defense successes were under development, planners saw a fly in the oint-ment that could render such traditional air defense concepts obsolete. This was the ballistic missile. Military gurus had feared these deadly warbirds since at least the days of the German V-2. By the early 1950s, American intelligence had learned that the Soviets were working on intercontinental ballistic missiles capable of carrying nuclear payloads to the United States. The reports pointed darkly to a time when bomber defense meant little. ICBMs would fly at around 24,000 miles per hour, descending on their targets at twenty-degree angles remindful of line drive home runs. Lincoln Lab's own 1953 and 1954 studies of estimated speeds and trajectories showed that not even the DEW Line would provide adequate warning of their arrival.

In response to the terrifying threat, a number of studies advocated massive civil defense programs, and several fanciful ideas gained surprising credence. A vast shelter carved into the granite bedrock under Manhattan was touted as capable of swallowing up the island's entire population in twenty minutes. At one point, the anthropologist Margaret Mead suggested keeping some percentage of newlyweds underground at all times to protect the breeding population.

On the military side, since most experts held out little hope of de-fending against missiles, the chief result was to extend the country's commit-ment to massive retaliation, both via Strategic Air Command bombers and American ICBMs. The term "mutually assured destruction" came into vogue. Still the specter of a missile race kicked up enough additional defense spending for Lincoln Lab to pursue the idea of a Ballistic Missile Early Warning System. Giant radars, vastly more powerful than anything in the DEW Line, would peer out thousands of miles to catch Russian drones as early as possible. These might afford perhaps fifteen minutes' notice of attack, not enough time to evacuate cities but ample to ensure SAC strike planes got off the ground.

A small effort to build a prototype of this revolutionary radar got under way in 1955. Slowly, as SAGE and its cohorts came into being, Lincoln devoted more of its computing, electronics, and radar know-how to missile defense. Unlike the Rad Lab, the facility would probably never attract any-one of the caliber of I. I. Rabi, Luie Alvarez, or the various other once-and-future Nobelists. Nor were there any Taffy Bowens to inspire employees with tales of real-world tribulation. But talent abounded, both in the still-sharp wartime radar veterans, and in the fresh crop of bright engineers and physi-cists who took up the mantle as their forerunners retreated to management.

The arriving generation formed a kind of mirror image of the pioneers before them. Men like Rabi, Alvarez, and Bowen had left science to apply scientific techniques to warfare. Lincoln Laboratory's rising stars often launched their careers in a military research facility, then slyly appropriated its highly specialized equipment for science. In no small way, they brought radar home to its roots.

19 • The Carpetbaggers

"The moon was such an easy target for this big radar. The echoes came back and they just shouted at you."

GORDON PETTENGILL

"We were a bunch of carpetbaggers."

PAUL GREEN JR.

MILLSTONE HILL RISES a few hundred feet above the northeastern Massachusetts plains to look down on a woodsy panorama dominated by scrub oaks and a few birch and evergreens. In the late 1960s the towns of Groton, Tyngsboro, and Westford swelled outward and encroached upon it, littering the landscape with power lines, new housing, and convenience stores. A decade earlier, though, it was a perfect place to build an experimental missile-watching radar. It sat in the heart of Native American farmlands and old colonist pastures, away from radio interference, yet only twenty miles from Lincoln Laboratory—and with a clear view toward the horizon in all directions. MIT set about to secure the hilltop and many surrounding acres. But land titles in the area went back generations, often without records. Officials worried at first that some unwritten claim would throw a kink into their plans. Nothing materialized, however, and construction got under way late in the summer of 1956.

With the rise of the intercontinental ballistic missile, Millstone Hill formed a focal point in the struggle to keep pace defensively with the Cold War's terrifying offensive escalation. Although the first U-2 surveillance flight over the Soviet Union had taken place on Independence Day 1956, the age of spy satellites and near-constant monitoring of America's most powerful enemy still lay several years off. In the interim, early warning radars arguably constituted the most critical element in the mutually assured destruction scheme, for they offered the best chance of preventing the United States from being taken by surprise and left unable to retaliate. It was a very thin line, though. Missiles, as opposed to bombers, could streak thousands of miles to their targets in under a half hour. Radars had to catch Soviet warbirds early—and reliably. A missed or spurious signal could cost the lives of millions.

When he first contemplated missile warning not long before the Millstone groundbreaking, Herbert Weiss, head of Lincoln's Group 31, Radar Systems, was not even sure radar was up to the job. The ballistic missile detection plan dreamed up in a lab study group called for three strategically placed radars—in Alaska, Greenland, and on the European side of the Atlantic—scattered across the top of the world to watch likely approaches to North America from the Soviet Union. These would not peer around the Earth's curvature—or over the horizon—to actual launching areas. However, with a line-of-sight range of 5000 kilometers, or 3000 miles, the radars could spot U.S.-bound missiles in time to provide perhaps twenty minutes' warning. That, anyway, was the ideal. Weiss likened the challenge to finding a trash can moving at supersonic speeds from a few thousand miles off. Nothing capable of such a feat existed. In fact, the very best warning radars of the day typically couldn't find something as compact as a missile from more than 150 miles away. Weiss was charged, therefore, with coming up with a little matter of a 20-to-1 improvement in the state of the art in range detection—an improvement in sensitivity of closer to 100,000 to 1.

It was just Herbie Weiss's sort of challenge. A short, round-faced man with a passion for overcoming obstacles, he considered himself lucky that life had positioned him to tackle some big ones. During World War II the New Jersey native had interrupted graduate studies at MIT to join the Rad Lab. Early on, he had helped develop airborne interception radar for the Black Widow: when a prototype conked out, he had once ad-libbed a demonstration before Navy czar Ernest J. King by peeking out the test plane window and hand-manipulating centering knobs to emulate the target aircraft on the cockpit scope. Weiss wound up the war as acting project engineer for Project Cadillac, and Jerrold Zacharias later recruited him to Los Alamos to help develop instrumentation for atomic tests in the Bikini atoll. Weiss had next joined Raytheon's pioneering Hurricane digital computer program. He then moved to the newly formed Lincoln, where he worked on radar detection studies associated with Project Charles. Another big task came in developing the DEW Line and its Fluttar radars. His success on that job had rendered him a natural to build the lab's missile detection prototype.

Like most radar designers, Weiss often viewed problems in terms of decibels, a catchall term for describing gains or losses in system performance: in general, the more decibels added to a radar, the greater its sensitivity and range. In the case of ballistic missile detection, he was looking for an improvement of just about fifty decibels. No single breakthrough like the cavity magnetron was going to carry him to the goal. So he began by setting his Group 31 associates, perhaps twenty engineers, to surveying the best available technology and trying to determine how far the limits could be pushed. The conclusion: for a radar to handle the missile detection job would require the world's highest-powered transmitter, most sensitive receiver, best processing, and biggest feasible antenna dish. "And then," Weiss recalled, "maybe you might do it."

The most straightforward approaches to gaining decibels lie in boosting transmitter power and increasing aperture area — a fancy way of saying build a bigger dish. On the antenna end, Weiss wanted something big and steerable — that could spot and follow a streaking missile. The largest reflector used for scatter communications, and it stayed fixed in position, was a sixty-foot-wide behemoth built by D. S. Kennedy, a small Cohasset firm. Weiss determined the same design could be stretched to eighty-four feet before things became too fragile, then set Division 7 engineers to devising a mount to render the antenna fully steerable, servodriven in both the elevation and azimuth axes. For the transmitter, the engineer looked to industry to develop monster klystrons, distant cousins of the original Varian brothers' creation that were a hundred times more powerful than those currently available. These monolithic transmitter-amplifiers, as tall as a man, would spit out huge amounts of power at around 440 megahertz. Lincoln was already experimenting with lower-powered klystron radars, and Weiss and Lincoln contracted with the Eimac Corporation in California to provide their giant counterparts for Millstone Hill.

Antenna and transmitter together held the potential to slash the decibel gap to almost nil. To get the rest of the way, Weiss plotted to do all the "little" things, such as minimizing receiver noise, scanning more slowly, integrating the radar signal over a longer period, and optimally processing the received signal energy. To achieve these goals, he meant to take full advantage of the lab's expertise in digital computer technology. Among its innovations, Millstone Hill would house one of the world's first entirely solid-state computers — the CG-24 — for real-time data processing.

At Lincoln in the 1950s, the red tape that would later plague big projects was still largely absent. Almost as soon as Weiss had his ducks in order, the lab began clearing trees and moving earth at its Millstone Field Station. By July 1957, construction was nearing completion. A circular drive topped the knoll. The big antenna hovered atop a ninety-foot-tall, milk bottle–shaped tower; concrete counterweights kept the dish balanced as it was pointed in varying angles and directions. In front of it stood two corrugated metal buildings, linked as one. The main structure, some three stories tall with a sloped roof, would hold the transmitter and power supplies. The data-processing and receiver area, along with the computer room, took up residence in the second structure.

Everything seemed designed on a gigantic scale. The waveguide, the biggest Weiss had ever heard of, ran almost the size of a heating duct; he tried to bet a secretary she could crawl through it. The whole system promised to be so huge, its powers so great, that the Lincoln staffers worried about harmful effects of the surging electromagnetic radiation. To try to extrapolate how much energy might be absorbed by the human body, the men ground up turkeys and chickens, added thermocouples, then used the radar to boil the concoction in front of the antenna feed. Meanwhile, Weiss set up a smaller radar to warn of aircraft that needed to be directed out of the main beam.

Throughout the summer, amidst intelligence reports that the Soviets were up to something on the missile front, an underlying sense of urgency permeated the Millstone project. That August, with the radar itself still several months from completion, Nikita Khrushchev, then first secretary of the Soviet Communist Party Central Committee, announced that the USSR had successfully test-fired an ICBM. Then, on October 4, Russia backed up its boasts by launching a satellite: *Sputnik*.

Since the rocket carrying *Sputnik* aloft could conceivably transport a nuclear warhead across continents, the stunning debut of mankind's first orbiter altered the geopolitical landscape overnight. Not only were the Soviets for real, they had upstaged the techno-powered United States in the satellite race. Millstone Hill was still incomplete. Among other things, it lacked a conical scan antenna feed for tracking, a shortcoming that was not taken care of until early the next year. Dan Dustin, leader of Group 312, Systems Research, was vacationing on his family farm up in New Hampshire. He got a frantic call from someone at the Naval Research Lab: "Can you track it, can you track it?" Dustin telephoned Glen Pippert, his number-two man, and got him on the job. Pippert joined Weiss and several others in a rush bid to cobble something together.

It took until the twelfth, eight days after *Sputnik* went into orbit, for the radar to pick up the satellite's echo. The big Eimac transmitters were on hand, but not fit for use. So the Lincoln team made do with a much lower-powered "exciter" klystron intended to help start its larger counterparts. Even with this limitation, the giant eye easily boasted the highest sensitivity of any radar ever built. With anything close to automatic tracking still out of the question, the men went low-tech. *Sputnik* emitted a regular radio signal, and by monitoring the Doppler-shifted "bleeps" it was possible to rough out the orbit parameters. From there, using slide rules, a protractor, and string on a desktop globe, they pinpointed the window of opportunity. As Weiss related, "We knew where the launch range was, and where *Sputnik* would break through the horizon." The radar was pointed in the right direction and placed in search mode inside the parcel of sky in which the rocket had to appear.

Perhaps a half-dozen people crammed into the tiny antenna control room for the event. In later years Lincoln's radar systems would produce an actual image, nearly as sharp as a photograph, of any object orbiting Earth. For *Sputnik,* nothing so dramatic existed. Echoes splashed across a standard A-scope that depicted range from left to right: the farther along the tube face a blip appeared, the greater an object's distance from the radar. To provide a hard copy record of the digitized data from the radar's filter banks, an Analex printer had been set up in the processing area across the room and rigged to switch on if returns broke a certain threshold.

Millstone caught *Sputnik,* dubbed Alpha I, when it orbited to within about six hundred miles of the dish. Someone called out, "We got it!" Or, "There it is!" From the screen's far left rose the large stamp of the outgoing pulse. It was followed by flat nothingness more than halfway across. Then, the satellite appeared as a fuzzy wide bar or plateau. As an image, it was hardly inspiring, but Gordon Pettengill, a young Group 312 physicist, found it high drama, a reminder of the awesome possibilities of the nuclear age. The rat-a-tat-rat-a-tat of the printer, like a machine gun, accentuated the feeling.

Herb Weiss felt none of that. He got on the phone and called his bosses to let them know, with just a trace of satisfaction, that Millstone worked as intended.

In the wake of *Sputnik,* which plunged toward Earth after less than two months aloft, and especially after the launch of *Sputnik II* on November 2, Millstone flew into life. The radar did its job beautifully, proving able to follow the orbiters in range, azimuth, and elevation; it seemed to Herb Weiss that every military officer and electronics expert involved in defense planning descended on the field station for a look-see.

The commotion dovetailed perfectly with a major brouhaha already under way in Washington. *Sputnik* lent credence to the perception of an ominous missile gap between the United States and Russia, with the Soviets far ahead. An odd coalition of hardliners, think tank gurus, and startled members of Congress, among them Massachusetts Senator John F. Kennedy, immediately used the event as fodder in a drive to step up the pace of U.S. defense projects. Perhaps the most influential of these efforts was *Deterrence and Survival in the Nuclear Age,* a white paper considered at a packed National Security Council meeting less than a week after *Sputnik II* went up. The dire report, stressing a rapid Soviet military buildup, was put out by a high-level government consulting body chaired by H. Rowan Gaither Jr.

and including in its ranks Jerry Wiesner, Al Hill, I. I. Rabi, James Killian, and the industrialist and MIT alumnus Robert C. Sprague. Fearing it would spur calls for outlandish spending, Eisenhower tried to put a lid on the study, but members leaked the general details, fanning the simmering fires.

The prime outcome, not just of the Gaither report but of all the assorted studies and pressure coming to bear at the end of 1957, was an acceleration of the U.S. missile program, a hardening of silos to survive a nuclear assault, and a diversification of Strategic Air Command bases. As for defensive measures, the Gaither Committee had concluded: "If, as it appears quite likely, an effective air defense system can be obtained, this will probably be a better investment than blast shelters." The United States was already soliciting proposals for a new radar warning network to do what the DEW Line didn't — watch for Soviet missiles. At the end of 1957, on the heels of the Gaither advice, the country moved to finalize contracts to build what was called the Ballistic Missile Early Warning System.

For Lincoln Laboratory, BMEWS more clearly delineated the end of bomber defense. The change would not come all at once. Until missiles entered mass production, bombers remained the best bet to deliver nuclear weapons. Therefore, SAGE, the Texas Towers, and the DEW Line were still deployed. But the writing was on the wall. Of forty-six planned operational SAGE centers, only twenty-two were built — and after twenty-eight crewmen were lost when a storm wiped out a Texas Tower off the New Jersey coast in January 1961, the warning platforms were abandoned altogether. Within a few years Lincoln would emerge as the chief U.S. center of space tracking and surveillance. Its radars, laser radars, and deep-space surveillance posts ultimately proliferated outward from Millstone Hill to Maui, Kwajalein Island, South Korea, and various points in between. These would follow, image, and catalogue virtually every object launched into Earth orbit. By the same token, the laboratory took the lead in developing military communications satellites, as well as electronic systems and techniques used to evaluate so-called penetration aids — decoys and other measures designed to increase the probability U.S. warbirds would reach their targets.

The missile warning system proved the first real step in this evolution. Dan Dustin was helping evaluate the proposals. In August 1956 he and Gordon Pettengill had authored an influential technical report, *A Comparison of Selected ICBM Early-Warning Radar Configurations*. This memo, part of a Group 312 series examining ballistic missile detection, outlined the basic look of BMEWS, as it came to be adopted. The two deemed an experimental unit like Millstone Hill unfeasible as the nation's prime missile watchdog. Among other things, to keep large regions of the sky under continual surveillance, something other than a pedestal-mounted antenna would be needed. Instead, the choice for the main radars was a network of four massive fixed arrays, each 440 feet wide and 165 feet tall. While nothing this size had been built, it represented a scaling up of well-known technology. Millstone-like trackers, less vulnerable to jamming, would take over once an ICBM had been spotted.

RCA won the prime BMEWS radar systems contract in January 1958. Heading the effort was Dyer Brainerd Holmes, a thirty-six-year-old engineer who would later serve as NASA's first director of manned spaceflight and then president of Raytheon. Under RCA's aegis Goodyear would build the Millstone-ish trackers, while General Electric, which had also competed for the prime contractor position, took day-to-day responsibility for the big surveillance radars. Sylvania won the signal and data-processing subcontract. Western Electric got the enormous job of handling logistical support and establishing the rearward communications network.

With a strong technical foundation already laid, things moved quickly. A survey study focused on three host sites — Clear, Alaska; Thule, Greenland; and Fylingsdale Moor in Yorkshire, England. Of these, Thule took top priority, since it could be built a few miles from an existing Air Force base and also afforded the best view of the most likely pathway for Soviet attack — unless, of course, its mere construction altered Russian plans.

Dustin and Herb Weiss joined a small Pentagon committee chaired by Hector Skifter, a Gaither Committee veteran and Airborne Instruments executive, to oversee the project and ensure its smooth operation. The real job, in Dustin's view anyway, was to keep an eye on RCA and make sure the maverick company, known for pushing the limits of feasibility, stuck to the plans. "They needed some hard watching," he explained. As it was, things went almost incredibly smoothly. What Herb Weiss called a "Chinese copy" of a portion of one of the big arrays was built in Trinidad, to test the concept on Cape Canaveral missile trials. Meanwhile, Brainerd Holmes and RCA got busy organizing the construction of the real Soviet watchers.

Even though the missile detection network involved only a few outposts, as opposed to the DEW Line's multitudes, it shaped up as another massive construction job. Each site was a vast complex with control rooms, communications posts, workshops, and living quarters. Thule, spanning more than a nautical mile, and required to withstand 160-knot winds, posed a special challenge. To avoid the effects of heat being conducted into the permafrost and unsettling the foundation, a refrigeration plant was built to keep the ground frozen year-round. Engineers jury-rigged a port to allow some 10,000 tons of electronics to be shipped in during the four-month weather windows. Fuel farms had to be constructed, electric lines laid. The surreal conditions stirred men's souls. As Holmes recalled: "Glaciers were visible. The landscape was one of an eternal dull twilight; the impression was desolate."

Much like Clear later on but on a bigger scale, Thule housed a small army of radars. Dominating the landscape were the four gigantic aerials, each larger than a football field standing on end. Covered tunnels, able to accommodate firetrucks, connected the antenna banks. Each aerial was fed by four separate radars—making for sixteen in all—assigned to a selected quadrant of space. Each of these radars, in turn, relied on a pair of monster klystrons to put out two parallel fan-shaped beams, one several degrees above the other. The beams probed different parts of the sky depending on which of a multitude of feed horns were connected: organ-pipe scanning. A rising missile would pass through the lower one first; and by measuring range, angle, and Doppler shift on both, a computer could calculate the trajectory solution. The swivel-mounted trackers, sheathed in radomes, would take over from there. Ultimately, the barrage of radar data was beamed by scatter communications systems back to Cheyenne Mountain near Colorado Springs for processing. Inside the mountain, behind blast doors, sat NORAD, the North American Air Defense Command, an upgrade of CONAD that included the Canadians.

Fylingsdale Moor was a special case. The British eschewed the organ-pipe technology and opted solely for Herb Weiss's tracking units, which had a range as great as the main warning arrays. The radar, like the other BMEWS trackers, was identical to Millstone Hill in almost every detail. However, in lieu of a conical scan, it employed a monopulse feed—an upscale, electronic method of balancing target echoes to find their precise origin. Conical scanning requires several pulses to pinpoint a target in azimuth and elevation. Any fluctuations between pulses, such as a slight tilt in an aircraft that causes its radar "cross section" to alter, reduces accuracy. In fact, Millstone Hill had occasionally been jammed by reflection lobes from *Sputnik II*'s long rocket body rotating end over end. With monopulse radar, the same data is obtained from a single pulse, as echoes from two or more beams are compared simultaneously. If the target is exactly on the main beam axis, the amplitudes—or phases, depending on configuration—of the return signal will be equal. If a difference appears, the angle of arrival can be computed and the target's location nailed down. At that point the antennas automatically adjust to track their prey.

Brainerd Holmes thought the British restricted themselves to the smaller dish to preserve the scenic beauty of the surrounding moors. It also provided more protection against shorter-ranged, sea-launched missiles. But Herb Weiss and other engineers quipped that the tracker's swivel mount meant it could be easily rotated to keep an eye on the French.

• •

In the waning days of 1957, the missile warning network yet in its infancy and *Sputnik II* streaking overhead, a handful of Lincoln Lab staff members began putting Millstone Hill through its paces. One of those assigned to the big instrument was Gordon Hemenway Pettengill, the young Group 312 member who in 1956 had joined Dan Dustin in evaluating missile-watching schemes and, only a few weeks earlier, had studied the original *Sputnik* echoes with a deep sense of foreboding.

The son of a stockbroker, Pettengill had been born in Providence but grew up in Dedham, Massachusetts, where the family moved just a month before the great market crash of October 1929. His father survived the onslaught well enough to start Pettengill out in private schools. But as the Depression deepened, the boy had to switch to public education. He found himself far ahead of the pack, and in 1942, at the age of sixteen, graduated from high school early and enrolled at MIT to study physics. Had he been a bit older, Pettengill might have found a home at the Rad Lab. Instead, he was too young to join the war effort. Once he did turn eighteen, the youth was drafted into the Army, serving first in the infantry, then for several months with a Signal Corps company stationed in Austria. He returned to the school in fall 1946 and earned his degree two years later.

Pettengill's first professional job was at Los Alamos, developing techniques for monitoring nuclear radiation levels. But after two years, feeling trapped in a dead-end position, he entered the University of California at Berkeley to study for a doctorate in high-energy nuclear physics. There, the young graduate student supported himself partly by working as a night cyclotron operator, and he had many encounters with campus legends E. O. Lawrence and Luie Alvarez. His enduring memory of Alvarez was as a dapper dresser who habitually arranged to be paged a half hour into the weekly physics colloquium. "This served two purposes," Pettengill recalled. "One, it put his name before everyone so they knew he was there, and secondly, it gave him a chance to leave. So he was credited with being, you know, how should I say it, a little more sophisticated than many of the other scientists."

As Pettengill advanced in his studies, and grew more sophisticated himself, he took stock of all the big names in nuclear physics. "I knew if I stayed in it I was probably gonna be a small fish in a very large pond," he confessed. As a youth, Pettengill had passed many hours lovingly repairing old junk radios, so when a job possibility at Lincoln Lab materialized in 1954, the freshly minted Ph.D. figured he stood a better chance of making a name for himself in his old boyhood domain of radio and electronics than particle physics. The chance to be near his family sealed the decision. Before joining Group 312 he had worked in Group 35, Solid State Physics. One of his first assignments involved Project Ragmop, a Lincoln effort to design the signal-processing system for the AN/FPS-17 radar, a highly classified General Electric set secretly installed near Diyarbakir, in southeastern Turkey. The radar's main job was to watch Tyuratam, a few hundred miles northeast, where Soviet missile tests were being conducted — and it had covered *Sputnik's* launch from the same area.

At Millstone Hill, while Herb Weiss's group fretted over the system hardware, Pettengill studied the surveillance techniques and geometries that could milk the best performance out of the beast. Besides watching *Sputnik II*, a world of things, he remembered, needed to be checked out. One task involved using the big radar to watch the ionization trail as streaking meteorites punched holes in the atmosphere; these objects had to be characterized, the radar footprint well established, so that they could not be confused with Russian missiles. The moon posed another concern. A stream of radar pulses reflecting off the disk might bring into play a situation called range ambiguity, where the system could not readily discern whether an echo was the result of a recent pulse hitting a close-by object such as a missile or a much earlier pulse bouncing off the moon. As it was, a few days after the pivotal BMEWS site at Thule went on-line in 1961, the rising moon disturbed the twin beams and triggered an alarm; special software was later developed to take account of lunar movements, as well as the various satellite debris collecting in space.

In looking at meteorites and the moon, Gordon Pettengill found himself conducting studies that basically amounted to radar astronomy. He soon crossed the line completely. At the time only a smattering of people in the world — Bernard Lovell's group at Jodrell Bank and maybe a few Russians — were capable of probing such objects by radar. Thrilled at being part of this select body, Pettengill and a handful of fellow devotees, some from Lincoln, some from MIT, began turning up at odd hours to look at whatever struck their fancy. The moon was a particularly appealing target. The orb was so easy to pick up, Pettengill liked to say it shouted.

Things only promised to get better. Monster klystrons, once successfully installed, would up average power to around one hundred kilowatts. The Lincoln-built CG-24 computer was due to be hooked up at Millstone for real-time data processing. On the receiver end, the facility was slated to host the first maser amplifier put to practical use: Charles Townes himself didn't undertake a communications experiment with his creation until a few months later. By lowering the radar's inherent noise levels, the maser alone afforded a fivefold improvement in sensitivity.

Early in 1958, Pettengill was approached by Robert Price and Paul

Green Jr., both members of Group 34, Communications Techniques. The two had been following Millstone's progress with great interest. In a burst of inspiration over lunch in the lab's cafeteria, they had calculated that at its full powers Millstone might be able to detect Venus. The planet was fast approaching inferior conjunction, when the sphere aligned itself directly between the Earth and sun and was closer than any other point in its orbit. Since Pettengill was as familiar as anyone with Millstone Hill's operation, the men wondered, would he like to join a small group trying to probe Earth's nearest planetary neighbor by radar?

Pettengill jumped at the chance to become one of the "carpetbaggers," as Green later called the team. A great irony of radar's development to that point lay in the fact that while the technology had proven instrumental in establishing radio astronomy, enabling scientists to tune in star noise millions of light-years away, radar itself remained confined chiefly to terrestrial activities. In the dozen years following John DeWitt Jr.'s detection of the moon, no planet had been reached by radar. At its closest approach Venus, Earth's nearest neighbor, is still a hundred times farther away than the satellite. So, although the planet offers ten times the moon's reflecting surface, the return echo from Venus is ten million times fainter than one bouncing off the lunar landscape. Even as late as 1958, nothing had overcome that handicap.

It was frustrating because radar promised a rich payoff for planetary studies. Unlike radio astronomy, which entailed listening for the natural radio creakings and moanings of the heavens, radar offered the chance to interact with the planets — to reach out and touch their mysterious surfaces. Long before Millstone Hill was built, it had been proposed by several astronomers, including Frank Kerr and Joe Pawsey in Australia, that planetary radar measurements could pay scientific dividends. Chief among the rewards, as the Lincoln men knew full well, radar contact with Venus held the key to an accurate determination of the astronomical unit by which stargazers gauge all solar system distances. This celestial ruler, generally defined as the mean distance from the sun to the Earth, in 1958 was known by optical measurements to an accuracy of about a tenth of a percentage point. Such an error range, trifling in small-scale studies, looms large when extended to space. It left astronomers in the awkward position of being able to draw a precise map of Earth's neighborhood through well-established angular observations of the planets, but without a definitive scale.

Correcting that shortcoming was important both for general interest and for the coming age of space probes. The key lay in determining the exact distance to Venus, in essentially the same way an aircraft's range is calculated from the time interval between the outgoing pulse and the return echo. Since every planet's orbital period around the sun was known, scientists armed with the range figure and schooled in Kepler's laws could almost instantly improve the Earth-sun distance calculation by several orders of magnitude.

Looking back, the group gathered around Price and Green marveled at their freedom to conduct such exercises, which had nothing to do with Lincoln's mission. The lab hierarchy, though, looked on such adventures benevolently, on the grounds that anything that increased understanding of the big radar would be helpful in the long run. As Green once enthused, "Lincoln was an outfit that had a lot of money, and it was sort of generally understood that if you turned smart people loose, it paid off in strange directions — like radar astronomy."

The men labored anxiously to get everything ready by the inferior conjunction of January and February 1958. It would be a close call. "If it hadn't been for the maser we would not have attempted it," Price reflected years later. Even then, luck had to be on their side. Since Venus loped around the solar system cloaked perpetually in clouds, astronomers had never been able to glean much about it. So, the only hope of getting even a faint echo lay in the gamble that the planet was almost a perfect sphere, its hidden surface not so riddled with mountains and valleys that the radar signal would ricochet in a myriad of directions, casting too small a portion back toward Earth to detect. Similarly, the group trusted that Venus was not spinning too fast on its axis, since a rapid rotation would so smear out the echo in frequency that it would again be too faint to pick up.

These factors remained out of the carpetbaggers' control. In the meantime, a host of basic technical problems had to be addressed. Even with monster klystrons and the maser, finding Venus and retrieving a reliable signal would depend on exacting computer programming and processing. Here, Price, as the effort's enthusiastic leader, blazed the trail. In boyhood, the native Philadelphian had been entranced by the magic of plucking radio waves out of the air. Everything he had done since seemed aimed at improving his skills in this art. He had gone across the Delaware River to college, studying physics at Princeton, where Robert Dicke supervised his senior thesis. Next, he pursued a doctorate at MIT in electrical engineering. Hunting for a dissertation topic, he had approached Al Hill, then head of the Research Lab of Electronics.

"What do you want, the classified or the unclassified?" Hill had queried.

"I'll take classified," an intrigued Price had shot back. That snap decision had led him, by a circuitous route that included a Fulbright year studying radio astronomy with Taffy Bowen's Radiophysics gang, to Lincoln Lab's radio communications group, where he was considered an expert in statistical communications theory. From Australia he had brought back a book called *Radio Astronomy*, co-written by Joe Pawsey and R. N. Bracewell, in which it was speculated that astronomers might one day look beyond the moon with radar. He had lent the volume to Paul Green—who seized on the idea and brought it up over lunch the day they approached Pettengill. Before the meal was over, infected with enthusiasm, the pair had rounded up a team of colleagues willing to try and detect the mysterious planet.

For the Venus hunt, Price took it upon himself to make certain Millstone's antenna was pointed as directly at Venus as possible and programmed to stay with the target, despite the relative orbital motions of the planet and Earth, as well as the rotation of the Earth on its axis. A good approximation of planetary positions at different times was laid out in the standard ephemeris used by optical astronomers. Roland Silver at the lab showed him how to write a Fortran program to take the information into account. Price then printed out the product, which showed the proper settings for the handwheel that adjusted Millstone's antenna in azimuth and elevation.

It would take roughly five minutes for a radar pulse to cross the twenty-eight million miles to Venus and return to Earth. Therefore, Price's group planned to generate pulses for about four and a half minutes, then turn off the transmitter for five minutes and listen. Any radio noise gathered would first be reduced to a relatively low frequency, then filtered so that only signals from the expected spectrum of Doppler-shifted returns would be passed on. To avoid range ambiguities, Price borrowed from spread spectrum techniques coming into widespread use in communications and radar. These allowed him to code, or modulate, sequences of outgoing pulses by randomly changing the frequency or phase of the transmissions. That way he could easily separate echoes that had traveled approximately the right distance from those that had bumped into something along the way. Remaining data would be digitized and stored on magnetic tape for processing in an IBM 704 computer programmed to pluck extremely faint signals from the mush of background noise.

Almost from start to finish, radar to computer, it was a tricky job that depended on pushing the state of the art. Everything finally came together on February 10. By chance, it was Gordon Pettengill's thirty-second birthday, though he didn't remember drawing the connection. Other team members included Thomas Goblick Jr.; Leon "Jake" Kraft Jr., like Pettengill a radar man; Robert H. Kingston, who handled the maser end of things; programmer Silver; and William B. Smith, another processing whiz. Price remembered gazing up at the planet, visible as a bright light in the sky. "I looked at it as a star, and it filled me with a sense of wonder that we might reach that with the radar beam."

That day, and again on the twelfth, the carpetbaggers made a total of five runs, one of which proved so garbled it was not deemed worthy of processing. Even then most of the crew gave up the task as hopeless. But Price worked off and on for several months, coaxing the computer along, until it finally conjured up barely detectable returns on two of the runs — one from each day. On both, the planet showed up as a large peak, or spike, a bit closer to Earth than optical measurements predicted.

It was not a joyous finding. The presumed echoes were, as Pettengill told it, "marginal, marginal, marginal." One alone would have been cast out without a second thought. Yet, the fact that they had two, extremely similar, showing a range difference of only 2.2 milliseconds — or two hundred miles after correcting for the planet's movement away from Earth — was tantalizing. It seemed highly unlikely, almost impossible, that spurious noise could have caused two so identical returns. Even so, Price wondered whether he had forced the data together. Viewed from the other end, getting only two returns, which were not perfectly matched, was a sign of trouble, not success. But a Lincoln statistical expert okayed the computations. So the carpetbaggers went with their findings, hurriedly writing up the results for the March 20, 1959, issue of *Science*, where they proclaimed, in a professionally muted way, the age of planetary radar astronomy.

"It is evident," the men reported, "that the success of the entire undertaking depended critically on the simultaneous application of significant advances in the state of several arts: (i) a maser operating in the 400-megacycle-per-second range; (ii) radar equipment combining high transmitter power and a large, precision antenna; (iii) digital recording and nonreal time digital processing of signals. It is felt that the advantages of such processing of received signals have been established for both passive and active radio and radar astronomy observations."

Lincoln Lab staged a press conference pegged to the article's appearance. *The New York Times* ran the story on its front page. Price posed for television cameras in front of Millstone Hill. Everything was happy madness and glory and the carpetbaggers would wince remembering it all. Some good news, though, preceded the bad.

•

Venusian close encounters are rare. The solar system's grand dance card only permits an inferior conjunction waltz with Earth every nineteen months or so, limiting opportunities for study but building anticipation. The planet's next close approach came during August and September 1959, and Millstone's radar astronomers jumped at the chance to score a second, more definitive hit.

Only Gordon Pettengill, Bob Price, and Bill Smith were left of the original eight. This time around Pettengill led the effort. His two colleagues enjoyed the processing challenge. But he alone sensed a calling in radar astronomy, which offered a perfect meeting place for his seemingly innate bent for electronics and chosen career in physics. When Millstone first came into being, Pettengill had shared a two-bedroom Cambridge apartment with Irwin Shapiro, an imaginative Lincoln theorist who also loved musing about the planets and the mysteries of the heavens. But Millstone's attraction proved so great that Pettengill had recently moved northwest to Concord to be nearer the big telescope.

The main goal of the latest Venus hunt was a solid contact to prove the first detection had been real, and in a sense the men got what they wanted. Their own impressive efforts, 138 transmission runs in all, drew a complete blank. But better news arrived from Jodrell Bank, where Bernard Lovell's giant 250-foot steerable telescope, mired throughout most of the decade in financial and design difficulties, was finally humming along. There, John V. Evans and G. N. Taylor had picked up a faint echo at about the same range as the initial Millstone detection. The Lincoln team, still concerned over their shaky earlier finding, heaved a collective sigh of relief.

As Venus pulled out of reach for another nineteen months, Price and Smith returned to their normal jobs. Pettengill, though, stuck more with astronomy. The lab still looked benevolently on bootleg experiments that pushed the state of the art. Besides, celestial studies took up very little time — the last Venus hunt had occupied just a few hours spread over a month — and Lincoln benefited from the positive publicity such sidelights afforded. Pettengill led a small but growing pack of Lincoln and MIT researchers booking time on the dish. With Venus vanished in the distance, the main place people turned was toward the moon.

Although astronomers had been gazing on the lunar surface for centuries, optical telescopes still could not resolve craters, ridges, or other details much below a third of a mile in size. Specialized polarization measurements and photometric examinations yielded additional clues about the microstructure. But whether the surface was sandy, layered in dust, solid as a rock, or riddled with craters and boulders too small to see from Earth — anything aside from being made of green cheese — remained open to question. The answers were not only of fundamental scientific value, but also crucial before a landing could be attempted. This was, after all, the dawn of the space age.

Radar stepped in to fill the gap. On one level it offered an unprecedented look at surface makeup. From the intensity of echoes received, investigators could infer a landscape's reflectivity. High values indicate a dense surface covered by a good conductor such as water. From lower results, a dielectric constant could be deduced, and by consulting elaborate tables of the absorption and permittivity properties of different Earth materials, it was possible to determine whether the radar beam was reflecting off soft basalt or hard rock, or even to identify the specific rock type. At the same time, the changing polarization of the returned signal indicated whether portions of the radar beam were reflected or absorbed as they penetrated into the lunar surface. That gave clues about the existence and depth of a layer of sediment or dust.

Even more intriguing to Gordon Pettengill, as he turned Millstone's eye toward the moon late in 1959, radar often afforded a better view of lunar surface form and structure: morphology. A radar beam can be envisioned as a spotlight. When energy from the beacon strikes a smooth surface lying perpendicular to the antenna, the echo is mirror-like, or specular. "The phenomenon can be likened to sunlight glinting from the sides of wavelets on the ruffled surface of a body of water," Pettengill and two colleagues once wrote. The steepness of the waves, or surface features in the case of the moon, can be determined by tilting the antenna to one side and charting how quickly the reflections fall off in power.

Before Pettengill got into the act, this type of study had been conducted by several researchers. The most notable was James Trexler, who headed Project Moonbounce, a modestly successful Naval Research Lab effort to co-opt the moon into the U.S. intelligence network either as a relay point for communications signals or to pick up reflected Soviet transmissions and determine the frequencies and locations of enemy warning radars. Trexler used a big Navy paraboloid dish in Stump Neck, Maryland, to examine what was called time dispersion. Because the satellite is a sprawling orb consisting of various peaks and valleys, a radar signal strikes different parts of the landscape at slightly different moments, smearing out reflected energy in time compared to the sharp transmitted pulse. The amount of spreading, therefore, is a good indication of surface roughness. In 1956, after finding that extremely short pulses aimed at the lunar midriff came back in tight bunches, Trexler concluded the surface consisted of reasonably gentle slopes and didn't fit the popular picture of a craggy landscape marred by steep cliffs and sheer ridges.

Millstone Hill's extreme sensitivity allowed Pettengill to carry Trexler's kind of experiment all the way out to the lunar limbs, which the Navy radar could not see. At the same time he went beyond the scope of any previous study by establishing novel techniques for the emerging branch of science. One aspect of his investigations hinged on the fact that rough or irregular surfaces typical of small features — boulders, say — reflect energy far differently than the large prominences Trexler and others had examined. Instead of producing mirror-like returns, these objects scatter energy diffusely over a broad front. When he arranged for his searchlight to illuminate the moon at wide angles of incidence, Pettengill found these returns tended to dominate over specular echoes. In a landmark paper that appeared in the *Proceedings of the Institute of Radio Engineers* in May 1960, he produced an elaborate plot of the moon's radar brightness that showed the landscape was covered with rough features about the size of his wavelength — 68 centimeters — that had remained hidden to both optical and radar astronomers before him.

This work alone would have attracted significant attention. But as part of the same study, Pettengill unveiled a stunning mapping technique that went far beyond the ambiguous topographic images then in existence. Such maps provided fairly accurate estimates of surface roughness at specific ranges. But because the moon is a sphere, knowing the range meant only that an echo had originated somewhere in a halo that ran completely around the disk's face. There was no way of telling where in the circle — left, right, top, bottom — the return originated.

The same held true for similar studies of frequency dispersion, the Doppler-shifted spread of a return pulse after it strikes a moving target. In the case of the moon and Earth, where rotation rates were well established, scientists could predict what the frequency dispersion would be at a given range if the moon were a perfect sphere, and compare it to the actual signal to enhance their crude vision of terrain slopes and overall cragginess. In this case, the band of constant Doppler shift ran up and down one side of the moon, like a slightly bent stripe. Once again, no way existed of pinpointing the exact origin of any echoes.

The core of the new idea — a marriage of these two techniques — had been conceived by Paul Green. Although part of the communications group, the North Carolinian, whose pleasant drawl accentuated his keen intelligence, made little distinction between radar and radio. In radar, a transmission reflects off something and is generally detected back at its point of origin. In radio, energy is beamed to another point, but might reflect off something, such as the troposphere, along the way. It is the same physics.

Specifically, Green was interested in a class of objects — radar targets and communications channels — physicists characterized as being nonideal. That is, after being struck by electromagnetic waves, they produce an incident signal spread out in time, frequency, or both. In radar, a missile spinning on its way to a target is a prime example of a nonideal object. The moon is another. In communications, nonideal surfaces were getting particular attention at Lincoln as part of Project Needles, a plan to enable worldwide communications by placing thousands of hair-like copper dipoles into Earth orbit: if their properties were well known, signals could simply be bounced off these artificial reflectors.

Project Needles proved unpopular and unnecessary, giving rise to the epithet Project Needless. Radio astronomers raised Cain about interference with their studies. And although the slender shards were successfully deployed and tested in space communications, the advent of satellites rendered the matter moot. Before that point, however, Green busied himself working out the physics of scattering—the loss of echo power as a function of both time and frequency spreading. "I was sitting in my living room wondering what the relationship was between the two of them," he recalled. Suddenly, he realized that by plotting the change in both variables backward, he arrived at exactly two small areas on any given sphere where they overlapped.

Green mentioned his idea to Pettengill over lunch one day in fall 1959, before presenting it more formally a bit later at a pioneering radar astronomy meeting held at MIT's Endicott House conference center about twenty miles southwest of campus in Dedham, Massachusetts. Pettengill immediately moved to bring the concept to reality. The basis of the idea, as applied to the moon, was that by selecting only a portion of the returned pulse for analysis, astronomers could isolate zones of equal range and Doppler shift that intersected at the desired twin points—making it possible to map radar reflectivity even if the antenna beam itself was too spread out to resolve the area. It was like placing a radar magnifying glass on specific areas a few hundred kilometers square, instead of thousands of kilometers long and wide.

As it turned out, the technique proved remarkably similar to synthetic aperture radar, which was developed at the University of Michigan in the late 1950s but remained highly classified and unknown to those at Lincoln. Synthetic aperture systems are used to image a small piece of the fixed ground from a high-speed aircraft or satellite, while in the astronomical incarnation the heavenly target moves in relation to the observer. Both hinge on coherent range and frequency measurements—with transmitter and receiver precisely synchronized by an ultra-stable crystal oscillator—to simulate the aperture of a longer antenna. Again, as radar experts like to say, the physics are the same.

Merely configuring the equipment and writing software code took Pettengill hours of sweat. He then laboriously plotted hundreds of thousands of echoes, gradually building up an electroencephalogram-like mosaic of the lunar landscape. Flat or gently sloped features appeared as tranquil brain waves, mountains and valleys as the sharp spikes of firing neurons. Owing

RADAR MAPPING OF THE MOON

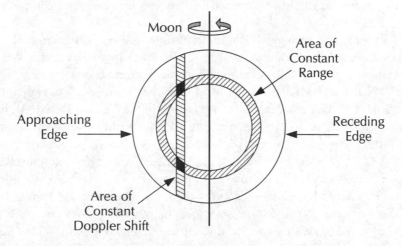

Source: Pettengill and Shapiro, "Radar Astronomy," p. 385.

to the ambiguity from twin areas of identical range and frequency shift, the map amounted to a snapshot of the full face superimposed onto half the face. Even with that caveat, it showed with startling clarity a number of locations where the moon was very rough.

The unprecedented lunar canvas created a stir in the astronomical community. The technique, soon known as delay-Doppler mapping, came to be widely employed, and after a time its purveyors learned to get around the inherent twin-image ambiguity by setting up two antennas in interferometer fashion and nulling out one of the points. In the case of lunar studies, delay-Doppler mapping headed the phalanx of powerful weapons in the radar arsenal, which together enabled researchers to build a picture of a relatively smooth surface marred only sporadically by craters and covered with a layer of firm sediment not unlike an ocean beach. These details proved instrumental in helping interpret results from the first successful American moon lander, *Ranger 4*, launched in April 1962.

Gordon Pettengill emerged from the study as the dean of America's rising generation of radar astronomers. But it was almost as if the fates were setting him up for a fall. A few months after the historic *Proceedings* paper appeared, Venus swung into position for another inferior conjunction. On this go-round, as events transpired, his old nemesis was coming back to haunt him.

Sullen and inscrutable, cloaked in clouds and immune to the probings of optical astronomers, Venus formed the archetypical radar target. The planet was set to line up between the Earth and sun on April 10, 1961, and this time around the hunt turned wild and merry as a host of competitors joined the fray.

By early that spring, Millstone's transmitter power had been vastly increased. Jodrell Bank, operating at a slightly lower frequency and with less power, had improved both its transmitter and receiver. They were dogged by RCA's prototype BMEWS tracker in Moorestown, New Jersey, as well as the Jet Propulsion Laboratory's Goldstone facility in the Mojave Desert. All had their own strengths and weaknesses, but all possessed the sensitivity to pick up Venus.

A Goldstone team, including JPL engineers Walter K. Victor and Robertson Stevens and a California Institute of Technology doctoral student named Richard M. Goldstein, won the race. The radar featured twin eighty-five-foot reflectors — one for transmission, the other for maser-enhanced reception — spaced six miles apart. Since it was designed for continuous wave operation, the set could not send pulses in the usual fashion. Instead, the transmitter was toggled on and off in varying, but predetermined intervals, thereby creating alternating periods in which the receiver would pick up radio noise with and without a signal. By comparing these on-off periods it was possible to cull out the radar echo, and from the known time factor determine the length of the round-trip journey. The first reliable planetary echo arrived on March 10. As more signals were captured, Duane Muhleman, a Jet Propulsion Lab colleague, calculated the astronomical unit at 149,598,600 kilometers, plus or minus 250 kilometers.

Pettengill was right behind. Strong signals were streaming into Millstone Hill by the twenty-second, and within two days he had enough to announce his own value for the astronomical unit of 149,597,850 kilometers, give or take. The rest of the Venus watchers — at Jodrell Bank, RCA, and eventually, a powerful Soviet radar in the Crimea — were in remarkably close agreement. The average value was 149.6 million kilometers, with none more than 2500 kilometers off that point. The finding immediately improved the old optical calculation by more than two orders of magnitude, and was afterwards adopted by the International Astronomical Union. Even before then, it contributed greatly to the *Mariner 2* Venus flyby mission launched in August 1962.

While good news for astronomy, the vastly improved estimate of the Earth-Venus distance released the Damoclean sword hanging over the carpetbaggers since their initial 1958 detection. Armed with the new figure, Bill Smith began sifting through the earlier data stored on magnetic tape. His efforts showed that neither the 1958 Lincoln detection nor the Jodrell Bank report the following year could have been real; it was apparently a remarkable, if unfortunate, coincidence that the two sets of spurious echoes agreed so closely.

As a junior author of the Millstone account, Gordon Pettengill was insulated somewhat from any fallout. But he and Price had compounded the initial mistake by publishing a second paper, based on the erroneous results, in which they had claimed an even better value for the astronomical unit. "It certainly taught me a lesson," Pettengill winced several decades later. "Although technically, they ought to have a chance of one in a thousand of being wrong, there are, you know, systematic effects." Pettengill bore the scars of his misjudgments the rest of his career. But he learned from them and he wasn't done with radar astronomy. Or Venus, for that matter.

•

After Venus again passed out of hailing distance in spring 1961, Pettengill was drawn increasingly into a long-running project to build what would remain for decades the world's most powerful telescope — a 1000-foot-diameter steel mesh reflector situated in the limestone hills of Puerto Rico.

The Arecibo Ionospheric Observatory was a Cornell University undertaking, with participation from the University of Sydney. The facility had been conceived chiefly by Cornell electrical engineering professors William Gordon and Henry Booker, who had sold the megaproject to the Advanced Research Projects Agency as a tool for conducting ionospheric research. During the Cold War such studies were deemed militarily crucial, both in understanding forward scattering for communications purposes and for predicting the likely effects of the flash of radiation from nuclear blasts in space.

From early on, however, it had been clear the best use of the Arecibo dish lay in astronomy. Overseeing much of Cornell's end of things, first as dean of physics, then of engineering, was Dale Corson, Taffy Bowen's old mate from the Rad Lab airborne interception project. Around the time of *Sputnik*, Corson had led the school's effort to beef up Cornell's lagging radio astronomy program. His search committee had looked to Harvard and spirited away Tommy Gold, a brash Austrian-born cosmologist famous for his continuous creation view of the universe. In fall 1958, seeking help on the radar front, Gold had detailed the Arecibo plans at a Millstone Hill luncheon seminar and brought Pettengill into the fold. The Lincoln man had started out as an informal consultant. But by June 1960, when earth-moving began, he was a regular visitor to the Puerto Rican site.

The big bowl was being set in a natural depression formed where water had long ago leached out the limestone, causing the terrain to collapse. Over the next three years, with the project snarled in construction delays and resulting lawsuits, Pettengill visited several times a year for lengthy stays. In July 1963, when the facility was nearly ready, he signed on as associate director and moved in completely, taking up residence in the nearby town of Arecibo.

As he helped prepare the facility for making observations, Pettengill grew anxious to get his own time on the dish. In the previous two years, radar astronomy had revitalized the study of the planets and solar system mechanics, a field many astronomers had abandoned only a decade or so earlier. After Bill Smith first raised the possibility, based on more forages into the magnetic tape archives, Venus's slow retrograde rotation had been established by radar astronomers at the Jet Propulsion Laboratory: unlike any known body except Uranus, the planet spun on its axis opposite to the direction of its motion around the sun. Things were exploding on other fronts, too, as transmitters and digital processors grew more powerful, receivers more sensitive, and new telescopes came on-line. Arecibo, churning out 140 kilowatts of average power and boasting the latest in receiving equipment, was the best of the best. As soon as Venus moved into range early in February 1964, Pettengill took up his long-running investigations of the planet.

The hunt resumed on the tenth, six years to the day from the carpetbaggers' initial foray. This time there was no ambiguity — Pettengill was stunned at the clarity of the echos. "The first time we turned on the thing, there was Venus just loud and clear," he marveled. All the thrills of the pioneering Millstone Hill days came gushing back. And this time, too, the birthday connection stuck. He always treasured the gift.

Pettengill continued probing the planet well into the summer. His right hand was Rolf Dyce, a Cornell-trained physicist on leave from the Stanford Research Institute, where he had conducted radar studies of the moon and the aurora borealis, as well as classified radar-based analyses of the atmospheric effects of nuclear tests. With Dyce, Pettengill's Arecibo experiments confirmed the retrograde rotation of Venus, while greatly improving values for the axis of rotation and tilt. The two, joined by Cornell graduate student Tommy Thompson, also probed the moon, honing delay-Doppler maps of the Northern and Southern hemispheres.

All this, though, turned out to be a warm-up act. Pettengill got busy refining his electronics and computing techniques, waiting for an opportunity to really put the radar through its paces. The chance came in the spring of 1965, when Mercury swung within Arecibo's field of vision. What he found surpassed even his wildest expectations.

Mercury was another of the solar system's enigmas. Its travels bring the tiny world only a bit farther from Earth than Mars, but the orbit is so close to the sun that optical astronomers had trouble penetrating the brightness to identify the faint surface landmarks needed to judge rotation rate and other basic characteristics. What observations had been conducted over the seventy-five years before Pettengill and Dyce set up shop at Arecibo held few surprises. The planet seemed to rotate on its axis every eighty-eight days, in perfect sync with its orbital motion, so that it always held the same face to the sun. Unusual, but easy to grasp. One troubling issue hinged on radiometric studies that indicated the sphere's dark side was somewhat warmer than it should have been, given the fact it had never seen the sun's rays. But even solid scientists sometimes ignore the trees for the forest.

The Arecibo team got its first good echoes on April 6. Although farther away than Mars, Mercury forms an easier radar target, partly because of the relatively unfavorable oppositions of the red planet, but also because its rotation is so slow returns aren't smeared out by the Doppler effect into a messy sludge that makes it hard to tell one from the next. It had first been spotted by radar in June 1962, some eight months before a Mars detection, by V. A. Kotel'nikov and his group at the Soviet Institute for Radio Engineering and Electronics. But neither that faint initial contact nor the few subsequent observations at Millstone Hill and the Jet Propulsion Laboratory had done anything to shake the optical findings.

Pettengill didn't expect to find anything significantly different. But he knew that Arecibo could help determine the planet's rotational axis as a clue to how the seasons unfolded, as well as pick up echoes from previously unglimpsed features. He and Dyce began their runs late each morning, when Mercury rose to within Arecibo's field of vision: the big eye could tilt just twenty degrees from the zenith. They would transmit pulses for roughly ten minutes, then go into listening mode. Returns were stored on magnetic tape and processed by a Control Data 160-A digital computer — it carried four kilobytes of memory and looked exactly like an office desk — housed in a big room behind the transmitter. Part of the work involved plotting the Doppler-shifted echoes over time, as the planet moved in toward Earth then back out on its path around the sun. By focusing on returns from a specific range-delay zone, Pettengill could calculate rotation rate more precisely, and, hopefully, divine the spin axis.

It was all standard stuff. Pettengill even used the same computer program he had for Venus, modified for Mercury's predicted leisurely eighty-eight-day rotation. He could envision the expected plot in his head. On the ends of each pulse his data would show two peaks, corresponding to the echoes from points of opposite — plus and minus — Doppler shift on different sides of the planet. The distance between the peaks would not be too great because Mercury's slow rotation, in perfect harmony with its trip around the sun, dictated little frequency spreading. However, when Dyce brought over the first data, the peaks were significantly farther apart than he had figured. "Almost immediately when I saw the data, I knew it didn't fit the one-to-one model," Pettengill recalled.

"Wait a minute," he told Dyce, excitement swelling inside him. "Things aren't making sense here." Unless he had fouled up his range calculations, and was therefore looking farther out on the limbs than he had supposed — where the frequency spreading should be greater — the unexpectedly large shifts meant the planet was spinning faster than everyone thought. Pettengill trusted his range data implicity; he had undertaken many similar studies of the moon and Venus. Yet, he had been burned with the Venus detection, so he suppressed his excitement while sitting down to calculate what the real rotation rate must be. From the initial runs, it was impossible to determine which direction the planet was spinning. But one thing was certain: it wasn't taking eighty-eight days to make a complete revolution. He concluded, therefore, that Mercury was either in a direct, or prograde, rotation about once every fifty-nine days or it moved in a retrograde orbit every forty-six days. In either case, after checking his figures several times, Pettengill felt he had the discovery of a lifetime. He called Tommy Gold at Cornell with the tentative news, then set about checking his results again.

Additional runs, on April 10, 1965, and again on the twelfth, confirmed the initial finding. There was still some ambiguity. The planet had reached the zenith of its inferior conjunction with Earth; and he had to wait until it moved farther away to nail down the spin direction. When the chance came on the twenty-fifth, he could see that Mercury was in a prograde rotation of fifty-nine days, plus or minus five days. He called Gold again. The Cornell man told Pettengill to start writing it all up; he could get a note published in a few weeks. The account appeared in the June 19 issue of *Nature*, accompanied by a theoretical explanation offered by Gold and Stanton J. Peale, who postulated how the gravitational pull of solar tidal torques could have nudged Mercury into a different rotation.

In one fell swoop the Pettengill-Dyce paper brought all the old confusion into focus. Thermal temperatures on the planet's presumed dark side were warmer than expected because the newly confirmed rotation meant that opposite faces of the planet alternately faced the sun at each close approach. Still, flashing back to Venus in 1958, Pettengill kept his self-congratulations to a minimum. "You don't want to provoke the gods by celebrating when it could be wrong, you know. . . . There's always that one chance in ten you've overlooked something — either in the equipment or your own calculations. So I didn't celebrate."

Even without angering any deities, Pettengill and Dyce missed one thing. Included within the margin of error for the planet's rotation rate was a value of 58.6 days, which would mean Mercury pirouetted on its axis in exactly two-thirds the eighty-eight days it took to journey around the sun. If true, Mercury would be unique in having its rotation locked to a multiple of its orbital rate, unleashing a torrent of intriguing possibilities for solar system and gravitational studies.

As it turned out, other scientists soon proved this to be the case, and Pettengill kicked himself for failing to see the implications of his own observations. But he was not a theoretician, and even Gold and Peale, who were, had missed it in the excitement of the main discovery.

•

The discovery of Mercury's true rotation rate marked the third major piece of gold — after the improved value of the astronomical unit and the retrograde rotation of Venus — mined by planetary radar. In its wake, behind the muscle of facilities like Arecibo, Goldstone, and others, the upstart astronomers continued along their auspicious path. In just a few years Earth-bound radars peered out to Jupiter, pinned down the masses and radii of Mercury and Venus, and added a few decimal points to the astronomical unit. That, in turn, yielded an even more exact solar system model.

Venus still called to Gordon Pettengill. He left Arecibo in 1965 and returned to Lincoln Lab as associate leader of the new Haystack Observatory. Another Herb Weiss creation, Haystack perched on an adjacent nob of the same hill that hosted Millstone, taking over the lab's radio and radar astronomy duties after 1962, when its predecessor's mission shifted to ionospheric physics and space surveillance. After three years there, Pettengill again took up residence at Arecibo — a fourteen-month stint as director — then joined the MIT faculty as a professor of planetary physics. At all these places, while he looked also at Mercury and the moon, he never lost the thrill for Venus. Haystack's fully steerable, 120-foot paraboloid and startling resolution pinned down the planet's radius. From Arecibo, Pettengill produced detailed maps that showed, when combined with data from other astronomers, gentler slopes on average than on the moon, but with large tracts of exceptionally elevated areas, and a curious accumulation in certain zones of matter remindful of terrestrial rocks and hydrocarbons.

It was an especially satisfying time. Pettengill was married late, in 1967, at the age of forty-one, to Pamela Ann Wolfenden, an Englishwoman he had met at Lincoln's ski retreat in Woodstock, Vermont. The couple moved down to Arecibo a year later with the first of their two children and returned to Cambridge after that time in the tropics. Pettengill's long-running association with former roommate Irwin Shapiro also thrived. Beginning in the 1960s, a formal collaboration evolved — Shapiro as theorist, Pettengill as experimentalist — that would produce nearly two dozen papers on various aspects of radar astronomy. Eventually, Pettengill would head MIT's Center for Space Research, while a bit down the road, Shapiro directed the Harvard-Smithsonian Center for Astrophysics.

There was one last frontier. As the space age became a reality, Pettengill began to think seriously about putting radar on a planetary voyager. Almost upon joining the MIT faculty in December 1970, he gained a spot on a National Aeronautics and Space Administration study group examining the feasibility of sending spacecraft to Venus. The mission's chief concern lay in gathering ionospheric, atmospheric, and meteorological data. Pettengill also proposed, however, mapping the planet's surface with a small radar altimeter.

The rationale for such a proposal was strong. Venus is similar to Earth in mass, diameter, and age. More than for any other solar system body, a radar map of the mysterious planet promised to shed light on Earth's own complicated mechanisms of existence. NASA listened. As it turned out, officials looked on the idea as a precursor to a more sophisticated, dedicated radar mission several years down the road. So when the initial Venus mission got the go-ahead, Pettengill's project was accepted.

The planning, design, and engineering of what came to be called *Pioneer Venus* stretched out seven years. Four times annually Pettengill flew out to NASA's Ames Research Center in Palo Alto, which managed the undertaking. As a radar man, he found himself a bit of a pariah among the atmospheric-minded. "So my experiment, which involved only the surface, was a little bit of an outcast," he recalled. "It was treated sort of as a black sheep that they wished wasn't there because it was heavy and took a lot of power and used a lot of bits to send the data. . . . They were very reluctant to let this big tiger in the cage."

Still, with NASA headquarters backing it, the radar flew. *Pioneer Venus* consisted of two spacecraft. The first, an orbiter carrying Pettengill's altimeter and various atmospheric equipment, was launched on May 20, 1978. Nearly three months later, on August 8, its sister craft lifted off laden with a series of probes that would be dispatched straight into the atmosphere to frenetically measure pressure, temperature, chemical composition, wind velocities, and wind directions at various altitudes.

The two spacecraft arrived at the planet that December: *Pioneer Venus 1*, the orbiter, which took a more indirect route, beat its sibling by only five days. Both were extremely successful. The probes showed that the Venusian atmosphere consisted almost entirely of carbon dioxide and was capped by a cloud layer high in sulfuric acid. Infrared radiation had been trapped in the atmosphere, bringing surface temperatures to a steamy 470 degrees centigrade, like a gigantic self-cleaning oven. The atmospheric pressure was ninety times that of Earth at sea level, like being nearly three thousand feet under the sea.

Pettengill's radar did its own pioneering. It had limitations. The altimeter charted topography by looking straight down and measuring round-trip pulse times. Although accurate to within about a thousand feet in gauging surface heights, it was nearly blind when it came to the lateral resolution that showcased slopes. In this dimension, features had to stretch out at least thirty miles just to be noticed. Still, the spacecraft's highly eccentric orbit, dictated largely by atmospheric studies, ran from 74 degrees north latitude to 63 degrees south latitude and encompassed more than 90 percent of the surface. Given this unprecedented global view of the planet, *Pioneer* proved a treasure trove.

Venus gazers on Earth had already spotted a vast highland area, taller than Mount Everest and towering some seven miles above its surrounding plain, that dominated the region at 65 degrees north latitude. This was a suspected volcanic caldera dubbed Maxwell Montes—for the physicist James Clerk Maxwell—and it turned out to be one of the few areas of prominence. *Pioneer*'s altimeter revealed the rest of the surface to be remarkably spherical and flat, a canvas of rolling plains where 60 percent of the landscape fell within a .6-mile range in height—a probable result of a slow rotation rate that didn't, by centrifugal force, create a midriff bulge like that adorning fast-spinning Earth. The radar also picked up signs of venerable tectonic activity in the planet's crust, especially in a long rift-like valley just north of twenty degrees south latitude, between 150 and 175 degrees east longitude. That was intriguing because scientists could find no sign of anything like the Earth's plate tectonics or sharp ocean-continental boundaries to indicate dynamic motion.

A last discovery inflamed the public's imagination. Permeating the higher altitudes, *Pioneer* found areas of low thermal emissitivity, as seen on its radiometer, but high radar reflectivity as measured back on Earth. The strange substance, unlike anything previously known, except water, was soon dubbed the "mystery material" in the press. "If it had been California, as I kept saying," recalled Pettengill years later, "it would have been water." But everyone knew there wasn't any water on Venus.

Mapping ceased in March 1981, when *Pioneer 1*'s fuel supply dwindled too low to keep the craft from drifting into a higher orbit and out of radar range. Long before then, the effort's clear success helped convince NASA to advance its idea of sending up a far more sensitive Venus mapper. The project was called VOIR, for Venus Orbiting Imaging Radar. It had been conceived specifically to exploit the capabilities of synthetic aperture radar. Long used by the military as an all-weather supplement to optical surveillance, the technique had been adapted for civilian observations of the Earth from space and deployed in the Seasat satellite launched by the Jet Propulsion Laboratory in 1978. It produced strikingly clear images, and NASA wanted to bring the technology to Venus before the Russians did.

Pettengill led the project's science study working group, which roughed out mission objectives and design parameters. In 1979, after the group submitted its report, and buoyed by *Pioneer's* mapping feats, he won the competition for what was ultimately called principal investigator of the Radar Investigation Group, giving him the main responsibility for the craft and its experiments.

As unforeseen snags materialized, and plans changed over the next several years, the endeavor would be scaled back and altered in scope several times. Reflecting the shifting circumstances, the project was renamed Venus Radar Mapper. Then it was christened, more simply, Magellan. Pettengill's own role diminished some as well. But anyway you sliced it, a detailed radar map of Venus remained the chief objective, and where radar was concerned, Gordon Pettengill held dominion. Magellan was really his baby.

20 • A Distant View

"Astronomy compels the soul to look upwards and leads us from this world to another."

PLATO, *The Republic*

LINGERING CLOUDS BROKE and a strong crosswind mercifully died down on an otherwise blustery overcast day at Cape Canaveral, Florida. It was around 2:45 the afternoon of May 4, 1989, and the space shuttle *Atlantis* had been on weather hold at pad 39B for fifty-nine tense minutes. Finally, just five minutes left in the launch window, the countdown resumed. *Atlantis* rose slowly out of billows of exhaust steam and disappeared in a blue glow behind the haze. In its cargo bay sat Venus-bound *Magellan*.

The shuttle made five revolutions around the Earth at an altitude of 296 kilometers, or nearly 178 miles, before its precious freight was eased out. The 38-foot-long voyager was almost stark white, with copper and black trim — the only adornments to the main body neat black lettering spelling out "USA" above a miniature, full-colored replica of the American flag. *Magellan* gathered itself for an hour, enough to orbit two-thirds the way around the Earth and slowly extend solar panels, before controllers ignited two solid-fuel inertial boosters that whisked it gracefully on a path toward the sun. After thrusters were fired to fine-tune the spacecraft's course, the inertial upper stage separated and left the radar-laden core on its own.

If all went well, *Magellan* would circle a little more than one and a half times around the sun and lope inward for a rendezvous with Venus early in August 1990. NASA controllers and engineers were playing the role of a quarterback lobbing a long bomb pass, not to where a receiver was, but where he would *be*. Only this pass traveled at 71,000 miles an hour and covered 788 million miles of space.

Gordon Pettengill had flown down to Florida almost a week earlier, in time for the originally scheduled launch date of April 28. But a recirculation pump problem had forced the countdown to be aborted thirty-one seconds shy of liftoff, and other commitments dictated his return to MIT. He breathed an inward sigh of relief when word came that things had gone well.

Magellan was six years late leaving Earth, and Pettengill had endured many occasions when he wondered whether the craft were doomed. The Venus Orbiting Imaging Radar initially had been scheduled to blast off in 1983. But problems kept surfacing. Under the concept put forth in his original proposal, Pettengill oversaw what NASA called a principal-investigator-supplied experiment, which meant that the MIT Center for Space Research would custom-build the synthetic aperture radar system. However, as overall estimated mission costs ran up the job was farmed out to the Hughes Aircraft Company, a subcontractor to the Martin Marietta Astronautics Group, the mission's prime contractor. That left Pettengill as an advisor on the radar construction end, although he still presided over the twenty-six-person Radar Investigation Group that planned most of the mapping experiment.

On the heels of this disappointment, Pettengill learned that the Soviets were threatening to steal his thunder. At the time French space scientists maintained a close working relationship with the Russian space program, even supplying some experiments, and through them, in early 1980 he got wind that the USSR planned its own synthetic aperture mission to Venus. That June, in Budapest for a meeting of the Committee on Space Research, a group set up to facilitate international cooperation in space, he cornered Roald Z. Sagdeev, director of the Soviet Institute of Cosmic Research (IKI), and asked him point-blank whether the rumors were true.

Sagdeev, not the type to shy away from a direct question, responded by sitting Pettengill down in a conference room with a few aides and candidly outlining the entire venture. "I was a little amazed," Pettengill admitted. "He believed in openness. He was one of the early *glasnost* people, and I guess he felt secure in his position and was willing to tell us what they were going to do." On the evening of June 5, and again the next night, the Soviet scientist, who later married a granddaughter of Dwight D. Eisenhower and settled in the United States, laid out the basics of four missions his country planned to Venus—a lander, two atmospheric-type experiments, and the synthetic aperture radar mapper. Pettengill took it all in, scribbling notes as Sagdeev and his comrades briefed him.

As soon as he returned home a few days later, Pettengill fired off a memo on his Sagdeev conversations to Thomas "Tim" Mutch, NASA's associate administrator for space science and applications, who would be killed in a Himalayan mountain climbing accident early the next year. The radar mission, Pettengill wrote:

> . . . is obviously close to my heart and I tried to find out as much
> as possible about it. I learned that it was to use a highly eccentric

(48-hour period) orbiter, with a periapsis altitude of 1500 km. The inclination is such that a synthetic aperture radar can view the pole. Although he would not quote a precise value of inclination, I estimate it from the likely SAR geometry to be about 75° (the same as Pioneer Venus). The surface resolution of the images to be obtained is 2 km (eight times worse than VOIR global). An attempt to launch the SAR mission in the next opportunity (fall 1981) will be made if all is ready, but Sagdeev felt that chances of this were less than 50–50. More probable was a launch in late spring of 1983.

Sagdeev had hinted, without actually saying it, that the Soviet radar mapper was designed to upstage VOIR, which at that point was nominally slated for a 1983 takeoff but already under well-publicized duress because of cost increases. "When I heard about it, initially it was very depressing, because these guys were going to scoop us basically," the American related. Down the road several years, when the Soviets shared their exact plans, he would find ways to turn the Russian venture to his advantage by making sure the American mission outshone it. In the meantime, though, the sense of depression only deepened.

That fall Pettengill left the United States for a year's sabbatical at Sydney University as a Guggenheim fellow. A few months later, early in February 1981, he got a telephone call from MIT colleague Sean C. Solomon, in charge of the school's Venus Orbiting Imaging Radar efforts in Pettengill's absence. Solomon had bad news. He related the word of Tim Mutch's death, and told Pettengill that the month-old Reagan administration had scrubbed the *Pioneer Venus* expedition, whose costs, originally estimated at around $200 million, had reached the dumbfounding level of $700 million. The one ray of light, Solomon went on, was that the President had thrown NASA a bone, indicating that if the budget could be cut in half, to around $350 million, the Venus expedition might go back on the books.

The fiscal axe set off a scramble to resurrect the Venus mapper. Officials eliminated virtually all VOIR's nonradar projects — infrared and ultraviolet experiments, a microwave ionospheric sounder, and a sophisticated Dicke-type radiometer to chart the planet's thermal emissions. That done, radar plans were reexamined. The VOIR had been set to occupy a tight, circular orbit at a constant altitude above the planet, a tremendously expensive step that required huge fuel expenditures. Such an orbit was necessary because NASA had adopted the same synthetic aperture processing methods the Air Force used on its surveillance flights. Radar data was filmed and run through analog optical processors whose lenses had to be specially ground to handle precise, predetermined geometries. Any change in altitude, or even the angle of the radar beam, meant adding other focusing lenses, a complication the Jet Propulsion Laboratory was not willing to endure.

Since all this had been decided, though, spectacular advances in digital computer processing had obviated the need for the old analog ways. This had been driven home with JPL's Seasat, the first civilian satellite to use synthetic aperture radar. Launched in June 1978, it aimed to generate high-resolution images of the Earth's oceans to gauge such factors as surface winds and sea states. But after NASA released analog pictures, the Canadian areospace firm of MacDonald Dettwiler and Associates Ltd. had mortified the Americans by digitally processing the images, making the JPL product pale by comparison. With Seasat well in mind, engineers reexamining the Venus mission realized they could adopt a much cheaper elliptical orbit, which required less rocket fuel to achieve. Mapping would take place during perigee, when the orbiter was closest to Venus, with a digital computer adjusting for changes in elevation and angle simply by changing a numerical coefficient.

Beyond this, virtually every other aspect of the project was placed under a microscope. Instead of custom-building the main aerial, planners liberated an old *Voyager* mission spare from mothballs. To save design costs, the command and control system was copied from Jupiter-bound *Galileo*. All the trimming brought the price tag down to around $300 million; but, Pettengill claimed, JPL officials tacked on the additional $50 million allowed as contingency funding just to be safe.

Washington was enough impressed that in the spring of 1983, even after the Soviets launched two Venus radar mappers, *Venera 15* and *16*, the revamped mission looked like a done deal. By then, the Russians were sharing their plans in detail, and it was clear that the American venture stacked up well against it. For one thing, essentially as Sagdeev had revealed, the Russian spacecraft would send back images of only the Northern Hemisphere, while NASA's expedition covered the entire planet. For another, the U.S. radar, with its digital processing and other advantages, promised roughly four times the resolution of its Soviet counterparts — compelling reasons, Pettengill felt, to go ahead with the plan.

Still, the American endeavor was not officially on the books, so he and several JPL engineers looked for ways to further improve the mission and assure their bosses an ever richer return on investment. As an initial step Pettengill pushed hard to get a radiometer back on board. He also supported a daring improvement in an already innovative JPL-Hughes data compression scheme.

The overall data compression plan was conceived mainly by Ronald Kwok and William T.K. Johnson, the Magellan radar system chief engineer.

Known as BAQ, for block adaptive quantizer, it was an extremely loose equivalent of High Definition Television for video images, relying on the fact that the overall average intensity of the radar scene changed very little from pixel to pixel. Essentially, phase information from the radar signal was preserved, but advantage was taken of the limited variation in the amplitude signal to reassemble a detailed picture with less of this data than usual. At the time Johnson had decided on three bits of amplitude data, but the new scheme boiled things down to a bare minimum of two.

Slicing the number of amplitude bits allowed spatial resolution to be improved 1.5 times, enabling the synthetic aperture system to image features as small as 120 meters, or about 400 feet. Johnson and spacecraft engineers also found the tape recorders ran faster, and that the entire system performance improved as well. The BAQ strategy would not have worked out well on a high-contrast planet such as the Earth, filled with water and areas of multivaried terrain, but it was perfect for Venus. When Johnson brought the idea to Pettengill, the MIT man pronounced himself suitably pleased. "It was almost something for nothing," Pettengill asserted. Instead of being roughly four times better than the Russian radars, the American system promised something like ten times the resolution of its rivals. Even then, the numbers could be misleading: the enhanced BAQ actually brought back a hundred times more data from every area on the surface than the Soviet system.

All the changes were incorporated into the project. The scaled-down, supercharged mission, reincarnated as the Venus Radar Mapper to shed VOIR's gold-plated image, got the formal go-ahead in October 1983, the same month the *Veneras* reached their destination. Pettengill knew well before then that the mission had been saved. The mapper was officially set to fly in May 1988, but the January 1986 explosion of the space shuttle *Challenger* incapacitated the American space program for thirty-two months, creating a backlog that forced the expedition on the sidelines for another year. After the disaster, the Venus foray got a project name: Magellan. Even then, Pettengill fretted about the choice: Magellan the explorer had been killed in the Philippines before completing his round-the-world quest to find a western route to the Spice Islands in the East Indies.

As if to push his nerves to the limit, when the 1989 launch finally neared, there had been one last gamble. The most natural takeoff window came late in the year — October. That would have enabled a direct approach that placed the craft into Venus's orbit within four months, in February 1990. However, the heavy *Galileo* mission to Jupiter, which needed a gravity assist from Venus to whip out to its objective, depended on the same four-week opportunity. Since *Galileo* was also shuttle-launched, and NASA did not feel comfortable planning two flights within a few weeks of each other, *Magellan* planners had been faced with a quandary: either leave in May and take fifteen months to reach Venus—six months later than an October departure would have put it there—or wait some two years for the next ideal window and a four-month trip.

The longer the route, the more things could go wrong. On the other hand, the May date not only got *Magellan* to Venus a year earlier than the alternative, it held down the costs associated with the "marching army" of experts needed to maintain the equipment and run the mission. "Plus," Pettengill added, "space is a much nicer environment than Earth. You don't get wet, you don't get environmental shocks, things like that. In fact, you can spend some of the time up there learning about your spacecraft, which we did. And most important of all, once you're launched, they won't cancel the project."

Magellan reached Venus as planned on August 10, 1990, although controllers had needed to execute three midroute trajectory corrections to keep the date. During the long journey overseers had put the craft through its paces, checking star reference points, recalibrating the gyroscope, and even ordering the radar through a mock run to test the mapping software. Things had gone well, but the mood was still tense that day in the Magellan Mission Support Area at Martin Marietta's Waterton facility south of Denver. Scores of engineers and software experts—flanked by a bevy of reporters, photographers, and television cameras—crowded around the fourth-floor operations room to witness the long-anticipated rendezvous.

The first critical command called for an eighty-two-second burn of *Magellan's* solid rocket motor, a maneuver designed to brake the craft to its desired orbit. The mission marked the first time a solid rocket had been exposed to space for anywhere near the fifteen months *Magellan* had been aloft. And to heighten the suspense, the spacecraft had to disappear behind Venus, taking it out of contact with the command center for thirty-seven long minutes.

"There were a lot of butterflies floating around in the air," recalled Rick Kasuda, flight software leader. But as soon as the craft finally came around the planet, he threw both hands up in a gesture of triumph and the cheering started: *Magellan* had hit a picture-perfect orbit. The big voyager, originally 7612 pounds but slimmed down by nearly two-thirds from firing its propellant, didn't need even the slightest orbit trim.

With the craft safely on course, the Jet Propulsion Laboratory and Martin Marietta teams began a tiresome, week-long series of calibrations and checks of the radar systems. Back in Pasadena, at the Science Mission Support Area on the second floor of JPL's Building 230, Gordon Pettengill eagerly awaited the end of the fine-tuning. He vividly remembered the early morning hours of August 17, 1990, when the first test pictures were due back. He had roused himself from sleep and walked into the analysis center shortly before 4 A.M. A dramatic close-up image of Venus dominated the large screen, and Pettengill sidled up next to another radar team member, project scientist R. Stephen Saunders of JPL.

"That looks fantastic," Pettengill exclaimed.

"Yeah, that's the good news," Saunders replied curtly. "The bad news is we've lost contact with the spacecraft."

•

The mission plan called for *Magellan* to come over the north pole and travel down the length of the planet for some 16,000 kilometers, or 9600 miles, almost to the south pole. The mapping segment would last 37.2 minutes, during which time the radar beam would highlight a narrow swath of the surface about 20 kilometers wide. From that point on, the elliptical orbit would carry *Magellan* too far from the planet to achieve the resolution desired. It would take roughly another two and a half hours to complete the orbit and bring the craft back around to the north pole and close enough to resume mapping. In the interim the planet would have rotated gently on its axis, so that as the spacecraft came over the pole, it actually flew over a slightly different area. Just like mowing a lawn, where each swath overlaps barely to prevent missing any grass, *Magellan* was to image strip after strip of Venus. In this way, a complete mosaic of most of the surface could be pieced together in 1789 orbits, about eight months of mowing.

The beauty of the scaled-down mission lay in its elegant efficiency. Original plans called for two high-data-rate antennas, one for mapping and one for transmitting the data in a steady stream back to Earth. To cut costs, however, *Magellan* had been stripped to a single one of these eyes. So whenever the craft moved too far from the planet for mapping, the aerial would be used to beam the data homeward. The plan called for roughly one hour of transmission, followed by a fifteen-minute housekeeping interval during which *Magellan* would fix its position by the stars and update its gyro-reference system. Then the voyager would resume transmission for another hour.

Everything started out fine. During the first test run *Magellan* performed magnificently. The craft came around for a second test, mapped the planet perfectly again, and forwarded images back to Earth for an hour before going off as planned for the star scan. But it had not come back. It later turned out a computer anomaly had canceled the onboard sequence that led to the star search.

Magellan was programmed to go into a light, hazy sleep if something like this went wrong. From its dream-like state, according to programming instructions, the spacecraft would swing the big solar panels toward the sun to keep its batteries charged, then reacquire its position by locking onto Sirius and Rigel and begin broadcasting a kind of SOS in the general direction of Earth. In this way it should theoretically take no more than two hours to get back in touch. But two hours had gone by. Then four. Then eight. What had happened, controllers later learned, was that the orbiter found Sirius all right but had locked onto a spurious star signal instead of Rigel. As a result, it had pointed the antenna completely away from Earth. No one on the ground knew this at the time. They had simply lost a $600 million spacecraft without a clue.

It wasn't until a second problem sent the craft into a deeper level of somnolence, almost a coma, that *Magellan* switched to a backup system with a more foolproof method of finding help. Now, instead of beaming only to where it thought the Earth should be, *Magellan* began coning—broadcasting its status throughout the sky, one section of space at a time. This time, thirteen hours after the craft had disappeared, the message got through.

Once back in contact, the Martin Marietta team ordered *Magellan* to stop the sky sweep and download its memory to earth. The memory dump enabled the team to cobble together new commands to get things working again. On August 17 the errant spacecraft regained its full senses. But four days later the signal disappeared into the ether for a second time. And it stayed lost. "We were tearing our hair out," remembered Kasuda. The young software engineer dug up the memory readouts from the first shutdown, hunkered over a desk in one corner of the operations room, and began poring through them step-by-step. Sometime after midnight he noticed that a small portion of the backup computer's memory had been overwritten. It took awhile to figure out what this meant, but then Kasuda realized that the same area of memory was used by the spacecraft to orient itself from its deep sleep. If it really were null and void, *Magellan* might be out for good.

Kasuda's team spent several hours designing commands to get the spacecraft back on track. Around five in the morning on August 22, they went down to the systems verification lab on the ground floor, where Martin Marietta was simulating the entire *Magellan* mission. Racks of equipment cabinets and breadboard mock-ups stuffed with electronics filled the room, pretending they were gyros, star scanners, sun sensors, and the like. Re-creating the effect of the bad commands, Kasuda's heart sank. The simulation showed that the spacecraft "would just spin out of control and never regain control. Never. Ever."

Giving up was out of the question, so mission planners beamed up the fresh commands anyway. *Magellan* possessed a low-gain antenna — good for receiving messages only — with a much wider field of view than the main aerial used for mapping and transmitting data. As long as the spacecraft was pointed anywhere within 180 degrees of Earth, the updated instructions could get through. Not knowing the orbiter's position, though, controllers sent the commands over and over, praying for a miracle.

Kasuda himself had finally gone home to get some rest and was asleep when *Magellan* checked back in at 2 P.M. on August 22, some seventeen hours after its last signal had been received. In the control room, some of those on duty cheered, others cried. Remembered Jim Neuman, one of the group leaders: "It was unbelievable. There were tears to be honest with you, because we had no expectation of ever seeing that thing again."

•

Magellan finally began mapping for real on September 15, some two weeks and several near coronaries later than envisioned. Even though control had been firmly reestablished, not everything ran smoothly. Soon after maneuvering into orbit, *Magellan* had fired explosive separation nuts to jettison the spent fuel case. The event had inadvertently sent an electrical shock through the craft, damaging a portion of the backup memory.

Other glitches kept popping up. To prevent overheating when exposed to the sun, the spacecraft had been coated with optical solar reflectors. But these mirror-like materials absorbed more heat than expected, forcing controllers to rotate the craft twice during each orbit so that the antenna could shade the main body from the sun. During these periods the radar had to be inactivated, causing annoying gaps in the Venusian mosaic. A short time later one of the two tape recorders began to fail. Without its counterpart to pick up the slack, a small amount of data was lost whenever the remaining recorder switched from one track to another. These showed up as black pinpoints on the radar images. Finally, late in 1990, a transponder developed a whistle, akin to the interference when two radio stations overlap frequencies. To avoid losing more data, *Magellan* had to switch to the backup. The loss of a transponder would haunt the mission down the road.

If all this weren't enough, every three months or so the original problem that had twice sent *Magellan* into a near-coma kept rearing its ugly head. In November 1990 and again in February 1991, *Magellan* started drifting off into dreamland. The revamped software commands automatically brought the spacecraft back on-line, but they had not cured the malady outright. Each time it took several hours to get up and mapping again, resulting in still more gaps in the data.

Once again, Rick Kasuda scoured the memory readouts. By April, he had discovered only one thing that looked out of whack: a point where the central processing unit appeared to overrun its processing frame. Although Kasuda thought this *could* send the computer into Never land, where it would run an "infinite loop" and be unable to resume normal operations, it seemed improbable. The problem could arise only if two specific events occurred within one-thirtieth of a second of each other. Out of all the millions of operations *Magellan* performed each day, the odds of such an alignment were so slim the glitch might appear, well, it dawned on Kasuda, once every three months.

Kasuda returned to the simulator downstairs to test his theory. After several days the "*Magellan*-on-a-bench" failed just like the real thing. From there, it proved a relatively simple matter to clean up the software once and for all, although in early May, before the repair process was complete, the spacecraft conked out yet another time. Finally, just before the Fourth of July holiday, *Magellan* got its cure.

•

Even as Rick Kasuda and the engineers suffered through their trials and tribulations, scientists like Gordon Pettengill were having a better time. When *Magellan* opened its quest to map Venus, the planet had already been studied repeatedly from places like Goldstone and Arecibo — and visited by twenty-one Earth expeditions, fifteen Soviet and six American craft that had mainly probed the atmospheric structure and makeup. The three voyagers carrying radar mappers — *Pioneer Venus 1*, and *Veneras 15* and *16* — had provided some further hard details. Pioneer's global map had shown general features: volcano-ish mountains, flat lowlands, and continent-like highlands. The *Venera* twins, offering much improved resolution of the Northern Hemisphere, had turned up impact craters and abundant evidence of volcanism. But it was left to *Magellan*, as NASA explained in its official mission booklet, to examine "the hills and valleys, craters, and lava flows — the telling details of Venusian geology." From these, scientists hoped to address the fundamental roles, past and present, played by plate tectonics, volcanoes, asteroid impacts, and erosion — factors often critical to Earth's own evolution.

Magellan's radar promised, on average, a resolution between 120 and 300 meters, or 400 and 1000 feet, about eight times better than the *Veneras*.

Three key science functions operated from the main transmitter. A small antenna under the spacecraft pointed straight down for radar altimetry, building up a topographic map. The larger, dominant aerial looked off to one side. In passive mode, it served as a radiometer. But it was the active mode, when the antenna worked as part of the synthetic aperture radar, that formed the heart of the mission. Functioning as a kind of flash camera, the system used the energy reflected off each object encountered to develop a two-dimensional picture vivid enough to identify craters, canyons, and lava flows from different geological periods.

The prime mission called for mapping 70 percent of the planet. But despite the engineering glitches, the first eight-month cycle alone covered 84 percent of the landscape, and with it came some answers to questions that had long plagued scientists. One of the chief debates going into the mission centered around questionable observations — from ground-based radars and *Pioneer* — of Earth-style tectonics. However, while the close-up *Magellan* images showed evidence of tectonic activity, such as fault zones and volcanoes, they turned up no signs of the long chains of volcanoes or mountains that on Earth demarked a global network of underground plates. The clear conclusion was that Venus must depend on some other mechanism for venting its hot interior. The nature of that mechanism — even the degree to which Venus remained active — quickly became the hottest controversy in planetary astronomy.

Not surprisingly, the race to come up with a working model of Venus that would satisfy the latest findings engendered several hotly contested theories. At the core of the argument lay *Magellan's* high-resolution images of Venus's impact craters. Planetologists commonly use craters to estimate the surface age of solar system bodies. As a rule, the more craters pockmark a planet's surface, the longer that landscape has been around. And it is not surprising to find regions of widely varying age on the same planet. On Mars, for example, volcanism and erosion have erased the craters in some areas, leaving a young and clean surface compared to the rest of the planet. The extent of this modification itself offers valuable clues about a planet's workings, such as the level of volcanism and the method of venting heat from deep in its bowels.

Magellan spotted more than nine hundred craters, giving scientists two important and ultimately interrelated pieces of information. First, the craters are distributed randomly, indicating the entire surface is roughly the same age. Secondly, the number of craters points to a surface age of around 500 million years, while studies of the Earth and moon indicate the solar system has been around for almost ten times that long.

Right away, a few scientists came to the tempting conclusion that something happened a half billion years or so ago to wipe the surface clean. "It's a shame in a way, because the ancient era of Venus, the first three-and-a-half billion years or so, are basically lost," bemoaned *Magellan* research associate Jeffrey J. Plaut of JPL. Astrogeologist Gerald Schaber of the U.S. Geological Survey in Flagstaff, Arizona, and Robert Strom of the University of Arizona at Tucson developed a computer model that indicates that sometime between 300 and 500 million years ago the surface of Venus was reconfigured by volcanic eruptions and tectonism in a process of "global resurfacing." After this upheaval, the model goes on, the planet's level of activity slowed dramatically as it cooled, leaving most of the more recent craters largely untouched. This certainly does not mean Venus is geologically dead, but a more active planet would show greater wear and tear. "That is the simplest explanation for the cratering record on Venus," asserted Schaber.

The two key ideas of global resurfacing and a relatively quiescent planet found wide appeal. Donald Turcotte, a Cornell University geologist, built on Schaber's observations to come up with his own theory that resurfacing has happened periodically. Turcotte's Venus is characterized by a period of rapid and violent plate tectonics, during which massive volcanism covers the surface in lava several kilometers deep. Such a global venting of heat from the interior rapidly cools the planet, shutting down its tectonic engine. As this happens, the planet's outer shell, or lithosphere, thickens to prevent fresh magma from reaching the surface, ensuring that subsequent craters are left largely intact. In this way heat is again trapped in the interior, until after several hundred million years the rising temperatures overwhelm the lithosphere, setting the process off again.

Not everyone, though, is a fan of global resurfacing. One of the chief critics, Roger J. Phillips of Washington University in St. Louis, maintains that a close examination of specific areas shows a much wider age spectrum than compatible with a global catastrophe. Roughly a quarter of the planet, mainly regions of volcanic rises and rifts, exhibits a lower overall density of craters than the rest of the planet—but a higher proportion of modified craters. To Phillips, such evidence of a vigorous planet indicates that the surface in some parts of these regions may be less than 100 million years old. Other Venusian venues, more haggard and shopworn, show signs of being many times as venerable. "There's a plethora of surface ages," the researcher asserts. In his Venus, then, there is no single catastrophic event followed by a long period of inactivity. Instead, even to this day the planet continues to undergo periods of volcanism.

A marriage of these two viewpoints was put forth by Sean Solomon, who had left MIT to become director of the Carnegie Institution's Department of Terrestrial Magnetism. Early in Venus's history, he theorizes, the planet was so hot that it deformed easily, rapidly wiping out craters and volcanoes. But as the planet cooled it went through a transition from a time when the crust was malleable to a more modern period when the surface was much more rigid. This makeover would have happened globally, and almost simultaneously. So, he notes, "That would look like a catastrophe, and yet there would not be a true catastrophe in the sense of a great belching of magma." Highlands areas, riding on the thick crust, would have held their heat longer, so they continued wiping out evidence of craters after the main body of the planet had cooled.

In January 1992, at the beginning of the third eight-month mapping cycle, the debates over these theories were just heating up when trouble struck *Magellan* yet again. The good transponder went out, leaving only its counterpart with the interfering "whistle." Small, strategically placed on-board heaters were turned on — and the craft was rotated sunward — to increase the temperature of this vital component. This raised the whistle's frequency, lessening its interference with the radar's own signal. But it was a makeshift effort, affording much less detail than previous cycles. The spacecraft hobbled along, trying to fill in holes in the data base. But by that September the signal had deteriorated even further, and the radar was turned off for good.

Even with the disappointing last segment, all but one percent of the planet had been mapped at least once — far outstripping the original goal. At that point, considering that the Bush administration had already eliminated most funding for the coming year, *Magellan*'s mission planners might have thrown in the towel. Instead, by scaling down to a skeleton staff — fewer than fifty from a heyday of more than four hundred — and carrying over some surplus 1992 funds, JPL decided to attempt a grand finale.

On May 25, 1993, at thirty-one minutes and thirteen seconds after five in the evening Greenwich mean time, a battered spacecraft *Magellan* began the last phase of its mission. Acting on a command sequence timed to kick in just as the Venus orbiter reached its farthest point from the planet, *Magellan* fired all eight of its hydrazine thrusters for 762.895 seconds, about thirteen minutes. The force slowed the craft enough so that as it drew close to Venus it moved downward into the planet's thick, carbon dioxide–rich atmosphere, to about 147 kilometers, or 88 miles, above the surface. There, after 748 similar dips into the top of the atmosphere over the next seventy days, the drag gradually robbed *Magellan* of enough momentum to move it into a low-altitude, almost circular orbit.

The dramatic "aerobraking" maneuver, the first ever attempted away from Earth, milked nearly another seventeen months of scientific life out of the interplanetary warhorse. The procedure, which required a dozen more perfectly timed thruster firings over the next three months to control the drag working on the craft, was deemed too risky to be included in the original mission plan. But with the spacecraft having done all it could on the radar mapping front, NASA officials and scientists saw little to lose in attempting a last experiment.

Meanwhile, the potential rewards were great. Circling the Venusian poles, *Magellan* would be close enough to the planet for the pushing and pulling of gravitational fields associated with large masses—both on the surface and below it—to subtly affect its orbit. Charting these perturbations and producing a detailed gravity map of Venus would be the icing on the cake for the voyager, providing a leg up toward resolving the heated controversy over how Venus lived and breathed. Through a series of calculations based on Newton's laws, it is relatively straightforward to predict the gravitational effects of known topography. Therefore, any further perturbations of *Magellan*'s orbit would have to result from subsurface forces. Armed with that data, since gravity fields on Venus are related directly to density variations, scientists hoped to infer the thickness of the lithosphere and even derive clues about magma plumes and other convective forces at work in the mantle beneath it.

The early results, taking into account relativity effects, the motions of planet and spacecraft, atmospheric drag, and thousands of other parameters, confirmed that Venus enjoys a much higher correlation between gravitational forces and surface features than does Earth—an indication to some that the lithosphere is significantly thicker. At the same time, gravity mapping revealed the lithosphere to be maybe thirty miles thick, compared to the Earth's value of about eighteen miles. Such a bulbous surface layer is compatible with Turcotte's idea of an ever-thickening shield that prevents heat from escaping until the temperature becomes so great the whole thing breaks up. However, it is far from conclusive, and the debate rages onward toward the millennium.

One last radar effort was coaxed out of the spacecraft. After the aerobraking procedure, Pettengill and Richard A. Simpson, a Stanford radar astronomer, successfully pled the case to try a novel experiment. Although both *Magellan*'s telemetry modulators were inoperative, meaning no data could be beamed back to Earth, the telemetry transmitters were capable of broadcasting a continuous signal. By tilting the orbiter so that the main antenna pointed at Venus at just the right angle, the beam would strike the surface and skip off directly to Earth the way a flashlight beam might reflect off a piece of glass. By measuring the intensity, frequency shift, and polarization of the waves reaching Earth, the men hoped to answer questions about surface composition left ambiguous by all previous studies from terra firma and spacecraft. It was something like looking through a sheet of polarizing glass and rotating it to better see a particular point in a particular direction. To pull it off, the big Deep Space Network aerials already tracking the craft — from Spain, Australia, and California — had to be specially configured.

Only the last run, on June 5, 1994, aimed at the eastern mountain peak Maxwell, the planet's highest point, produced a dramatic result. The region is coated in mystery material, a substance of high radar brightness and low thermal emissivity that had aroused public and scientific imaginations for years. The best models explained the phenomenon as characteristic either of a surface material rich in a conductive mineral like pyrite or a far stranger situation in which some sort of substance, transparent to microwaves, surrounded lumps of another type of matter that readily scattered radar energy. However, from the way the polarization of the radio waves changed as the beam was reflected off Maxwell's surface, the second option was effectively ruled out. Moreover, the value of the mystery material's dielectric constant was shown to be higher than feasible with the pyrite-rich model — so high, in fact, it might dictate a surface coating of metal or ferroelectric material.

Geochemists didn't like either of these latest alternatives. The metallic option brought to mind a picture of the surface covered with something akin to aluminum wrap, which would be so reactive with the atmosphere it couldn't last long, explains John A. Wood, a *Magellan* investigator from the Harvard-Smithsonian Astrophysical Observatory. The second scenario, dependent on highly unusual combinations and concentrations of rare elements, seemed absurd outside a laboratory. The apparent contradictions quickly brought the focus back to pyrite, or something similar, but layered across the surface perhaps as much as several millimeters deep. Iron in the pyrite layer could have come from the low, hot plains level, reacting with atmospheric chlorine to form iron chloride, which would evaporate and drift up to the relatively cool high altitudes abundant in mystery material. There, it would plate out onto the surface as the garden-variety mineral pyrite, freeing the chlorine that rendered it volatile to return to the atmosphere and facilitate the process anew.

The debates over the mystery material and the workings of Venus will continue, in much the same way geologists and other scientists argue about the Earth. Theories are put out, to be shot down or supported. Science as a whole advances. Not long after the synthetic aperture radar part of the mission ended, after more than two decades' hard sweat on the project, Gordon Pettengill expressed a wistful regret about the engineering failures, the software bugs, component malfunctions, and other problems that didn't have to be. Yet he couldn't deny his pride in the mission, or his admiration for the imaginative ways in which a spacecraft millions of miles off in space had been repeatedly brought back from the dead to answer questions for the living.

Down the hall from Pettengill's sixth-floor office in the Massachusetts Institute of Technology Center for Space Research, principal research scientist Peter G. Ford patiently oversaw the vast job of processing and annotating *Magellan*'s altimetry and radiometry readings well into 1996. Combined with radar maps enhanced at JPL, and still other high-resolution, cohesive mosaics provided by the U.S. Geological Survey, the spacecraft by that time had fathered some three hundred CD-ROMs' worth of data that had become something of bestsellers in scientific circles. There had never been a mission like *Magellan*, either in terms of the amount of information gathered — some one trillion bytes, more than all previous NASA missions combined — or in the speed with which data was made available to the general scientific public.

On October 11, 1994, even as the collating, reducing, smoothing, and composing continued, the Jet Propulsion Lab terminated *Magellan* via a fiery entry into the Venusian atmosphere. Already at that point, the orbiter's solar panels had been badly damaged by thermal shock from its constant swings in and out of the sunlight, causing power output to drop precipitously. The batteries had nearly failed, the craft was down to its last gyro pair, and funding was clearly at an end. Preferring to perform one final experiment rather than simply walk away, controllers fired the thrusters in four bursts to drive the craft down into the upper atmosphere. On the way down, *Magellan*'s shopworn solar panels were deployed as propellers to create an aerodynamic torque on the spacecraft. This action was resisted by the jets, thereby enabling engineers to calculate the torque needed to prevent the body from rolling on its axis. From that came a very nice measurement of the aerodynamic properties of a spacecraft in the upper fringes of the atmosphere. The measurements continued until the twelfth, when *Magellan* could no longer maintain attitude and contact was lost for good.

Much of *Magellan* vaporized almost instantly. However, some of the quartz glass insulating layer, and possibly the batteries and a few other protected internal components, were expected to survive the burn-through and float down like leaves through the thick atmosphere to land on the hot, hidden surface. There they may have taken days to corrode away, unseen testaments to the long journey from peace to war and back again, and to the insatiable thirst for knowledge that leads, as Plato proclaims, from this world to another.

Epilogue: A Quiet Glory

"You did not invent the women!"

LORD COHEN, chairman,
Royal Commission on Awards to Inventors,
to Sir Robert Watson-Watt, July 1951

*"Memories crowd in on me; memories of acute anxieties and tragedies—
memories too of absent friends—and of the dismal prophecies of utter destruction of the cities of England which proved in the event so greatly exaggerated.
I wonder if the part that scientists have played will ever be faithfully and
fairly recorded. Probably not."*

last diary entry of SIR HENRY TIZARD

ON THE AFTERNOON of October 20, 1954, Robert Alexander Watson-Watt drove with second wife, Jean, along the north shore of Lake Ontario on his way from Toronto to Kingston. As he roared down the road just outside Port Hope in a Buick Century, much as he had motored along the Suffolk coast nearly two decades earlier in his beloved Daimler, bound with Taffy Bowen and Robert Hanbury Brown to investigate Bawdsey Manor as a possible home for the fledgling British radar project, Sir Robert's repose was punctured by a shrill police whistle. An officer of the Ontario Provincial Police stepped onto the road from some unseen spot and motioned for the car to pull over. Then the policeman informed the Scot he was being charged with speeding.

"How did you do it; was it by radar?" Watson-Watt queried.

"It was not by radar; it was by an electronic speed-meter," the policeman responded.

"You may be interested to know," Jean informed the officer, "that King George VI knighted my husband for inventing radar."

"I don't know who invented anything; I know you were driving at an excessive speed."

Watson-Watt tried to explain he was hurrying to speak before cadets at Canada's Royal Military College, but his golden tongue did no good. He was fined $12.50 Canadian, a fee paid on the spot. When he finally arrived at the academy, Sir Robert incorporated the episode into his speech, "Electronics and the Service Man." Suddenly, he heard a rustle and looked up to see several men rushing from the hall. They were reporters. That evening the story was widely carried on radio. On the morrow it made television, as well as newspapers in North America and Western Europe. By some fluke, after Watson-Watt appeared on a Chicago television show the next year, the story was regurgitated as if it had just happened, with the Canadian penalty transferred into American dollars, and everyone apparently insisting it was radar, not a speed meter, that had foiled him. Sir Robert was in heaven. After all, he noted, "what Scot could be insensitive to the world's record value in publicity at a total cost of $12.50, even in the more valuable of the dollar currencies concerned?"

Amidst the attention, an anonymous newspaper reader penned this ditty:

> Pity Sir Robert Watson-Watt
> Strange target of this radar plot.
> And thus, with others I can mention,
> The victim of his own invention.
> His magical all-seeing eye
> Enables cloudbound planes to fly.
> But now by some ironic twist
> It spots the speeding motorist.
> And bites, no doubt with legal wit,
> The hand that once created it.

Already by 1954, the legacy of radar radiated across a wide front. Yet, the episode nevertheless marked a rare public flare-up for a technology that remained largely in the background, especially compared to its big sister, nuclear energy. Radar was still the quiet revolution produced by World War II, and while well known in their field, its pioneers experienced a largely silent glory compared to the Oppenheimers, Tellers, and other nuclear bombmakers.

Fast-forward to near the turn of the century, some seventy years after the first radar systems were wired up, and the technology's reach had extended so far as to almost stymie any attempt to categorize it. The decades following World War II encompassed the Cold War, the Korean War, the Gulf war, the rise of civil airliners, the TV weather man, satellites, the space age. New radar techniques became widely familiar in name, though not theory— over-the-horizon, pulse compression, chirp, synthetic aperture—and there was the rise of Stealth technology to conceal aircraft from radar detection. The waves of the insurgence propagated further than any of the radar patri- archs could have envisioned. "Radars are like automobiles. There are coupes and Cadillacs. . . . It's hopeless to try to figure it all out," claimed William

Delaney, a vice president of Lincoln Laboratory in the 1990s. Delaney wasn't even addressing the technology's immense impact on a myriad of other arenas: television, cathode ray screens, telephone systems, microwave communications, solid-state electronics, quantum electronics, computing. Beyond that, à la Charles Townes and even more unquantifiable, lurk the musings and creations of the many minds so wonderfully prepared by the wartime radar drive.

There was at least one, extremely limited attempt to put a value on some of the World War II–vintage radar developments. It opened in July 1951, three years before Watson-Watt got his speeding ticket. That month the Royal Commission on Awards to Inventors met to consider claims for contributions to the invention and development of British radar. Taking credit for various advances before and during the war, more than twenty people put themselves up for consideration. The leading candidates belonged to a ten-person syndicate led by the Orfordness–Bawdsey Manor crew, including among them Watson-Watt, Taffy Bowen, Arnold "Skip" Wilkins, and Robert Hanbury Brown. Sir Robert had apparently started the ball rolling, galvanizing Bowen, who was determined that the Scotsman alone should not get the reward for their labors, and before long a host of early recruits were in on the action. To present a united front, the men submitted their claim as a team, agreeing in advance on relative percentages by which they would divvy up any payout.

It was generally a bleak scene, at least to Brown. The hearings lasted forty-four days. Yet, he writes, "I only went to two meetings of the Commission. I found them dull and sad, like a funeral service for the spirit of Bawdsey Manor. In the early days of radar we never worried about who had invented what, and we never imagined any reward other than seeing the thing work."

Times, though, had changed. The proceedings opened on July 12, a Thursday. Watson-Watt presented an opening statement, then continued his six-hour testimony through Friday morning and into that afternoon, recounting the early history of British radar in excruciating detail. As Brown recalls, "It was after lunch and Sir Robert was in full polysyllabic flood telling the Commission about the many important things which he had done."

Watson-Watt was being questioned by the Crown's P. J. Stuart Bevan. Partial transcripts of the proceedings indicate he had given himself credit for bringing in members of the Women's Auxiliary Air Force to help operate the Chain Home network, and that Bevan was trying to rank that achievement in importance. "I have no doubt that you were responsible for the training of the first women operators and so forth, but that is very much a sideline, is it not?" the Crown's man queried.

"No. I would not accept that," countered the Scot. "I think that my suggestion for the introduction of women operators and for their training, and the demonstration that they could be very quickly trained to be very remarkable operators, was a notable contribution by myself, and one which was personal to myself."

Brown records that by this time Lord Cohen of Walmer, the commission's chair, seemed to have fallen asleep. But at the mention of the WAAFs, Cohen exclaimed: "You did not invent the women!" That brought down the house. But Sir Robert managed to continue: "That was before the period we are discussing; but I did in fact make to the Air Ministry the first suggestion that was made derogatory to the then male operators, and I put forward a reasoned argument on which I based my belief that women operators would in fact do as remarkably well as they did in the operation of radar."

On the last day of the hearings, evidently part of a four-day marathon closing speech, Sir Robert estimated that the British government had spent some £500 million on radar. He compared himself to a stockbroker seeking a commission. "If I accept my own figure then I have not a right to press my client very hard on my brokerage rate and so I offer my total radar team one eighth per cent, the very lowest form of brokerage life." He meant one-eighth of one percent, or £625,000. "I put that forward as a very modest estimate," he told the commission, adding, though, that if every British citizen contributed the smallest coin of the realm to his award, the figure would be £112,000.

In the end, the total granted to his syndicate came to less than £90,000 — roughly $250,000 U.S. at the time and $1.5 million in 1995 value. The commission made individual recommendations, but Brown and a few others distributed the money on the basis of the group's previous agreement. Sir Robert got about £52,000, Wilkins and Bowen £10,400, and the rest £2,080 — apparently on the theory that the work done by Bowen and Wilkins was five times more valuable than the main group's, and Watson-Watt's was five times more important than that of Bowen and Wilkins.

It was more than they had dreamed of when starting the work. Even two decades later Watson-Watt's share was presumed to have been the largest ever recommended by the commission, and Brown, for one, put his own slice toward a down payment on a house. Still, in absolute terms it wasn't all that much given the scope of the ultimate payoff. Perhaps had the radar men known where their invention was headed, they might have pressed for more.

•

Airplanes, ships, boats, cars, trucks, buses, satellites, spacecraft. From monolithic military installations like the triangular Raytheon-built phased arrays deployed in the 1990s as upgrades to the Ballistic Missile Early Warning System to tiny portable devices for the police officer, a survey of radar in the 1990s shows that the technology has done its share for the post–World War II economic globalization. Even setting aside those other fields it has critically influenced, such as microwave ovens and solid-state physics, radar has carved out a key role for itself in a wide range of products and endeavors from the obvious to the obscure.

Predictably, the military continues to drive the technology. *Jane's Radar and Electronic Warfare* annual lists more than a thousand systems designed for virtually all aspects of military endeavor. In the United States, Lincoln Lab remains the closest thing to the Rad Lab. Although its people often exude TLA, or Typical Lincoln Arrogance, the lab's overall expertise in digital communications and radar techniques is unsurpassed. In specific areas, though, others reign. Hughes and Westinghouse, the latter part of Northrop Grumman Corporation since 1996, own the airborne intercept business. General Electric bought RCA, then sold the resulting combined defense electronics unit to Martin Marietta, which in turn merged with Lockheed—but GE's radar child remains a big player in large search or early warning systems. Raytheon is the king of phased arrays. Bell Labs might have been part of the list, but by the mid-1970s, after years of designing radars for planes, ships, and submarines, as well as the Nike series of guided missiles, it had grown disgusted with the government and abandoned the field to stay focused on its core communications business.

Over the postwar decades, radar systems benefited both from a steady amalgamation of improvements in everything from solid-state and semiconductor components to digital processing and a deeper understanding of the theory behind the technology. During World War II models were often designed by brute force and rushed into service when it seemed as though they worked. Peacetime engineers unveiled a second generation of radar techniques — imaginative manipulations of the basic concept of sending out a pulse and waiting for an echo. Among other things, just as a baseball pitcher can extend his arsenal by putting a snap on a straight fastball to turn it into a hard-breaking slider, electromagnetic pulses can be modulated in phase or frequency to extend the powers of a given transmission—a field known as pulse compression or spread spectrum technology. With it and other innovations, the radar set has become almost a magician's bag, from which purveyors conjure up novel beam shapes or pull enemy rabbits from heavy clutter.

Virtually all the latest twists and turns came into play during the Gulf war. A million people squared off in the desert following Iraq's seizure of Kuwait in August 1990. And when the U.S.-led coalition finally moved to counter the aggression, its onslaught was marked by the greatest concentration of electronics skulduggery since D-day—and possibly ever. On the first night of the war, at 1 A.M. on January 17, 1991, Pave Low helicopters fitted with night-flying and navigational equipment took off from Al Jouf airfield in western Saudi Arabia, headed for two clusters of enemy mobile early warning radars positioned sixty miles apart in the open desert, north of the Iraqi-Saudi border. Behind them, divided into teams assigned to each radar outpost, came Apache attack helicopters. As the Pave Lows closed on the targets, crew members dropped green phosphorescent lights to mark key points, then pulled back and let their heavily armed comrades navigate onward with the help of forward-looking infrared systems, or FLIRS, which created images from contrasts in heat emitted by objects on the desert floor. Arriving at their targets nearly simultaneously, the Apaches let loose a cannonade of laser-guided Hellfire missiles, then raked the sites with rockets and cannon fire, opening a hole for other aircraft taking part in attacks on Baghdad, as well as suspected Scud missile sites in western Iraq.

In concert with the desert assault, all hell began to break loose around the permanent radar net known as Kari, the French word for "Iraq" spelled backward. Missile crews in Saudi Arabia launched BQM-74 target drones designed to simulate a combat plane's radar cross section, and Navy A-6s fired unguided Tactical Air Launched Decoys. As Iraqi radars lit up to track the invaders, other aircraft, mainly Air Force F-4G Wild-Weasels, fired high-speed anti-radiation missiles like lethal rain. That first night the Weasels let loose 118 HARMs. At roughly the same time came Tomahawk cruise missiles, some bound for such targets as the Presidential Palace, others carrying the latest descendants of Window—carbon filament spools designed to short out electrical grids when they fall across power lines. Meanwhile, F-117 Stealth aircraft, covered by jamming planes, handled the main bombing runs on Baghdad. The whole assault worked so well the only coalition loss of the night was a HARM-firing Navy FA-18 that collided with or was shot down by a MiG. Of the forty-nine F-117 bombing attacks, intelligence reported twenty-seven hits. These were not pinpoint attacks. But they apparently proved far more accurate—and safer to both pilots and civilians—than could have been managed with conventional forces.

The rest of the six-week conflict saw the appearance of a host of other important electronics systems. Three AWACS planes stayed up at all times to handle the huge volume of air traffic. Lesser known was J-STARS, the Joint Surveillance and Target-Attack Radar System, a sophisticated airborne platform for tracking ground traffic in much the same way AWACS follows air activity. During the Gulf war two prototypes hied to the Middle East. The planes, modified B-707s, carry two types of radar in their underbelly. In one mode, a Moving Target Detector can spot trucks, tanks, even small columns of soldiers. In the other, a synthetic aperture radar provides detailed still imagery of the ground situation. During Desert Shield and Desert Storm the J-STARS aircraft flew forty-nine missions and by most accounts lived up to their promise. On January 25 one tracked an Iraqi armored column and called in warplanes: fifty-eight of seventy-one vehicles were reported destroyed. Four days later, when Saddam Hussein launched a surprise four-pronged assault against the Saudi border town of Khafji, J-STARS again played a role in spotting the attackers, though the planes' exact contribution is not clear from published accounts. While the thrust caught the coalition off guard and allowed a portion of the Iraqi troops to briefly occupy Khafji, the air-led counterattack thwarted the main body before it really got started. Coalition forces soon overwhelmed Saddam's remaining fighters.

Noncooperative target recognition, a trick of advanced processing which can "fingerprint" enemy aircraft through the characteristic modulation imposed on radar signals by their jet engines, also made an appearance in the desert war. The conflict marked as well a test of Firefinders — counter-battery radars designed to locate the source of enemy artillery and mortar fire. Such sets are not exactly new. At the Rad Lab back in the spring of 1944, Ed Purcell was working on a counter-mortar radar when the water vapor mystery cropped up. But the latest incarnations employ advanced digital processing and can feed positional data to multiple rocket launchers for precise and rapid counterfire. Launcher shells house cluster bombs capable of destroying an area roughly three football fields long and two wide. In theory, with the Firefinders a retaliatory blow can be off before the first enemy round hits the ground. According to one account, "If the enemy guns do not move immediately after firing one round, they are doomed." Counterbattery fire was extremely effective during the Gulf war. On February 27, 1991, during the opening of the ground campaign, U.S. troops of the 24th Mechanized Division of the XVIII Airborne Corps came under artillery fire from an estimated five Iraqi battalions. Radar located the enemy artillery positions and directed a reportedly devastating counterassault. "We'd get a sensing round from them and the Iraqis would get forty-eight in return," Colonel Paul Kern told the *Army Times*.

The civilian sector, of course, has utilized advances in military techniques and processing ever since the veil of secrecy was lifted from the World War II radar labs. Approaching the dawn of the twenty-first century, airways worldwide are controlled via radar. The technology seems omnipresent during the TV weather report. The maritime radar market has reached everything from ocean liners to pleasure boats. Giant telescopes at places like Arecibo and Goldstone generate delay-Doppler maps capable of resolving features as small as 120 kilometers, or 72 miles, wide on Mercury. Continuing in the Seasat vein, high-flying satellites and space shuttles gaze down on Earth with the latest in radar systems, charting wind patterns, storm fronts, and oceanic currents. A recent incarnation, the Spaceborne Imaging Radar SIR-C/X-SAR, was first carried aloft on April 9, 1994, aboard *Endeavour*. It can operate simultaneously on up to three distinct frequencies. Finally, following the tradition of the *Veneras* and *Magellan*, the spacecraft *Cassini*, due for takeoff in October 1997, will harbor a sophisticated radar system to map Titan, the mysterious big moon of Saturn hidden from optical astronomers under dense organic haze.

Beyond these more or less mainstream niches for radar have come a host of more fanciful spin-offs. Radar routinely gauges the speed of baseball pitches. It has been used on Greyhound buses to warn of impending collisions. Low-frequency ground-penetrating radars are able to peer up to a hundred feet beneath the surface, depending on such factors as soil conductivity and the presence of nearby power lines. They have been deployed to hunt for hidden mines, underground tombs, North Korean tunnels penetrating the demilitarized zone, old pipes and rebar that may pose construction hazards, and even murder victims. In the mid-1990s extremely low-powered broadband radars are under development for such uses as home burglar alarms, fighting crib death by noting disruptions in a child's breathing pattern, and auto warning systems that mount in a tail light or bumper: fender-defenders.

While the myriad faces of radar cut widely across geographic boundaries, few routinely touch everyday existence like the air traffic systems, weather watchers, and police speed gun Watson-Watt thought might have tagged him. Air traffic control, the most visible aspect of the technology, dates back to World War II, as exhibited in the Battle of Britain and the Blitz. Long-distance air route sets track planes between destinations. Airport surveillance systems, looking out only twenty to fifty miles, control denser traffic around busy airfields. Still others watch runway activity to avoid ground collisions. But despite all the promise and experience coming out of World War II, radar's incorporation into modern air travel came slowly. The first radar put to general use guiding civilian airliners over air routes was the AN/CPS-1—otherwise known as MEW—which could monitor traffic to about two hundred miles. By 1955, although the number of domestic airline passengers had shot up from 6.7 million to 38 million annually in the first postwar decade, only two air route systems were in place, along with thirty-two airport surveillance radars. Three of Luie Alvarez's Ground Controlled Approach systems, donated by the Air Force after World War II to airports in New York, Chicago, and Washington, were likely also up and running— though GCA's main employment would come through the military.

In the face of increasing numbers of near misses, the sparse radar coverage came up during several congressional hearings, adding impetus to wider use of the technology. However, air route surveillance was still not in general service when a United Air Lines DC-7 and a Trans World Airlines Super Constellation collided in uncontrolled airspace over the Grand Canyon on June 30, 1956, killing all 128 aloft. The disaster sparked a media and political firestorm that dramatically accelerated plans for nationwide installation of air traffic control radars. Within a month Congress appropriated funds for the purchase and installation of eighty-two additional air route sets. The first large order, for twenty-three systems, went to Raytheon.

As 1995 opened, the critical role of radar in air traffic control could be seen in several major projects, plus a grab bag of minor ones. The country fielded 116 air route radars and 221 airport surveillance systems. Backing up these primary units are Secondary Surveillance Radars, in which pulses trigger transponders aboard aircraft, and from the coded reply controllers get more complete details of the plane's identity, pressure altitude, and the like. Such a capability can be traced to Watson-Watt's 1935 memo describing the need for identifying friend from foe, but did not emerge in civilian air traffic until the mid-1950s. The latest version, still being installed in 1996 after a money crunch and a variety of technical obstacles kept it in check for more than two decades, is known as Mode S. Although a product of collaborations between top laboratories in England and the United States, much of the impetus came from Lincoln Laboratory's Herb Weiss.

In the late 1960s, risen to head of Lincoln's Radar Division, Weiss flew to Washington frequently on government business, becoming exasperated at the long delays caused by an unprecedented buildup in air traffic. Current FAA radars only provided sparse information that was passed to controllers over poor communications systems: the coded response of an airborne transponder triggered by radar told whether the craft was military, commercial, or private — but identifying the specific flight, or even the aircraft's altitude, for good computerized tracking was impossible. The inaccuracy of the setup required a great deal of manual attention and forced controllers to space planes out farther than should have been necessary. Since designing better radars, digital processing, and communications had been Lincoln's forte since the SAGE days, it struck Weiss that the lab could help end air travelers' misery. "I just saw all this fancy stuff we were doing at Lincoln," he related. "And then you look at what the FAA was using and they were back in World

War II — and that really got me riled up." It took some eighteen months to overcome inertia, both within his own shop and the FAA. Ultimately, though, Weiss succeeded in bringing Lincoln into the fold to develop a more reliable, automated system for accurately identifying, tracking, and controlling aircraft.

What came to be called the Mode S system proved compatible with plans to equip planes with sophisticated electronics to warn of impending collisions and guide pilots toward the proper evasive maneuvers. The latest incarnation of these, TCAS, for Traffic Advisory and Collision Avoidance System, started as a cooperative effort between Lincoln and the MITRE Corporation. It draws on features of aircraft transponders such as Mode S, tapping into these systems to derive altitude information and other data about the approaching plane. By 1995, TCAS was standard on all large carriers flying into U.S. airports.

Part and parcel with airport traffic control is the monitoring of weather conditions by radar, another practice dating back to World War II. Again, a number of interrelated efforts were afoot by the 1990s. Out front in the public eye come models to warn of wind shear, gust fronts, and microburst downdrafts — hazards associated with at least eighteen aviation accidents and 575 fatalities in the United States since 1970. Seeing embodiment as the FAA's Terminal Doppler Weather Radar, these systems depend on digital moving target detectors and Doppler processing to paint a constantly updated picture of weather conditions. In 1988, Raytheon won the contract to build forty-seven systems, eventually drawing on software algorithms transferred from Lincoln Lab. The radars are encrusted in balloon-like radomes mounted on towers a few miles outside major airports. By spring 1996, after many delays, about three-quarters of the forty-seven systems were installed, and ten had been officially commissioned.

The Terminal Doppler program arose from a much broader effort to improve weather prediction and tracking by radar. The biggest effort of the 1990s is NEXRAD, the much-maligned Next Generation Weather Radar program begun in 1980 as a key to providing better warnings of the violent thunderstorms and tornadoes that yearly lash across the United States. The intent is to keep a keen eye on all storms and weather systems — except the wind shear and microbursts covered by Terminal Doppler Weather Radar — at ranges up to about 140 miles. In 1991, the first NEXRAD system was set up in Norman, Oklahoma — an early step in the installation of 116 radars to be delivered by prime contractor Unisys for about $850 million. By March 1996, the number of radars had crept up to 120, with nearly all delivered and 89 commissioned. Costs, though, had swelled to slightly more than $2 billion, and the effort was seen by some as a boondoggle brought about by needless adaptation of the latest technology, when a much less expensive modernization of existing systems would have served as well. At the same time, the system seemed to earn rave reviews for its ability to pick up previously undetectable small storms on the verge of transforming into tempests.

A final way in which radar has impinged on the everyday psyche is through the police speed gun. "For the Lord's sake, don't let the cops know about this," one of Alfred Loomis's Tuxedo Park employees uttered in 1940 when their crude microwave radar tracked cars on a nearby highway. But within a few years the technology was catching speeders, and it continues to do so more than half a century later. The principle behind most police radar guns is continuous-wave Doppler. A steady stream of energy is transmitted, and from the frequency shift in reflected waves the speed of a moving object can be determined. By the early 1990s, it was estimated that upward of 200,000 traffic measurement devices were operating in the United States. Of these, a few hundred employed laser technology, which works in the same manner as pulsed radar, but with light waves instead of radio energy. Virtually all the rest were microwave systems, usually on the X-band around three centimeters, or K-band in the one centimeter region.

Early radar guns were more like World War II military installations, mounted at strategic points along oft-traveled roadways. However, in the 1970s, the integrated circuit brought easily portable devices into patrolmen's hands. Watson-Watt may not have envisioned it, but eventually the presence of police radars led to a market for detection devices — fuzzbusters — in much the same way air-to-surface-vessel radars brought Metox to U-boats. Like that early high-stakes cat-and-mouse game, the battle between police and speeders has seen measure met by countermeasure and counter-countermeasure. Law enforcement has unveiled the Stalker, which can shift between frequencies and operates at extremely low power levels that make it hard to pick up. The latest detectors incorporate digital signal processing and are designed to spot these low-level police radars, as well as laser guns. With an estimated fifteen to twenty million fuzzbusters in American cars, the Wizard War has long since been extended to the interstate.

●

And the radar pioneers? The pace of the revolution they engendered proved so swift that even those who were middle-aged at the dawn of World War II lived to see the technology's sweeping effects on society and science.

Sir Robert Watson-Watt drew on his reputation as the father of radar long after the war. Once establishing his consulting business, he continued as a part-time government advisor — among other things leading British delegations to international forums on radio aids to marine navigation — before disrupting everything sometime around 1952 to be with Jean Drew Smith in Canada.

In his memoir, *The Pulse of Radar*, Watson-Watt writes that he and Jean had discovered "a community of thought, taste, and spirit which neither of us had conceived to be attainable between two human beings; that life apart from one another could no longer have any worth-while meaning for us." For a time his wife, Margaret, apparently would not grant a divorce. But money from the inventor settlement in late 1951 assured Margaret's future, and the next year, childless, they dissolved their union. A few months later, in a November civil ceremony in London, the Scotsman married Jean. She had been wedded twice before — once divorced and once widowed — and had a son and a daughter by her first husband. And so Sir Robert became a stepfather.

The union seems to have stirred a rebirth in Watson-Watt. He writes, evoking the radar theme: "We are all immersed in a field of radiation from these values of the everyday. She carries a more sensitive receiver than do most of us. With it is a secondary transmitter, so intensifying the effects of the incoming radiation that they can be clearly perceived within our less sensitive mechanisms. . . . Among the many repair jobs she has done on my own radar system is the cure of a terrible defect of my IFF equipment, which often made the error opposite to her own, of identifying as foes those whose return signals were merely slightly distorted indications of 'friend,' or, at worst, 'unidentified.' "

The couple bought an old church in Quebec city, dismantled it, and used the bricks to help build a stunning cranberry red home overlooking a creek in the Toronto suburb of Thornhill. Sir Robert was busy in those days, traveling the world to scientific conferences, lecturing, and managing his consulting business. For a time, he kept the London shop, headquartered in a small house at 7 Gayfere Street, behind Westminister Abbey, and also started a Canadian counterpart, Adalia, Ltd. From Montreal, Adalia expanded to Toronto and New York City before collapsing in bad debt. Sir Robert was a scientist, a kind of philosopher, even a poet. But he was a bad businessman. As his stepson, Tony Drew, remembered, the family spent years paying off obligations. "I don't know if it was because he was a Scotsman or not, but he didn't declare bankruptcy."

Hard times forced the Watson-Watts to sell their beloved cranberry home. The couple moved to a cozy house in Sterling Forest, New York, near Tuxedo Park, where Sir Robert could keep up his business association with the Axe family of mutual fund fame, and where he grew active in community affairs. In December 1964, Jean and Sir Robert attended a local

Christmas buffet. They had loaded their plates and sat down to eat when Jean collapsed from a cerebral hemorrhage and died. "I honestly feel that Robert was never the same after that," said Drew. "He worshipped my mother. He really was in total shock." Drew and his wife, Stella, flew down from Canada and brought Sir Robert back for a bleak Christmas.

After the holidays Watson-Watt returned to London. Eventually he resumed an old acquaintance with Air Chief Commandant Dame Katherine Jane Trefusis-Forbes, director of the Women's Auxiliary Air Force during World War II. They were married in 1966. The last years of Watson-Watt's life proved tragic. His memory flagged. By the late sixties or early seventies, he was in the more advanced stages of what Drew felt certain was Alzheimer's disease, though it was not diagnosed. Jane, younger and spryer, took care of him. But in 1971 she died unexpectedly, and Sir Robert was sent to a Royal Air Force nursing home in Inverness, Scotland. He died there on December 5, 1973, at age eighty-one.

"For those who worked with him," eulogized Sir Robert Cockburn, the World War II countermeasures wizard, "it is difficult to realize that nearly forty years have passed since he unleashed the ferment of invention, imagination and initiative which gave us an overwhelming superiority in the new technique of radar. More than any single endeavour, it was radar that brought us through the war to ultimate victory. . . ." Tony Drew savors a different kind of memory. In 1955, when he was barely twenty-one, Drew and his sister, Dennie, were living in England. Jean and Robert had come over on holiday, and after the visit the children escorted the couple to Southampton to board the *Queen Elizabeth I*. They had arranged to stop by the ship early, and Drew found the top officers gathered in anticipation of meeting Sir Robert. "From the captain on down they were all waiting at the top of the gangplank to shake his hand. . . . A number came up and said, 'Do you realize what this man did during the war?' And I remember how proud I was as a kid."

In the heyday of British radar before and during World War II, the country's two central figures on the science policy level above Watson-Watt were Sir Henry Tizard and Frederick Lindemann, after 1941 Lord Cherwell. Though mutual friends attempted to engineer their reconciliation, the one-time chums never settled their quarrels. After his great success with the mission to America, which engendered the Radiation Laboratory and significantly aided the development of the proximity fuze and other innovations, Tizard found himself on the outskirts of government decision-making. In 1943 he left Whitehall for the presidency of Oxford's Magdalen College. As the European war drew to a climax, his stock rose again. In 1946, Tizard was called back to London with Clement Attlee's Labour government, serving for five years as the country's top science advisor. He finished his career at Oxford.

As Tizard resumed academic life, Cherwell, who had returned to Oxford himself as a professor at Christ Church, arrived back at Whitehall with a resurgent Winston Churchill. He stayed until 1953 and was instrumental in establishing the Atomic Energy Authority, inaugurated the next year. He remained active on that body, while resuming duties at Oxford, until his retirement. Even then he did not withdraw from public life, among other things speaking before the House of Lords on science and technology policy. Lord Cherwell was struck ill and died suddenly the night of July 2–3, 1957.

Sir Henry lived two more years, until the morning of October 9, 1959. As far as their influence on Britain's war effort goes, the two made similar contributions. Tizard managed the development of the radar chain that proved so critical during the Battle of Britain. Cherwell unmasked the woeful ineffectiveness of the British bombing campaign, then led the fight to bring radar to bear offensively, despite his severely flawed reasoning of the effects of the ensuing onslaught on German morale.

History, though, has afforded Tizard by far the greater legacy. A defining stamp was placed by Sir Charles Snow, the British scientist and philosopher, who contrasted Tizard and Cherwell in vividly entertaining lectures at Harvard University in 1960. His portrayals were turned into two small books, *Science and Government* and *Appendix to Science and Government*. In the treatises Snow characterizes Cherwell as almost always wrong on key science policy questions, citing as an example a "wildly impracticable" inclination toward infrared detection over radar. "He believed, as much as any man of his time, that he could solve any problem by his own a priori thought," says Snow. "This is the commonest delusion of clever men with bad judgment."

This account, while brilliant, should be taken with a grain of salt. As Snow asserts, Cherwell didn't listen well to others. However, he was often right in his views, suffering more from being off-putting than wrong. R. V. Jones, the Wizard War's intelligence guru, tried to set the record straight. In three articles for the London *Times*, Jones pointed out that Cherwell had been among the first to see the threat posed by Hitler and promote the idea that science could offer some defense to the bomber. Cherwell's early thoughts on locating planes through their infrared emissions, vilified by Snow, in one sense proved ahead of their time. When Jones penned his articles, at least three homing missiles — Firestreak, Sidewinder, and Falcon — found their quarries through infrared detection, though the technology is still unreliable for air warning. Jones did, however, see Tizard as the better scientific leader, and he did not quibble with Snow's admonition that Cherwell illustrated the dangers of placing a scientist alone among nonscientists in a position of power. However, he concluded, "an injustice" had been done to Cherwell.

As for Jones himself, he received two American awards, including the Medal of Merit, and was as far as he knew the only British scientist twice decorated by the United States. Jones left the Air Staff on September 30, 1946, having served exactly ten years as a civil servant, to become professor of natural philosophy at the University of Aberdeen. He first considered a career in radio astronomy, but decided Martin Ryle and Bernard Lovell already possessed too strong a head start, and so decided to concentrate on making small-scale improvements in instruments of measurement. Just before undertaking the move north, he received a phone call.

"This is Winston Churchill speaking from Westerham. Mr. Jones, how long are you going to be in the southern regions of our land?"

The former leader invited Jones to visit, and the physicist found Churchill sick in bed with a cold, wanting to talk over the war, particularly the role played by scientific intelligence. Jones stopped in again less than a week later, rating the two days as "among the most interesting of my life." Alone, he spoke to Churchill candidly. "I told him that I did not like the way the country was going, with strikes and the clamour for a 40-hour week, and he replied, 'I could have given them a 40-hour week — if they would work for 40 hours!' And he wept as he told me that he never thought that he would see the British Empire sink so low. . . ."

Churchill remained impressed by his young devotee, asking Jones back into government as director of scientific intelligence in the Ministry of Defence when he reascended to the prime ministership in 1951. He wanted Jones to strengthen the entire process of applying science to warfare. But Jones, who took up the position in 1952, resigned after only a year. Churchill, having already suffered a stroke, was far from the height of his powers, and without his leader's firm backing Jones found the peacetime bureaucracy like quicksand.

Jones returned to Aberdeen, where he finished his career. He wrote several books on modern arms issues and intelligence, and his World War II memoir — published as *Most Secret War* in Britain and *The Wizard War* in the United States — is a classic. A follow-up more than a decade later revealed the identity of the Oslo report author as Hans Mayer, which Jones had deduced before writing the earlier account but kept secret for fear of endangering the man's position.

Since Jones ranked as an inveterate practical joker, it was possible he could shed light on a wartime story Luie Alvarez had heard but had never been able to confirm. This held that the oft-repeated bit of folk wisdom that carrots were good for one's eyes had been originated — or at least given wide audience — by British agents during the Blitz. In order to keep the Germans ignorant of airborne radar and Ground Controlled Interception, the story went, the spies let it slip in bars around air bases that RAF pilots were being fed carrots to hone their night vision. This sounded a bit like an R. V. Jones operation. However, the retired intelligence ace noted that, sadly, he could not claim credit. He did recall, though, that in 1941 Air Vice Marshal Sir Charles Medhurst had mentioned being the originator of the hoax. Even without that arrow in his quiver, Jones stands out in a colorful profession. In 1994, the U.S. Central Intelligence Agency named a medal in his honor — the R. V. Jones Intelligence Award — with its namesake the first honoree. Inscribed on the medallion is the legend "Awarded for scientific acumen applied with art in the cause of freedom."

Other influential figures of British radar also built successfully on their wartime base. Martin Ryle and Bernard Lovell, whose ventures into radio astronomy dissuaded Jones from pursuing the same line, entrenched themselves as pioneers in that field. For widespread contributions, especially the application of interferometry to radio studies, Ryle won a Nobel Prize in physics in 1974. Lovell was knighted in 1961. Robert Hanbury Brown, after refusing to leave Britain with Watson-Watt, spent a dozen years with Lovell at Manchester, then another twenty-six at the University of Sydney. The quintessential "boffin," the term given by the British military on the eve of World War II to nerdy but vitally important scientists and radar engineers, Brown also earned a distinguished reputation in radio astronomy and served in the early 1980s as president of the International Astronomical Union. As his last major Australian project, he largely oversaw the fund-raising and design of the Sydney University Stellar Interferometer, a 640-meter-long optical telescope built in Culgoora, New South Wales. Brown was retired and back in his native Britain in 1995, as the big interferometer moved into full-scale operation.

Denis Robinson, a critical span in the radar bridge who helped convince the Americans to push ahead with microwave sub-hunters, and kick-started Allied research into semiconductor crystals as radar detectors and mixers, made America his adopted home. High Voltage Engineering Corporation, which he co-founded, proved successful, though after the Englishman's retirement the company fell victim to the 1980s leveraged buyout game. Robinson, ever gracious, gave several interviews for this book, growing misty-eyed while reminiscing about the tight bonds and lifelong rewards that arose from the World War II radar effort, and continuing to help even as disease slowly claimed his life. In February 1994, during a telephone conversation to check certain facts and recollections, he noted: "You're lucky you got to me, I won't be around in a year." Denis Robinson died on August 25, 1994, at the age of eighty-six.

Alfred Lee Loomis, who once greeted Robinson in his skivvies, maintained his powerful, behind-the-scenes presence for years after the war. His Tuxedo Park laboratory was closed in 1940, as the Rad Lab opened, and later divided into a handful of luxury apartments. Loomis himself caused something of a scandal in the Park by divorcing Ellen in 1944 and then marrying Manette Hobart, until roughly the same time the wife of his partner and friend Garrett A. Hobart III, who also served at the Radiation Lab. As *Fortune* noted in a rare feature about Loomis, "As often happens in such circumstances, some old friendships have been strained." After that, pens Luie Alvarez, Loomis and his second wife enjoyed "an extraordinarily happy time together during the final 32 years of Alfred's life, until he died at his home in 1975, a few months before his 88th birthday. His lifestyle underwent a dramatic change from one of multiple homes staffed by many servants to a very simple one, in which he and Manette cooked dinner every evening in East Hampton, side by side in the kitchen."

Alvarez called Loomis, who was elected to the National Academy of Sciences at the age of fifty-three, not long before America entered World War II, "the last of the great amateurs of science." During a protracted postwar legal battle with several independent inventors, the modern-day Ben Franklin first won, then lost on appeal, the claim to the Loran invention, though it was through his efforts, and of those at the Rad Lab, that the idea became a practical navigational tool. Loomis conducted, along with son Farnsworth, some biological research, but largely enjoyed more leisurely pursuits, hosting golfing visits to the Pebble Beach area in Monterey, California, and wintering in Jamaica. His three sons, to whom he reportedly gave a million dollars each when they turned fourteen, also became accomplished, though their stories are not universally happy. Alfred Lee Jr. emerged as a successful lawyer, venture capitalist, and sportsman who won an Olympic gold medal in sailing in 1948 and nineteen years later managed the *Independence-Courageous* yachting syndicate that triumphantly defended the America's Cup. Farney became a physician and a professor of biochemistry at Brandeis University, but later killed himself. Henry, after serving as a Navy radar officer in World War II, then working for years in physics, rose to be president of the Corporation for Public Broadcasting. He sat for many years on the MITRE Corporation board.

Lee DuBridge, the soft-spoken Radiation Laboratory director, lived a long and vigorous life. He remained president of Caltech for some twenty-three years, until January 1969: he had planned to retire two years earlier, at the age of sixty-five, but stayed on for a pivotal fund-raising campaign. His retirement was finally precipitated in November 1968, when he flew to New York to see President-elect Richard Nixon at the Hotel Pierre and agreed to serve as presidential science advisor. Some eighteen months later, approaching the age of seventy, DuBridge stepped down from that position. However, he continued to sit on the President's Science Advisory Committee.

DuBridge saw Caltech through a period of expansive growth, while maintaining its flavor as a small and elite institution. He presided over the birth of the Jet Propulsion Laboratory, as well as the dedication of the 200-inch Hale Telescope facility, yet throughout this impressive career retained the unfeigned humility and graciousness that characterized him during the Rad Lab days. In 1954 he testified in Washington on behalf of J. Robert Oppenheimer during a security hearing looking into the former A-bomb boss's reported communist sympathies. The testimony pitted him against Edward Teller, his old Radiation Lab cohort David Griggs, and, to some degree, Luie Alvarez, who backed off from stating that Oppenheimer was disloyal but asserted that his fellow physicist's influence against the hydrogen bomb seemed to be pervasive. "It was a rigged hearing," DuBridge once said of the case, which ultimately saw Oppenheimer stripped of his security clearance. "So it was a very sad business. . . . And I regarded the charges against him as trivial, if not false." DuBridge and Robert Bacher, another Rad Lab man who had gone on to Los Alamos, visited Oppenheimer several times after it was all over.

On the personal side, during his undergraduate days at tiny Cornell College in Iowa, DuBridge had noticed an attractive young woman and asked her out. But Arrola Bush responded that they had not been properly introduced and turned him down. A bit later he arranged to meet Doris May Koht, who did accept a date and impressed him by catching frogs in a local pond. They married in 1925 and remained together until she died in her sleep shortly after their forty-eighth anniversary. After a year, when DuBridge remarried, it was to Arrola Bush. She had also married and been widowed—all the while staying periodically in touch with the DuBridges, visiting the couple several times in Doris's last days. Later, Arrola wrote inviting DuBridge to visit her in Massachusetts. He took her up on the offer in January 1974 and proposed before the evening was over. They married that November and honeymooned aboard the *Queen Elizabeth 2*, on its first round-the-world cruise. The couple returned to Pasadena and lived there, chiefly, until DuBridge's death in January 1994. He was ninety-two years old.

Vannevar Bush, America's Tizard and Cherwell rolled into one, then spiced up with a dash of Churchill and Franklin Roosevelt, died June 28, 1974, in Belmont, next door to Cambridge, at eighty-four. Through much of his life, the patriarch of scientific warfare greatly influenced political and scientific affairs, though his clout waned steadily after World War II. He continued as president of the Carnegie Institution for another decade, retiring in 1955 at age sixty-five. Not long before he stepped down, in April 1954, he joined DuBridge in launching a staunch defense of Robert Oppenheimer. Other pro-Oppenheimer testimony came from James Conant, Mervin Kelly, Hans Bethe, and I. I. Rabi, but Bush was especially vociferous. He railed that the hearing smacked of "being interpreted as placing a man on trial because he held opinions, which is quite contrary to the American system. . . . I think no board should ever sit on a question in this country of whether a man should serve his country or not because he expressed strong opinions. If you want to try that case, you can try me. I have expressed strong opinions many times. They have been unpopular opinions at times. When a man is pilloried for doing that, this country is in a severe state. Excuse me, gentlemen, if I become stirred, but I am."

The revocation of Oppenheimer's clearance perhaps illustrated that Bush no longer ranked as omnipotent. Still, he did not go away. He served for years as an AT&T director, and in 1957 began a five-year stint as chairman of pharmaceutical giant Merck & Company, where he was also a director. In the late 1950s, too, he became chairman of the MIT Corporation and remained the honorary chair until 1971. Entrenched in his Belmont home, he reveled in gadgeteering, creating a basement machine shop where he worked on hydraulic pumps, piston-free engines, and valve inserts for the heart and brain. The straight-shooting engineer wrote widely on science policy issues and was sought out by journalists for advice and comments on contemporary events.

Bush left the most enduring legacy of the radar pioneers. His tremendous success directing civilian military science during World War II heavily influenced the way the United States funds research, both for the military and otherwise, in the postwar age. Even before World War II ended, far-thinking statesmen like Bush looked ahead to the transition to peacetime and saw a series of pitfalls on the horizon. For starters, with some two thirds of national defense contract R&D dollars going to sixty-eight corporations, and an astonishing 40 percent to only ten companies, Bush was among those who fretted that a few large concerns might be thwarting the entrepreneurs and independent inventors so crucial to long-term economic prosperity. In fall 1944 the engineer maneuvered to get Franklin Roosevelt to request a report on postwar scientific policy. Roosevelt died before Bush completed the document. But the next July the OSRD boss sent *Science — The Endless Frontier* on to Harry Truman.

It was a landmark document that formed a focal point in a lengthy congressional debate over the best method for distributing federal research funds and ensuring the scientific, technical, and economic prosperity of the nation. Bush's vision included increased funding of basic scientific research and training, while placing fundamental defense studies in a new civilian-run agency. In this his plan paralleled that of chief rival Senator Harley M. Kilgore of West Virginia. But the senator, an unabashed New Dealer, wanted sweeping overhauls that included distributing money on a geographical basis and assigning patents resulting from federal financing over to Uncle Sam. Bush not only favored protecting the basic patent situation, he also was adamant about adopting a best-science approach to determining who got funded — meaning only those most likely to succeed received financing, no matter where they hailed from. Lastly, unlike Kilgore, who wanted the National Science Foundation — Bush called it the National Research Foundation — strictly accountable to the commander in chief, Bush envisioned the upstart agency directed by a professional board chosen by the President, but largely free of political controls.

Although "an instant smash hit" in newspaper editorial circles, Bush's document could not carry the day. In 1947, after Congress approved a largely Bush view of the National Science Foundation, Truman vetoed the bill, objecting to the agency's insulation from presidential controls. The stalemate continued until a compromise was reached in early 1950. In the end, Bush got his way on the best-science approach and salvaged a power-sharing arrangement between the President and a private board. Military research, contrary to the *Endless Frontier*'s recommendations, remained outside the NSF purview — though Bush later came to feel this was for the best. Capitalizing on their five-year head start, the armed services, led by the Office of Naval Research, became the main supporters of academic research, and NSF's strength was diluted further by the emergence of the Atomic Energy Commission and NASA as additional funding sources.

For such reasons, Bush is typically viewed as the goat in the debate. Nathan Reingold, the distinguished science historian, summed it up: "I do not see Bush as the progenitor of a new post–World War II order but as a Moses who pointed out the way to a promised land, remained behind while others moved into the new domain, and realized to his dismay that the new promised land did not match his vision. This is a sketch of Bush as a loser, as a believer in certain old-fashioned virtues others chose to ignore. This is a sketch of a very impressive human being in whose losing we may have all sustained wounds."

But the flip side of the equation is that without Bush's haranguing and effort, NSF almost assuredly would have looked completely different. Whether that is good or bad can be debated, but the fact remains that Vannevar Bush exerted more influence than anyone over the shape of American scientific policy. In a postwar world that saw even Churchill stripped of power, it is hard to imagine how the engineer could have retained more influence than he did.

Moreover, Bush's style and influence extended to the military, even as the armed forces moved beyond his control, *for he had shown the services how to manage civilian research.* This was clear in Army Chief of Staff Dwight D. Eisenhower's April 1946 memo to his commanders, in which he stated that a clear lesson from the war was that the armed services could not have won without the techniques and weapons developed by scientists and industrial concerns. Scientists, he felt compelled to note, are "more likely to make new and unsuspected contributions to the development of the Army if detailed directions are held to a minimum." The Navy took a similarly openhanded view toward the eggheads, clearing red tape and leaving individuals to do their own thing, which was often basic science with only a tenuous link to military applications. This was the stuff, after all, that led Charles Townes to the maser. By 1949, "Even Bush noted that American scientists were decidedly happy with ONR, happy enough, possibly, not to mind the absence of a National Science Foundation," writes the historian Daniel J. Kevles. To a large degree, that state of relative bliss arrived because the Navy learned much from the way Vannevar Bush ran the OSRD, avoiding the World War I practices that had spurred him to get involved in mobilizing civilian science for war in the first place. So the crafty engineer may have lost the battle but come out on top after all.

Throughout World War II, Bush worked closely with trapper-turned–Pentagon advisor Edward Lindley Bowles, who for his wartime services was awarded the Distinguished Service Medal, the U.S. Presidential Order of Merit, and the Order of the British Empire. Bowles stayed on at the Pentagon for a time after the war. He wore many hats for the rest of his career, advising the Air Force until 1957. He played a significant role in establishing the RAND Corporation as that service's research and development offshoot. He served as a Raytheon consultant for some twenty years and chaired a congressional committee that proved influential in allocating television channels in the UHF range.

The wily Missourian, however, was not welcomed with open arms back at MIT, and his relationship with the school never fully blossomed again. After the war it was clear that President Karl Compton didn't really want Bowles back: the best job offered was his old professorship, with nothing in the way of a leadership position. Ultimately, when he left the Pentagon in 1947, Bowles signed on as a consulting professor to the electrical engineering department. Five years later he switched over to the Sloan School of Industrial Management in the same capacity — where he helped bring Jay Forrester into the fold — and finally received a consulting professor (emeritus) designation when he retired in 1963. Over the course of his life Bowles contributed $400,000 to the school, yet was miffed enough at his treatment that upon his death in September 1990, at the age of ninety-two, he left his papers to the Library of Congress and not MIT, greatly disappointing school authorities.

Bowles's relationship with Vannevar Bush retained its love-hate flavor. In the early days after the war, when the Department of Defense was being formed and it seemed possible Bush might be named the first secretary of defense, Bowles was approached by his old mentor about becoming Air Force secretary, although nothing came of either appointment. In 1948, apparently on the heels of receiving his national awards, Bowles wrote his former advisor, acknowledging past debts:

"I want to express my gratitude to you for the formal recognition that has come to me.

"Your support of my activities during the war I shall not forget. At the time when things were going not too well you were largely responsible for giving the indorsement that enabled me to work for colonel Stimson. Had it not been for this sustaining and inspiring episode — the opportunity to serve under a man who had confidence in me — I don't know where I would have landed or whether I would ever have recovered my spirits."

At the same time, though, the long-standing tension and competition persisted. Bowles had this to say for an official MIT history published in 1985: "Bush was brilliant, I wasn't; it made me work all the harder."

Many Rad Lab members rose to positions of great prominence. Edward Purcell, Norman Ramsey, and R. V. Pound became professors emeritus at Harvard, with offices on the same floor. In 1989, partly for his work in developing atomic clocks, Ramsey won the Nobel Prize in physics, becoming the last Rad Lab member to claim the award. Until his death on March 7, 1997, Purcell lived on Wright Street, his home since before the war.

George Valley Jr. served as Air Force chief scientist for a year, until 1958, then spent the remaining twenty years of his career as a professor of physics at MIT, becoming professor emeritus. Retired to Concord, Massachusetts, he remains at once friendly and irascible — and very sharp.

Dale Corson, Taffy Bowen's old mate whose adventures included the 1941 showdown with the British over airborne interception and co-invention of the Low Altitude Bombing radar set that devastated Japanese *marus*, became president, then president emeritus, of Cornell University.

Jerry Wiesner, after serving as John F. Kennedy's science advisor, earned the same titles at MIT. He died in October 1994.

One of the most versatile scientists emerging from the radar days was Luie Alvarez, whose father had reared him to regularly devote time to pure thought. In summer 1943, to demonstrate Ground Controlled Approach to the British, he spent six weeks at the Royal Air Force airfield at Elsham Wolds. "We landed every type of plane the RAF owned, every rank of pilot, from sergeant to air chief marshal, and on several occasions the entire bomber squadron returning from a mission over Germany," the GCA inventor boasted. When he left that August, Alvarez turned his creation over to Coastal Command for use in bringing home bombers from anti-submarine patrol. George Comstock, a Rad Lab staff member who later sat on George Valley's ADSEC Committee and Project Charles, took up the reins. Comstock passed operational control to a young officer named Arthur C. Clarke, who told the GCA story in his novel *Glide Path*. In the book, which was dedicated to the two Americans, Clarke predicted a physics Nobel by 1958 for the character modeled after Alvarez. The prediction proved ten years off the mark. But when Alvarez did receive the prize in 1968 for his liquid-hydrogen bubble chamber and the host of short-lived, subatomic particles his research team discovered, Clarke wired congratulations and reminded his old friend of the forecast.

Alvarez never stopped his creative thinking. After John F. Kennedy was assassinated, he studied pictures of the Zapruder film in *Life* magazine and from such obscurities as the rate at which a bystander clapped his hands was able to see a host of things the FBI had missed, debunking many conspiracy buffs in the process. In the late 1970s, in conjunction with eldest son Walter and Berkeley geochemists Frank Asaro and Helen Michel, he advanced the theory that asteroids or comets caused the extinction of the dinosaurs.

In 1987, Alvarez underwent surgery for a benign brain tumor, seemed to be on the road to recovery, then was diagnosed with cancer of the esophagus. His health declined steadily, and he died on August 31, 1988, at seventy-seven.

Alvarez's great versatility was nearly matched by that of Robert Dicke who died in March 1997—three days before former colleague Ed Purcell. Despite being beaten out in the measurement of the cosmic background radiation, Dicke established a wide reputation as a scientific virtuoso, adept experimentally, technically, and theoretically. He held more than fifty patents, and before the war ended filed key disclosures in two important forms of radar — monopulse and chirp. He made valuable contributions to cosmology, sparring with Princeton colleague Albert Einstein over the origins of the universe. In April 1992 he treated the author to a wild car ride through Princeton in which the future of this book seemed briefly in doubt.

Gordon Pettengill retired from MIT in March 1995, but remains a senior lecturer and Lincoln Laboratory consultant, and is still turning out original research papers based on *Magellan* data and other work. He is a member of the laser altimeter team of the *Mars Global Surveyor*, launched on November 6, 1996.

Unique among the radar pioneers was Edward George "Taffy" Bowen, who remained a good friend of Luie Alvarez and many other World War II scientists, and whose far-reaching influence on the radar story spanned three continents. A pioneer of Watson-Watt's upstart program that largely won the Battle of Britain, he personally shepherded the cavity magnetron to North America and proved instrumental in providing the Radiation Laboratory's initial direction — even motivation. After World War II he seized the initiative in utilizing radar for peacetime science, overseeing the rise of Australia, previously a scientific backwater, as a preeminent center of radio astronomy.

Down Under, people still remember him well — an avid sailor and cricketer standing ramrod straight, eyes twinkling, face lit up by an infectious and devilish grin. Women of his laboratory place him in the old school, flirting outrageously.

As an administrator, Bowen stayed true to his word and kept out of the day-to-day affairs of his researchers, guarding their interests but not taking credit for scientific successes. In the beginning, at least, he and Joe Pawsey complemented each other perfectly. Pawsey was the quintessential scientist, thinking things over interminably. Then Bowen, the dynamic man of action, stepped in to make the decisions.

In World War II's wake, the Council for Scientific and Industrial Research, reconstructed in 1949 as the Commonwealth Scientific and Industrial Research Organization, launched a broad-based attack aimed at bringing Australia up to speed scientifically and technologically. Bowen's Radiophysics Division led the way. Its distance-measuring equipment enabling airplanes to tell by radio how far they were from an airport rated as an unqualified success. A solid-state program produced the first Australian-made transistors. Radiophysics also built one of the world's first digital computers. It was Bowen, to a large degree, who started these programs, then decided to drop almost everything to concentrate on radio astronomy and his own pet cloud-seeding project.

Some fault this decision, arguing that the abandonment of emerging fields like computer science slowed Australia economically. The alternative view is that by concentrating on a few things, the country made its name internationally instead of being swamped by the far greater accumulation of experience, talent, and funding available in the United States and Western Europe. Bowen's radio astronomy program proved so successful that the International Union of Radio Science held its 1952 General Assembly in Sydney, home of the Radiophysics Lab's pioneers. It marked the first time a meeting of any international scientific union had been held outside Europe or the United States.

Bowen ran Radiophysics with zeal and charismatic energy until his 1971 retirement. A monument to his efforts can be seen in the Parkes Telescope, which remains the largest radio astronomy observatory in the Southern Hemisphere. Bowen's tireless efforts, as well as his salesmanship and contacts with powerful friends in England and America, got the telescope built.

In a way, Parkes was Taffy Bowen's last big dream. By the early 1950s, radio astronomy had become a major field inside astronomy, with Radiophysics perhaps the world's preeminent force. From the ranks of Joe Pawsey's researchers had come legends in the field—Bernard Mills, John Bolton, Gordon Stanley, and W. N. "Chris" Christiansen. Already, though, economics was changing the game. Radio astronomy could no longer be conducted with scuttled wartime receivers and antennas. Staying on top of stiff international competition required larger arrays and state-of-the-art processing, as exhibited by Bernard Lovell's plans to build a giant 250-foot-diameter radio dish at Jodrell Bank.

Bowen wanted a massive, steerable radio telescope like Lovell's for southern skies, but was thwarted initially by a lack of funding. Late in 1951, during one of his frequent trips to the United States, he revealed his hopes to various Rad Lab companions he looked up whenever he returned to the States. Among those he saw were Lee DuBridge, Vannevar Bush, Karl Compton, and Alfred Loomis, then a trustee of both the Carnegie and the Rockefeller foundations. All, as Bowen put it, were "men of enormous stature in American science." All, too, were concerned by their country's general absence from such a seemingly important field of science as radio astronomy; the Ed Purcell–Doc Ewen discovery earlier that year of the twenty-one-centimeter line of interstellar hydrogen had not belonged to any formal program. They asked Bowen for his views on how to bring the United States up to speed, and it was through this back door that the Welshman ultimately paved the way for Parkes.

Bowen's advice was for the United States to do what he couldn't—

build a big radio telescope to compete with Jodrell Bank. Caltech, together with the Carnegie Corporation, already operated the world's largest optical scope, the 200-inch Palomar instrument in the hills outside San Diego. At DuBridge's invitation, in the spring of 1952, Bowen drew up plans to construct a 300-foot-diameter radio dish to complement Palomar. Bush and Loomis pledged to seek funding and install Bowen as head of the new observatory.

After his return to Sydney, Bowen got a surprise. Bush and Loomis decided that the telescope could just as well be erected Down Under with U.S. financial backing. Bowen didn't quite understand the turnabout. "Perhaps," he later speculated, "it was the feeling that America already had too many of the world's good things, and that it was time to come to the help of other parts of the world." However, the genesis for the shift in thinking seems to have been a specialized Carnegie Corporation fund created to finance science in the United Kingdom's colonial empire. Allocations had been cut off during the war years, so by the early 1950s the fund was flush with money it had to distribute inside the British Commonwealth. In April 1954 the trustees, including Bush and Loomis, announced they would contribute $250,000 toward the construction of Bowen's big dish. Late the next year, after Bowen toured the United States with hat in hand, the Rockefeller Foundation put up another $250,000, with the proviso that the Australian government equal it. Within a few weeks, the match was approved. Not long afterwards, Caltech went ahead with plans to build its own radio facility, two 90-foot antennas at Owens Valley that drew on Office of Naval Research funds. Bowen's staff members John Bolton and Gordon Stanley became the director and assistant director, respectively. But the Welshman stayed in Australia to see his dream to completion.

It took six years to get the dish built. Three were spent perfecting the novel design, which eventually consisted of a steerable, 210-foot-diameter paraboloid, not the 300-foot array Bowen first envisioned. Meanwhile, a site was selected a few miles outside the small ranching town and former gold mining enclave of Parkes, perched in the flatlands 250 miles west of Sydney. Unlike optical telescopes, radio eyes are not hindered by clouds and do not need to rest on a mountain or peak. But they do require a clear view in all directions, well away from the interfering radio noise of large cities. The flat township fields, though ripe with blowflies and bush flies, seemed ideal.

The Parkes Telescope no doubt met most of Bowen's expectations. Almost amazingly, given troubles at Jodrell Bank and elsewhere, the big dish was dedicated on October 31, 1961. John Bolton was brought back from Caltech to direct the facility, which for years ranked as the world's best place to pursue radio astronomy. Besides charting the southern sky with unprecedented accuracy, it became a leading outpost in the hunt for quasars. By the early 1970s, close to 350 had been detected from the fly-ridden astronomical mecca.

For all its successes, however, Parkes broke the hallmark Radiophysics cohesion. Financially, the division could not support both it and ongoing interferometry efforts led by men like Bernard Mills and Chris Christiansen. The pair left for the University of Sydney. At the end of 1961, Joe Pawsey decided to take a job directing the four-year-old National Radio Astronomy Observatory in the United States. Early the next year, while visiting the West Virginia site, he suffered a partial paralysis of his left side, and a few months later a malignant brain tumor was discovered. Pawsey died in November 1962, and with him went much of the heart and soul of Australian radio astronomy. The early 1960s saw an exodus of Radiophysics personnel, many to upstart American programs. However, by the early 1990s, Parkes was in better standing, discovering pulsars and deep space masers, and forming a vital leg of the Australia Telescope, a network of eight antennas spread across New South Wales that can be linked to form one of the world's premier radio telescopes.

After retiring from CSIRO in 1971, the sixty-year-old Bowen became chairman of the Anglo-Australian Telescope, a large optical facility being built jointly by Britain and his adopted land. He left in 1973 to serve as scientific counselor to the Australian embassy in Washington, D.C., where he renewed old Rad Lab friendships and pursued his love of sailing out on the Chesapeake in a thirty-two-foot yacht named *Sosie*.

Bowen didn't return to Australia until 1976. For years his home situation had deteriorated: Vesta had declined to move to Washington, and not long after her husband came back, feeling forever estranged, returned to England. The couple were divorced in 1979, after forty-one years of marriage. Afterwards, Bowen tried to grow closer to his sons. David had studied electrical engineering, gotten a master's degree in business administration, and become a management consultant. Edward worked as a geophysicist, while John began a career as a computer analyst before turning much later to sugarcane farming.

The senior Bowen continued to pursue rainmaking and cloud physics, though his studies were always regarded with suspicion in the scientific mainstream. Just after his retirement, disgusted with the clunky CSIRO computers, he had decided to put his data on a personal computer. He felt uncomfortable buying one, so he presented a check to David, who purchased an HP-85 with sixteen kilobytes of random access memory. His son then helped write a software program that Bowen used to track rainfall patterns for years.

Shortly before Christmas in 1987, Bowen dropped a bottle of champagne by David's house, then went to Edward's but arrived complaining of numbness in one arm. Later that day, back at his Sydney home, he suffered a severe stroke that kept him hospitalized for nearly a year. Afterwards, Bowen walked with a stick, but retained much of his mental sharpness. He moved to a nursing home and resumed his rainmaking studies, finally shedding his old Hewlett-Packard computer for an IBM clone. A memoir, *Radar Days*, was published in 1987 and filled many vital holes in the record, even as its author's health deteriorated. Edward George Bowen died on August 12, 1991, at the age of eighty.

His sons mailed notices of Bowen's death to Britain, America, and around Australia. Old friend Robert Hanbury Brown wrote the biography for the Royal Society, to which Bowen had been elected in 1975. Despite playing a prominent role in Australian science, Brown noted, his comrade had refused to become a citizen of the country, sacrificing the possibility of honors in favor of keeping his British nationality. Bowen reveled, particularly, in his Welsh heritage, and delighted in the nickname Taffy, which he had kept since it was given to him in America in the early 1940s and savored as a lifelong reminder of those radar days.

Notes

ABBREVIATIONS

Archives

AIP	Niels Bohr Library, American Institute of Physics, College Park, MD
CAC	Churchill Archives Centre, Churchill College, Cambridge University, England
CAR	Carnegie Institution of Washington Archives, Washington, DC
CSL	Cabot Science Library, Harvard University, Cambridge, MA
CTA	California Institute of Technology Archives, Pasadena, CA
DRMC	Division of Rare and Manuscript Collections, Cornell University, Ithaca, NY
IWM	Imperial War Museum, London
LC	Library of Congress, Washington, DC
LLA	Lincoln Laboratory Archives, Lexington, MA
MIT	Institute Archives and Special Collections, Massachusetts Institute of Technology Libraries, Cambridge, MA
NA	National Archives, Washington, DC. Note: Unless otherwise specified, documents are from RG 227, Records of the Office of Scientific Research and Development
NAEST, 61	National Archive for Electrical Science and Technology, Institution of Electrical Engineers, London
NANER	National Archives New England Region, Waltham, MA. Note: Unless otherwise noted, all documents here belong to Series: Records of the Office of Historian; Subgroup: MIT Radiation Laboratory, RG 227
NRL	Historian's files, Naval Research Laboratory, Washington, DC
PRO	Public Record Office, Kew Gardens, Britain
TBL	The Bancroft Library, University of California, Berkeley

Papers and Collections

AVHL	Archibald Vivian Hill, CAC
CKFT	John D. Cockcroft, CAC
EGBN	Edward G. Bowen, CAC
ELB	Edward Lindley Bowles, LC
EOL	Ernest O. Lawrence, TBL
HAHB	Henry A. H. Boot, NAEST, 61

HSTP   History of Science and Technology Program, TBL
LAD   Lee Alvin DuBridge, CTA
ROHO   Regional Oral History Office, TBL
VB   Vannevar Bush, LC; MIT
WONU   Arthur Edgar Woodward-Nutt, CAC

Terms

CCN   Cape Cod Network
CH   Chain Home Network
LL   Lincoln Laboratory
OHI   Oral History Interview
PC   Project Charles
RG   Record Group
RL   Radiation Laboratory
WW   Whirlwind

PROLOGUE TO NOTES

In constructing this book's narrative, I have pieced together information from a wide array of sources, often making it impossible, given space constraints, to match each item with its exact source. In such cases, I have written overviews to particular subjects or sections, listing all sources but not by item. Should a reader need more, I can supply additional detailed notes through the publisher.

Boldfaced page numbers indicate prime sources drawn on throughout a chapter or section. In a given entry, sources generally appear in alphabetical order, although if one or two heavily outweigh the others, they are listed first.

Preface

13 Building 20 history: Beam (1988); Brand (1994), pp. 26–28; Garfinkel (1991).
13 "The edifice . . . campus": Quoted in Brand (1994), p. 26.
14 "Ah, Building . . . procreative": Quoted in Garfinkel (1991), p. 11.
14 scheduled for demolition: Dunbar interview, June 7, 1996.
15 more than 300 radars: *FAA Administrator's Fact Book*, July 1995, p. 29.

1 • The Most Valuable Cargo

Several PRO documents are vital to tracing activities of the British Technical Mission to the U.S. in fall 1940. Even when not yielding specific information, these often served as invaluable yardsticks: "British Technical Mission," vol. 1, items 1–100, AVIA 10/1; "Activities of British Technical Mission from September 9, 1940," pp. 1–15, AVIA 10/2; various minutes of daily meetings, AVIA 10/4.

27 "When the . . . shores": Baxter (1947), p. 142.
27 Contents and description of black box are pieced together from Bowen (1987), pp. 152–53; Watson-Watt (1959), p. 261; R. Clark (1962), p. 138; R. Clark (1965), p. 259; Alvarez (1987), p. 84; Bryant (1991), p. 74; Baxter (1947), p. 120; Cockcroft (1985), p. 329; Blackett (1960), p. 649; Bowen OHI with David K. Allison, side 5, NRL; "Minutes of Daily Meetings of Technical Mission in America," AVIA 10/4, PRO. Hartcup and Allibone (1984), p. 98, say the box was purchased at Army & Navy Stores, Westminster.
27 Bowen's movements: Key is Bowen (1987), pp. 150–54.

27 desperate times in Britain: Churchill warned of invasion Sept. 11, 1940. Lovell (1991), p. 42; cf. Churchill (1949).

28 magnetron most prized treasure in black box: Widespread reading, cf. Bowen (1987), pp. 152–53.

28 CH capabilities: Mainly Swords (1986), pp. 195, 227.

28 Magnetron technical capabilities, reasons for bringing it to U.S.: Wide reading, esp. Bowen (1987); Guerlac (1987).

30 "Out. Don't . . . reserved?": Bowen (1987), p. 154.

30 "The would-be . . . interruptions": Ibid.

30 "I was . . . it": Ibid.

31 TM preparations: For pre-mission wrangling, see R. Clark (1965), pp. 248–71. Some Hill actions from "Memories and Reflections," chs. 1, 5, memoir in AVHL, 5/4, CAC; Hill, "RDF in Canada and the United States," Canadian Radar folder, Box 49B, NANER; cf. Bowen (1987), p. 213.

31 Items withheld from black box: R. Clark (1965), p. 259; Woodward-Nutt, "Memories — Mainly Aeronautical," WONU 1/6, p. 26, CAC.

32 Tizard Mission makeup, member backgrounds, general movements: R. Clark (1965), pp. 258–60; Woodward-Nutt, WONU 1/6, p. 21; Bowen (1987), p. 152; Guerlac (1987), p. 253; Kevles (1979), p. 230.

32 Voyage of *Duchess of Richmond*, mission journey to D.C.: Key are Woodward-Nutt, WONU 1/6, pp. 21–25, CAC; Bowen (1987), pp. 155–56; cf. Cockcroft diary notes, CKFT, 21/6, CAC; Guerlac (1987), pp. 253–54; Hartcup and Allibone (1984), p. 99. Early Tizard Mission movements are convoluted. Bowen, p. 156, says Cockcroft stayed in Canada. However, Cockcroft's diary has him in D.C. Also, Guerlac, p. 254, has the *Duchess* arriving in Halifax Sept. 7, a day later than was the case.

33 "submachine guns . . . orifice . . .": Bowen (1987), p. 155. Woodward-Nutt is more authoritative.

33 "I was . . . arrived": Woodward-Nutt, WONU 1/6, p. 25, CAC.

33 "No administrative . . . rather annoyed": R. Clark (1965), p. 263.

33 Tizard's movements in D.C.: R. Clark (1965), pp. 262–63. Description of Tizard's physical characteristics and manner is from Snow (1961), pp. 6–7.

33 Cosmos Club: Kevles (1979), p. 50, names the club. A club official confirmed the location. The club has since moved to Massachusetts Avenue, west of Dupont Circle.

33 Bush background and actions, Carnegie background: Bush OHI, Reel 1-A, p. 1, Reel 2-A, p. 73A, MC 143, MIT; Bush (1970), p. 74; Kevles (1979), pp. 67–69, 296–300; Morse (1977), p. 121; Wildes and Lindgren (1985), p. 84.

34 "Of the men . . . or third": Loomis OHI, May 21, 1943, Loomis folder, Box 49A, NANER.

34 "That's okay . . . on it": Kevles (1979), p. 297. Events leading up to and including VB's meeting with FDR come from Bush (1970), pp. 35–36; Kevles (1979), pp. 296–97. Meeting date is from FDR Library; cf. Rhodes (1986), p. 338.

34 "That, in . . . was": Bush (1970), p. 32.

34 "excel . . . be necessary": FDR to VB, June 15, 1940, NDRC folder, Box 1, E-3, NA.

35 early NDRC: Baxter (1947), p. 17; Kevles (1979), pp. 296–98; Guerlac (1987), pp. 246–47; cf. Wilson OHI, May 6, 1943, Interviews 1943–1945 folder, Box 58C, NANER.

35 Alfred Loomis contributing to MIT: Wildes and Lindgren (1985), p. 196.

35 early Bush-Tizard meetings: Bush OHI, Aug. 20, 1944, Interviews — Final Drafts folder, Box 49A, NANER; R. Clark (1965), pp. 264–65. That Bush was unaware of the magnetron, but had been tipped off to British radar, see Bush OHI, NANER.

35 "behind the barn": Bush OHI, Aug. 20, 1944, Interviews — Final Drafts folder, Box 49A, NANER.

36 Wardman Park description is from EOL (72/117c), C27, f5, which has hotel stationery. Loomis's suite is described by Woodward-Nutt, WONU 1/6, p. 25, CAC; cf. "British Technical Mission," AVIA 10/4, PRO.

36 events surrounding magnetron disclosure: Key is EGBN, CAC, esp. "The Tizard Mission to the USA and Canada — Sept 1940," April 22, 1985, 4/1; and EGBN 4/3, which includes Bowen's diary excerpts; Bowen (1987), pp. 88, 159–62, 193.

Also needed to complete the puzzle: Bowen OHI, April 29, 1943, Bowen folder, Box 49A, NANER; Bowen OHI, Side 6, NRL; Cockcroft notes, AVIA 10/1, PRO; "Activities of British Technical Mission from September 9, 1940," AVIA 10/2; R. Clark (1965), p. 268; Cockcroft (1985), p. 329; Guerlac (1987), pp. 249–54, 369.

For U.S. point of view: Baxter (1947), p. 120; Guerlac (1987), pp. 249–50, 369. More details, including Harold Bowen's drinking, come from Compton OHI, Aug. 20, 1943, K. T. Compton folder, Box 58C, NANER. U.S. radar state of the art is from Allison (1981), p. 149 ff.

37 Bush often delegated authority: Kevles (1979), p. 297.
37 Wilson as Bush's alter ego: Baxter (1947), p. 16.
37 "I still . . . table": quoted in R. Clark (1965), p. 268.
37 "that . . . do next": Newton et al. (1991), p. 12. The unattributed source is undoubtedly Bowles; cf. OHI with Norberg, May 6, 1977, Tape 2, p. 51, 13, 6, ELB: "Before, we didn't know what to do and when you don't know what to do, you write a report."
37 small wooden box . . . thumbscrews: Hartcup and Allibone (1984), p. 99.
37 "It was . . . Compton": Bowen OHI, side 5, NRL.

2 • Radiation Laboratory

38 "the greatest . . . world": Newton et al. (1991), p. 19.
38 Tuxedo Park mansion: Physical description mainly from visit to estate and Tuxedo Park library real-estate documents. Many details of lab in the 1930s are from Alvarez (1983); Alvarez (1987), pp. 78–81, and a longer version of the Alvarez manuscript, pp. 650–80, provided by Peter W. Trower. Also helpful were interviews with Loomis, April 15, 1993; Degnan, Sept. 20, 1994; Compton OHI, Box 58C, NANER; Lewis OHI, May 5, 1943, Lewis folder, Box 49A, NANER; Guerlac (1987), pp. 222–24.
39 "Hello, Ed . . . over?": Trower-Alvarez manuscript, p. 680.
39 "For the . . . this": Lewis OHI, Box 49A, NANER. A slightly abridged version is in Guerlac (1987), p. 224.
39 German air raids on Britain: I rely throughout chapter on Gilbert (1989).
39 Tuxedo Park meeting: Key are Bowen (1987), pp. 159–62; and Cockcroft, "Notes on Visit to Loomis Laboratory, Tuxedo Park, 28 and 29 September, 1940," British Scientific Mission folder, Box 43, NANER. Also important is Bowles OHI with Romanowski, March 17, 1982, ELB. Cf. Alvarez (1980), p. 329; Baxter (1947), p. 16; Hartcup and Allibone (1984), p. 97. Bowles's background is in Wildes and Lindgren (1985), pp. 106–23.
40 "The atmosphere . . . cause": Bowen (1987), p. 162.
40 "All we . . . gasp": Bowles OHI with Romanowski, p. 12, 13, 10, ELB.
40 luncheon meeting: Guerlac (1987), p. 255.
40 Tizard leaves for England: Date is in dispute. Bowen (1987), p. 162, says he and Tizard left D.C. on Oct. 2. R. Clark (1965), p. 269, puts Tizard's departure on the 5th, and has details of bet with Cockcroft. Tizard left the 5th, as confirmed in "Activities of British Technical Mission from September 9, 1940," p. 7; AVIA 10/2. However, Tizard may have passed several days in New York.
40 "for all . . . worth": Bowen (1987), p. 163.
40 events surrounding magnetron's demonstration: Bowen (1987), pp. 162–66. Some technical help from Smullin interview, Sept. 8, 1994; also Brittain (1985), p. 67; Fagen (1978), pp. 25–30. Physical description of Whippany comes from Bell Laboratories archives photo, Townes interview, Feb. 13, 1996.
41 "Very gingerly . . . amazement": Bowen (1987), p. 166.
41 Loomis excitedly told his cousin: Kevles (1979), p. 303.
42 acute sense of urgency: Factors leading Bush, Compton, and Loomis to create a central lab are gleaned from Bush and Loomis OHIs, Box 49A, NANER.
42 second Tuxedo Park meeting, formation of RL: Central are Bowen (1987), pp. 171–174, 215–17; Guerlac (1987), pp. 248, 257–59, 284; "Notes on Meeting of Micro-wave Sub-committee of N.D.R.C. Held Friday, Oct. 11, 1940 at the Carnegie Institution," AVIA 10/1, PRO; "Meetings of Microwave Sub-Committee of N.D.R.C.," AVIA 10/1, PRO. This

document, apparently by Bowen, summarizes the Tuxedo meetings of Oct. 12–13, and the NYC meeting the 14th.

42 call from Mervin Kelly: Bowen (1987), pp. 166–68; cf. Fagen (1978), p. 114.

42 "Oh my . . . up": Ibid., p. 166.

42 "There was . . . say": Ibid., p. 167.

43 "We agreed . . . worked": Ibid., p. 168. The 7-hole magnetron didn't work. Also, Megaw was a pioneering engineer who made important improvements in the original device. See Swords (1986), pp. 265–66.

43 Lawrence summoned by Bush: Notes, presumably draft of Bush telegram to EOL, Oct. 3, 1940, Lawrence file, Container 64, VB.

43 "Our course . . . shores": Gilbert (1989), p. 131. He Anglicized "defense."

43 highest priority . . . AI radar: Previous histories omit a fourth item, a microwave search-light director: "Notes on Meeting of Micro-wave Sub-committee of N.D.R.C. Held Friday, Oct. 11, 1940 at the Carnegie Institution," AVIA 10/1, PRO. Also, the priorities of items 2 and 3, the navigation system and gunlaying radar, were later flopped. See "Meetings of Microwave Sub-Committee of N.D.R.C.," AVIA 10/1, PRO.

44 "This changes . . . British": Lewis OHI, Box 49A, NANER.

44 agreement hammered out: Contract details from Bowen (1987), p. 175; L. Brown, The Tizard Mission, draft manuscript chapter.

44 Bowen imploring them on: Bowen's role is described by Bainbridge OHI, May 1, 1943, Box 49A, NANER. Details of Oct. 18 meeting and subsequent budget figures are from A. Loomis to Bush, Oct. 24, 1940, AVIA 10/2, PRO.

44 "We were . . . them": Romanowski OHI, Tape 2, pp. 44–45, 13, 10, ELB.

45 "I protested . . . scotch": Bush OHI, Box 49A, NANER.

45 "I am . . . MIT": Biographical historical note, June 20, 1988, RL folder, ELB. Several versions of this story exist. Most authoritative are Guerlac (1987), p. 258, and Bowles OHI, Aug. 21, 1943, Box 49B, NANER. Bowles himself tells story in at least three slightly different ways — in the quoted note, the 1943 OHI, and Romanowski OHI, Tape 2, pp. 45–46, 13, 10, ELB.

45 sandbagged Compton: Guerlac (1987), p. 258; Bowles, Romanowski OHI, Tape 2, p. 45, 13, 10, ELB; Compton OHI, Box 58C, NANER.

46 christened the RL: Rationale behind name comes from Guerlac (1987), p. 261; Kevles (1979), p. 303.

46 primarily nuclear physicists: Reasons for staffing RL with physicists are from Bowen (1987), p. 173; DuBridge interview, May 17, 1992; cf. Alvarez (1980), p. 330.

46 EOL's role in forming RL: Trower-Alvarez manuscript, p. 698; Johnston (1987), p. 55; Bowen (1987), pp. 174–75; Kevles (1979), p. 271.

46 Bainbridge's experiences and background come from OHI, Box 49A, NANER; Rhodes (1986), p. 652.

46 DuBridge's early RL recruitment, except where noted, comes from DuBridge, "Memories," Part I, unpublished memoir, CTA; DuBridge interview, May 17, 1992; DuBridge to EOL, Nov. 1, 1940, EOL (72/117c), C 27, f2; Rigden (1987), pp. 131–35; Guerlac (1987), pp. 260–62; Kevles (1979), pp. 303–304; handwritten notes, presumably of an OHI with Rabi, dated May 27, 1943, Rabi folder, Box 49A, NANER.

46 "I can't . . . important": Rigden (1987), p. 131.

46 "that was . . . in": Guerlac (1987), p. 260; cf. Kevles (1979), p. 304.

47 series of lab visits: DuBridge to EOL, Nov. 1, 1940; Barrow's role in Bainbridge OHI, Box 49A, NANER; Barrow's background, Wildes and Lindgren (1985), p. 119 ff.

47 Algonquin Club visit: H. Guerlac, notes on conference with A. Loomis, Jones, and DuBridge, Loomis folder, Box 49A, NANER; cf. Wilson OHI, Box 58C, NANER; DuBridge to EOL, Nov. 1, 1940. Rigden (1987), p. 132, refers to British radar specialists being present. But it could only have been Bowen, since Tizard Mission notes, AVIA 10/2, show Cockcroft in Canada.

47 "It was . . . it": DuBridge to EOL, Nov. 1, 1940.

47 "So the . . . fun": Ibid. Parens his.

47 American desire to fight Hitler: Cf. Kevles (1979), pp. 281, 289.

48 "The war . . . it' ": Rigden (1987), p. 132.
48 Rabi joined RL on Nov. 6: Rabi to W. Loomis, May 2, 1942, Rabi folder, Box 49A, NANER.
48 Zacharias joins: J. Goldstein (1992), p. 50.
48 Ramsey joins: Ramsey interview, March 30, 1992.
48 Street, Getting, Purcell join: Purcell interview, March 26, 1992; cf. Purcell OHI, IEEE Center (1993), pp. 241–52.
48 McMillan and Alvarez arrive: Trower-Alvarez manuscript, Alvarez (1987), p. 87; Alvarez OHI, May 18, 1943, Alvarez folder, Box 49A, NANER.
48 "I never . . . thought": Trower-Alvarez manuscript, p. 701.
48 Pollard leaves Yale: Pollard (1982), pp. 38–39.
48 Bainbridge and Eastham make arrangements: Bainbridge OHI, Box 49A, NANER; Guerlac (1987), p. 261.
49 Karl Compton overview: W. G. Tuller, "Microwave Radar in Its Infancy," Miscellaneous verbal information folder, Box 59A, NANER.
49 "We chose . . . team": Rigden (1987), p. 134.
49 "A sudden . . . asking": Alvarez (1987), p. 83.
49 "It's simple . . . work?": Kevles (1979), p. 304; Condon's background: *American Men of Science* (New York: Jaques Cattell Press, 1960).
49 magnetrons locked in safe, windows painted black: Rigden (1987), p. 134; Newton et al. (1991), p. 12.
49 Bowen drives to Cambridge, Lawrence leaves: Bowen (1987), p. 177; Guerlac (1987), pp. 261–62; Lewis OHI, Box 49A, NANER; Tuller, "Microwave Radar in Its Infancy," Box 59A, NANER.
51 Baker background: Baker OHI, IEEE Center (1993), pp. 33–39.
51 Loomis scrawls on chalkboard: Alvarez OHI, Box 49A, NANER.

3 • Beginnings

52 "The flying . . . fly": Quoted in Bruce-Briggs (1988), p. 22.
52 "that period . . . Peace": Churchill (1948), p. 37.
52 WWI figures: Kinsey (1983), p. 4; Harrison (1976), p. 23.
52 Gothas, strategic bombing: Rhodes (1986), p. 98; cf. Hastings (1979), p. 31.
52 early air defense: Swords (1986), pp. 48–49, 178–80; Rhodes (1986), pp. 98–99; Bruce-Briggs (1988), p. 19; Fennessy (1985), pp. 1–3; cf. "Committee of Imperial Defence—Progress of Balloon Barrage Scheme," Oct. 5, 1936, CAB 3, 6, PRO.
53 "I think . . . yourselves": Hastings (1979), pp. 38–39; cf. Bruce-Briggs (1988), p. 21.
53 1934 air exercises: R. Clark (1965), p. 106.
53 "cursed, hellish . . . air . . .": Churchill (1948), p. 84.
53 Lindemann-Churchill relationship: Churchill (1948), p. 72; Birkenhead (1961).
53 "That there . . . exhausted": R. Clark (1965), pp. 107–108.
53 British search for air defense technologies: Rowe (1948), pp. 4–6; R. Clark (1965), pp. 108–13. Rowe's personality: Bowen (1987); R. Brown (1991). Bowen and Watson-Watt (1957a, 1959) provide additional tidbits.
54 "to consider . . . aircraft": R. Clark (1965), p. 111.
54 The idea quickly accepted: For invitations to join the air defense committee, see 2/4481, PRO.
54 "damaging radiation": Watson-Watt (1959), p. 51.
54 W-W background: Ibid., pp. 5, 17–22; cf. Bowen (1987), p. 7.
55 "bubbled out . . . fountain": R. Brown (1991), p. 11.
55 failure to appreciate good food: F. S. Barton OHI, Jan. 20, 1944, General Intelligence folder, Box 43A, NANER.
55 "Please calculate . . . 1 km": Quoted in A. Wilkins (1981), p. 131. He uses "RF," not "radio frequency." Abridged for clarity, since W-W's note was lost, and quote was from memory. Wilkins erroneously dates memo Feb. 1935.
55 Wilkins's role: A. Wilkins (1981), pp. 131–35; Watson-Watt (1959), pp. 52–53.

55 "Well, then . . . them": A. Wilkins (1981), p. 132.
56 "Meanwhile attention . . . required": Watson-Watt (1959), p. 53.
56 committee pounced . . . facts and figures: Rowe to CSSAD, Feb. 4, 1935, EGBN, 5/46. Tizard Committee meeting minutes: general intelligence folder, Box 43A, NANER.
56 W-W again turned to Wilkins: Watson-Watt (1959), p. 54.
56 planes 10 miles distant . . . 20,000 feet: Wilkins's conclusions in W-W's famous memo, "Detection of Aircraft by Radio Methods," Feb. 12, 1935, Watson-Watt (1959), p. 427 ff.; cf. Watson-Watt (1957a), p. 470 ff.
56 "It turns . . . re-checking": Watson-Watt (1959), p. 54.
56 "radar" coined, accepted by British: Guerlac (1987), p. 5; Swords (1986), p. 1.
56 how radar works: Example from Guerlac (1987), p. 15.
57 "the whole . . . Station": Watson-Watt (1959), p. 430.
57 Scotsman proposed: Ibid., p. 427 ff.
57 CSSAD response to W-W memo: R. Clark (1965), pp. 117–18; cf. Watson-Watt (1959), p. 63.
57 Dowding and sound mirror: Fisher (1988), pp. 23–24. Other sound mirror details: Watson-Watt (1959), pp. 60, 63; R. Clark (1962), p. 26; Wood and Dempster (1990), p. 80.
57 Daventry test: Except where noted, I rely on R. Clark (1962), pp. 36–37; Watson-Watt (1957a), pp. 110–11; (1959), p. 64; A. Wilkins (1981), pp. 136–38.
58 "atmospherics": Watson-Watt (1959), p. 64.
58 Heyford's communications aerial: Barton OHI, Box 43A, NANER.
58 "Britain . . . once more": Clark (1962), p. 37.
58 "It was . . . mirrors": Watson-Watt (1959), p. 65.
58 "all the . . . reason". R. Clark (1965), p. 118.
59 £12,300 allocated: Watson-Watt (1959), p. 114.
59 never dreaming . . . same trail: That British felt they alone had radar is evident from wide reading: cf. Watson-Watt (1959); R. Jones (1978).
59 Einstein's theory languished: Kevles (1979), pp. 86, 159.
59 radar as simultaneous invention: See Süsskind (1988, 1994).
59 radio background of radar: I draw mainly on Guerlac (1950); Swords (1986). Important for numerous clarifying details: Appleton and Barnett (1925); Brittain (1990); L. Brown (1992b); Burns (1988); Kevles (1979), p. 137; Rawlinson (1985); Seitz (1996); Süsskind (1994); Swords (1994); Tucker (1978), pp. 1232–33; Tuve (1974); Smullin interview, Sept. 8, 1994.
62 crucial refinements in CRTs: the critical ones were the electrostatically focused CRT and the multielectrode electron tube, ready in 1930; L. Brown correspondence, March 1, 1996.
63 U.S. radar program: I rely mainly on Allison (1981); Guerlac (1950, 1987); Page (1962a, b); Swords (1986). Cf. van Keuren (1994); Guerlac to Baxter, Oct. 12, 1944, Historical Project folder, Box 47A, NANER.
64 Orford and Orfordness: Physical description of town from visit of March 14, 1993. Main details of "island" life drawn from Bowen (1987), esp. pp. 11–21; R. Brown (1991); A. Wilkins (1981). Swords (1986) has valuable technical details, pp. 181–83, 202–207, 292. A few minor holes filled by Kinsey (1981), pp. 1, 34; Guerlac (1987), pp. 135–44; Rowe (1948), pp. 14–17, 31; Watson-Watt (1959), p. 114; Bowen OHI, April 27, 1943, Bowen folder, Box 49A, NANER.
65 "It was . . . birds": R. Brown (1991), p. 6.
65 "benign influence": Bowen (1987), p. 8.
65 hanged by . . . extinct: Ibid., p. 9.
65 "Many were . . . breakfast": Ibid., p. 16.
66 "a marvel . . . crudity": Guerlac (1987), p. 135.
66 showed range . . . horizontal scale: Best source for early Orfordness technical progress may be Guerlac (1987), pp. 135–38.
66 stone's throw . . . receiving corps: A. Wilkins (1981), p. 141, describes setup, generally confirmed by R. Brown correspondence, Sept. 29, 1995. Parallel receiving aerials were for determining aircraft height.

67 "We were . . . excited": Bowen (1987), p. 16.
67 early height finding: First Orfordness radar worked at 50 m. Wilkins strung half wave dipoles between two towers 30 m apart, probably separating aerials by 25 m, or half operating wavelength. By comparing signals in two horizontally separated antennas he could measure elevation angle of the incoming signal: Bowen (1987), p. 21; A. Wilkins (1934); A. Wilkins (1981), p. 141; cf. R. Brown correspondence, Sept. 29, 1995.
68 Bawdsey Manor purchase: Quilter's background: Kinsey (1983), pp. 1, 187. For price: Kinsey (1981), p. 1; A. Wilkins (1981), pp. 150, 154; Bowen (1987), p. 22, says £23,000.
68 manor description: March 1993 visit; Kinsey (1983); R. Brown (1991); Savills real estate brochure.
69 Bawdsey life: Mainly Bowen (1987), pp. 22–47; R. Brown (1991), pp. 4–22; Guerlac (1987), pp. 140–48. Exceptions noted.
69 close to 20 staff: G.Touch OHI, June 9, 1944; Touch folder, Box 49A, NANER.
69 first-name basis: Carter (1988), p. 157; cf. Touch OHI, Box 49A, NANER; Kinsey (1983), p. 39.
70 CH accuracy: Swords (1986), p. 188; Guerlac (1987), p. 145.
70 "To say . . . stations": Guerlac (1987), p. 172, n.38.
71 Sgt. Naish: full name from F. Mason (1990), p. 439.
71 "Let's go . . . look": Bowen (1987), p. 43.
72 "Signal lights . . . *Courageous*": Bowen (1987), p. 43. Story from ibid., pp. 42–45. Guerlac (1987), p. 149, mistakenly puts episode a year later.
72 "Yes, that . . . reporting?": Bowen (1987), p. 45.
73 "We had . . . capabilities": Ibid., p. 45.
73 RAF took active interest: Rowe (1948), p. 23; cf. Watson-Watt (1959), p. 122.
73 service work at Bawdsey: R. Clark (1962), p. 42; Kinsey (1983), p. 32; Watson-Watt (1959), p. 116.
73 "Rabbits . . . vermin": R. Brown (1991), p. 22.
73 "standard of . . . servants": Ibid.
73 Rowe and staff: Bowen (1987), p. 50; Guerlac (1987), p. 154; Touch OHI, Box 49A, NANER.
73 Army progress: Rowe (1948), p. 38; Watson-Watt (1959), p. 238.
74 CH progress: Guerlac (1987), pp. 143–44; R. Brown (1991), p. 35; Rowe (1948), p. 26; Swords (1986), pp. 193–95.
74 Bowen's AI project: Bowen (1987), pp. 45–73, except where noted. Wood interview, June 10, 1996, helped get ASV antenna configuration straight.
75 "Where is . . . it": Ibid., p. 70.
75 "My God . . . close": Ibid.
75 "Tea, tea . . . one": Ibid., p. 73; cf. R. Brown (1991), p. 28.
75 crash order for AI radar: Bowen (1987), p. 78; R. Clark (1965), p. 189; Guerlac (1987), p. 150, attributes order to Dowding's test flight.
75 Third Reich growth: Churchill (1948); cf. Rhodes (1986), pp. 244–46; 309.
76 *Plutôt mourir . . . changer*: R. Brown (1991), p. 4; Alexander interview, Sept. 7, 1994.

4 • A Line in the Ether

77 "The odds . . . infinite": Churchill (1949), p. 296.
77 Bawdsey evacuation: Bowen (1987), pp. 84–85; Guerlac (1987), p. 153.
77 Zeppelin flights: Some confusion exists about flight dates. Wood and Dempster (1990), p. 10, say the first took place in May. However, a complete flight log held by historian Alfred Price reveals only one flight besides that of Aug. 2–4, 1939. It began July 12, 1939, but despite unconfirmed reports of its detection by CH stations stayed well out over the North Sea and does not appear to have been an espionage flight. Price correspondence, April 17, 1995, and June 1, 1995; L. Brown correspondence of March 1 and 13, 1996.
77 move from Bawdsey: except where noted, Bowen (1987), pp. 80–97; Rowe (1948), pp. 42–43, 53–56.

77 Chamberlain's radio address: Churchill (1949), pp. 362–63.
77 3 million Londoners: Kennett (1982), p. 103.
78 CH technical capabilities: Swords (1986), pp. 236–39. Technical help from Barton interviews.
78 PPI development and discussion: Guerlac (1987), pp. 153–56; Dale R. Corson correspondence, Oct. 16, 1995.
78 Rowe recruits from prepared lists: Guerlac (1987), p. 153; D. Robinson interview, July 30, 1992.
80 move to St. Athan: Date was Oct. 26, 1939. "Progress Reports Dr. Bowen 1939–40," AVIA 7, 344, PRO.
80 "The sight . . . life": Bowen (1987), p. 93.
80 Bowen's feud with Rowe well known: See Bowen (1987), pp. 56–57, 138; R. Brown (1991), pp. 59–62; Lovell (1991), p. 15; Rowe (1948), pp. 55, 69; V. Bowen interview, May 18, 1994.
81 Bowen stopped filing status reports: Rowe to Bowen, April 6, 1940, AVIA 7, 344, PRO.
81 airborne radar figures: Estimate of 100 installed comes from fact 50 were in by Feb., 200 in July: Bowen (1987), p. 95; also see A. Price (1973), pp. 54–56. Production order numbers come from Bowen (1987), p. 135.
81 move to Swanage: Chiefly, Bowen (1987), pp. 137–39; Lovell (1991), pp. 25–28. A few details from Rowe (1948), pp. 57–66; R. Brown (1991), p. 28; Pringle (1985), p. 340.
82 Dunkirk: Churchill (1949), pp. 102–104; Wood and Dempster (1990), p. 130.
82 ". . . we shall . . . surrender . . .": Churchill (1949), p. 104.
82 TRE reaction to invasion of France: Bowen (1987), p. 137; Lovell (1991), p. 29; Rowe (1948), p. 81.
82 Bowen's increasing isolation: V. Bowen interview, May 18, 1994; Guerlac (1987), p. 232; R. Brown (1991), p. 61; Lovell (1991), p. 26; Denis Robinson OHI, June 8, 1944, Robinson folder, Box 49A, NANER.
82 first magnetron test: Boot (1946), HAHB; Guerlac (1987), p. 229; Miller (1976); Randall (1946b); Shearman and Land (n.d.); Swords (1986), pp. 265–66.
83 magnetron kept secret: Bowen (1987), pp. 144–49, indicates TRE did not learn of it until June 1940. Callick (1990), p. 64, has samples being brought to TRE that July.
83 early Bowen attempts to develop mw radar: Bowen (1987), p. 142; D. Robinson interviews.
83 early British mw efforts: Bryant (1991); Rowe (1948), pp. 34, 80.
83 Birmingham efforts leading up to magnetron's invention: Key sources: Boot OHI with Norberg, Aug. 9, 1977 (84/21c), HSTP; Boot (1946), HAHB; Bryant (1991); Miller (1976); Randall (1946a, b). Corson and Hartman interviews, April 17, 1995, and Corson correspondence were instrumental on technical end.
84 early mw transmitters: Brittain (1985); Watson-Watt (1959), pp. 224–25; Seitz (1995).
84 klystron: Norberg and Seidel (1994); Seitz (1996); Varian and Varian (1939).
84 Randall background: Chiefly, M.H.F. Wilkins (1987).
85 "substantial . . . work": Ibid., p. 495.
85 Boot background: Boot OHI with Norberg, Aug. 9, 1977 (84/21c), HSTP.
87 diagram: Corson and Hartman interviews, Corson correspondence; cf. anon. (1946b); Guerlac (1987), p. 226.
87 measure with Lecher wires: Boot notebook Feb. 24, 1940, B1, HAHB; Boot and Randall, draft book chapter, April 23, 1946, p. 9, A.4, HAHB. Notebook shows clearly the 9.5 cm figure, which would be perpetually misquoted, even by Boot, as 9.8 cm; cf. Watson-Watt (1959), p. 236.
88 magnetron development at Birmingham, GE: Boot and Randall draft chapter, p. 12, A.4, HAHB; Guerlac (1987), pp. 228–29; Randall (1946b); Swords (1986), pp. 265–66.
88 TRE learns of magnetron, Bowen joins Tizard: Except where noted, account is from Bowen (1987), pp. 149–53; Lovell (1991), pp. 12, 24–25, 41–42.
88 French magnetron work: Callick (1990), p. 63; cf. Seitz (1996).
88 "As we . . . tube": Lovell (1991), p. 42.
89 Tizard puts together mission: R. Clark (1965), pp. 202, 257–58.

89 Buildup for war: Wood and Dempster (1990), except where noted.

90 July 10, 1940. This date commonly taken as Battle of Britain opening: J. Mason (1989), p. 119; cf. Wood and Dempster (1990), p. 161.

90 CH configuration and deployment: Rowe (1948), p. 44; Swords (1986), pp. 236–37; Wood and Dempster (1990), p. 93.

90 series of air games: Early filter, operations rooms described in Guerlac (1987), p. 143–45; Wood and Dempster (1990), pp. 92–97, 105, 116–18, 258. For Bentley Priory, see Canfer et al. (1990). Author also relies heavily on visit of March 12, 1993, with escort by Norman Greig, and same-day visit to Fighter Command, Uxbridge, with operations room tour by Warrant Officer "Chris" Wren.

91 German efforts to unravel British radar: Wood and Dempster (1990), pp. 10–11. Also see Breuer (1988), pp. 47–48. Explanation for second LZ 130 flight's failure to find CH signals comes from A. Price, based on an interview with Ernst Breuning, who flew on both missions: Price correspondence, April 17, 1995.

93 CH towers hard to hit: Despite their great height, they made small targets: Price correspondence, April 17, 1995.

93 "It is . . . action": Wood and Dempster (1990), p. 209.

93 Hitler strikes back on London: Kennett (1982), pp. 118–21; cf. Rhodes (1986), p. 342; Wood and Dempster (1990), pp. 259–67.

94 September 15, 1940: Chiefly, Churchill (1949), pp. 293–97, with a few details from Wood and Dempster (1990), pp. 271–72. Physical description of the operations room is mainly from visit of March 12, 1993. Note: Churchill, p. 294, describes the rows of bulbs depicting fighter status in reverse order, so that what he remembered as at the bottom was actually on top.

94 "I don't . . . quiet": Churchill (1949), p. 294.

96 "What other . . . infinite": All from Ibid., p. 296.

5 • The Rooftop Gang

98 "In many . . . war": Bowen (1987), pp. 181–82.

98 Bowen . . . regaled RL members: Trower-Alvarez manuscript, pp. 705–706.

98 Bowen in U.S.: Bowen (1987), pp. 169, 178–82; Bowen OHI, May 16, 1979, side 5, NRL; "British Technical Mission," vol. 1, items 1–100, AVIA 10/1; "Activities of British Technical Mission from September 9, 1940," AVIA 10/2; Bowen progress reports, AVIA 10/3 — all PRO.

98 "The hell . . . Limeys": Bowen (1987), p. 182.

98 Bowen gets nickname Taffy: D. and V. Bowen interviews.

99 latest AI and ASV longwave sets: ASV Mark II and AI Mark IV.

99 early mw radars at U.S. industry: Allison (1981), pp. 105–107, 121–24; cf. Guerlac (1987), pp. 108, 204.

99 "There appears . . . England": Bowen, n.d. "Report on U.S. Commercial Detection Activities," AVIA 10/1, PRO.

99 Bowen efforts to bring Vesta to U.S.: See "U.S.A. liaison: correspondence resulting from Tizard Mission 1940–1942," AVIA 7/2796, PRO.

99 RL eased Bowen's loneliness: V. Bowen interview, May 18, 1994.

100 CH ineffective against night bombing, failure of early AI radars: Bowen (1987), pp. 125–27; Burcham (1985), p. 385; R. Clark (1965); Guerlac (1987), pp. 163–64; Orgel (1985), p. 412; Wood and Dempster (1990), p. 324.

100 Bowen rallied RL comrades: Alvarez (1987), p. 88. Note: The angular beam spread from a directional radar antenna is proportional to wavelength divided by antenna width. With latter limited by practical constraints, shortening wavelength from 1.5 m (as in ASV Mark II and AI Mark IV) to 10 cm decreased beamwidth 15 fold, making it far easier to distinguish between closely grouped targets: DuBridge and Ridenour (1945).

100 RL's early goals: Guerlac (1987), p. 262.

100 "We accepted . . . Britisher": Pollard (1982), p. 40.

100 Pollard's citizenship: Ibid., p. 37.
100 Yanks off to fast start: Scene-setting from Trower-Alvarez manuscript, p. 702; Newton et al. (1991), p. 13; Pollard interview, Aug. 29, 1994; various interviews and readings.
101 RL's inaugural radar: Description from Van Voorhis OHI, March 15, 1944, Van Voorhis folder, Box 49B, NANER.
101 Christian Science Mother Church: Alvarez to Guerlac, May 18, 1943, Alvarez folder, Box 49A, NANER, cf. Van Voorhis OHI, Box 49B.
101 Corson remembered: Corson interview, Oct. 4, 1994.
101 "ROOF OUTFIT . . . LEE": DuBridge to EOL, Jan. 6, 1941, EOL (72/117c), C27, f2.
101 DuBridge background and style: Alvarez (1987), p. 87; Kevles (1979), p. 301.
101 TR box: Smullin interview, Sept. 8, 1994; Purcell interview, March 26, 1992; cf. Trower-Alvarez manuscript, p. 702.
102 "If we . . . payroll": Alvarez (1987), p. 90.
102 Lawson description and TR success: Pollard (1982), pp. 172–73; Trower-Alvarez manuscript, pp. 712–13; Guerlac (1987), p. 262.
102 "HAVE SUCCEEDED . . . EYE": Guerlac (1987), p. 262.
102 enmity between MIT engineers and RL physicists: See Hazen (1976); Leslie (1993), p. 22.
103 rooftop radar set, Alvarez bet with EOL: Marshall OHI, April 29, probably 1943, Marshall folder, Box 49A, NANER; Guerlac (1987), pp. 262–63.
103 "What we . . . immense": Pollard (1982), p. 51.
104 crude telescopic sight: Rigden (1987), p. 136; Smullin interview, Sept. 8, 1994.
104 gloomy meeting: "Div. 14 radar committee, general correspondence 1941," folder in Box 5, E-112, NA; Guerlac (1987), p. 263.
104 "Ernest caught . . . away": DuBridge interview, May 17, 1992; cf. Rigden (1987), p. 136. Meeting minutes, E-112, Box 5, NA, show detection range of 2.5 miles was reported.
104 "We've done . . . boys": DuBridge interview, May 17, 1992.
104 call breathed fresh life: Guerlac (1987), p. 263.
104 "I HAD . . . LUNCH": Marshall OHI, Box 49A, NANER.
105 rising around four: March 7 and 10 events drawn mainly from Alvarez OHI, Box 49A, NANER; Corson interview, Oct. 4, 1994. Bowen (1987), pp. 131, 184, details March 10 British trials; cf. Lovell (1990), p. 61.
   About dispute over whether Americans got air-to-air echoes on March 10: Besides Corson, Baxter (1947), p. 146, says the AI set spotted planes 5 miles away. However, Bowen visited East Boston Airport later that same day. His diary entry states the U.S. radar did not get air-to-air echoes, Box 31, Guerlac papers, DRMC; cf. Bowen (1987), p. 184.
105 "the equipment . . . results . . .": Guerlac (1987), p. 263.
105 March 27 foray: Chiefly, Bowen (1987), pp. 184–85. Also important are Guerlac (1987), pp. 263–64; Alvarez OHI, Box 49A, NANER. Minor, mostly clarifying, details are from Alvarez (1987), p. 91; Pollard (1982), p. 49; and Pollard interview, Aug. 29, 1994.
105 previous month . . . Tuxedo Park: Guerlac (1987), p. 302, n.36.
106 "I remember . . . miles": Bowen, side 6, NRL; cf. Bowen (1987), p. 184.
106 several subs cruising: Bowen (1987), pp. 184–85, mentions one sub. Guerlac (1987), p. 264, and Alvarez (1987), p. 91, say several.
106 no one had yet detected a sub: Bowen, side 6, NRL. Lovell correspondence, Dec. 7, 1995, says first British detection of a sub with airborne mw radar likely didn't take place until April 1941.
106 "We returned . . . wildfire": Bowen (1987), p. 185.
106 ranks swelled toward 200: Guerlac (1987), p. 264.
106 New recruits from disparate fields: Newton et al. (1991), pp. 32–33.
106 Chance background: Chance OHI, IEEE Center (1993), p. 49. Chance's genius is from Corson correspondence, Oct. 16, 1995.
106 "a physicist's . . . physicists": Guerlac (1987), p. 297.
106 DuBridge initially understood: DuBridge, "Memories," Part I, p. 31, CTA.
106 "I think . . . war": Ibid.

106 Rad Lab's structure and management style, and Wheeler Loomis's background, are taken primarily from Kevles (1979), p. 307, and Guerlac (1987), pp. 265, 297; and supported by many interviews.

107 he claimed to be unversed: Wheeler Loomis's attitude is drawn from his interview with Guerlac, March 11, 1946, "Administration Problems (1946)," folder in Box 58C, NANER; Guerlac (1987), p. 285.

107 "a son . . . bitch": Kevles (1979), p. 307.

107 DuBridge said "Yes." Loomis said "No": A. Price (1984), p. 20.

107 Rabi group's location: Newton et al. (1991), p. 13, illustration.

107 Rabi's early work: Rigden (1987), p. 137; Guerlac (1987), p. 266.

107 "to develop . . . enemy": "Longhairs and Shortwaves," Fortune, Nov. 1945, p. 162.

107 "How many . . . kill?": Composite quote from many interviews.

107 German attacks on Britain: Churchill (1950a), pp. 38–39.

108 Dowding's RL visit: Bowen (1987), pp. 184–86; Guerlac (1987), pp. 268–71; Alvarez (1987), p. 89. Early AI status from Corson interviews.

108 Bowen hoped: Bowen to Cockcroft, May 15, 1941, AVIA 7, 2796, PRO.

108 Loran status: Bowen (1987), pp. 173–74; Guerlac (1987), pp. 284–85, 525 ff.; Kevles (1979), p. 306.

109 gunlaying project: Except where noted, Abajian, Davenport, and Getting interviews; Davenport correspondence, April 3, 1994, Dec. 8, 30, 1995; Getting (1989), pp. 106–13, 128; Guerlac (1987), pp. 276–78. A few details gleaned from Alvarez (1987), p. 95; Trower-Alvarez manuscript, pp. 727–28; Davenport OHI, IEEE Center (1993), p. 64.

109 conical scanning attributed to Loomis: Baxter (1947), p. 147. For British and German developments, I am indebted to L. Brown, personal communication.

110 "kept bouncing . . . echoes": Davenport interview, April 6, 1994.

112 "It was . . . magic": Getting interview, March 31, 1994.

112 ASV project launched: Bowen (1987), p. 185; Guerlac (1987), p. 273.

112 Semmes installation: Baxter (1947), p. 147; Guerlac (1987), pp. 273–74, 399–400; Pollard interview, Aug. 30, 1994.

112 "All the . . . it": Bowen to Cockcroft, May 15, 1941, AVIA 7, 2796, PRO.

112 "There is . . . once": Ibid.

112 RL's financial plight: Guerlac (1987), pp. 264, 285; Kevles (1979), p. 305; for role of Compton and MIT, see "Statement of the Radiation Laboratory at MIT," "Div. 14 radar committee, general correspondence 1941," folder in Box 5, E-112, NA.

113 America wavers long way from war: Cf. Alvarez (1987), p. 94.

6 • The Radar Bridge

114 "Our impudent . . . war": Hill, June 18, 1940, "RDF in Canada and the United States," Box 49B, NANER.

114 "The essential . . . naturally": Quoted in R. Clark (1965), p. 254.

114 Bush and OSRD formation: Baxter (1947), p. 124; Bush (1970), pp. 42–51; Kevles (1979), p. 299.

115 RL biggest enterprise in Bush's empire: See Baxter (1947), p. 456; Guerlac (1987), p. 658.

115 Rockefeller never paid a dime: Guerlac (1987), p. 246.

115 British order AI radars: Bowen (1987), pp. 185–86.

115 Semmes trials: Baxter (1947), p. 147; Guerlac (1987), pp. 273–74, 399–400; Pollard (1982), pp. 57–59.

115 AI performance: Bowen (1987), p. 186; Guerlac (1987), p. 341.

116 liaison offices opened: Baxter (1947), pp. 121–23.

116 Darwin background: Kevles (1979), p. 169.

116 Bainbridge trip: Bainbridge OHI, Box 49A, NANER; Bainbridge interview, 1994; Alvarez (1987), p. 3; Guerlac (1987), pp. 259, 296.

116 Ramsey trip: Ramsey interviews, March 30, 1992; Oct. 12, 1993.

116 RL-TRE comparison tests: Chiefly, Bowen (1987), pp. 186–88; interviews and correspon-

dence with Corson and Heath, Ramsey interviews. Also important: Bowen diary entries, Box 31, Guerlac Papers, DRMC. 14/17/2354; Bowen to Charles Darwin, July 19, 1941, War Effort OSRD letters April–June 1941, 47, 1, ELB; Guerlac (1987), p. 272.

116 Darwin grants Bowen trip home: Darwin to R. H. Fowler, July 1, 1941, AVIA 7, 2796, PRO.

117 Leeson House: R. Brown (1991), p. 96; Lovell (1991), p. 42.

117 "They had . . . offer": Bowen (1987), pp. 187–88.

117 "Immediately . . . away": Ramsey interview, Aug. 16, 1994.

117 early British crystal detector and TR work: Mainly, Guerlac (1987), pp. 232–35, 268, 586; Ramsey interviews; Torrey and Whitmer (1948), p. 19. On soft Sutton tube: Some do not think of it as a klystron since it had neither the electron gun nor reflector electrode of the reflex klystron. Also Pound, interview of July 3, 1992, noted the tube had to be modified with a "keep alive," which protected a crystal from a very short spike, too fast to be measured by oscilloscopes. The keep alive allowed for "a little gas discharge at all times" with DC current source, outside the TR tube's gap region, so ions were available to start the discharge more rapidly, greatly reducing burnouts from spikes.

118 "I guess . . . country": Ramsey interview, July 2, 1992.

118 state of U.S. crystal studies: Fagen (1978), pp. 44–45; Guerlac (1987), pp. 289, 587, 594; Ramsey interviews; Torrey and Whitmer (1948); Torrey interviews.

119 Robinson: Chiefly, Robinson interviews; his OHIs, June 6, 8, 1944, Box 49A, NANER; follow-up correspondence, June 22, 1944, British Scientific Mission (1941–1943); folder in Box 43A. Also important: Robinson (1983); a few details in Robinson OHI, IEEE Center (1993); Robinson (1988).

119 Rowe had dispatched: Robinson interview, July 30, 1992. Note that in OHI, IEEE Center (1993), p. 283, DR says he knew of magnetron in late 1939, but the device wasn't invented until Feb. 1940. He probably confused it with the generic mw project, which he helped initiate in 1939.

119 Seitz's investigations: Seitz (1996).

119 "You see": D. Robinson interview, July 30, 1992.

119 German bombing and effectiveness of British nightfighters: Churchill (1950a), pp. 39–42; Guerlac (1987), pp. 156, 163–64, 272; Orgel (1985), pp. 412–13; clarifying details from A. Price correspondence, April 17, 1995.

120 U-boats: Unless noted, van der Vat (1988).

120 losses exceeded 100: Churchill (1949), p. 638; (1950a), p. 697.

120 Enigma primarily for rerouting convoys: Though some have claimed otherwise, the authoritative story is Hinsley (1994), pp. 2, 6; Winterbotham (1974), p. 84; cf. van der Vat (1988), p. 198.

121 Radar . . . took on greater importance: R. Smith et al. (1985), pp. 367–68; cf. Churchill (1950a), p. 107.

121 *Bismarck*: Devereux (1994); van der Vat (1988), pp. 188–94. The prime role of the ASV Mark II in the hunt and the fact the *Catalina* carried the radar are from Hezlet (1975), pp. 205–58.

121 tracking U-boats by longwave radar: A. Price (1973), pp. 43–44, 176; R. Smith et al. (1985), pp. 367–68.

121 Bowen's stream of . . . accolades: Robinson to Guerlac, June 8, 1944, Box 49A, NANER.

122 "It's all . . . useful": D. Robinson interview, Oct. 7, 1993.

122 "After all . . . list": D. Robinson interview, July 30, 1992.

122 "You're it": D. Robinson interview, Oct. 7, 1993.

122 Corson holds Dee's daughter: Corson interview, Nov. 7, 1994.

123 "And these . . . years": D. Robinson interview, July 30, 1992.

123 "Come along . . . scotch?": Ibid.

123 "Anything . . . said so": Ibid.

123 "We'll . . . all": Ibid.

124 "I had . . . shortage": Ibid.

124 RL gains on British: The UK still led on many fronts. In Sept. 1941 word reached RL that James Sayers of Oliphant's lab had found a way to prevent the magnetron's tendency

to switch frequencies. "Strapping" entailed connecting every other resonator pole—those expected to be in phase—with copper strips. This displaced unwanted frequencies by increasing coupling of the magnetron's natural frequencies. Power efficiency jumped 20–40%; a similar discovery was made earlier by Percy Spencer of Raytheon but ignored at RL: Guerlac (1987), p. 564; Rolph (1991), p. 13.

124 Bowen returns from England: Minutes of Microwave Committee meeting of Oct. 24, 1941; "Div. 14 radar committee, general correspondence 1941," folder in Box 5, E-112, NA.

124 British and American AI: Corson and Heath interviews, Corson correspondence, April 25, 1995; Heath letter of Sept. 28 1995; cf. Guerlac (1987), pp. 24–25.

124 "The problem . . . was": Heath interview, April 29, 1995. In the British system, if a plane headed straight at a target, the radar screen showed a circle whose radius decreased proportionally as the gap between the craft narrowed. If the quarry flew to one side, the circle's intensity grew uneven, growing brightest on the side facing the target. The effect varied with target range and size, making it hard to judge direction. The U.S. system had an A-scope showing range along a horizontal time base. For ASV mode, it apparently also had a B-scope that showed range vertically, azimuth along the horizontal. The elevation scan was disconnected. The main screen for AI was a C-scope, which showed target bearing (horizontally) and elevation (vertically) relative to host plane.

124 DuBridge phases out Project I: Bowen (1987), pp. 188–89; Newton et al. (1991), pp. 16–17; Bowen's reports in "Div. 14 radar committee, general correspondence 1941," Box 5, E-112, NA.

124 designed with radar in mind: Heath correspondence, Sept. 8, 1994.

124 status of RL ASV projects: Guerlac (1987), pp. 275, 287–88, 322–23; D. Robinson interviews.

125 production radars: Guerlac (1987), p. 687; cf. Pollard (1982), pp. 105–106.

125 "Did you . . . Harbor": D. Robinson interview, July 30, 1992.

125 Pearl Harbor: Prange (1982); Watson-Watt (1959), pp. 291–307. Support from Guerlac (1987), p. 118; Morison (1963), pp. 47–48; Rhodes (1986), pp. 391–92.

126 Army paid a local radio station: Rhodes (1986), p. 391.

126 "To, To, To": Prange (1982), p. 503.

126 "Tora! Tora! Tora!": Ibid., p. 504.

127 status of U.S. Navy radar: Allison (1981), p. 181.

127 CXAM set sunk, resurrected: McNally (n.d.), Part II, pp. 3–5, 9; Guerlac (1987), pp. 88–89. The set was later installed on the carrier Hornet and lost at sea at Guadalcanal.

127 status of U.S. Army radar: Colton (1945).

127 "Lee and . . . convoys": Getting (1989), p. 118. Account spans pp. 117–18; clarifying details in Davenport correspondence, Dec. 8, 1995.

127 most dramatic turning point: Guerlac (1987), p. 288.

127 DuBridge . . . sold house: DuBridge, "Memories," Part I, p. 31, CTA.

127 military liaison at RL: Guerlac (1987), p. 299; Newton et al. (1991), p. 190.

127 RL relationship with industry: Guerlac (1987), pp. 298, 687–89; Newton et al. (1991), p. 82.

128 "endless trips . . . overcome": Pollard (1982), p. 72.

128 Raytheon recruited: Collins OHI, May 3, 1944, Box 49B, NANER; cf. Guerlac (1987), p. 563. Note: Other factors figured in Raytheon's building magnetrons. One was engineering genius Percy Spencer. In addition, said Bowles: "It turned out that Raytheon moved so God danged fast, compared to the Western Electric, that it was unbelievable." Then, too, Bush was a Raytheon founder, though he had resigned from the board June 3, 1938, after taking the Carnegie Institution job, while Bowles had been in a lawsuit with AT&T. Added Bowles, "Now I must confess that I closed my eyes and never thought of these things." Bowles OHI with Norberg, Tape 2, pp. 51–55, 13, 6, ELB, VB's resignation from Raytheon: Scott (1974), p. 91.

128 "wholesome jealousy": J. Goldstein (1992), p. 52. That the same held for other firms comes from several interviews, esp. Purcell, March 26, 1992.

128 demands of war: RL actions drawn from Guerlac (1987), pp. 288–92, 661, 668–71; Newton et al. (1991), p. 19.

129 Dicke's experience: Dicke interviews, April 1, Nov. 12, 1992.

129 "We are . . . research": Wiesner interview, May 22, 1992.

129 RL organization and security: Guerlac (1987), pp. 292–93; Newton et al. (1991) throughout.

129 civilian guards, many retired: Pound (n.d.), ch. II, p. 10.

129 "Forgotten": Pollard (1982), p. 96.

129 MPs with riot guns: Pound (n.d.), ch. II, p. 10.

129 Allen: Several accounts of episode exist, including DuBridge, "Memories," p. 32, CTA. Most authoritative is Corson, letter of Oct. 16, 1995.

129 RL's sense of community: Many interviews. Mixed sense of outrage and pride after Pearl Harbor is from Ramsey interview, Aug. 16, 1994.

129 For RL's daily operations and Bowen story, I rely on Newton et al. (1991), pp. 36–40; lunching off campus comes from various interviews.

130 ratio of men to women: Guerlac (1987), p. 668.

130 "It is . . . once": Newton, et al. (1991), p. 40.

130 university-like feel of lab: Various interviews; many particulars, including W. Loomis's views, are found in Loomis OHI, March 11, 1946, "Administration problems (1946)," folder in Box 58C, NANER.

130 explore brainstorms: Pollard (1982), pp. 104–105; many interviews confirm RL style.

130 "How many . . . kill?": Composite quote from many interviews.

130 Loran: Guerlac (1987), pp. 525–30; Newton et al. (1991), p. 27.

131 Gunlaying: Getting (1989), pp. 112–13; Getting and Davenport interviews.

131 Bell's first radar work: Fagen (1978), p. 24.

131 Parkinson predictor: Fagen (1978), pp. 24, 134–39; Warren (1975). Abajian, Davenport, and Getting interviews made technical details real. The key to Parkinson's innovation, separating it from conventional strip chart recorders, lay in its nonlinearity. A special wire-wound card caused the pen to move not with the exact voltage being applied but with the logarithm of the voltage. That made it a powerful electronic calculating device.

131 "It didn't . . . gun!": Fagen (1978), pp. 135–36.

132 Zacharias work at Bell: J. Goldstein (1992), p. 52; Guerlac (1987), p. 271.

132 XT-1 development picks up: Davenport interviews and correspondence. Getting (1989), pp. 115–27, supplements main story and provides Keller anecdote; RL organizational details from Guerlac (1987), pp. 293–95.

134 score of projects: Cf. Newton et al. (1991), p. 19.

134 Alvarez background and projects: Unless otherwise noted, see Alvarez (1987); Trower-Alvarez manuscript; Alvarez OHI, Box 49A, NANER; "The Eagle Story: How It All Began," Radar, No. 10, June 30, 1945: 28–33. Critical in understanding things were interviews and correspondence with Johnston, as well as Johnston (1987). Additional details gleaned from Guerlac (1987), pp. 387, 448–49, 497–99; with thanks to Corson interviews.

134 ". . . It reminded . . . it": Trower-Alvarez manuscript, p. 760.

137 rid Eagle of one hang-up: There still remained problem of scanning the beam left and right to build up a ground picture. Conventional mechanical scanning — rotating the antenna — was impossible. The already crowded B-29 did not have room. The dilemma set Alvarez thinking again, and he saw a way to scan the beam by introducing a waveguide of varying width. As successive dipoles along the guide shifted phase in response to the change, the beam angle also moved; each twist of the waveguide centered the beam over a different area.

137 "I don't . . . job": Alvarez (1987), p. 98.

137 "I'm convinced . . . unnecessarily": Ibid.

138 two precision antennas: Getting a good elevation angle reading — showing how close plane was to ideal glide path — required an accuracy of .5°. LA used a 12-foot-tall antenna that radiated a beaver-tail–shaped beam parallel to the horizon. The beam could measure elevation angle by being mechanically rocked up and down on its axis. It would be so

narrow vertically that when the beam pointed up, almost no energy would spill down to cause a repeat of Oceana. When it pointed into a runway, a mirror image appeared. But just as desert drivers can easily tell an oncoming car from its inverted image on the roadway, operators knew a plane was not trying to land from below the Earth. Johnston even found twin images helpful. As planes descended, the images merged, he noted. "At that point, you knew the pilot had better have his wheels down."

The second antenna, an 8-foot-wide arrray, operated similarly, only its beam stood on end, fanning out vertically, but scanning left and right. Loomis made the important suggestion that since each antenna rocked with simple harmonic movements, the phases could be 90° apart, enabling a single radar transmitter to switch between them and rapidly find a plane.

138 generate information to error meters: Although the Mark I was a prototype, its somewhat clumsy mechanical scanning feature worked fine and would have done for the production model. However, Alvarez incorporated Eagle's electrical scanning feature as part of an improved Mark II system. To rush the set to war, the men decided to skip the prototype stage and design it straight for mass production.

138 strategy elegantly simple: Buell (1980), pp. 282–83.

139 U.S. mustered almost no defense: Morison (1963), pp. 110–11; van der Vat (1988), pp. 237–39.

139 four attacks in 8000 hours: Kevles (1979), p. 306.

139 RL efforts: Guerlac (1987), pp. 319–24; Corson interview, Oct. 4, 1994.

139 ASD: Ramsey interviews; cf. Guerlac (1987), p. 346.

140 "Find us . . . over": Ramsey interviews, March 30, 1992, Oct. 12, 1993.

140 "This slowed . . . flight . . .": Ramsey interview, March 30, 1992.

140 "They all . . . work!": D. Robinson interview, July 30, 1992.

140 Dumbo I: D. Robinson interviews; Guerlac (1987), pp. 276, 322; a few tidbits from Robinson OHI, IEEE Center (1993).

140 300 planes with longwave ASV: A. Price (1973), p. 71.

140 radar kills: The first was Nov. 30. The other, the first radar night kill, took place in Straits of Gibraltar on Dec. 21; A. Price (1973), pp. 19, 76–78; Watson-Watt (1959), pp. 127–28.

141 "Robinson . . . these!": Robinson OHI, IEEE Center (1993), p. 291.

7 • Battle for the Atlantic

142 "The U-boat . . . bombed": Churchill (1950a), p. 107.

142 "Admiral King . . . being": Bush (1970), p. 91.

142 Bush maneuverings: Kevles (1979), pp. 308–309; Baxter (1947), pp. 29–31. See Bush (1970), pp. 51–52, and Stimson and Bundy (1947), intro., pp. 323, 465–68, for relative points of view.

143 "We want . . . one": J. Goldstein (1992), p. 51.

144 "The use . . . fire": Bush OHI, Aug. 20, 1944, Box 49A, NANER.

144 Bush and Bowles maintained: Bowles OHI with Norberg, Tape 1, pp. 36–48; cf. Tape 3, side 1, p. 6, 13, 6, ELB.

145 "Bush has . . . 'it':" Ibid., Tape 1, p. 47.

145 "And I . . . methods": Ibid., p. 43.

145 "I understand . . . methods": Ibid.

145 "he was . . . crooks": Ibid., p. 49.

145 "fraternity of physicists": Bowles, OHI with Romanowski, 13, 10, ELB.

145 "I am . . . others": Bowles to Compton, Feb. 3, 1942, Correspondence between KT Compton and ELB, folder in ELB, LC; cf. Compton to Bowles et al., May 5, 1941, both in 47, 3, ELB.

146 "pasted . . . him": Bush OHI, Box 49A, NANER.

146 Bowles arrived in Washington: Wolfgram (1980); Bowles OHI with Green, May 6–7, 1971, pp. 6–9, 13, 5, ELB.

146 "I was . . . discouraged": Wolfgram (1980), p. 37.

146 war secretary's office: Site from Goldberg interview, Nov. 22, 1994.
146 "I've been . . . Bowles": Bowles, Green OHI, p. 7, 13, 5, ELB. That Bowles took issue with Watson Watt's report is from Kevles (1979), p. 310.
146 The next Monday: Guerlac (1987), p. 699; Kevles (1979), p. 310.
146 Outfoxing opposition: Bowles background drawn from ELB, esp. Norberg OHI, pp. 1–18, 13, 6; and Green OHI, 13, 5, notably Tapes 1, 3. Also see Kevles (1979), p. 310, and Wolfgram (1980).
147 Stimson's move: Goldberg interview, Nov. 22, 1994; cf. Kevles (1979), p. 310.
147 "I spent . . . ones": Bowles, Norberg OHI, Tape 1, p. 9, 13, 6, ELB.
147 Bucket Brigade: Morison (1963) for naval war accounts.
147 "the red . . . tankers": Ibid., p. 121.
148 up in Dumbo II: Guerlac (1987), p. 322; Bowles, OHI with Green, Tape 3, side 1, p. 3, 13, 5, ELB.
148 April 22: Date derived from comparing events in Guerlac (1987), p. 322, and Stimson and Bundy (1947), p. 510.
148 "Lo and . . . them": Bowles, Green OHI, Tape 3, side 1, p. 3, 13, 5, ELB.
148 "That's good . . . home": Guerlac (1987), p. 322.
148 "I've seen . . . you?": Ibid. That note was tantamount to an order; see Stimson and Bundy (1947), p. 510.
148 Stimson's total conversion: See Stimson and Bundy (1947), p. 509.
148 DuBridge launched crash program: Guerlac (1987), p. 276.
148 Bowles's trapper instincts . . . : Sub-fighting plans from Bowles, Green OHI, Tape 3, side 1, pp. 8–9, 13, 5, ELB.
148 What distinguished. . . : Guerlac (1987), pp. 715–20; Kevles (1979), p. 311; cf. Bush (1968), p. 64.
148 Operations Research: Blackett (1962), esp. pp. 171–76; R. Jones (1990), pp. 188–89.
149 one chronicler claimed: R. Clark (1962), p. 216. General improvement in ASW is a fact, but with better training, equipment maintenance, etc., picking out one factor is difficult. Study is in Blackett (1962), p. 191.
149 American OR program: Morison (1984), pp. 222–23; Morse (1977), pp. 159–84.
149 3000 square miles: Bowles, "Preliminary Report on the Submarine Search Problem," May 1, 1942, Submarines B file, Box 7, E-117, RG 107, NA; cf. Kevles (1979), pp. 311–12.
149 Bowles harbored few illusions: Army-Navy infighting and Bowles's tactics leading to First Sea Search Attack Group (FSSAG) compiled from Bowles, Green OHI, pp. 4–7, 13, 5, ELB; Bowles, Memorandum for the Secretary of War, Aug. 7, 1942, Submarines General file, Box 7, E-117, RG 107, NA; Guerlac (1987), pp. 320–21, 701; Kevles (1979), pp. 306–12, footnotes; Morison (1984), pp. 243–44, footnotes. Corson interviews important to sorting out details of RL's B-18 group and early FSSAG days; as well as Radar, No. 9 (April 30, 1945), p. 27. Log for official 1942 submarine sinkings from Morison (1984), p. 415.
150 SADU: SADU-FSSAG relationship confused. AF History Offfice was no help. Author's interpretation drawn from wide readings, Corson and Fletcher interviews; cf. Thompson et al. (1957), p. 255; Radar, No. 5 (Sept. 30, 1944), p. 25.
150 "We merely . . . airplane . . .": Bowles, Green OHI, p. 7, 13, 5, ELB.
151 log 1274 hours . . . four sunk: Radar, No. 5 (Sept. 30, 1944), p. 25. Also see Analysis of Combat Activity, First Sea Search Attack Group folder, Box 49B, NANER. Twenty-two other sub sightings were due to MAD, while three were made visually.
151 "You couldn't . . . way": Corson interview, Nov. 7, 1994.
151 "Escort is . . . success": Buell (1980), p. 288.
151 U-boat situation deteriorated: Morison (1963), pp. 113–21, says Navy got 8 Us in this period, but one was Army's. Rohwer (1965), p. 271, tells how pleasantly surprised Germans were by the slow U.S. response.
151 gloomy month of June: Churchill (1950b), pp. 112, 860; van der Vat (1988), p. 255.
151 merchant losses surpassed WWI total: Kevles (1979), p. 306.
151 Optimists could find: German migration from U.S. shores is from Churchill (1950b), p. 110; Guerlac (1987), p. 717; Rohwer (1965), p. 272.

152 "simply . . . gadget," "ipso . . . weapon": Quoted in Kevles (1979), p. 311.
152 "each craft . . . leadership": Bowles, Memorandum for the Secretary of War, Sinking by Submarine — off Barnegat Light, June 1, 1942, 31, 7, ELB.
152 "the factual . . . submarine": Quote and overall account: Bowles, Memorandum for the Secretary of War: A. Submarine Destruction Program, Aug. 7, 1942, Submarines General, Box 7, E-117, RG 107, NA.
153 First Bomber Command: Activities from Morison (1984), p. 241.
153 "Scientific aids . . . 'click' ": Bowles memo, Aug. 7, 1942, Box 7, E-117, NA.
153 Arnold cut Bowles's request in half: handwritten note on ibid.; Bowles, Memo to the Secretary of War, April 2, 1943, 30, 1, ELB.
153 Army Air Forces Antisubmarine Command: Morison (1984), pp. 241–42, (1963), p. 129.
153 SCR-517: Details of the set's development, production, and deployment compiled from many sources: Fagen (1978), pp. 70–71, 95–98; Guerlac (1987), pp. 271–76, 717; Schoenfeld (1995); Thompson et al. (1957), pp. 249–53; Marshall, "Formation of First Anti-Submarine Army Air Command," memo to Commander-in-Chief, U.S. Fleet and Chief of Naval Operations, Sept. 14, 1942, Submarines General, Box 7, E-117, RG 107, NA.
   Guerlac says Bell began 517 project after Pearl Harbor, Fagen says in Oct. 1941. On number of sets built: Fagen shows 2000 models of 517 and its AI cousin, the SCR-520. Guerlac notes that only 100 520s were made. A smaller, lighter version, the SCR-720, came into service in 1943 and became the RAF standard until 1957 — Bowen (1987), pp. 188–89.
154 Antisub Command met roadblock: Stimson and Bundy (1947), p. 512.
154 Bay of Biscay strategic importance: A. Price (1973), p. 31.
154 U-boat recharging: Figures from Schoenfeld (1995), p. 45.
154 Joubert increasingly dependent on radar: Leigh Light introduction, mw ASV work, mainly from Lovell (1990), pp. 89–91, (1991), pp. 156–62; R. Smith et al. (1985), pp. 368–71. Some ASW operations from Roskill (1954), p. 358, (1956), pp. 88–89, 112, 205; van der Vat (1988), pp. 274–309. Bowen (1987), pp. 113–14, and A. Price correspondence fill gaps.
155 *Metox*: Ibid.; A. Price (1973), pp. 92–93. Price correspondence April 1995 helped explain how device was used.
155 "The German . . . States": Churchill (1950b), pp. 254–55.
156 Dönitz's forces: Roskill (1956), pp. 354–67; van der Vat (1988), pp. 311–27; cf. Bekker (1974), p. 312.
156 "As the . . . way": Bush (1970), p. 88.
157 reports streaming across VB's desk: See "Report to the Joint Chiefs of Staff from the Joint Committee on New Weapons and Equipment," Jan. 4, 1943, Submarines General, Box 7, E-117, RG 107, NA; also Kevles (1979), pp. 312–13.
157 mw radar topped list: See Bush (1970), p. 82.
157 2000 workers, 50-odd projects: Newton et al. (1991), p. 19.
157 1943 U.S. radar production: Guerlac (1987), p. 690.
157 NDRC reorganization: See J. G. Trump to EOL, Jan. 2, 1943, EOL (72/117c), C27, f2.
157 DMS-1000: D. Robinson interviews; Baxter (1947), pp. 42, 149; Guerlac (1987), p. 323; Roskill (1956), p. 205; R. Smith et al. (1985), p. 374; and *Radar*, No. 5 (Sept. 30, 1944), p. 27.
157 SCR-517: Schoenfeld (1995), esp. pp. 20–21. There were 21, but two lost en route; cf. Bowles, *The Acute Problem of Ocean Borne Transport and Supply*, Memorandum Report to the Secretary of War, March 1, 1943, pp. 18–24, 39, 6, ELB; Morison (1984), p. 244; *Radar*, No. 5 (Sept. 30, 1944), p. 28.
157 George: Guerlac (1987), pp. 323–25; R. Smith et al. (1985), pp. 371, 374.
157 Wellingtons with Leigh Lights: A. Price (1973), p. 118.
157 Casablanca Conference: van der Vat (1988), pp. 314–15.
158 VLR bomber numbers, capabilities: Morison (1956), p. 43; Roskill (1956), pp. 362–64; R. Smith et al. (1985), pp. 371, 374.
158 Perhaps worse . . . : Most U.S. details, including Sanford's exploits, from Schoenfeld

(1995), pp. 44–62. Some SCR-517 particulars from Fagen (1978), pp. 95–97; cf. A. Price (1973), p. 117. Note: Though some claim otherwise, the SCR-517 did not have an A-scope.

158 just as Bush feared: From wide study of VB, esp. OHI, MC 143, MIT; and Kevles (1979), p. 308.

158 Liberator crew based in Iceland: OR study from R. Clark (1962), p. 218.

158 Germany . . . defeated by bombing alone: Wide reading; see Bowen (1987), pp. 114–15; Lovell (1990), p. 88; Roskill (1956), p. 371.

158 King did not rank . . . Atlantic: Buell (1980); cf. van der Vat (1988), p. 334.

159 what to make of King: Admiral's background, beliefs, from Buell (1980).

159 shaved with a blowtorch: Kevles (1979), p. 312.

159 "I see . . . convoys": Quoted in ibid., p. 313; cf. Wolfgram (1980). For more convoy conference details see van der Vat (1988), pp. 325–26.

159 King could dig . . . : Figures from Morison (1956), p. 9.

159 officers waxed impatient: Ibid., p. 28.

159 *The Acute Problem* . . . : Memorandum Report to the Secretary of War, March 1, 1943, 39, 6, ELB; cf. Morison (1956), pp. 27–28.

160 Loran status: Guerlac (1987), p. 530.

160 "Were it . . . warfare": Memo to War Secretary, March 1, 1943, 39, 6, ELB. Parens his.

160 King unmoved: For Stimson, Marshall actions, Stimson and Bundy (1947), pp. 512–13.

160 "Mr. Secretary . . . King": Bowles, Green OHI, p. 14, 13, 5, ELB.

160 President monitoring situation: Roskill (1956), pp. 362–64; Stimson and Bundy (1947), pp. 512–13; van der Vat (1988), p. 326.

160 There remained the question: See Stimson and Bundy (1947), pp. 512–13; van der Vat (1988), pp. 353–54.

160 "The usual . . . scientific": Bush, testimony, pp. 147, 152, unknown document, CAR.

161 "There's too . . . radar": Bush (1970), p. 91; cf. Bush OHI, Reel 7-B, p. 467, MC 143, MIT; Kevles (1979), p. 312.

161 VB considered going straight to FDR: Bush (1968), p. 61, (1970), pp. 88–89; Bush OHI, Reel 7-B, p. 467, MC 143, MIT.

161 "I have . . . belief": Bush to King, April 12, 1943, Cooperation: National Research Council to Navy, Box 34, E-13, NA; cf. Kevles (1979), p. 314.

161 "Antisubmarine . . . method": Ibid.

162 "Recent . . . troubled": Ibid.

162 "I feel . . . somewhere": Bush to Jewett, April 14, 1943, Cooperation: National Research Council to Navy, Box 34, E-13, NA.

162 King-Bush maneuverings: Best overall source is Kevles (1979), pp. 313–15. Important King–10th Fleet details are in Buell (1980), pp. 226–27, 293–94; Farago (1962); Morison (1956), pp. 21–23.

For Bush's end, see Bush, Memorandum of Conference with Admiral King, April 19, 1943; Bush to King, April 20, 1943, Box 34, E-13, NA.

163 "Wouldn't our . . . him . . . ?": Bush, memo of conference with King, April 19, 1943.

163 "The headquarters . . . Bush": Quoted in Morison (1956), p. 22.

163 "to reinforce . . . groups' ": Ibid., p. 23.

163 By early 1943: For German maneuverings, except where noted, Bekker (1974), pp. 322–32. Prime Allied source is Roskill (1956), esp. pp. 364–69. Van der Vat (1988) fills gaps.

163 Dönitz twice shifted headquarters: See also A. Price (1973), pp. 120–21.

164 "almost certainly": Bekker (1973), p. 323.

164 code breaking: Hinsley (1994), pp. 6–7; van der Vat (1988), pp. 257–58, 310, 328–29.

164 similar undetected attack: Evidence of mw radar was mounting: cf. Rohwer (1965), pp. 283, 446n.

164 a dozen Wellingtons with Leigh Lights and mw radar: Lovell (1991), pp. 159–62; Morison (1956), p. 90.

165 strategy backfired: A. Price (1973), pp. 32–34; van der Vat (1988), pp. 333–34, 342; cf. Zuckerman (1975), pp. 479–80.

165 Radar robbed U-boats: Morison (1956), pp. 92, 99.
165 navigating by Loran: Baxter (1947), p. 44.
165 escort carriers: L. Brown (1994), p. 132; Morison (1984), p. 235 chart.
165 naval mw radar and Huff-Duff: Prime sources; Morison (1984), pp. 226–28, (1956), pp. 83, 89; A. Price (1973), p. 109; Watson-Watt (1959), pp. 165–73.
165 ONS5 attack: Bekker (1974), pp. 332–34; Morison (1984), p. 212, (1956), pp. 71–75; Rohwer (1965), pp. 301–302; van der Vat (1988), pp. 331–32.
166 "Enemy radar . . . located": Rohwer (1965), p. 301.
166 "The enemy . . . possible . . .": Bekker (1974), p. 335.
166 attack on SC130: Rohwer (1965), pp. 304–305; Roskill (1956), pp. 366–76.
166 One of Dönitz's sons: Morison (1956), p. 79n.
167 "Thus our . . . devices": Rohwer (1965), p. 306; Morison (1956), p. 83, has slightly different wording.
167 That same day Dönitz ordered: Morison (1963), p. 366; van der Vat (1988), p. 337.
167 May ASW figures: Morison (1963), p. 245; van der Vat (1988), p. 333. At least five other Us went down, but not by Allied action; cf. Guerlac (1987), p. 719n.
167 June–July figures: van der Vat (1988), p. 343.
167 sophisticated attack procedures: A. Price (1973), pp. 30, 34; Roskill (1960), p. 42.
167 479th Antisubmarine Group: Schoenfeld (1995), pp. 130–35.
167 480th Antisubmarine Group: Ibid., pp. 76–83, 104–106; cf. Radar, No. 5 (Sept. 30, 1944), p. 29.
168 Ernst Salm: Radar, No. 5 (Sept. 30, 1944), p. 29; Schoenfeld (1995), pp. 100–101.
168 "He deserves . . . us": Radar, No. 5 (Sept. 30, 1944), p. 29.
168 August figures: Morison (1956), p. 132; van der Vat (1988), pp. 343, 366.
168 four months that began May 18: Morison (1963), p. 376.
168 ratio of subs sunk to spotted: Radar, No. 5 (Sept. 30, 1944), p. 23.
169 Bush knew from intelligence reports: Bush, testimony to Congress, p. 152, CAR; German weapons descriptions from Morison (1956), p. 138; A. Price (1987), p. 143; van der Vat (1988), p. 350.
169 Aphrodite never worked: A. Price (1987), p. 143.
169 Foxer: Bush, testimony to Congress, CAR; Roskill (1960), p. 40.
169 "I think . . . done": Bush to Congress, p. 153, CAR.
169 Dönitz's counterattack failed miserably: Roskill (1960), pp. 37–47.
169 "the centimetric . . . visibility": Quoted in Burns (1988), p. 69.
169 "Microwave radar . . . 1943": Morison (1963), p. 127.
170 "The greatest . . . radar": Bush (1970), p. 82.
170 churning out vessels . . . faster than subs: van der Vat (1988), p. 270.
170 "from menace to problem": Quoted in Buell (1980), p. 299.

8 • The Will to War

171 "It is . . . war": Quoted in Kennett (1982), p. 54.
171 "The bombardment . . . prohibited": Ibid., p. 63 ff.
171 "It is . . . cities": Ibid., p. 65.
171 early British bombing policy and effectiveness: Except where noted, I rely throughout on Hastings (1979); Kennett (1982).
171 inclement weather . . . three days of five: Guerlac (1987), pp. 765–66.
172 Lindemann background: Chiefly, Birkenhead (1961); Snow (1961, 1962).
172 Churchill's friend and advisor: Churchill (1949), p. 338.
172 Tizard and Lindemann: R. Clark (1965), pp. 222–23, adds to prime sources.
172 Lindemann and CSSAD: R. Clark (1965), pp. 122–46, 172–75; Snow (1961); cf. Churchill (1948), pp. 133–34.
173 Prof challenged Sir Henry: Lindemann's focus on IR detection is often taken as a sign he was anti-radar. Not so: see R. Jones (1961a, b, c); cf. McElheny (1973).
173 Butt report: R. Clark (1962), p. 183; Hastings (1979), pp. 117–18; Kennett (1982), p. 129. For intelligence overview, see R. Jones (1978), p. 210.

174 Sunday, Oct. 26: Watson-Watt (1959), p. 382.
174 Sunday Soviets: Rowe (1948), pp. 84–86; D. Robinson interview, July 30, 1992.
174 "The rules . . . right' "; D. Robinson interview, July 30, 1992.
174 During the Oct. 26 soviet: Meeting, early TRE bombing radar work, drawn from Lovell (1991), pp. 87–94; Rowe (1948), pp. 110–16; Watson-Watt (1959), pp. 340, 382. For technical description of Gee and Oboe, chiefly Lovell (1991), pp. 87–88; Guerlac (1987), pp. 732–39.
175 Gee as Loran inspiration: See Bowen (1987), p. 173.
175 Bowen's bombing radar ideas: Watson-Watt (1959), p. 380; EGBN 2/3, CAC.
175 "This is . . . war": Lovell (1991), p. 94.
175 Rowe named Lovell head: Development of TRE's bombing radar: Lovell (1990) and (1991), except where noted.
176 "At last . . . bombers": Lovell (1990), p. 72.
176 "Nothing was . . . priority": Rowe (1948), p. 117.
176 "I hope . . . punctually": Quoted in Churchill (1950b), p. 252.
177 the magnetron . . . virtually indestructible: A. Price (1987), pp. 122–23.
177 Cherwell insisted on klystron: Rowe (1948), pp. 117–19; Lovell correspondence, Dec. 7, 1995; cf. Lovell (1991), p. 105.
177 Downing Street meeting: Lovell (1991), pp. 133–35.
178 "We don't . . . October": Ibid., p. 134.
178 "What does . . . breadboards": Ibid.
178 "We had . . . this": Rowe (1948), p. 149. For crash program story, see pp. 148–49.
178 Bomber Harris: Hastings (1979), pp. 148–50.
178 "You might . . . night!": Ibid., pp. 149–50.
178 "It has . . . workers": Quoted in Kennett (1982), p. 129.
179 Extrapolating wildly: Zuckerman (1975), pp. 470–71.
179 "In 1938 . . . people": Quoted in Hastings (1979), pp. 140–41. Parens his.
179 Tizard and Blackett saw: Blackett (1960), pp. 649–50; Snow (1961), pp. 49–52.
180 "We are . . . relentlessly": Hastings (1979), p. 172.
180 On the heels: TRE developments from Lovell (1991), pp. 135–60; Lovell correspondence, Dec. 7, 1995.
181 Thirteen Pathfinders: Guerlac (1987), p. 737.
181 "Heartiest . . . action": Lovell (1991), p. 153.
181 German technicians find magnetron: A. Price (1987), pp. 134–35.
182 George Valley: Chiefly, extensive interviews with Valley.
182 "I better . . . in": Valley interview, May 10, 1992. Exact trip date unknown: Guerlac (1987), p. 434, says GV went to England that fall. Wilson to DuBridge, Aug. 22, 1942, says GV about to leave, Radar 1941 thru 1942, Box 56, E-1, NA. There were only three major German air raids on England from September through December 1942 — none on London: Collier (1957), pp. 513–14. Valley witnessed either a minor attack, or planes from one of these raids that were off target.
183 "I didn't . . . bitch": Valley interview, Aug. 13, 1993.
183 "I came . . . birds": Valley interview, May 10, 1992.
183 "It didn't . . . to": Valley interview, Aug. 13, 1993.
183 projects on RL books: Guerlac (1987), pp. 380–90; Valley interviews.
184 "H$_2$S did . . . America": Lovell (1991), p. 146.
184 Rabi and Purcell in England: Guerlac (1987), pp. 731–72; Purcell interview, March 26, 1992; Purcell OHI, IEEE Center (1993).
184 "a pretty lousy radar": Purcell interview, March 26, 1992.
184 "What we . . . against": Ibid.
184 rankled Bernard Lovell: Lovell (1991), p. 146.
184 matter rested, tension lingering: Even Bowen found H$_2$S lacking. He had returned to England, meeting with Dee Aug. 17. "His boys were shaken by the detail shown on the American 3 c.m. A.S.V. pictures as compared with their own H$_2$S results," he wrote in his diary. "It was difficult to find anyone, apart from O'Kane, who consistently had confidence in H$_2$S." EGBN, 4/4, CAC.

184 When GV took over: Valley interviews; technical progress from Guerlac (1987), p. 380; Lovell (1991), p. 147.
185 "I stopped . . . better": Valley interview, Oct. 29, 1993.
185 8th Air Force: Structure and operational details from Freeman (1986), (1990); Guerlac (1987), pp. 763–70.
185 radar bombing conference: Meeting minutes, provided by GV; Valley interviews; Corson interviews, correspondence.
186 Griggs: Background from Kevles (1979), pp. 316–17.
186 Eagle: Specs from *Radar*, No. 10 (June 30, 1945), pp. 32–33; Thompson and Harris (1966), p. 487.
186 Valley charged antenna group: Some technical details from Guerlac (1987), pp. 348–49, 380–82.
187 "Wowie, zowie": Valley interview, May 10, 1992.
187 "When I . . . about": Ibid.
187 Special Mission on Radar: For U.S., Guerlac (1987), pp. 742–46. British view is in Lovell (1991), pp. 181–84.
188 Eagle mired in difficulties: *Radar*, No. 11 (Sept. 10, 1945), pp. 38–40.
188 Watson-Watt hyphenated name: Ratcliffe (1975), p. 563.
188 go into high gear: Guerlac (1987), pp. 381–82, 770–71; Kevles (1979), p. 316. For more on U.S.-UK problems over H_2S, see Lovell (1990), pp. 100–101.
188 "The success . . . size": Griggs, "Memorandum Report to the Assistant Secretary of War for Air," June 21, 1943, 40, 3, ELB.
189 "the radar . . . on": Guerlac (1987), p. 743.
189 "the progressive . . . weakened": Quoted in ibid., p. 761; Lovell (1991), p. 173; cf. Kennett (1982), pp. 135–36.
190 a pretty lousy radar: H_2S travails from Lovell (1991), pp. 173–76, 195.
190 Gomorrah: Besides Hastings (1979) and Kennett (1982), see Rhodes (1986), pp. 472–74. A. Price, personal correspondence, helped with general bombing strategies.

9 • Tangled Web

192 "A delusion . . . snare": quoted in Lewis C. Henry, ed., *Best Quotations for All Occasions*, (New York: Fawcett Premier, 1970), p. 54.
192 "It is . . . deceiver": Ibid.
192 The morning after: Hastings (1979), pp. 232–33; R. Jones (1978), p. 302.
192 Wizard War. Throughout, except where noted, R. Jones (1978), with key assistance from A. Price (1967) and Price follow-up correspondence. Both books are intelligence-countermeasure classics.
193 "Here's a . . . you!": R. Jones (1978), p. 68. Many Oslo Report details, including author's identity, are in R. Jones (1990).
193 "Aircraft warning device": R. Jones (1990), p. 334.
194 British worries about German jamming: Cockburn (1985), p. 426.
194 strong evidence of very high frequency signals: these were 30 MHz.
195 "Would it . . . would!": R. Jones (1978), p. 101.
195 "All I . . . files!": Ibid., p. 102.
195 "a source . . . power": Ibid., p. 107.
196 TRE starts countermeasures: In addition to R. Jones (1978), and A. Price (1967), see R. Clark (1962), pp. 113–14; Cockburn (1985).
196 W-W holiday, subsequent British thinking: Watson-Watt (1959), pp. 174–76; Cockburn (1985), p. 427.
197 *Seetakt*: Cockburn (1985), p. 426–27; A. Price (1967), p. 75; cf. Kümmritz (1994), pp. 31–42. There were two *Seetakts*, on same frequency, one for surveillance, one for naval vessels. Former chiefly deployed after 1942.
197 "curiosities": R. Jones (1978), p. 190.
198 Detectors set up: Some details in Cockburn (1985), p. 429.
198 "What good . . . successfully": R. Jones (1978), p. 192.

198 shots taken Nov. 22: Date from *Radar*, No. 6, (Nov. 15, 1944), p. 14.
199 photos taken Dec. 5: Date from Churchill (1950b), p. 248.
199 Bruneval raid: Chiefly, R. Jones (1978), esp. pp. 225–45; Breuer (1988), pp. 125–36. Clarifying details in A. Price (1967), pp. 6, 125–36; Priest OHI, Aug. 16, 1944, general intelligence folder, Box 43A, NANER.
200 "I can ... sir": R. Jones (1978), p. 238.
201 "Cabar Feidh!": Breuer (1988), p. 133.
202 Hitler did not push defensive radars: Beyerchen (1994), p. 268.
202 Germans knew radar: Overviews are in Kümmritz (1994); A. Price (1967), pp. 55–70; Swords (1986), pp. 91–100. Term *Funkmessgerät* is in Guerlac (1987), p. 5n.
203 Germany had drawn first blood: Hastings (1979), pp. 7–26; A. Price (1967), pp. 61–62.
203 night defense problem to Kammhuber: Chiefly, R. Jones (1978), pp. 264–69; A. Price (1967), pp. 63–70; with some *Helle Jagd* bits from Kennett (1982), p. 137. A. Price correspondence helpful in resolving discrepancies.
203 By late 1941, Kammhuber had divided: In initial incarnation, line included searchlights. These withdrawn by early 1942.
204 *Scharnhorst-Gneisenau* breakout: Churchill (1950a), pp. 104–105, (1950b), pp. 98–100, 229; Giessler (1961); Watson-Watt (1957b), (1959), pp. 356–75.
206 one more raid: Nissen (1987). Main Dieppe invasion from Churchill (1950b), pp. 457–59.
206 detected 20 miles offshore: L. Brown (1994).
206 Nissenthal plunged into sea: Nissen (1987), pp. 192–93, tells how he cut Freya phone lines before leaving. He claims this forced radar crew to radio plots in plain language, enabling British to divine key information about the radar's function. However, A. Price, personal correspondence, notes Freya's role was already well known; L. Brown, personal correspondence, notes Canadian witnesses doubt any phone lines were cut.
206 first airborne jammers: Mandrel chiefly from Cockburn (1985), pp. 428–32; cf. *Radar*, No. 6 (Nov. 15, 1944), pp. 18–19. Moonshine from A. Price (1984), pp. 37–39; R. Jones (1978), pp. 290–91.
207 RRL: Guerlac (1987), pp. 287–90; McMahon (1984), pp. 200–203; A. Price (1984), pp. 19–32; Suits et al. (1948), pp. 10–14. Terman's appointment on Feb. 12, 1943, is from Chalfant (1992).
208 "Mr. Jones ... here": R. Jones (1978), p. 297.
208 "Very well ... Window": Ibid.
208 *Düppel*: It means dipole.
208 "In the ... work": Quoted in R. Jones (1978), p. 297.
209 8th AF first dropped chaff: A. Price (1984), pp. 83–85, 99; cf. Watson-Watt (1959), p. 397.
209 RRL at Great Malvern: A. Price (1984), pp. 86–89; cf. Guerlac (1987), p. 819.
210 H₂S not ignored: Hastings (1979), pp. 288 89, 299–304. Sortie figures from Lovell (1991), p. 211.
210 Freeman Dyson: Dyson (1981), pp. 28–31.
210 "this crazy ... murder": Ibid., p. 30.
210 "failed utterly ... where": Ibid., p. 29.
210 "See, banks ... rain": Ibid., p. 31.
210 "I have ... device": A. Price (1967), pp. 134–35.
210 Leo Brandt: R. Clark (1962), p. 196; Kümmritz (1994), p. 44.
211 "We must ... race!": A. Price (1967), pp. 136–37.
211 German radars based on the transmitter: Kümmritz (1988), pp. 222–26.
211 "One may ... defence ...": R. Clark (1962), p. 198.
211 Americans gearing up: Guerlac (1987), pp. 742–53, 777–84, 799, 817. Some Griggs background from Corson correspondence, Oct. 16, 1995.
212 equipment woes and lack of training: *Radar*, No. 1, April 1944, p. 5.
212 "... Tests prove ... group ...": quoted in Guerlac (1987) , p. 783.
212 *"The availability ... overcast"*: "Division 14 fiscal contract budget and program — Jan. 5, 1944," folder in Box 2, E-112, NA.

212 TRE soon rushed: Lovell (1991), pp. 183–88, 195–96; Lovell correspondence, Dec. 7, 1995.

212 H₂X circular probable error: Kennett (1982), p. 160; Corson, personal correspondence, Oct. 16, 1995, explained term.

213 "Doolittle is . . . later": Corson notes, supplied by him, with clarifications in correspondence of Oct. 23, 1995. Generals' backgrounds from *Who's Who in America, 1946–7*. Doolittle was promoted that year to lt. gen.

213 APS-15: Guerlac (1987), esp. pp. 785, 799, 812 n.64, 813 n.84; 1084–91.

213 Hitler's defenses did not break: Rest of section chiefly from R. Jones (1978) and A. Price (1967). Some Wittgenstein and Meurer details from Hastings (1979), pp. 270, 275.

214 Eisenhower curtailed area bombing: Ibid., p. 308.

214 D-day: R. Jones (1978), esp. pp. 400–411; Price (1967), esp. pp. 125–26, 199–211. Also important were R. Jones (1990), pp. 124–25; A. Price (1984), pp. 122–31; Burns (1995); Pringle (1985), pp. 346–54; Watson-Watt (1959), pp. 414–16. Cf. Hastings (1979), pp. 69–80.

214 Rebecca-Eureka: Guerlac (1987), pp. 840–46, *Radar*, No. 3 (June 30, 1944), p. 26; No. 4 (Aug. 20, 1944), p. 11.

216 "No conceivable . . . then": Price (1987), p. 210.

216 Mickey on D-day: DuBridge to Bush, June 21, 1944, "Division 14 division files Jan.– June 1944," folder in Box 16, E-1, NA. Also Guerlac (1987), p. 843; *Radar*, No. 3 (June 30, 1944), p. 2.

217 air bombardment effectiveness: Ambrose (1994), pp. 245–53, 577; Guerlac (1987), pp. 842–43.

10 • Victory

218 "So while . . . old": Kennett (1982), p. 177.

218 buzz bombs: R. Jones (1978), esp. pp. 336–428. Davenport interviews helped bring home horror of attacks. Technical details: Collier (1957); Longmate (1981); Breuer (1988), p. 223; "Secrets of the Flying Bomb and the Rocket Bomb Revealed," *Illustrated London News*, n.d., reproduced by Temple Fortune Press. Copy supplied by J. K. Maynard.

219 MEW deployed: Guerlac (1987), pp. 846–52; Pollard (1982), pp. 92–94. Additional details from Pollard interview, Aug. 29, 1994; *Radar*, No. 3 (June 30, 1944), pp. 3–12; DuBridge to Bush, June 21, 1944, Box 16, E-1, NA.

220 beating other radars by 20 miles: Coordination committee meeting notes, July 19, 1944, Box 47A, NANER.

220 A typical patrol: Guerlac (1987), pp. 857–58; Bowen (1987), p. 134.

220 shifted hundreds of A-A guns: Guerlac (1987), pp. 858–59; R. Jones (1978), pp. 427–28.

220 imperiled maiden: Guerlac (1987), p. 859.

220 Into the turmoil: Getting (1989), pp. 129–34; Guerlac (1987), pp. 485, 859. I rely extensively on interviews with Abajian, Davenport, and Getting.

221 proximity fuze: L. Brown (1993); Baldwin (1980), pp. 211–12; Baxter (1947), pp. 35, 221–42.

221 fuze formed shell nose, emitted oscillating signal: A. Price (1984), p. 242.

222 fuze restricted to Pacific: A fuze fired from light cruiser *Helena* first claimed a Japanese plane on Jan. 5, 1943 — Baldwin (1980), pp. 233–34.

222 584s at Anzio: Baxter (1947), pp. 114–15; Guerlac (1987), pp. 484, 853; Morison (1963), pp. 358–60.

222 "One night . . . air": Papian interview, Jan. 30, 1995.

222 584 claims 37 planes: Baxter (1947), p. 115.

222 Churchill's plea shook loose: Guerlac (1987), pp. 485, 859. A 584 was first fired operationally in Britain on Feb. 4, 1944, from Lippets Hill, London, Col. W. R. Goodrich, CAC, "Report on operational employment of SCR-584," E-V 1944 folder, Box 92B, Office of Publications, NANER.

223 "Seven . . . them": Davenport interview, April 6, 1994.

223 "the curve . . . London": Pile to Marshall, Aug. 12, 1944, included in Bush to Loomis, Aug. 31, 1944, "Div. 14 July–Dec. 1944," folder in Box 16, E-1, NA.
223 buzz bomb figures: Aug. 28, 1944, from Baxter (1947), p. 235. Overall figures: Guerlac (1987), p. 859. Collier (1957), pp. 384, 523, has slightly different numbers but trend is same.
223 "the almost . . . weapon": Guerlac (1987), p. 689.
224 U.S. and RL radar production figures: Guerlac (1987), p. 690; Newton et al. (1991), p. 77.
224 RL's major projects reached front lines: Newton et al. (1991), p. 55.
224 astonished DuBridge: DuBridge to Bush, July 19, 1944, "Div. 14 July–June 1944," folder in Box 16, E-1, NA.
224 "Los Alamos . . . weeks": Rigden (1987), p. 170. Story on pp. 167–70.
224 RL ran field stations: Guerlac (1987) and Newton et al. (1991).
225 "Where do . . . fired?": Fortune, Nov. 1945, p. 163.
225 MEW crated, waterproofed, shipped: DuBridge to Bush, June 21, 1944, "Division 14 division files Jan.–June 1944," Box 16, E-1, NA; Notes on coordination committee meeting, July 19, 1944, Box 47A, NANER; Radar, No. 5 (Sept. 30, 1944), pp. 30–31.
225 "The Huns . . . tactic": Griggs to Bowles, Aug. 30, 1944, "Div. 14 radar divisions July–Dec. 1944," Box 16, E-1, NA. Historian A. Price is skeptical of Griggs's account, noting among other things DG could not have known German intentions, and that enemy fighters were badly outnumbered, making such an attack risky: Price correspondence, April 17, 1995.
225 Patton screamed profanely: Griggs to Bowles, Aug. 30, 1944, "Div. 14 radar divisions July–Dec. 1944," Box 16, E-1, NA; and Griggs to Bowles, Oct. 17, 1944, "synchronous bombing (1944–1947)," Box 92B, NANER.
226 SCR-584, more mobile: Guerlac (1987), pp. 485–89, 892–93; some details from Radar, No. 5 (Sept. 30, 1944), p. 3 ff. Abajian interview, May 11, 1995, provided details of how the 584 worked against ground targets.
226 584 in Belgium: Baldwin (1980), pp. 273–74; Getting (1989), p. 134. Antwerp importance from Churchill (1953), pp. 165–79.
226 "literally cried . . . fuzes": Baldwin (1980), p. 274.
226 rankled VB: Bush OHI, Reels 7-A, 7-B, pp. 439–47, MC 143, MIT.
227 foxhole tests: Baxter (1947), p. 233.
227 Navy almost lived by fuze: During 1943, 75% of rounds fired by 5-inch guns were time fuzed, the rest proximity fuzed. However, VT fuzes accounted for 51% of enemy planes shot down, outdoing conventional fuzes 3–1. This increased to 6–1: Baldwin (1980), p. 245. Navy's financial role in VT fuze development is in Baxter (1947), p. 229.
227 Combined Chiefs approval: Date from Baxter (1947), p. 236.
227 210,000 rounds: Cole (1965), p. 655.
227 double the previous success rate: Baldwin (1980), pp. 273–74.
227 fuze and Battle of Bulge: Baldwin (1980), pp. 277–91; Cole (1965), pp. 360–63, 374–75, 500–504, 626–27.
227 Hitler managed to assemble: Churchill (1953), p. 238.
228 "The funny . . . us": Baldwin (1980), p. 279.
228 "The number . . . measured": Story, quote in Radar, No. 9 (April 30, 1945), p. 12.
228 AI radars for rare tactical missions: Radar, No. 8 (Feb. 20, 1945), p. 8.
228 a like tale unfolded: Chiefly, Guerlac (1987); cf. Baldwin (1980), pp. 277–78, who notes how devastating VT fuze was against aircraft. Not clear from any account is the number of 584s involved.
228 Allied strategic bombing and German oil output: Hastings (1979), pp. 318–19, 381–82, 395–96.
228 Bomber Command relied on H_2S: Ibid., pp. 335–36; A. Price (1967), p. 215.
229 8th AF dependence on H_2X: Guerlac (1987), pp. 785, 799, 1084–91.
229 A number of technical innovations: Radar, No. 3 (June 30, 1944), p. 15; Radar, No 5 (Sept. 30, 1944), p. 17 ff., 34; Corson interviews.

229 "From the . . . flag": *Radar*, No. 5 (Sept. 30, 1944), p. 17.
229 Getting within a mile . . . rare: *Radar*, No. 10 (June 30, 1945), p. 12; at one stretch, near war's end, only 5.6% of bombs dropped in total overcast came within a mile.
229 290,000 tons: Guerlac (1987), p. 799 table.
229 "Every mission . . . mind": Ridenour story and quote: "Notes on coordination committee meeting, Nov. 7, 1944," Box 47A, NANER.
229 two air forces . . . overwhelmed Hitler's air defenses: Chiefly, R. Jones (1978), pp. 466–71; A. Price (1984), pp. 168–71, 187; cf. Hastings (1979), pp. 345–49.
229 killed up to 100,000: Hastings (1979), p. 397.
230 "My saddest . . . Germany!": quoted in A. Price (1967), p. 228.
230 vast palls of fire and smoke: See Hastings (1979), pp. 392–95, 402.
230 U-boats managed brief resurgence: See van der Vat (1988), pp. 346–48, 373–77; Morison (1963), p. 559.
230 ASD: Figures from Guerlac (1987), p. 394.
230 Reich falls: See Churchill (1953), pp. 445–70.
230 inventory of German radar: R. Jones (1978), pp. 470–71.
230 POST MORTEM: A. Price (1967), pp. 240–42, (1984), pp. 195–96.
231 DuBridge: "Memories," Part I, pp. 36–37, CTA.
231 a few RL radars found way westward: Guerlac (1987); cf. Fagen (1978), p. 70, for SCR-720 production details.
231 CXAM at Coral Sea and Midway: Guerlac (1987), pp. 913–16, 925–30, 944–45.
231 thousands of ASV units: Exact figures hard to find. Bowen (1987), p. 180, says 7000 Mark IIs ordered from Philco by U.S. Navy, 10,000 more from Research Enterprises Ltd. in Toronto; Skolnik (1985), p. 184, says 26,000 ASB sets made from 1942 to 1944: ASB had a duplexer and seems to have been derived from Mark II. Cf. OSRD (1946), pp. 13–14.
231 SJ radar: Fagen (1978), pp. 75–76; Guerlac (1987), p. 916. Set went into production early in 1942, with about a dozen in service that October. The SJ, used both to find targets and avoid detection, wreaked havoc on the Japanese: cf. Devereux (1994), p. 169; Morison (1963), pp. 495–96, 504.
231 Bell fire-control radars: Fagen (1978), pp. 25–26, 49, 55–56; McNally (n.d.), p. 3. The Mark IV was first U.S. application of the cavity magnetron, which scaled linearly with wavelength: L. Brown, unpublished manuscript.
232 nagging sense of more to accomplish: See Pollard (1982), p. 97; cf. Pound interview, Feb. 6, 1996.
232 Taffy Bowen in Australia: Bowen (1987), pp. 196–99; Australian radar details in Guerlac (1987), pp. 1106–1109.
232 "a wild ride": Bowen (1987), p. 198.
232 "I got . . . pilots": Ibid.
232 "By mid-1943 . . . devices": Ibid., p. 196. Note: "than" reads as "then" in Bowen's text and was changed for this account.
232 Low Altitude Bombing: Much of LAB's development and production from Guerlac (1987), pp. 377–78, 997–1013; Fagen (1978), pp. 100–102. Critical to understanding early stages, and how it worked, were extensive interviews with Corson; cf. Zann interview, March 29, 1996. Also important were *Radar*, No. 1 (April 1944), p. 17 ff.; *Radar*, No. 6 (Nov. 15, 1944), p. 3 ff. LAB first saw combat in fall 1943: OSRD (1946), p. 78.

The brilliant Havens took over project after Corson went to Pentagon in fall 1942. He later developed a high-altitude version with a PPI that showed true ground range rather than image-distorting slant range to target. It might have displaced H_2X had the war lasted much longer.

233 perfect complement to subs: For overview of U.S. sub operations and importance of radar, see Morison (1963), pp. 495–97, 501–507.
233 "We flew . . . know": Carson interview, March 27, 1996.
234 SCR-584 in Pacific: Guerlac (1987), pp. 1018–25. Abajian's account from interview of March 22, 1994; Abajian OHI, IEEE Center (1993), pp. 1–12.
234 world's largest radar army: *Radar*, No. 11 (Sept. 10, 1945), pp. 26–28.
234 "That was . . . us": IEEE Center (1993), p. 8.

234 "This is . . . shooting": Quoted in Guerlac (1987), p. 1021.
234 "Anytime we . . . beam": Abajian interview, March 22, 1994.
235 "A-battery . . . fire": Ibid.
235 "Sure enough . . . bogey": Ibid.
235 H₂K: RL details from Guerlac (1987), pp. 508–22; cf. Newton et al. (1991), pp. 26–27. Actions of key participants, state of scientific thinking: interviews with Dicke, Pound, and Purcell; Purcell's videotaped lecture, Radar and physics, March 18, 1986, *The Morris Loeb Lectures in Physics*, CSL.
235 Merchant's Limited: Pound interview, July 3, 1992.
236 theory already advanced: See Van Vleck (1942).
236 Singing at right frequency: Analogy from Pound interview, Nov. 10, 1992.
236 wherever effect peaked, it should be weak: Van Vleck (1942), pp. 1, 21.
237 few aces up sleeves: Guerlac (1987), pp. 790–95.
237 Lane Bryant Building: *Radar*, No. 11, (Sept. 10, 1945), p. 42.
237 Pacific war escalated: Morison (1963) for general events; Kennett (1982) for strategic bombing campaign.
238 The Coral Sea and Midway: Morison (1963), esp. pp. 144–63; radar role and CXAM problems: Guerlac (1987), pp. 913–16, 925–30, 944–45.
238 Raytheon SG: Guerlac (1987), pp. 399–400.
238 Combat Information Center: Ibid., pp. 925–34, 990n.
239 Battle of Philippine Sea: Ibid., p. 967; Morison (1963), pp. 333–45.
239 VT fuze picked up slack: Baldwin (1980), p. 244.
239 vice admiral: Rigden (1987), p. 163.
239 Musashi works: Kennett (1982), p. 168; Craven and Cate (1953), pp. 554–55.
239 friends called him "Sir": In correspondence from Jerome E. Schroeder, NORAD historian, Nov. 3, 1995.
239 APQ-13: Fagen (1978), pp. 102–106. All told, 7902 APQ-13s and slightly modified 13As shipped by V-J Day. Cf. Guerlac (1987), pp. 1043–50. Chart detailing various sets and manufacturers supplied by Valley.
240 "scopogenic": *Radar*, No. 4 (August 30, 1944), pp. 4–5.
240 Okinawa and kamikazes: Morison (1963), pp. 542–54; *Radar*, No. 10 (June 30, 1945), p. 36. L. Brown, correspondence of March 1, 1996, notes that the proximity fuze was not nearly as efficient against kamikazes as against buzz bombs, since naval-fire control radars and gun directors had great difficulty tracking the low-flying, piloted craft.
240 Japan . . . a hair's breadth behind: Primarily, Nakajima (1988) and R. Wilkinson (1946a, b). Also helpful is Swords (1986), pp. 130–35. For Japanese magnetron work, see Brittain (1985). Russian scientists and a Swiss engineer had also invented the cavity magnetron independently and ahead of the British, in the mid-thirties, about the same time as the Japanese, but made little use of it: L. Brown. cf. Swords (1986), p. 142.
241 "a few . . . engineering": Talbot II. Waterman, "Personal narrative FE AF Trip," Box 47A, NANER.
241 Midway defeat . . . might have been averted: Cf. Churchill (1950b), esp. pp. 221–24; Morison (1963), esp. pp. 149–63. L. Brown, correspondence of March 1, 1996, notes that two Japanese battleships at Midway had recently been fitted with radar. However, these were positioned well behind the carriers and inconsequential to the outcome.
241 Imperial Navy planned radar upgrade: R. Wilkinson (1946b); Guerlac (1987), pp. 917–18.
242 second-rate technicians: Japanese Wartime Military Electronics and Communications, Section I, 1 April 1946, U.S. Army Forces, Pacific, Techn. Liaison E/SWMEC Sec I, IWM.
242 technological gap with Allies: Lt. Col. S. K. Wolf, n.d. Operations Analysis—FEAF, document in "Report on OAS-FEAF," folder in Box 47A, NANER.
242 largely unchallenged: Cf. A. Price (1967), p. 250; L. Brown (1994).
242 defending against enemy air raids: For Japan's air defenses, except where indicated, R. Wilkinson (1946b); D. R. Hutchinson, Sept. 28, 1945. "Radar and early warning networks of the Japanese Army and Navy": Report based on personal interrogations of Japanese officials, IWM.

242 Army controlled A-A guns: Ibid.; A. Price (1984), p. 226; R. Wilkinson (1946b).
243 lacked good GCI system: Cf. L. Brown (1994). Some magnetron details in Talbot H. Waterman, Personal narrative FEAF trip, Box 47A, NANER.
243 losses rarely surpassed 3%: A. Price (1984), pp. 227–33.
243 Eagle: Guerlac (1987), pp. 1050–53; *Radar*, No. 11 (Sept. 10, 1945), p. 36 ff. Some bombing details from Kennett (1982), p. 174; for Eagle's scanning features, see Skolnik (1980), p. 184.
243 Dolan's death: *Radar*, No. 11 (Sept. 10, 1945), p. 43.
243 "the most . . . date": Guerlac (1987), p. 1052.
243 postwar analysis: E. H. Sharkey, "Comments on operation of AN/APQ-7 in 315th Wing —Case 24839," restricted memo of Sept. 18, 1945, copy supplied by Corson. Sharkey was a Western Electric engineer.
244 Little Boy and Fat Man: Rhodes (1986), esp. pp. 701–704, 733–42; Alvarez (1987), pp. 139–46. Some Little Boy details from A. Price (1984), pp. 232–33.
245 approach via APQ-13: Guerlac (1987), pp. 1053–55.
245 Alvarez never believed Behan's story: Alvarez (1987), p. 146. Note, Guerlac (1987), p. 1053, casually states: ". . . the second bomb was dropped by radar. . . ." However, this statement is not borne out by Beser (1988), pp. 133–34.

11 • Emergence

246 "There was . . . tackle": Davenport OHI, IEEE Center (1993), p. 74.
246 "They can . . . can": *Aeneid*.
246 atomic bomb only ended war, radar won it: Oft-repeated phrase, usually slightly modified. See DuBridge, "Memories," Part I, p. 38, CTA; Kevles (1979), p. 308; Rigden (1987), p. 164.
246 V-J Day convocation: Newton et al. (1991), p. 57. Site is misidentified in inside photo as Eastman Court.
246 Bush months earlier sent word: DuBridge, "Memories," Part I, p. 37, CTA.
246 "On this . . . lives": V-J Day speech, folder 131.1, LAD.
247 "It has . . . me": Ibid.
247 "This war . . . pledge": Ibid.
247 Abajian . . . was ecstatic: Abajian interview, May 11, 1995.
247 Edythe Baker: Baker interview, May 12, 1995.
247 "I got . . . it": *Nova*, "Echoes of War," Aug. 29, 1991.
248 "Is now . . . management": Rollefson OHI, IEEE Center (1993), p. 318. Rollefson background and anecdote, cf. pp. 308–18.
248 *had* soared to tremendous heights: Guerlac (1987), pp. 668, 690–91; "Longhairs and Shortwaves," *Fortune*, Nov. 1945, p. 169. The 431 figure cited in *Fortune* is for non-engineers only. Number of radar type designations: Personal communication from Stephen Johnston, International Radar Directory.
248 One of top priorities: DuBridge interview, May 17, 1992; Rigden (1987), pp. 142, 164; Newton et al. (1991), p. 6.
248 released for public consumption: Don Bennett to DuBridge, Sept. 2, 1945, and DuBridge to Bennett, Sept. 7, 1945, 131.1, LAD.
248 piecing together story for press: DuBridge, "Memories," Part I, p. 37, CTA; Newton interview, May 17, 1992; Newton et al. (1991), p. 6.
249 "a great . . . scientists . . . ": *Time*, Aug. 20, 1945, p. 78.
249 "We expected . . . one": DuBridge interview, May 17, 1992; "Memories," Part I, pp. 37–38, CTA.
249 *Technology Review*: DuBridge and Ridenour (1945).
249 staff occasionally invited: Bennett correspondence, 131.1, LAD.
249 "Oppenheimer became . . . MIT": Rigden (1987), p. 164.
249 did nothing to diminish: Evident from many interviews.
249 No single technology more versatile: Cf. L. Brown (1994).

250 "Surely radar . . . bomb": Bethe interview, Nov. 7, 1994.
250 "legacy to posterity": Quoted in Coordination Committee meeting notes, Sept. 13, 1944, Box 47A, NANER.
250 gist of his concern: Rigden (1987), pp. 164–65; cf. DuBridge and Ridenour (1945); *Fortune*, Nov. 1945.
250 20 years of normal progression . . . : Oft-repeated statement from wide reading, many interviews; cf. Garfinkel (1991), p. 10.
250 ". . . I realized . . . Laboratories": Rigden (1987), p. 164.
250 Rabi's plan: Ibid., p. 165; cf. John L. Danforth to John E. Burchard, April 12, 1946, "Division 14 — general 1946," folder in Box 15, E-1, NA; DuBridge to division heads and group leaders, Aug. 11, 1945, "DuBridge, L.A. LNR 1942–46," folder in Box 85A, NANER.
250 "People came . . . dying": Rigden (1987), p. 165.
251 $495,024.07: Document dated June 27, 1946, "Final Report of the Editorial Board [1946]," folder in Box 92B, NANER.
251 250 RL staffers: Newton et al. (1991), p. 58.
251 occupational bible: Rigden (1987), p. 165.
251 "the biggest . . . Septuagint": ibid., p. 164.
251 "ambassadors of . . . preparedness": Quoted in Kevles (1975), p. 44.
251 "The lessons . . . counterpart . . .": Quoted in Leslie (1993), pp. 24–25.
251 ONR: Kevles (1979), pp. 354–55; excellent general source on federal funding of research. Cf. Kevles (1975); Allison (1981), p. 184; J. Goldstein (1992), pp. 72–74.
251 deeply influenced research style: Wide reading; cf. Forman (1995), p. 411.
252 "We were . . . Alamos": Ramsey interview, March 30, 1992; cf. J. Goldstein (1992), pp. 78–80.
252 "We had . . . science": Lovell (1990), p. 7.
252 radars for ships and . . . air traffic control systems: Scott (1974), pp. 184, 229; cf. Bowen (1987), pp. 200–201; McNally (n.d.), p. 18.
252 GCA: Alvarez (1987), p. 159: *Radar*, No. 7 (Jan. 1, 1945), p. 19 ff.; No. 10 (June 30, 1945), p. 38.
252 Loran: Bowen (1987), p. 200; Guerlac (1987), p. 531.
252 communications and video revolution: Cantelon (1995).
253 radar in meteorology: RL work in Guerlac (1987), pp. 641–42. Generalities from Donaldson (1990), pp. 115–16; Fletcher (1990), p. 6; Fletcher interview, June 1, 1994.
253 mw radar provided . . . multitude of pathways: See DuBridge and Ridenour (1945); Forman (1995).
253 placement program: Newton et al. (1991), p. 57.
254 "A couple . . . said": Kyhl OHI, IEEE Center (1993), p. 170. Schwinger was a 1965 physics Nobelist.
254 telegram still in one pocket: Rigden (1987), p. 133.
254 even *thought* about things differently: Widespread interviews; cf. DuBridge and Ridenour (1945); Forman (1995).
254 "I didn't . . . sort": Davenport OHI, IEEE Center (1993), p. 74.
254 "open sesame": Robinson OHI, IEEE Center (1993), p. 294.
254 RL Nobelists: Rabi, McMillan, Purcell, Schwinger, Bethe, Alvarez, Samuelson, Van Vleck (lab consultant), Jack Steinberger, Ramsey. Presidential science advisors: DuBridge, Killian, Wiesner. TRE Nobelists: Hodgkin, Kendrew, Ryle.
255 "There was . . . ideas": quoted in Garfinkel (1991), p. 10.
255 Basic Research Division becomes RLE: J. Goldstein (1992), pp. 70–77; Leslie (1993), pp. 7, 14–15, 23–25; Wildes and Lindgren (1985), pp. 207, 245–47; cf. *RLE Currents*, vol. 4, no. 2 (Spring 1991); Valley interviews.
255 mw spectroscopy: Bleaney, Dicke, Strandberg, and Townes interviews; Townes correspondence, Oct. 16, 31, 1995; Forman (1995).
256 Lamb shift: Ramsey interview, March 30, 1992; Forman (1995), pp. 426–28; J. Goldstein (1992), p. 86.
256 Lovell: Hey (1973), pp. 17–37; Lovell (1990), pp. 110–17.

256 High Voltage Engineering: Corson interview, April 17, 1995; H. Robinson interview, Sept. 12, 1995; Robinson (1988). Good overviews on radar's application to accelerators, etc., are in Forman (1995); Alvarez (1987), pp. 153–56.

256 microwave oven: Scott (1974), p. 180; cf. Behrens (1978); Krim interview, May 20, 1994.

12 • Snow on the Doorstep

258 "His soul . . . dead": Quoted in Hugh Rawson and Margaret Miner, eds., *The New International Dictionary of Quotations* (New York: Signet, 1988), p. 243.

258 "If winter . . . behind?": Quoted in ibid., p. 244.

258 Purcell invariably walked: Main sources for chapter's narrative are interviews with Dicke, Pound, Purcell, and Torrey; periodic correspondence with Pound; and two chapters from Pound's unpublished memoir, copy supplied by him.

258 "one is . . . alone": Ridenour (1947), p. 36.

259 "Louis, 'betting'. . . quit": Purcell interview, March 26, 1992. Purcell used "bet," but he wrote "betting."

259 Purcell had grown up . . . : Background and McNair story from Purcell OHI, June 8, 1977, pp. 3–4, AIP.

259 Miss Edwards: Given name Helen, according to Mattoon High School Yearbook, *The Riddler*.

259 "Well, we . . . rope": Purcell OHI, pp. 3–4 AIP.

259 "And then . . . that moment": Ibid., p. 4.

260 in Hitler's Germany: A few of Purcell's background details added from his OHI, IEEE Center (1993).

260 hazel eyes: Description from photos, B. Purcell interview, Dec. 10, 1995.

261 probably late that summer: Dicke rarely dated notebooks, so exact chronology is not clear. Much on RL's K-band program is in Guerlac (1987), pp. 508–17.

262 "It just . . . trick": Dicke interview, Nov. 12, 1992.

262 suggested as much to Dicke: This contradicts Guerlac (1987), p. 519, who states Dicke suggested idea to EP. Dicke notes Purcell at first suggested a different means of conducting the experiment than was actually used. Dicke pointed out an improved method.

262 late in 1944: Time element fixed partly by Dicke to Van Voorhis, Feb. 6, 1945. He states one radiometer is complete, two afoot, Correspondence folder, Box 791, Fundamental Developments Group, NANER.

262 several more radiometers: Methodology and findings are in Dicke et al. (1946); cf. Dicke (1946).

263 more accurate measurements at Columbia: Guerlac (1987), pp. 519–21.

263 "Given the . . . anyway": Purcell interview, May 11, 1995.

263 "the more subtle part": Purcell interview, Nov. 4, 1992.

263 "This difference . . . emission": Ibid.

264 Charting this effect . . . : NMR historical background, Hahn interview, Feb. 13, 1996; Rigden (1985); cf. Gerstein (1994).

264 analysis of magnetic cooling: Purcell and M. Hebb. 1937. "A Theoretical Study of Magnetic Cooling Experiments," *Journal of Chemical Physics* 5: 338–50.

265 Gorter: Gorter (1967); Rigden (1985). A series of chance events had worked against Gorter. Among other things, his samples were later shown to be of too high a purity for NMR to be readily observed—an almost unheard-of condition in science, where pure samples are generally a kind of Holy Grail.

266 Hennessy's: William Hennessy Candies Inc., 493 Massachusetts Avenue.

266 "What's your . . . cheese": Pound interview, July 3, 1992.

266 "My first . . . no": Torrey interview, Nov. 12, 1992.

266 "That night . . . calculations": Ibid.

266 "I believe . . . do it": Ibid.

267 several weeks passed: In addition to participant interviews, I draw on Rigden (1986) for confirming details and clarification.

268 steady snowfall: *Boston Globe* records show snow for that night, .17 inch of total precipitation.
268 "Talk of the Town": "Fateful Night," Aug. 18, 1945, p. 18. Pound tells story in memoir, and related it at Golden Jubilee of NMR, Harvard University, Dec. 10, 1995.
269 Gulfwax brand: A like one-pound package exhibited at NMR Jubilee.
270 going seven hours: It is not true, as has been reported, that the experiment finally succeeded because field homogeneity was achieved by using appropriate shims for the magnet. The shims had been tested and modified before the first attempt on Dec. 13. The difference was turning up the generator: Pound correspondence, July 19, 1995.
270 "Dec. 15 . . . magnitude": Pound's notebook, in his possession.
270 "The method . . . nuclei": Purcell, Torrey, and Pound (1946), p. 37.
271 Stanford team: Cf. Bloch, Hansen, and Packard (1946). Some Bloch background from index to his papers, SC 303, Stanford archives. For comparison of Purcell-Bloch methods, see Rigden (1986).
271 Like Purcell, Bloch knew Rabi: Rigden (1986), pp. 441, 446.
271 teams first interacted: Ibid., pp. 445–46.
271 "You should . . . NMR": quoted by Purcell interview, Nov. 4, 1992.
272 "great heaps . . . field": Purcell (1964), p. 219; cf. Rigden (1986), p. 447.

13 • The New Astronomers

273 "At the . . . community": Lovell (1990), p. 130.
273 moon's detection by radar: T. Clark (1980); Gould (1946a, b, c).
274 "Man had . . . planet": Quoted in T. Clark (1980), p. 44.
274 Signal Corps trotted out: Besides main sources, see Stodola (1988).
274 "I thought . . . rockets": DeWitt interview, Dec. 29, 1992.
275 As Evans director: Ibid. Apparatus description and actual detection from main sources, plus DeWitt interview, Oct. 9, 1995; DeWitt and Stodola (1949); Stodola (1988). The men had been trying for a detection since the previous Sept.; this was their second system.
275 listened over loudspeakers: The speakers were hooked to an IF amplifier.
275 "because there . . . moon": Quoted in Gould (1946a).
276 "If intelligent . . . own:" Quoted in Gould (1946b).
276 Harry Edward Burton: Ibid.; background from *Who's Who in America*.
276 next to nothing of scientific interest: For early days of radar astronomy, Hey (1973), esp. pp. 7–37; Lovell (1990), esp. pp. 104–30, 166–80. Clarifying details from Forman (1995), pp. 410–14; Lovell (1983), and Lovell correspondence, June 23, 1994.
278 Ryle hit upon the idea: Ryle's ignorance of optical interferometers is spelled out in Lovell correspondence, June 23, 1994.
279 "I began . . . wavelength": Lovell (1990), p. 110.
279 "By the . . . community": Ibid., p. 130.
280 no Ph.D. . . . awarded: W. T. Sullivan III, paper given at American Physical Society meeting, March 25, 1993, Seattle; Home (1988), pp. 247–48.
280 Bowen in Australia: Bowen's actions drawn primarily from Bowen (1984), (1987), p. 196 ff.; interviews with Bowen family; and Sullivan's OHI with Bowen, Dec. 24, 1973, transcript provided by Sullivan. For early CSIRO days, Bowen (1984); Sullivan (1988); Bowen-Sullivan OHI; Sinclair and Slee interviews, April 22, 1994. Also helpful: R. Brown (1991), R. Brown et al. (1992).
282 "the whole . . . era": Bowen (1984), p. 86.
282 "After a . . . come": Ibid.
282 plans for research programs: Besides prime sources, see F. Kerr (1984).
282 Watson-Watt consulting firm: R. Brown (1991), p. 88.
282 "I was . . . work": Bowen OHI with Sullivan, p. 27.
282 Joe Pawsey: Background from Lovell (1964); cf. Bowen OHI with Sullivan, pp. 11–13; F. Kerr (1984).
283 early Pawsey solar studies: Beyond main section sources, see Pawsey (1953); Wild (1972).

R. Brown, correspondence of Sept. 29, 1995, was extremely helpful in explaining work's significance.

283 good fix on point of origin: To determine a radio source's size with the sea-cliff interferometer, one had to study the ratio of the intensity of the fringe's sinusoidal pattern of maximum and minimum.

284 Dover Heights: Main sources; Bolton (1982); visit of April 24, 1995; interviews with Ekers, Minnett, Slee. Also helpful are Pawsey (1953), esp. pp. 147–48; Robertson (1992), pp. 44–49; Wild (1972).

285 Hey accidentally discovered an oddity: Lovell (1990), pp. 173–74.

287 associated with optical objects: Ibid., p. 174.

288 early successes . . . begat more success: Bowen OHI with Sullivan, June 22, 1978, Part II, p. 42, transcript from Sullivan.

288 "Each morning . . . huts": Christiansen (1984), pp. 113, 115.

288 "With the . . . time": Wild (1972), third page of unnumbered reprint.

289 catalogue of . . . stars: Lovell (1990), p. 174.

289 Scotsman met Canadian woman: R. Brown (1991), p. 90; Watson-Watt (1959), pp. 422–23; Drew interview, Aug. 28, 1995.

289 Lovell built 218-foot-wide antenna: For Jodrell Bank; R. Brown (1991), pp. 90–100; Lovell (1990), pp. 168–76; cf. Hey (1973), p. 35.

290 two distant galaxies in collision: Pawsey (1953), p. 148.

14 • Twenty-one

291 "Here was . . . instrument": Purcell, videotaped lecture, Radar and physics, March 18, 1986, *The Morris Loeb Lectures in Physics*, CSL.

291 Purcell not an astronomer: Background from Purcell OHI, p. 39, AIP.

291 word of a remarkable prediction: Chapter based on interviews with Purcell and Ewen, and extensive correspondence with latter. Vital to filling in details: "The Discovery of the Twenty-One Centimeter Line," a chapter from W. T. Sullivan III's unfinished radio astronomy history.

291 meeting at Yale: Despite yeoman efforts, Yale astronomy department librarian Kim Monocchi has not been able to find meeting record. Sullivan, correspondence of May 31, 1996, notes that it could have been the American Astronomical Society meeting of December 1948, which would support Purcell's version of events.

292 jumped at the chance: Few more details from Purcell OHI, p. 37.

292 "There was . . . it": Ewen interview, Oct. 24, 1995.

292 "If you . . . magazine": Morrison and Morrison (1987), p. 240.

292 smuggled copies of *Astrophysical Journal*: Sullivan manuscript; Hey (1973), pp. 23–24.

293 optical astronomers . . . studying telltale spectral lines: Morrison and Morrison (1987); for general spectroscopy description, see Rhodes (1986), pp. 72–75.

293 interstellar matter remained impenetrable: Ewen (1953).

293 war-scarred Holland: Sullivan manuscript; Hey (1973), pp. 23–24.

293 hyperfine transition: Good explanation in Morrison and Morrison (1987), p. 240.

294 "The amount . . . Earth": Ibid.

294 The two papers Ewen dug up: I. S. Shklovsky, "Monochromatic Radio Emission from the Galaxy and the Possibility of Its Observation," which appeared in the Russian *Astronomichesky Zhurnal* in 1949; van de Hulst (1945a), "Origin of the Radio Waves," translated from the Dutch at Cornell University, and including van de Hulst's original 1945 work, various notations added three years later, and a two-page summary, "*Radiogolven uit het Wereddruim*," or "Radio Waves from Space Summary." A more recent translation of Shklovsky's paper appears in Sullivan (1982), pp. 318–24. The van de Hulst paper was supplied by Ewen, although a version (1945b) appears in Sullivan (1982), pp. 302–16.

As it turned out, Sullivan relates, Shklovsky had read a brief write-up of van de Hulst's work in a 1947 review of radio astronomy. Shklovsky did not have access to the Dutchman's paper and so worked out expected signal strengths himself. In the process, he calculated the probability of the hyperfine transition taking place as a factor of four too low. After Beth

Purcell translated Shklovsky's work, her husband must have picked up on this error, because it was brought to Ewen's attention. However, the mistake had no real bearing on the decision to try and detect the line: Ewen interview, April 25, 1996; B. Purcell interview, April 26, 1996; Sullivan (1982), p. 317.

294 "lick your . . . hydrogen": Ewen interview, Oct. 24, 1995, modified in correspondence of April 22, 1996.

294 "Would the . . . undetectable?": Ewen correspondence, Oct. 19, 1995. Also see Hey (1973), pp. 24–25; Piddington (1961), p. 75; and Robertson (1992), pp. 80–82, for discussion about the difficulty of the task.

295 "So I . . . was": Ewen interview, Oct. 24, 1995.

295 "I don't . . . Bob": Ewen interview, July 27, 1994.

295 Dicke returned to Princeton: RD's actions, except where noted, from interviews.

296 pair observed the sun: Dicke and Beringer (1946).

296 upper threshold of 20 degrees K: Dicke et al. (1946); Dicke interviews; cf. Hey (1973), p. 27.

296 Russell about to retire: Dicke interviews, Russell's role confirmed by Lyman Spitzer Jr. interview with Alex de Ravel, Oct. 5, 1995.

296 "I thought . . . research": Dicke interview, Nov. 12, 1992.

297 fateful decision: For cosmic background radiation detection, chiefly Weinberg (1977), pp. 39–49; Wilkinson and Peebles (1983); cf. Wilson (1983). The fact that Penzias and Wilson found it accidentally does not negate their work's importance. It took remarkable scientists to methodically rule out other radiation sources.

297 Southworth: Hey (1973), pp. 16–17, 26; Southworth (1945); technical help from Ewen.

298 Dicke's vastly more accurate technique: Radiometer description pieced together from Dicke, n.d., copy supplied by Dicke; Dicke (1946), almost, but not quite, a direct copy of the RL report; Dicke et al. (1946); with technical help from Dicke, Ewen, and Richard Muller. Note: Although an excellent method for detecting broadband noise sources, Dicke's technique does not reduce the internal receiver noise — or the statistical level of noise fluctuations present on the output signal. If a receiver is gain stable, it will be two times more sensitive than a Dicke approach. But even today receivers are not gain stable. So Dicke sacrificed a factor of two in theoretical sensitivity to be able to achieve a practical result.

Also, Dicke's lock-in amplifier by itself proved a valuable scientific tool in the postwar era and formed the heart of his own spectroscopy studies.

299 Pawsey stopped by: Dicke interviews.

299 Bowen had known about the radiometer: Bowen (1984), p. 88.

299 Piddington and Minnett: See their 1949 paper.

299 virtually all the action . . . at meter wavelengths: Wide reading, cf. Forman (1995). Mills and Minnett interviews revealed thinking of the day.

299 Purcell's brainstorm: Interviews with principals; various Pound communications.

300 second oscillator performed frequency switching: The account does not do justice to the complicated workings of Ewen's radiometer. In a true Dicke-switching scheme, switch is located forward of all active receiver circuits, including mixer. This meant the first local oscillator would have to be switched in frequency at 30 Hz — a difficult problem. Ewen's hybrid is called a "total power" and frequency switching receiver. He had to temperature stabilize all front-end components. With the second oscillator and IF amplifer, he obtained a bandwidth of 16 kHz through a minor modification of an existing filter. The frequency shift was about 10 kHz, far less than desired. However, when his switching system proved able to move only between frequencies 10 kHz apart, he could not guarantee he could get off the search bandwidth to find a reference point.

By far the best source for experiment's hardware details is Sullivan's unpublished chapter. He notes William Mumford had shown such gas discharge tubes provided an extremely constant noise source over a broad range of mw frequencies, with output varying little from tube to tube. Shortly after Ewen's experiment, their use in radio astronomy became standard.

301 "Papa" Friis, "Moms" Mumford: Pound interview, June 22, 1992.

301 "Right here . . . Laboratory": Ewen interview, July 27, 1994.
301 became women's lounge: Ewen interview, July 27, 1994; Pound correspondence, July 10, 1995.
301 plywood antenna, lined with copper sheeting: Main sources; Morrison and Morrison (1987); Purcell videotaped lecture, CSL; photos of experimental setup supplied by Ewen.
301 Rumford fund: Purcell to Harlow Shipley, Jan. 12, 1950; Shipley to Purcell, Feb. 28, 1950. Copies supplied by Ewen.
302 Purcell left that summer: Pound interview, June 29, 1994; Ewen correspondence, Oct. 19, 1995.
303 "Tell me . . . this": Ewen interview, July 27, 1994.
303 "Well, go for it": Ibid.
303 van de Hulst arrived: Sullivan's unpublished chapter; cf. F. Kerr (1984), pp. 137–8.
303 unrelated experiment in same frequency: Pound correspondence, July 10, 1995.
303 a Friday night on Easter weekend: Ewen and Purcell (1951); Ewen correspondence, Oct. 19, 1995.
303 trained south, toward Aquila and Ophiucus: Sullivan correspondence, May 31, 1996.
304 "It's just . . . it' ": Ibid.
304 "I think . . . thesis": Ibid.
304 "And we . . . that": Ibid.
304 "One of . . . right": Purcell interview, Nov. 4, 1992.
304 "What are . . . 1420?": Ewen interview, July 27, 1994.
305 called in van de Hulst and Kerr: Besides main sources, see F. Kerr (1984), pp. 136–38. Note: In his taped lecture (CSL), Purcell says he was called in on Saturday, April 7. However, this was not the first detection. F. Kerr (1984), p. 137, relates that he was notified the morning after the March 25 discovery. Ewen, correspondence of April 22, 1996, maintains this was not possible: he had not yet told Purcell. Kerr was probably brought in the day after Ewen told Purcell.
305 very similar equipment: Ewen correspondence, Oct. 19, 1995; Pawsey and Bracewell (1955), p. 267.
305 Purcell insisted: Robertson (1992), p. 83; Purcell interview, Nov. 4, 1992.
305 Bowen had been urged: Bowen (1984), p. 95; Robertson (1992), p. 82.
305 Christiansen and Hindman: Sullivan, unpublished chapter; cf. Sullivan (1982), p. 335.
305 enhanced picture of universe: Cf. Sullivan (1982), pp. 325–27.
305 Purcell always considered: Purcell interview, March 26, 1992.
305 whole new universe awaited: See Hey (1973), pp. 126–28; Piddington (1961), p. 75.
306 first radio equivalent to Hale: Hey (1973), p. 35.
306 Lovell redesigned reflector: Robertson (1992), p. 138.
306 revealing galaxy's long, spiral shape: Hey (1973), esp. pp. 127–30, 145–46; cf. Millman (1983), p. 152.
306 banks of filters instead of radiometers: A dual filter technique is two times more sensitive than the switch frequency technique. However, frequency switching was fast, easy, and cheap. Writes Ewen, ". . . for that piece of genius, we are forever indebted to Ed Purcell." Letter of Oct. 19, 1995.
306 "the method . . . measurements": Hey (1973), p. 29.
306 latest generation Dicke radiometer: Author indebted to Richard Muller for explanation of the radiometer's impact on radio astronomy.
307 "that measly little line": Purcell videotaped lecture, tape 1, CSL.
307 "So hurray . . . line . . . .": Ibid., tape 2.

15 • The Magic Month

308 "Most important . . . month. . . .' ": Shockley (1974), p. 47.
308 "The initial . . . involved": Ibid., p. 61.
308 piled into the auditorium: Account taken from Bell Labs press release, July 1, 1948,

supplied by AT&T archives; *The Transistor*, a description of the event, with photos, in *Bell Laboratories Record.*, vol. XXVI, no. 8 (Aug. 1948): 321–24; Smits (1985), pp. 15–16.

309 "A device . . . Laboratories . . . .": Quoted in Bernstein (1984), pp. 77–78.

309 "However, aside . . . popular": Quoted in Smits (1985), p. 16.

309 "destined to . . . applications": quoted in Ibid.

309 demonstrations . . . drew gasps: Shockley (1974), p. 62.

309 a resounding thud: Bernstein (1984), p. 80. He also gives price.

309 not until 1959 did solid-state overtake tubes: Macdonald and Braun (1977), p. 1063; cf. Smits (1985), p. 16.

309 these features made clear: Ibid., p. 1062; Smits (1985), p. 16; cf. Weiner (1973), p. 33.

310 wagering heavily something vitally important: Impact of war on Kelly's thinking spelled out in Hoddeson (1981). Also helpful: Townes interview, July 19, 1995.

310 No organization could rival it: Cf. Bernstein (1984), pp. 8–9; Hoddeson (1977), p. 63.

310 Bowen lunched with Darrow: Bowen (1987), p. 165.

310 "He had . . . person": Ibid.

311 economical way to amplify: Hoddeson (1977), pp. 62–65; (1981), p. 45.

311 Bell's fledgling radar effort: Allison (1981), p. 123; Fagen (1978), p. 7.

311 Semiconductors: For a lovely discussion of semiconductors and rectification, see Bernstein (1984), p. 80 ff. For early crystal rectifiers: Basalla (1988), p. 44 ff.; Olson (1988), p. 59; Torrey and Whitmer (1948), pp. 5–6.

311 hunt for transistor: Except where noted, I rely on Hoddeson (1977, 1981).

311 Mervin Kelly: Background from Bell press release noting Kelly's death on March 18, 1971, AT&T archives. Personality from wide reading, but especially von Mehren interview, Oct. 27, 1994.

312 Shockley's first assignments: Shockley (1963), pp. 3–4.

312 "When Kelly . . . me' ": Quoted in Hoddeson (1977), p. 65. Author has changed placement of dashes only. Also helpful in drawing out study group details: Townes interview, July 19, 1995; Shockley (1974), p. 51.

313 first stab at a semiconductor amplifier: Shockley (1963), pp. 4–6; (1974), pp. 51–52.

313 small-scale radar project: Allison (1981), p. 129; Fagen (1978), pp. 7, 19–25, 63–65, 90.

313 crystals offered way to reduce noise: See Callick (1991), p. 18.

314 Southworth rummaged around the dusty offerings: Hoddeson (1981), p. 49. The site is usually identified as Cortland Street in lower Manhattan. But interviews indicate it was probably Cortland Alley, close to Bell headquarters.

314 Ohl: Background and actions chiefly from his OHI, AIP. Also important for transistor's prehistory: Hoddeson (1977), (1981). Some general details of semiconductors and purification process are from Millman (1983), pp. 417–18; Olson (1988); Scaff (1970); Scaff and Ohl (1947); Torrey and Whitmer (1948); cf. Bernstein (1984); Braun (1992).

315 Becker, witness to WS's amplifier disclosure: See Shockley (1974), p. 51.

315 "we were . . . laboratory": Brattain OHI with Holden and King, pp. 20–21, AIP.

315 detected odor of acetylene: Millman (1983), p. 419; Brattain OHI with Holden and King, p. 23 AIP; cf. Ohl OHI, AIP; Scaff (1970).

315 p-n junction arose from segregation of impurities: Smits (1985), p. 23; Brattain OHI with Holden and King, p. 23, AIP; Scaff, Theuerer, and Schumacher (1949), p. 383.

316 "Well, I . . . is": Scaff OHI, p. 19, AIP.

316 Brattain joined Shockley in revised disclosure: Shockley (1974), pp. 51–52.

316 "The early . . . War II": ibid., p. 52.

316 Ohl, Scaff, and Theuerer assigned: Ohl OHI, p. 45 AIP; Scaff OHI, p. 24, AIP.

316 Wooldridge and Townes shunted to radar: Townes interviews.

316 Brattain . . . airborne detectors: Brattain OHI with Holden and King, AIP.

316 Shockley to Navy: Shockley papers, SC222, Stanford.

317 "Nature abhors . . . tube": Quoted in Bernstein (1984), p. 85.

317 Robinson browsed: D. Robinson interview, July 30, 1992; Seitz (1996).

317 stumbling blocks in building mw radar receiver: Callick (1991), pp. 18–19; Seitz (1995).

317 "Everything about . . . doing": D. Robinson interview, July 30, 1992. Ibid., for anecdote, with some Skinner background from H. Jones (1960).
318 "That one's . . . just right": D. Robinson interview, July 30, 1992.
318 hosts made significant strides: Callick (1991), pp. 19, 54; Guerlac (1987), pp. 232–35.
318 Ramsey at first greeted with skepticism: Ramsey interviews, March 30, July 2, 1992. Status of Bell's crystal program at time confirmed by Seitz, correspondence of Dec. 20, 1994.
318 pair set up a rooftop demonstration: Ramsey, Corson interviews.
319 "That started . . . electronics": Ramsey interview, March 30, 1992.
319 By fall, crystal investigations ongoing: RL crystal work from interviews with Bethe, Nov. 7, 1994, and Torrey, Nov. 12, 1992, Oct. 25, 1994; Torrey and Whitmer (1948), p. 9; Seitz (1995).
319 Bethe and Schwinger: Bethe interview, Nov. 7, 1994.
319 RL became focal point: WWII crystal work from Guerlac (1987), pp. 289, 587, 594; Olson (1988); Pearson and Brattain (1973); Scaff and Ohl (1947); Seitz (1995); Torrey interviews; Torrey and Whitmer (1948); Weiner (1973); correspondence with Seitz.
319 Bell a major part of attack: Above sources, but chiefly Fagen (1978), pp. 44–45, and Scaff and Ohl (1947), pp. 8–14; cf. Scaff, Theuerer, and Schumacher (1949), p. 383; Hoddeson (1981), p. 50.
320 Lark-Horovitz: Background from Henriksen (1987).
320 before RL temporarily abandoned crystal research: Seitz correspondence.
321 Crystal Crackin' Papa: Seitz (1995).
321 "We sort . . . great' ": Torrey interview, Oct. 11, 1993.
321 many . . . reached same conclusion: Ibid.; Pearson and Brattain (1973), p. 1800.
322 Bell entered conflict with 2300 members: Fagen (1978), esp. pp. 9–19; Guerlac (1987), p. 691; Hoddeson (1981), p. 51.
322 right hand to Buckley: M. Sparks interview, June 12, 1995.
322 odd-looking towers: Brooks (1975), pp. 214–16; Cantelon (1995); Mabon (1975), p. 55 ff.
322 near and dear to Kelly's heart: For MK's solid-state plan, I rely mainly on Hoddeson (1981).
322 visited Bell's crystal program weekly: Scaff OHI, p. 26, AIP.
322 couldn't see final payoff clearly: Townes interview, July 19, 1995; cf. Hoddeson (1981).
323 With Buckley's blessing: Bernstein (1984), p. 91.
323 "Subject: . . . solids": Quoted in Weiner (1973), pp. 25–26.
323 headquarters spilling out: See Hoddeson (1981), p. 51; Bell Labs News, Sept. 22, 1986. Townes interview, July 19, 1995, for timing of move.
323 Murray Hill: M. Sparks and E. Sparks interviews; Hunt (1943); plot plans and other design documents supplied by AT&T archives.
324 William Shockley: Wartime activities from Morse (1977), pp. 183, 187; Shockley papers, SC222, Stanford.
324 Kelly wooed him back: Shockley (1974), p. 52; cf. Weiner (1973), pp. 29–30.
324 hand-picked to complement each other: For personalities of Shockley and Morgan, I rely on E. Sparks interview, June 10, 1995.
324 rewired MIT elevators: Fine interview, 1994.
324 tape-recorded all conversations: Shockley papers, SC222, Stanford.
324 organizational chart: Physical Research setup, primarily from Hoddeson (1981), pp. 54–55; Bernstein (1984), p. 91; Weiner (1973), p. 31.
325 Bardeen and Brattain: For Bardeen's background I rely on his OHI, AIP; and Physics Today, April 1992, with articles by Holonyak Jr., Lubkin, and Pake. Brattain's life is drawn from OHI with Holden and King, AIP; OHI with Weiner, AIP; Shockley (1974), p. 89; Weiner (1973), pp. 27, 33. Also helpful: interviews with Holonyak Jr., Nov. 23, 1994, and E. Sparks, June 10, 1995; Seitz correspondence, June 9, 1995.
325 "if they . . . Gulf": Bardeen OHI, p. 3, AIP.
325 "Say, John . . . living?": Lubkin (1992), pp. 23–24.
326 Bardeen . . . had to share an office: Herring (1992), p. 30; Hoddeson (1981), pp. 54–55, 62.

326 "Come in . . . this": E. Sparks interview, June 10, 1995.
326 "I cannot . . . do it": OHI with Holden and King, p. 33, AIP.
326 Shockley's manner would drive a wedge: Brattain OHI with Weiner, pp. 32–33, AIP. As he says, "for reasons of Shockley."
327 "Walter, I . . . this?": Scaff OHI, p. 37, AIP; M. Sparks and E. Sparks confirm Shockley's strong sense of humor.
327 Shockley's first instinct: For Shockley team's work, I rely, except where noted, on Shockley (1963, 1974, 1976) and Hoddeson (1981); cf. Bernstein (1984), esp. pp. 93–97; Weiner (1973).
327 first workable transistor design: Shockley was foiled in his attempts to produce the first field-effect transistor, explained Fred Seitz, "because of the presence of special energy levels associated specifically with the surface. With advancing technology, one learned to neutralize those states." Letter of June 9, 1995.
328 "Reduce a . . . necessary": Bardeen OHI, p. 12, AIP.
328 "Our group . . . germanium": Shockley (1974), p. 55.
328 "For these . . . materials": Ibid.; cf. Brattain OHI, Holden and King, p. 24, AIP.
328 "We were . . . problems": Quoted in Weiner (1973), p. 31.
329 "In showing . . . zero": Shockley (1974), p. 64.
329 "will to think": Ibid.
329 "the will . . . act": Ibid., p. 65.
329 ". . . these tests . . . semiconductor": Ibid., p. 66.
329 switch from silicon to germanium: Seitz correspondence, June 9, 1995, for benefits of germanium.
330 rather than decreasing surface conductivity: Herring (1992), p. 31.
330 gold contact . . . drew electrons out: An excellent treatment is in Bernstein (1984), pp. 93–94. Also helpful: Hoddeson (1981), pp. 72–74; Herring (1992); Holonyak interview, Nov. 23, 1994.
330 ". . . and how . . . block": OHI with Holden and King, p. 31, AIP.
330 "I found . . . collector . . .": Ibid.
331 "It was . . . flop": OHI with Weiner, p. 21, AIP.
331 "Mr. Watson . . . you": See Shockley (1974), p. 73.
331 "The circuit . . . quality": Hoddeson (1981), p. 74.
331 "laboratory confidential": Ibid., p. 75.
331 "Bardeen and . . . see' ": Bernstein (1984), p. 95.
331 "transresistance, transistor . . . it!": Ibid.
332 "Frankly, Bardeen . . . inventors": Shockley (1974), p. 75.
332 "I told . . . everybody' ": OHI with Weiner, p. 25, AIP. For more on discord caused by Shockley, see Herring (1992), p. 32.
332 something better was needed: Cf. Bernstein (1984), p. 96.
332 pair of p-n junctions: For junction transistor, I rely primarily on Bernstein (1984), pp. 97–105; Shockley (1974), pp. 80–85; with help from M. Sparks interview, June 12, 1995; Seitz correspondence.
333 prestigious *Proceedings* . . . : Brittain (1990), p. 19.
333 "I got . . . me": OHI, p. 6, AIP.
333 "That morning . . . mind": Ibid., p. 5.
333 "Yes, I'd . . . that": Ibid.

16 • A Sense of God

Note: By far the bulk of this chapter is drawn with much appreciation from extensive interviews with Charles Townes, July 15–16, 1994, and Sept. 14, 1995, as well as several phone conversations and letters. The other prime source, repeating many of the same details, is Townes's rich oral history (1994), ROHO, TBL. Unless otherwise noted, I use these sources.
334 "And so . . . lifetime": Townes (1992), copy supplied by Townes.
334 transistor . . . mass production: Bernstein (1984), p. 79. In October 1951, Western Electric began manufacturing primitive devices. They appeared in phone system the next year.

The first commercial application outside communications came in 1953, in Sonotone hearing aids.

335 The committee, due to meet . . . : Cf. Townes (1984), p. 547; (1992).

335 sources of high, sharp frequencies: Cf. Bertolotti (1983); Townes (1984).

336 if one believed in God, He was everywhere: Townes (1992) discusses faith and background.

336 "Some of . . . God-made": Townes (1992).

336 Latin, Greek, Anglo-Saxon: Bertolotti (1983), p. 75 ff.

337 "The fact . . . things": Townes (1994), p. 21, ROHO.

337 learned about electromagnetic theory: Some details in Townes (1992).

337 extraordinary community of scientific pioneers: Bertolotti (1983), p. 75 ff.; Kevles (1979), p. 219, for Pauling; Townes (1992).

337 arrived in Pasadena: Some Caltech details from Townes (1992). Note: He did not receive Duke degree until 1937; it was held up because requirements had been completed so quickly.

338 just-launched company program: Carroll (1970), p. 78. For other Bell Labs details, see Nebeker (1993), pp. 68–69.

339 "On Monday . . . assignment": Townes (1992). Some punctuation changed.

339 "This work . . . job": Nebeker (1993), p. 68.

339 Overall, the bombing radar . . . : Some details from ibid., pp. 67–74.

340 report written two years earlier: report dated April 27, 1942.

340 "I couldn't . . . possible": Townes interview, July 15, 1994.

340 voiced his concerns: Townes interviews; Nebeker (1993), p. 75.

341 "And I . . . home": Nebeker (1993), p. 75.

341 a way to transport himself: Townes (1992).

341 Spectroscopy: general description from Rhodes (1986), p. 72. Physics applications mainly from Forman (1995), pp. 404–409; with help from Townes; cf. Millman (1983), pp. 151–52.

341 Two central factors: Forman (1995), pp. 406, 410; Townes interviews.

342 Cleeton-Williams experiment: Forman (1995), pp. 408–10; Guerlac (1987), p. 191; Bleaney (1984), p. 466. Townes, letter of Oct. 16, 1995, vital in understanding work's importance.

342 "a spectacular . . . spectroscopy": Townes (1992).

342 precise high-frequency beams: Main sources; Townes (1992); cf. Forman (1995), pp. 422–24.

342 "Microwave radio . . . engineering": Quoted in Nebeker (1993), p. 76; cf. ibid., p. 90n.

342 Had Bell turned down: Main sources; Nebeker (1993), p. 76.

343 he and various competitors derived: Townes interviews; Forman (1995), pp. 422–28; Millman (1983), pp. 151–52.

343 "It was . . . period": Townes interview, Sept. 14, 1995.

344 offered up an unsolicited job: Main sources; Townes (1992), (1968), p. 700. For Rabi's doings, see Rigden (1987), p. 180.

344 "You've made . . . company": Townes (1994), p. 92, ROHO.

344 where radio frequency spectroscopy flourished: Cf. Bertolotti (1983), p. 76.

344 led Townes to Franklin Park: Date from Townes's notebook. Cf. Bertolotti (1983), p. 76; Townes (1992).

345 several researchers . . . already achieved population inversion: The Harvard work appeared in *The Physical Review* in Jan. 1951, just a few months before Townes's insight. There were three papers. The key one, dealing with negative temperatures, was by Pound and Purcell. For more on how this influenced Townes's thinking, see Townes (1994), pp. 144–45, ROHO. Until the Franklin Park incident, Townes felt that because of the second law of thermodynamics he could not put enough molecules in high-energy states to produce useful radiation—Townes correspondence, Oct. 16, 1995. Cf. Townes (1984), pp. 547–48.

346 Nothing central . . . hadn't been around 25 years: Townes (1983), p. 7681; Bertolotti (1983), p. 7.

346 Townes came to believe: Townes (1983), p. 7681, (1984), p. 547.

346 Lending credence to his view: Other maser work from Bertolotti (1983), pp. 74, 84–85; Carroll (1970), pp. 75–76; Townes (1983), p. 7681.

346 "I'm not . . . work": Quoted in Carroll (1970), p. 83; Gordon background and general events from Gordon interview, July 6, 1994.

346 Zeiger: Background, general events from Zeiger interviews.

347 Townes decides to build first device at 1.25 cm: Townes, Gordon, and Zeiger inteviews; Bertolotti (1983), pp. 77–81, has several details, including time factor of switch to K-band in first quarter of 1952. See Townes (1984), p. 547, for what he envisioned as fruits of the work.

348 main physics offices ringed Pupin: Department layout from Novick interview, on or about Oct. 26, 1995; Alex de Ravel interviews with Jerry Packer and Irene Tramm, Oct. 18, 1995.

348 Gordon set up shop there: Experiment details from Gordon, Townes, Zeiger interviews.

348 plowed through $30,000: Some details from Bertolotti (1983), p. 80.

349 Danos bets a bottle of scotch: Townes (1984), p. 548.

349 "They sort . . . it!' " Quote and story in Townes (1992).

349 "No, I . . . work": Ibid.; cf. Townes interviews.

349 afternoon of April 8, 1954: Date from Townes's notebook. Subsequent events chiefly from Gordon and Townes interviews; a few details from Bertolotti (1983), p. 80; Carroll (1970), p. 83.

350 physics community reacted quickly: The seminal paper is Gordon, Zeiger, and Townes (1955a); see also (1955b). Other details of announcement and its impact drawn from Bertolotti (1983), pp. 80–81.

350 Townes took sabbatical: Townes interviews; Townes (1968), p. 701; Carroll (1970), pp. 84–85.

350 similar train of thought: Strandberg, Bloembergen, and Townes interviews; Bertolotti (1983), pp. 90–91; Carroll (1970), pp. 84–88; Townes (1968), p. 701. A few Bell Labs details from Millman (1983), pp. 151–52.

351 maser counter to common view of invention: Townes (1968), pp. 699–701; (1983), p. 7679, (1984), p. 549. Bertolotti (1983), pp. 83–84, spells out impact on spectrometry.

352 "I played . . . waves": Townes (1992).

352 setting stage for first laser: Bertolotti (1983), pp.104–106; Carroll (1970), pp. 94–98; Townes (1992).

352 Prochorov and Basov: Bertolotti (1983), pp. 84–89, 115–117.

353 selection committee headed by Bush: Besides the fact that Townes was not an MIT insider, he attributed Bush's rejection in part to disagreements over NASA funding Townes helped secure to pursue rocket and space research. Bush felt the government should not support such work, and that MIT should not take the money: Townes interview, Sept. 14, 1995.

353 "Now, that . . . research": Townes (1992).

353 "You may . . . done": Ibid.

17 • Rad Lab Redux

354 "For the . . . power": "Problems of Air Defense," *Final Report of Project Charles*, Aug. 1, 1951, vol. I, p. xvii, LLA.

354 "Our restrained . . . place . . .": Ibid., p. vi.

354 unbearably bright light: Holloway (1994), pp. 196–219. For U.S. monitoring of the Soviet A-bomb test, I rely on Ziegler and Jacobson (1995); cf. Ziegler (1988).

355 took off from Misawa: Usually, the craft left from Yokote. Engine problems had forced change.

356 Bush: For Bush doings, see Kevles (1979), pp. 356–58.

356 "We have . . . U.S.S.R.": Quoted in *New York Times*, Sept. 24, 1949, p. 1.

356 "ATOM BLAST . . . DISCLOSED": Ibid.

357 "Johnny, quit . . . days": Rigden (1987), p. 201n.; Rigden interview with Alex de Ravel, Nov. 2, 1995.

357 debate over H-bomb: See Getting (1989), pp. 237–38; Dyson (1985), p. 138.

357 radar better technically, worse operationally: Valley interview, Dec. 19, 1994; Jacobs (1986), pp. 2–3.

357 George Valley: Valley's story, except where noted, is from lengthy interviews; Valley (1985).

357 "clip level": Valley interview, May 10, 1992.

358 "No, I . . . it": Ibid.

358 May-Johnson bill: A good summary is in Kevles (1979), pp. 349–50.

358 "I had . . . listen": Valley (1985), p. 197.

358 Bob Robertson: Some background from Kevles (1979), p. 318.

358 bare-bones warning net: For the state of air defense, mainly Schaffel (1990), esp. 70–73, 90–98; cf. Getting (1989), pp. 345–46.

359 "So when . . . all": Valley interview, May 10, 1992.

359 "Well, what . . . do?": Ibid.

359 CONAC: Founding from *History of Air Defense*, n.d., Air Defense Command press release, LLA.

359 "The site . . . hats": Valley (1985), p. 198.

359 AN/CPS-5: Characteristics in Fagen (1978), p. 85; Guerlac (1987), p. 460. Cornett (1973) has set at Otis AFB on Cape Cod, but photos show it at Truro, as Valley said. Deduction the CPS-5 was in Lashup net is derived from Schaffel (1990), p. 91, showing station map from the next year. Author thanks Dave Barton for discussion of benefits of longer wavelengths. Note that for optimum MTI, a radar needed to hit a quarry with as many pulses as possible in a single antenna scan, dictating some combination of relatively slow scanning and wide beamwidth, a function of antenna aperture area. Choosing too low an operating frequency meant building huge, costly radars. The UHF band around 600 MHz, or 50 cm, would have been ideal. But TV removed that option. That left L-band.

359 "guys fit . . . grass": Valley interview, Dec. 19, 1994.

360 "He started . . . attack": Valley (1985), p. 198.

360 "to find . . . problem": Quoted in ibid., p. 199.

360 "Regarding the . . . weeks": Quoted in ibid.

360 AFCRL: Location from Everett interview, Jan. 20, 1995; Liebowitz (1985).

361 Marchetti: Some background from Getting (1989), p. 351.

361 like a computer mouse: Cf. J. Pearson (1992), p. 7.

361 Lashup rapidly taking shape: Bruce-Briggs (1988), pp. 76–77; Fagen (1978), pp. 652–57; Schaffel (1990), p. 123. Disposition and type of early Lashup radars: Cornett (1973).

361 Plexiglas maps: Jacobs (1986), p. 2.

362 not peer much below 5000 feet: Chiefly, Valley (1985). Specifics on CPS-5 and 6B: Barton interviews; Barton (1975), p. 4; Guerlac (1987), pp. 461–65, 615–23; Schaffel (1990), pp. 120–21, 142–43, 159–60. Note that the GE-built 6B was another RL legacy. Dubbed V-beam due to shape of its two-beam pattern, it was the first "3-D" radar, capable of determining an aircraft's range, azimuth, and height. The L-band set also had an MTI. However, the radar relied on five distinct transmitters and receivers to cover various elevation sectors, which could render data ambiguous. All told, V-beam offered little range or low-altitude benefit over the CPS-5 and its trusty height finder.

362 "To attack . . . flyers . . .": Valley (1985), p. 204.

362 "to an . . . idiot-like' ": Quoted in Schaffel (1990), p. 150.

363 "The individual . . . time": Valley (1985), p. 205.

363 "Say, I . . . there": Quoted in ibid., p. 206.

364 "Now you're . . . George": Quoted in ibid., p. 207.

364 ENIAC et al.: Status of early digital computer projects in Redmond and Smith (1980), p. 166.

364 Wiesner had made reputation: JW's background and view of matters from Wiesner interview, May 22, 1992; Wiesner OHI, IEEE Center (1993), p. 433. Note that Valley is called "Elling," and Bowles, "Voles." This is transcriber's error probably due to JW's stroke-induced slurred speech.

364 Project Cadillac: Guerlac (1987), pp. 537–49.
365 estimated cost . . . $3 million: Redmond and Smith (1980), p. 166. For blow-by-blow account of funding woes, see pp. 126–28.
365 steady drone of WW's fans: Description from Everett and Valley interviews; Jacobs (1986), pp. 6–12; Redmond and Smith (1980), p. 184, and photos throughout.
366 Forrester background: Forrester (1992); Jacobs (1986), p. 41; Redmond and Smith (1980), pp. 14–15.
366 Everett background: Everett interviews; Redmond and Smith (1980), pp. 14–15.
366 somewhat infamous at MIT: Everett interview, Jan. 20, 1995; confirmed by other accounts.
366 "Neither one . . . lecturer": Ibid.
366 backing of Brown: Ibid.
366 "high signal . . . ratio": quoted in Redmond and Smith (1980), p. 136. Details of JF-RE management style spans pp. 132–36.
366 WW had started life: Forrester (1992), pp. 8–10; Jacobs (1986), pp. 8–9; Redmond and Smith (1980), pp. 1–15; cf. Bruce-Briggs (1988), p. 54.
367 party about to end: Everett interview, Jan. 20, 1995.
367 Friday, Jan. 27: Redmond and Smith (1980), pp. 174–75.
367 short for a big processor: As things turned out, it took decades before many computer needs outstripped a sixteen-bit register.
367 "They could . . . style": Valley interview, May 10, 1992.
368 "Horrifying": Ibid.
368 Counterbalancing discouraging first impression: WW's capabilities and Valley-Forrester encounter from Everett, Forrester, Valley interviews; Redmond and Smith (1980), esp. pp. 174–77; Valley (1985). JF's version is slightly different than GV's, but facts are generally the same. Valley (1985) indicates more than a few days elapsed before his return, but Redmond and Smith, pp. 174–75, fix the date from JF's notebook.
368 "So, how . . . need?": Valley interview, May 10, 1992.
369 "George, do . . . No": Ibid. Note that "kludge" was coined expressively in 1960, but Valley swears Marchetti used it.
369 WW still not a functioning computer: Redmond and Smith (1980), pp. 179–91; Valley (1985), p. 210.
369 "ambiguity in . . . out": Valley (1985), p. 210.
369 "appalled": Quoted in Schaffel (1990), p. 119; ibid. for Time account.
369 start construction of permanent radar system: air defense details from Schaffel (1990), pp. 113–15, 124–29.
370 managed Sandia: Fagen (1978), p. 656.
370 "probably Bell . . . spec": Valley interview, Dec. 19, 1994.
370 CONAC was run . . . by Whitehead: CONAC evolution from "History of Air Defense," press release, LLA. ADC was formed in March 1946, absorbed into CONAC in Dec. 1948 and reestablished Jan. 1, 1951; Schaffel (1990), p. 140.
371 balding man barely 5 feet 7: Background in Schaffel (1990), pp. 102–103.
371 "Darkter, my . . . mercy!": Valley (1985), p. 202.
371 "General, that . . . Clausewitz": Ibid.
371 "Well, Dark . . . metcha": Ibid.
371 deal would be done quickly: Valley (1985); cf. "Operational Plan Semiautomatic Ground Environment System for Air Defense," LLA. Western Electric accepted the task in Dec. 1950, though actual contract date is unclear; Fagen (1978), p. 546 ff.
371 "George, what . . . stupid": Ibid.
371 AFCRL-Harrington doings: Main sources; Harrington interview, Feb. 29, 1996; Harrington correspondence; Redmond and Smith (1980), pp. 179–92. Best description of DRR is Harrington (1983).
372 Ridenour pushed new air defense development center: Besides Valley's accounts, events are reconstructed from Bruce-Briggs (1988), pp. 49–50, 55; Getting (1989), pp. 225–33, 357–60; J. Goldstein (1992), p. 114; Schaffel (1990), pp. 81, 121–22, 141–46. AF chief scientist details from Dr. Walton Moody, AF History Office.

373 "It is . . . proposed": Quoted in Valley (1985), p. 212.
373 "willing to . . . want": Quoted in Schaffel (1990), p. 141.
374 MIT turned down proposal: MIT infighting over new lab drawn from Valley interviews; Valley (1985); Getting (1989), pp. 230–33; J. Goldstein (1992), pp. 71–75, 113–14; Leslie (1993), p. 33.
374 "a nasty . . . bitch": Valley interview, May 10, 1992. For Zacharias's view, see J. Goldstein (1992), pp. 114–15.
374 "What stuck . . . it": Quoted in R. Goldstein (1993), p. 113.
374 "It will . . . electronics": Ibid., citing 1982 Alfred P. Sloan Foundation video of PC participants, deposited at Smithsonian Institution.
375 "This was . . . reality": Valley interview, May 10, 1992.
375 recently acquired Lever Bros. building: Sloan Foundation grant for purchase came in late 1950 — management school public relations.
375 PC participants: PC, vol. I, p. xxi; vol. II, pp. 3–4, LLA.
375 first month spent in . . . : Ibid., vol. 1, p. v ff.
375 "Had Margaret . . . Physicists": Valley (1985), p. 213.
376 "Having Johnny . . . yes": Valley interview, May 10, 1992.
376 April 20 turning point: Everett, Valley interviews; Redmond and Smith (1980), p. 193; Valley (1985); Wieser (1983), p. 365; cf. PC, vol. 1, p. v ff; Leslie (1993), pp. 27–35.
376 "If I . . . too": Valley interview, Dec. 19, 1994.
377 "homeless": PC, vol. III, VII, 1–13, LLA.
377 PC recommendations: Scattered throughout volumes, but see "Summary of Major Conclusions," vol. 1, pp. xxiii–xxxii. For ground observer corps details, see Bruce-Briggs (1988), p. 64; Schaffel (1990), p. 128.
378 A site for facility: For early LL, I rely chiefly on E. Freeman (1995), pp. xiii–13; Getting (1989), pp. 233–34. Valley interviews fill in details, with LL organizational diagram, Aug. 1, 1952, supplied by lab. RL members listed in Newton et al. (1991), p. 203 ff. Also helpful: J. Goldstein (1992), pp. 115–16; Leslie (1993), pp. 33–35; Redmond and Smith (1980), pp. 198–200.
379 carried a sense of urgency: Many interviews; cf. E. Freeman (1995), intro.
379 PC members had been briefed: Valley interviews; PC, vol. I, pp. 3–4; cf. Schaffel (1990), p. 130.
379 "If not . . . him": PC, vol. 1, p. 4.

18 • The SAGE Age

380 "We tend . . . it": Everett interview, June 6, 1995.
380 "We invented . . . problem": Everett interview, Jan. 20, 1995.
380 Late afternoon of April 16: The incident Next to the Real Thing is from Schaffel (1990), pp. 169–71; cf. p. 148.
381 "We have . . . over": Quoted in ibid., p. 170.
381 "Tell 'em . . . thing": Quoted in ibid., p. 171.
381 "the very . . . evidence": Quoted in ibid., p. 171.
382 exodus from Cambridge to Lexington: Route 128 history from Lampe interview; LL specifics, unless otherwise noted, are from E. Freeman (1995), though organizational diagram, Aug. 1, 1952, is helpful. For LL's style, early division of labor, I also have drawn on Everett and Valley interviews; Jacobs (1986), pp. 126–27; Leslie (1993), p. 35.
382 Zacharias apparently had no interest: Cf. J. Goldstein (1992), p. 116.
382 Hill and Valley: Backgrounds from interviews; Who's Who in America entries. Styles, personalities, reasons for choosing Hill as LL director, from many interviews, readings; cf. Jacobs (1986), p. 88.
382 "would have . . . concerned": Hill interview, Jan. 12, 1993.
383 Valley hurt by the decision: GV's story, again except where noted, is from lengthy interviews; Valley (1985).
383 Hill saw GV as loose cannon: Hill interview, Jan. 12, 1993.
383 "Hill 'n Valley": From Bruce-Briggs (1988), p. 57. For more: Jacobs (1986), pp. 88–89.

383 CCN: Jacobs (1986), pp. 22–27; E. Freeman (1995), pp. 17–21; Harrington (1983); Valley (1985).
384 magnetic cores: Redmond and Smith (1980), pp. 183–85, 204–207; Pugh (1984), pp. 68–81; Everett interviews; Forrester interview, Jan. 30, 1995; Papian interview, Jan. 30, 1995; Olsen interview, March 15, 1996.
385 Olsen background: Olsen interview, March 15, 1996.
385 ADIS vs. LL: Valley interviews; Valley (1985), pp. 220–24; E. Freeman (1995), pp. 25–26. AF side: Schaffel (1990), pp. 199–202.
386 "If a . . . flounder . . .": Valley (1985), p. 221.
386 "But, instead . . . university": Ibid.
386 laboring through holidays: Everett interview, June 6, 1995.
387 "we stand . . . withdraw . . .": Quoted in E. Freeman (1995), p. 26; slightly different version in Schaffel (1990), p. 201.
387 spewed forth over loudspeakers: E. Freeman (1995), p. 26; Forrester interview, April 24, 1996.
387 "I was . . . Lincoln": Valley (1985), p. 222.
388 "Lincoln's power . . . effort": Schaffel (1990), p. 202. Valley (1985), p. 224, says ADIS, though mortally wounded, was not dead until late that year.
388 CCN operational trials: Lincoln Laboratory Review, Dec. 1954, pp. 3–5; E. Freeman (1995), pp. 18, 27; Jacobs (1983), p. 328.
389 MTC and magnetic cores: Everett interview, June 6, 1995; Jacobs (1983), pp. 324–25; Redmond and Smith (1980), pp. 206–207.
389 "We took . . . charm": Everett interview, June 6, 1995.
389 "Young kid . . . now": Olsen interview, March 15, 1996.
389 feast of leading-edge technology: Key is Annals of the History of Computing Special Issue: SAGE, vol. 5, no. 4 (Oct. 3, 1983): Astrahan and Jacobs (1983), pp. 342–43; Harrington (1983); Wieser (1983), pp. 366–68. Also helpful: E. Freeman (1995), pp. 18–22; Jacobs (1986), pp. 23–25, 74–75; LL Review, Aug. 1955, p. 3. Barton, Everett, Forrester interviews made it all clearer.
391 "This was . . . profession": Forrester interview, Jan. 30, 1995.
391 CCN operational: Astrahan and Jacobs (1983), pp. 342–43; Freeman (1995), pp. 22, 27–30; Jacobs (1983), p. 328; LL Review, Dec. 1954, pp. 3–5, Feb. 1955, p. 11, Aug. 1955, p. 4; Valley (1985), p. 215.
392 GV and AF figured: Valley interviews.
392 Building F construction: LL Review, Aug. 1955, pp. 5–7.
392 selection of IBM: Jacobs (1986), pp. 28–29, 34, 43–60; Everett interviews; time line in Jacobs (1983), p. 328. Exceptions noted.
393 "Oh, Karl . . . mills?": Pugh (1984), p. 76, anecdote and quote.
393 contract for FSQ-7: IBM press release, June 27, 1958, LLA; cf. E. Freeman (1995), pp. 22–25.
393 old necktie factory: Astrahan and Jacobs (1983), p. 344; cf. Pugh (1984), p. 95.
393 Manufacturing in Kingston, NY: IBM press release, June 27, 1958, LLA.
393 honeymoon with LL ended: Everett, Forrester interviews; cf. Pugh (1984), p. 102, for IBM perspective.
394 "You run . . . plant": Everett interview, Jan. 20, 1995.
394 IBM-LL trade workers: Jacobs (1986), pp. 60–62, 70–73.
394 "intolerable": Story and quote, Everett interview, Jan. 20, 1995.
394 XD-1 moves to LL: General details from E. Freeman (1995), pp. 23, 27–28; Jacobs (1986), pp. 51–58.
395 SAGE: Valley (1985), pp. 224–25; cf. E. Freeman (1995), p. 27.
395 "Where did . . . from?": Ibid., p. 225.
395 North American Van Lines: from photo, LL Review, Aug. 1955, pp. 6–7.
395 Experimental SAGE Subsector: General details from wide reading, interviews. See E. Freeman (1995), pp. 30–31.
395 three layers of radar coverage: General air defense plans from Schaffel (1990), pp. 152–56, 187–94, 209–20.

395 DEW Line: E. Freeman (1995), pp. 35–44; Fagen (1978), p. 455; cf. Bruce-Briggs (1988), pp. 56–57.

396 SAGE-CONAD integration: Schaffel (1990), pp. 246–53; "History of Air Defense," n.d., ADC press release; similar untitled ADC release, both from June 27, 1958, LLA; CONAD formed in Sept. 1954. In 1957, NORAD, the North American Air Defense Command that included Canada, came into being. Both coexisted until 1975, when CONAD was abolished: Alex de Ravel interview with NORAD assistant historian Jerry Schroeder, Oct. 26, 1995.

396 Red Book: Jacobs (1983), p. 328, (1986), p. 82; cf. E. Freeman (1995), p. 28.

396 identical FSQ-7s, direction center plans: Jacobs (1986), pp. 28, 130 ff., diagram opposite p. 98; Fagen (1978), pp. 574–80, diagram, p. 548.

396 JF initially estimated: Valley (1985), p. 220.

396 actual prototype held: Jacobs (1986), pp. 96–97; E. Freeman (1995), p. 31; Everett interviews.

396 excruciating workup: Everett interview, Jan. 20, 1995, including switch anecdote.

397 LL embroiled in controversy: Everett, Forrester, Valley interviews; E. Freeman (1995), p. 28; Jacobs (1986), pp. 86–89, 114.

397 Hill resigns: E. Freeman (1995), p. 28. Hill background from RLE Currents, vol. 4, no. 2 (Spring 1991): 14–15; Who's Who in America. SHAPE stands for Supreme Headquarters, Allied Powers in Europe.

397 JF quits: Forrester interview, Jan. 30, 1995; cf. Valley interview, Sept. 6, 1995.

397 "We had . . . indispensable": Ibid.

397 industrial dynamics: Forrester (1992); Forrester interview, Jan. 30, 1995; Pugh (1984), p. 127.

398 GV fired: Valley interviews. His official AF title was Chief Scientist to the Chief of Staff, USAF. For LL management shuffle, see E. Freeman (1995), p. 28; Jacobs (1986), pp. 86–89, 114.

398 "Both Whirlwind . . . it": Everett interview, Jan. 20, 1995.

398 SAGE specs: IBM press release, SAGE fact sheet, both June 27, 1958, LLA; E. Freeman (1995), pp. 23–25; Jacobs (1983), pp. 326, 329; cf. Pugh (1984), p. 126.

399 liked to mock SAGE: Barton interviews.

399 air games rigged: Cf. Bruce-Briggs (1988), p. 96.

399 FSQ-7 downtime: E. Freeman (1995), p. 25.

399 last SAGEs: Jacobs (1983), p. 329. Everett interview, June 6, 1995, for why SAGE lasted so long.

399 SAGE innovations: E. Freeman (1995), p. 33; Jacobs (1986), pp. 74–75, 168; J. Pearson (1992), p. 7; Redmond and Smith (1980), pp. 215–16.

400 SAGE contractors, subcontractors: IBM press release, June 27, 1958, LLA; Jacobs (1986), p. 54; Wieser (1983), p. 366.

400 first major digital network: Fagen (1978), pp. 546, 574–75, 579; Jacobs (1986), pp. 90–93.

400 Long Lines: Bruce-Briggs (1988), pp. 87–88.

400 SDC: E. Freeman (1995), p. 33; Leslie (1993), pp. 36–37; Everett interview, June 6, 1995.

400 DEC: J. Pearson (1992). Doriot background: Harvard Business School Bulletin, Oct. 1987, p. 13; Olsen interview, March 15, 1996.

401 IBM: SAGE figures mainly from Pugh (1984), p. 126; cf. Jacobs (1983), p. 329; Leslie (1993), p. 36. SABRE: Leslie, pp. 36–37; Pugh (1995), pp. 219, 326.

401 cores basis of . . . computer storage: E. Freeman (1995), p. 17.

401 manna from heaven: For importance of cores to IBM, I rely on Pugh (1984) throughout, esp. pp. 2–3, 76–77, 112–59.

401 "all the . . . tooling": Olsen interview, March 15, 1996.

402 JF-IBM-RCA patent fight: Ibid., pp. 81–92, 208–12; Forrester interviews.

403 MITRE formed: MITRE (1979), pp. 1–20; E. Freeman (1995), pp. 32–33; Jacobs (1986), p. 143. Leslie (1993), p. 39, suggests MIT REsearch; the Robert Everett joke: Bruce-Briggs (1988), p. 89.

404 AEW: E. Freeman (1995), pp. 36–40; Weiss interview, Feb. 9, 1995.
404 DEW: Fagen (1978), pp. 455–60; E. Freeman (1995), pp. 35, 41–42; Lindgren (1966); Weiss interview, April 5, 1995.
405 ICBM problem and civil defense responses: Bruce-Briggs (1988), pp. 101–18, 201; a few LL details in E. Freeman (1995), p. 47 ff.

19 • The Carpetbaggers

407 "The moon . . . you": Pettengill interview, Jan. 22, 1993.
407 "We were . . . carpetbaggers": Green interview, March 7, 1995.
407 Millstone Hill: Except where noted, I rely on Weiss interviews, esp. Feb. 9, April 14, 1995. Visit of April 14, 1995, and old photos help physical descriptions.
408 three strategically placed radars: Dustin interview, April 28, 1995.
409 "And then . . . it": Weiss interview, Feb. 9, 1995.
409 approaches to gaining decibels: In addition to Weiss, author is indebted to Dave Barton and Gordon Pettengill for help with technical details and calculations.
409 monster klystrons: General details, Shearman (1991), pp. 78–82; LL work: *LL Review*, Feb. 1955, p. 5; cf. E. Freeman (1995), pp. 48–49.
409 CG-24: see E. Freeman (1995), p. 48.
409 Millstone Hill circa 1957: From photos lining MH corridors.
410 *Sputnik*: For events surrounding launch, I rely on Bruce-Briggs (1988), p. 133; Lovell (1990), pp. 255–57.
410 "Can you . . . it?": Dustin interview, April 28, 1995.
410 LL watches *Sputnik*: Dustin, Oldel, Pettengill (April 5, 1995), Pippert, Weiss interviews set scene. Kraft et al. (1958) give tracking particulars. *Sputnik I* was captured by radar on Oct. 12, 15, and 17. Only screen photo found is from the 17th, showing the craft 590 miles away.
410 "We knew . . . horizon": Weiss interview, Feb. 9, 1995.
411 "We got . . . is!": Consensus of interviews.
411 ominous missile gap: For how wrong the perception was, Kaplan (1983).
412 Gaither report: Bruce-Briggs (1988), pp. 128–33; Halperin (1961); Kaplan (1983), pp. 125–49; Killian (1977), pp. 96–101. Rabi and EOL were on an advisory panel.
412 "If, as . . . shelters": Bruce-Briggs (1988), p. 132.
412 For LL . . . end of bomber defense: See E. Freeman (1995), p. xiv.
412 SAGE curtailed: Bruce-Briggs (1988), pp. 138–40.
412 Texas Towers wiped out: See R. Johnston (1961); anon. (1962).
412 Dustin and Pettengill, 1956: A *Comparison of Selected ICBM Early-Warning Radar Configurations*. LL Technical Report No. 127, LLA. For help understanding thinking of the day: Barton, Dustin, Weiss interviews.
413 BMEWS contract and development: E. Freeman (1995), p. 48. Scott (1974), p. 339. Dustin, Holmes, Weiss interviews critical for understanding a poorly documented story.
413 "They needed . . . watching": Dustin interview, April 28, 1995.
413 "Chinese copy": Weiss interview, Feb. 9, 1995.
413 "Glaciers were . . . desolate": Scott (1974), p. 339.
414 Fylingsdale Moor: Besides main sources, Dave Barton helped much with technical details.
414 monopulse radar: Some aspects were conceived at NRL as early as 1940. The essential patent, though, was filed by Robert Dicke soon after World War II. Several companies, notably RCA, pursued the technique, and it was widely applied in missile guidance and fire-control radar well into the 1990s: Dicke patent, 2,830,288; Barton (1991); Barton interviews; Skolnik (1980), p. 160, (1985), p. 185.
415 Pettengill: GP's story throughout chapter, except where noted, is from extensive interviews.
415 "This served . . . scientists": Pettengill interview, Jan. 2, 1993.
415 "I knew . . . pond": Ibid.
416 Project Ragmop: Ward (1992). Some details from Pettengill, Lerner, and Siebert inter-

views. FPS-17 employed pulse compression, or spread spectrum, technology—another technique patented in one form by Dicke after WWII. William Siebert, GP's boss, shared a prestigious Pioneer Award for helping develop the technique—see Cook and Siebert (1988).

416 surveillance techniques: Pettengill, Green interviews.

416 rising moon triggered alarm: Weiss interview, April 14, 1995; Bruce-Briggs (1988), p. 160, for Thule service date.

416 promised to get better: Green, Pettengill, Weiss interviews; specs in E. Freeman (1995), p. 48; R. Price et al. (1959). A klystron helped track *Sputnik II* in Nov. 1957, but problems forced removal: Kraft et al. (1958).

416 first maser amplifier put to practical use: Townes interview, Feb. 13, 1996.

417 might detect Venus: For hunt, I rely mainly on Green, Pettengill, R. Price interviews.

417 "carpetbaggers": Green interview, March 7, 1995.

417 10 million times fainter: The problem is with inverse square law: signal returning to receiver decreases as fourth power of the target's distance.

417 radar promised rich payoff: General discussion from Hey (1973), p. 122; F. Kerr (1952); Lovell (1990), pp. 277–79; Pettengill (1966), (1968), p. 275; R. Price et al. (1959). Cf. Shapiro (1968); J. Thomson et al. (1963), for astronomical unit.

418 "Lincoln was . . . astronomy": Green interview, March 7, 1995.

418 "If it . . . it": R. Price interview, April 21, 1995.

418 "What do . . . classified": Ibid.

419 *Radio Astronomy:* The book was Pawsey and Bracewell (1955).

419 take roughly five minutes: Experiment details from Pettengill, R. Price interviews; Pettengill (1966), (1968); R. Price et al. (1959); R. Price (1959); cf. LL press release, March 20, 1959, LLA.

419 "I looked . . . beam": R. Price interview, April 21, 1995.

420 "marginal . . . marginal": Pettengill interview, April 5, 1995.

420 "It is . . . observations": R. Price et al. (1959), p. 753.

420 front page: *New York Times*, March 20, 1959.

421 latest Venus hunt: Pettengill, R. Price interviews; Lovell (1990), esp. pp. 220–66, 278–79; Hey (1973), p. 122; Pettengill (1966).

421 turned toward moon: General discussion from Pettengill interviews, wide reading, esp. Evans and Hagfors (1968); Green and Pettengill (1960); Pettengill (1966); Pettengill et al. (1980).

422 "The phenomenon . . . water": Pettengill et al. (1980), p. 57.

422 Project Moonbounce: Hey (1973), p. 87; A. Price (1989), pp. 47, 89–90, 160–61, 278.

422 establishing novel techniques: GP's innovations drawn from Green, Pettengill interviews. Key paper is Pettengill (1960). Also important: Butrica (1996), pp. 332–33; J. Evans (1968); Green and Pettengill (1960); Pettengill (1966); cf. E. Freeman (1995), pp. 101–105.

424 Project Needles: Green interviews; E. Freeman (1995), pp. 65–67. Renamed Project West Ford, with last orbiting dipoles nearly gone by 1966.

424 "I was . . . them": quoted in Butrica (1996), pp. 332–33.

424 synthetic aperture radar: Barton, Green, Pettengill interviews helped explain basic difference between SAR and delay-Doppler Mapping. See Skolnik (1980), p. 185; cf. Cutrona (1985), 21.1–3.

425 Sullen and inscrutable: Third Venus hunt from Goldstein, Pettengill, R. Price, Renzetti interviews; Hey (1973), pp. 98, 122; Lovell (1990), pp. 279–80; Pettengill (1968), pp. 284–89; Pettengill et al (1962); Pettengill and Shapiro (1965), p. 383; Shapiro (1968), pp. 143–44, 166–68, 181–82; Thomson (1963). The JPL method of detecting Venus—involving comparisons between an object of interest and general background noise—is the same basic principle used by Robert Dicke and others in making radiometery measurements, and in fact was suggested by Goldstein's brother, Samuel, a radio astronomer.

427 "It certainly . . . effects": Pettengill interview, Jan. 22, 1993.

427 Arecibo: background: Corson interviews, April 1995; Dyce interviews, June 1995; Pettengill interviews, Jan. 1993.

428 radar astronomy revitalized study of planets: For overview, see Pettengill (1966); Pettengill et al. (1980); Shapiro (1968).

428 "The first . . . clear": Pettengill interview, Jan. 22, 1993.

428 Mercury a solar system enigma: For what was known, Hey (1973), p. 122; Pettengill (1968), pp. 275, 289–90, 300; Shapiro (1968), pp. 144, 180.

429 first good echoes: Mercury detection from Dyce, Pettengill interviews; Pettengill and Dyce (1965); cf. Hey (1973), pp. 122–23.

430 "Almost immediately . . . model": Pettengill interview, Jan. 22, 1993.

430 "Wait a . . . here": Quoted in Dyce interview, June 29, 1995.

430 solar tidal torques: Peale and Gold (1965).

430 "You don't . . . celebrate": Pettengill interview, Jan. 22, 1993.

431 Mercury locked to orbital rate: See Shapiro (1968), pp. 144, 180.

431 continued on auspicious path: See Pettengill (1968), p. 318; Pettengill et al. (1980); Shapiro (1968), pp. 169–70, 181–82.

431 Haystack: Landry, Weiss interviews, April 15, 1995; E. Freeman (1995), pp. 49, 101 ff.

432 "So my . . . cage": Quotes from Jan. 22, Feb. 1, 1993, interviews. Merged with GP's permission.

432 Pioneer Venus: Mainly, Pettengill interviews, Pettengill et al. (1980); planetary details also from J. Mason (1989).

433 "If it . . . water": Pettengill interview, April 5, 1995.

433 VOIR, Seasat: Murray (1989), pp. 125–29.

20 • A Distant View

435 "Astronomy compels . . . another": Quoted in Hugh Rawson and Margaret Miner, eds., The New International Dictionary of Quotations. (New York: Signet, 1988).

435 Magellan launch: Young (1990), esp. pp. 101–103; Magellan fact sheet, Nov. 9, 1990; "Magellan: Revealing the face of Venus," JPL 400–494, March 1993.

435 GP had flown to Florida: Pettengill's story throughout from interviews.

436 Magellan six years late: Besides Pettengill interviews, Magellan history and team makeup from Murray (1989), pp. 128–29; Young (1990), pp. 85–86.

436 "I was . . . do": Pettengill interview, April 5, 1995.

436 Mutch: Title from NASA public relations.

436 "is obviously . . . 1983": GP to Mutch, June 11, 1980, supplied by GP. Parens his.

437 "When I . . . basically": Pettengill interview, April 5, 1995.

437 scrubbed Pioneer Venus: Pettengill interviews; Young (1990), p. 9, notes mission canceled in 1982, but GP says word was out long before.

438 mortified Americans: Pettengill interview, April 5, 1995; correspondence from John Macdonald, Nov. 20, 1995, Michael Desandoli, Nov. 21, 1995. Some Seasat details: Murray (1989), p. 125; Cantafio (1990), pp. 22.1–4.

438 placed under microscope: Pettengill interviews; Young (1990), pp. 9–11, 52.

439 BAQ: W. T. K. Johnson interviews, April 7, June 16, 1995; e-mail correspondence with Johnson; Pettengill interviews; cf. Young (1990), p. 7.

439 "It was . . . nothing": Pettengill interview, Feb. 1, 1993.

439 mission renamed: Young (1990), pp. 10–11, 81–83; Johnson interview, June 16, 1995; Pettengill interview, April 5, 1995.

439 one last gamble: Pettengill interviews, Feb. 1, 1993, April 5, 1995.

440 "Plus, space . . . project": Pettengill interview, Feb. 1, 1993.

440 midroute corrections, tests: Young (1990), pp. 107–11.

440 Mission Support Area: Kasuda, Neuman interviews, April 13, 1993.

440 "There were . . . air": Kasuda interview, April 13, 1993.

440 Magellan weight before and after firing solid rocket motor: Neuman interview, March 13, 1996.

441 GP eagerly awaited: Pettengill interviews.

441 "That looks . . . spacecraft": Pettengill interview, April 21, 1993.

441 *Magellan* mission plan: Pettengill, Kasuda, Neuman interviews. NASA/JPL fact sheets supply supporting figures.
441 Everything started out fine: For stories of lost contact, I rely, except where noted, on Kasuda, Neuman interviews, April 13, 1993; Kasuda interviews, April 19, 1993, April 18, 1996.
442 "We were . . . out": Kasuda interview, April 13, 1993.
443 "would just . . . Ever": Kasuda interview, April 19, 1993.
443 "It was . . . again": Neuman interview, April 13, 1993.
443 mapping for real: Pettengill interview, April 21, 1993.
444 21 Earth expeditions: Young (1990), p. 4. She has 20, but *Galileo* had not yet reached Venus.
444 three voyagers carrying radar: Summarized in ibid., pp. 4–7; details filled in with many interviews.
444 "the hills . . . geology": Ibid., p. 6.
445 Three key science functions: Pettengill interviews; NASA/JPL fact sheets.
445 The prime mission: Discussion of scientific debates surrounding Venus comes from interviews with the principals; wide reading.
446 "It's a . . . lost": Plaut interview, April 15, 1993.
446 "That is . . . Venus": Schaber interview, April 23, 1993.
446 "There's a . . . ages": Phillips interview, March 25, 1996.
447 "That would . . . magma": Solomon interview, April 21, 1993.
447 trouble struck yet again: Kasuda, Neuman, Pettengill interviews.
447 "aerobraking": Technical discussion: Lyons interviews, correspondence. Science aspects come from many interviews, esp. Plaut, Sjogren, Solomon, Turcotte. Also helpful: *Magellan* fact sheet, Sept. 30, 1994; press release, Oct. 12, 1994, 9462.
448 One last radar effort: Ford interviews; Pettengill interview, April 5, 1995.
449 coated in mystery material: discussion from Ford, Wood interviews and correspondence, chiefly in April 1995.
450 vast job of processing: Ford interview, April 5, 1995; Pettengill interview, April 23, 1995; NASA/JPL press release, Oct. 12, 1994, draft.
450 Magellan's last days: Neuman interview, March 13, 1996; Pettengill interviews, April 5, 23, 1995; NASA/JPL press releases, Oct. 12, 1994; *Magellan* fact sheet, Sept. 30, 1994.

Epilogue: A Quiet Glory

451 "You did . . . women!": Cohen, transcripts of Commission on Awards to Inventors, EGBN, 5/2, p. 28, CAC.
451 "Memories crowd . . . not": quoted in Clark (1965), p. 366.
451 W-W drove with Jean: Watson-Watt (1959), pp. 229–30; *Toronto Globe and Mail*, Oct. 21, 1954, p. 1; correspondence from Tony Drew.
451 "How did . . . speed": Watson-Watt (1959), pp. 229–30.
452 "what Scot . . . concerned?": Ibid., p. 230.
452 "Pity Sir . . . it": Anon., courtesy of Tony Drew.
452 "Radars are . . . out": Delaney, videotaped lecture filmed Sept. 1991, LL.
453 Royal Commission: EGBN, 5/2, 5/44, CAC; R. Brown (1991), pp. 91–92; Ratcliffe (1975), p. 566; also see W-W to V. Bush, May 7, 1951, Guerlac papers, 14/17/2354, DRMC.
453 "I only . . . work": R. Brown (1991), p. 91.
453 "It was . . . done": Ibid.
453 "I have . . . myself": EGBN 5/2, transcript, p. 28, CAC.
454 "You did . . . women!": Ibid.
454 "That was . . . radar": Ibid.
454 "If I . . . estimate": EGBN, 5/44, transcript, p. 21, CAC.
454 W-W's share presumed largest: London *Times*, Dec. 7, 1993, p. 21.
455 *Jane's:* 1995–96 edition.
455 Bell Labs: Cf. Fagen (1978); much on missiles in Bruce-Briggs (1988).

455 second generation of radar techniques: wide reading; Barton interviews; Barton (1984); Skolnik (1985).

456 Gulf war: I rely, except where noted, on M. Gordon and Trainor (1995); supplemented with A. Price, "Spoof and punch," version of article for *Asia Pacific Defence Review*, Nov./ Dec. 1993. Cote interviews, July 19, Aug. 8, 1995, invaluable for making sense of things.

456 Three AWACS planes: Dunnigan and Bay (1992), pp. 160, 167.

456 J-STARS: Ibid., pp. 196–97; M. Gordon and Trainor (1995), p. 266–68; cf. Friedman (1991), pp. 175, 187–88.

457 Noncooperative target recognition: Cote interviews.

457 Firefinders: Ibid.; Dunnigan and Bay (1992), p. 286; Cote interviews.

457 "If the . . . doomed": Dunnigan and Bay (1992), p. 286.

457 "We'd get . . . return": quoted in Blackwell (1991), p. 16.

458 120 km wide on Mercury: See Bond (1992), pp. 37–39; Paige (1994).

458 in Seasat vein: Wide reading, esp. *Eos*; NASA/JPL SIR-C/X-SAR fact sheet, Dec. 1994.

458 *Cassini*: Johnson interviews; Young (1990), pp. 170–71; NASA/JPL fact sheet, May 1995.

458 ground-penetrating radars: Wide reading, interviews; cf. Colorado School of Mines news release, "CSM Tunnel Detection Expert Seeks to Save Time, Money, Lives," Nov. 21, 1991; Cohn (1993).

458 low-powered broadband radars: See Ajluni (1994); Flanagan (1995); Stover (1995).

458 Air traffic control: General history and discussion from Rochester (1976); *FAA Administrator's Fact Book*, July 1995; cf. Whitnah (1966).

459 Secondary Surveillance Radars: Weiss interviews; E. Freeman (1995), pp. 235–45; Trim (1988), (1994); cf. *FAA Fact Book*; R. Kent (1980), pp. 189–90.

459 "I just . . . up": Weiss interview, Aug. 23, 1995.

460 Terminal Doppler Weather Radar: Wide reading; cf. E. Freeman (1995), pp. 240–41; "Terminal Doppler Weather Radar (TDWR): A briefing paper," FAA July 1, 1988; Phillips (1994); Raytheon news releases. For 1996 status, FAA communication of March 18, 1996; Michelson interviews.

460 NEXRAD: Wide reading, esp. *EOS*; cf. R. Kerr (1993); Skolnik (1991), pp. 13–14. S. Johnston interview, March 11, 1996. Latest status from National Oceanic and Atmospheric Administration.

461 "For the . . . this": Lewis OHI, Box 49A, NANER.

461 radar guns: Press materials for Cincinnati Microwave Inc., courtesy Fran Dym Communications Inc., NYC.

461 Watson-Watt: Drew, Reburn interviews; also Watson-Watt (1959), pp. 422–25. For details on postwar career: R. Brown interviews, correspondence; Ratcliffe (1975), p. 552; *Who's Who in America*.

462 "a community . . . us": Watson-Watt (1959), p. 422.

462 "We are . . . 'unidentified' ": Ibid., pp. 424–25.

462 "I don't . . . bankruptcy": Drew interview, Aug. 28, 1995.

462 Sterling Forest days: A bit from Joan and John White interviews, Sept. 22, 1994.

463 "I honestly . . . shock": Drew interview, Aug. 28, 1995.

463 W-W last days: Beyond main sources, see McElheny (1973); Bowen (1987), p. x, foreword by Lord Bowden.

463 "For those . . . victory . . .": Quoted in Ratcliffe (1975), p. 549.

463 "From the . . . kid": Drew interview, Aug. 28, 1995.

463 Tizard and Cherwell: For HT's last days, I rely, except where noted, on R. Clark (1965), esp. pp. 386–87, 417. For Cherwell's: Birkenhead (1961), pp. 267, 295–337; R. Jones (1957).

464 "wildly impracticable": Snow (1961), p. 34.

464 "He believed . . . judgment": Snow (1962), p. 35.

464 set the record straight: R. Jones (1961a, b, c); cf. McElheny (1973).

465 "an injustice": R. Jones (1961b), p. 15.

465 R. V. Jones: I rely, except where noted, on R. Jones (1978), esp. p. 520 ff.; cf. R. Jones (1990); a few details from *Who's Who*.

465 "This is . . . land?": Quoted in R. Jones (1978), p. 521.
465 "among the . . . life": Ibid.
465 "I told . . . low . . .": Ibid., p. 522.
465 Alvarez carrot story: Related by Richard Muller.
466 sadly, he could not claim credit: Jones interview, March 8, 1993. Medhurst's role is from Jones correspondence of May 6, 1996.
466 "Awarded for . . . freedom": Inscription and other award details from Jones interview, April 23, 1996; CIA public information.
466 Hanbury Brown: R. Brown interview, Aug. 25, 1995; R. Brown correspondence, Sept. 29, 1995; Who's Who.
466 "boffin": Etymology related in R. Clark (1962), pp. vii-viii.
466 Denis Robinson: D. Robinson, H. Robinson interviews; a few details on HVEC from Fortier interview, Aug. 23, 1995.
466 "You're lucky . . . year": Robinson interview, Feb. 19, 1994.
467 Alfred Loomis: Alvarez (1980), (1983); Who's Who in America.
467 something of a scandal: Conversations with George Boynton, James Degnan, Kathy Norris, Tuxedo Park, Sept. 20, 1994.
467 "As often . . . strained": Fortune, March 1946.
467 "an extraordinarily . . . kitchen": Alvarez (1983), p. 33.
467 "the last . . . science": Alvarez (1980), p. 309.
467 AL lost Loran claim on appeal: Many documents found in 47, 8, ELB, LC. Most pertinent is U.S. Court of Customs and Patent Appeals, Appeals Nos. 6300, 6301, decision of Jan. 31, 1956. However, the patent was issued to Loomis, April 28, 1959, no. 2,884,628, cf. Loomis Papers, Box 1, MC 264, MIT. Repeated attempts to find record at U.S. Patent Office have been fruitless.
467 Loomis sons: Alvarez (1980), pp. 313–14; Farnsworth obit, New York Times, Aug. 12, 1975, p. 32; Alfred Jr. obit, New York Times, Sept. 13, 1994, p. B11; Henry entry in Who's Who in America; Quillen interview, May 18, 1996.
467 Lee DuBridge: I rely mainly on "Memories," CTA. Also helpful, Caltech press release/ obit, Jan. 23, 1994.
468 pitted him against Teller, Griggs, Alvarez: Goodchild (1981), pp. 255–56.
468 "It was . . . false": DuBridge OHI with J. Goodstein, Feb. 19, 1981, Session 1, Tape 1, p. 67, CTA.
469 Vannevar Bush: A good overview of VB's life is Zachary (1991a, b). The best source on Bush's impact on science policy is Kevles (1979), esp. pp. 343–66. These should be supplemented with Bush (1945), (1968), (1970); VB's lengthy OHI, MC 143, MIT, and bio, in MC 78, MIT.
469 staunch defense of Oppenheimer: Goodchild (1981), pp. 212–14, 236, 250.
469 "being interpreted . . . system": Quoted in ibid., p. 25.
469 "I think . . . am": Quoted in Zachary (1991b), p. 103.
470 "an instant . . . hit": Quoted in Kevles (1979), p. 347.
470 "I do . . . wounds": Quoted in Zachary (1991b), p. 102.
471 clear lesson: Quoted in Leslie (1993), p. 24.
471 "more likely . . . minimum": Quoted in Zachary (1991b), p. 100.
471 "Even Bush . . . Foundation": Kevles (1979), p. 364.
471 Edward Bowles: Accomplishments from résumé, 5, 11, ELB; "Misc. Biographical Info. — Edward Lindley Bowles," MIT memo, Sept. 7, 1990; G. Fowler (1990).
472 ELB relationship with MIT: See K. Compton to ELB, March 7, 1946, ELB to KC, July 2, 1946; J. Killian to ELB, May 14, June 12, 1947; ELB to Bush, July 2, 1947 — all in Bowles file, Container 13, VB. The $400,000 contribution is from ELB to P. Gray, Sept. 22, 1980, 57,7, ELB.
472 "I want . . . spirits": ELB to VB, July 12, 1948, Bowles file, Container 13, VB.
472 "Bush was . . . harder": Wildes and Lindgren (1985), p. 109.
473 Luis Alvarez: Alvarez (1987), pp. 103–10; Muller (1988); Hurley (1988). For more on the Nobel, see Stevenson (1987).
473 "We landed . . . Germany": Alvarez (1987), p. 106.

473 JFK assassination: Alvarez (1976); Garwin (1987).
473 dinosaur extinctions: See Angier (1985).
473 Robert Dicke: Dicke interviews.
474 Taffy Bowen: Bowen's life from R. Brown et al. (1992); interviews, correspondence, with family and friends. Early CSIRO days from Home (1988), pp. xv–xvi; Robertson (1992), pp. 29–32; Sullivan (1988), pp. 311–13; interviews with Bowen colleagues.
475 first time a meeting . . . outside Europe or U.S.: Sullivan (1988), p. 330.
475 Parkes: Central are Bowen (1984), pp. 98–103; Robertson (1992). Also helpful: visit to Parkes, April 20, 1994; interviews with Bowen colleagues, Bowen OHI with Sullivan, Dec. 24, 1973, pp. 29–36; CSIRO publications, "The Parkes Radio Telescope," and "CSIRO's Australia Telescope."
475 "men of . . . science": Bowen (1984), p. 99.
476 "Perhaps, it . . . world": Ibid., p. 100.
477 Pawsey decided: For JP's life, see Lovell (1964).
477 After retiring from CSIRO: Bowen's late life primarily from interviews with David and Edward Bowen, R. Brown; R. Brown correspondence, Sept. 29, 1995.

Glossary

AI	Airborne Interception radar.
AI Mark IV	Most common and successful British longwave airborne interception radar, operating at a wavelength of 1.5 meters.
Airborne Interception	Intercepting an aircraft from another plane while in flight. In radar, refers to successfully detecting and locating intruder.
Air-to-Surface-Vessel	Airborne radar designed for spotting ships and submarines.
AN	Army-Navy equipment.
AN/FSQ-7	IBM-built digital computer, modeled after Whirlwind, that formed the brains of SAGE.
A-Scope	Radar screen designed to show range only, from left to right, with targets appearing as horizontal blips.
ASD	"Dog." 3-cm Philco-built ASV developed at Rad Lab.
ASG	"George." Philco 10-cm ASV radar developed at Rad Lab.
ASV	Air-to-Surface-Vessel radar.
ASV Mark II	Most common and successful British longwave ASV set. Operated at 1.5 meters. Used successfully with Leigh Lights to find U-boats in Bay of Biscay in summer 1942.
Azimuth	Angle of a target in the horizontal plane as measured clockwise, usually from true north.
Bearing	Angle of target in the horizontal plane in relation to some understood reference direction.
BMEWS	Ballistic Missile Early Warning System. Network of long-range warning and tracking radars for watching for missile attacks along northern approaches to United States.
B-Scope	Radar screen presenting range and bearing in rectangular coordinates.
Cape Cod Network	Trial radar and communications network to serve as testing ground for Whirlwind and what became the SAGE concept.

Cavity Magnetron	Microwave radar transmitter.
Chaff	Metal strips with high radar reflectivity designed to confuse enemy radars. Known to British as Window.
Chain Home	Network of British meter wave radars employing separate transmitting and receiving towers rising as high as 300 feet. Used for air warning in the Battle of Britain and throughout World War II.
Chain Home Low	British radar net used to spot planes flying below the coverage zone of Chain Home system.
Clutter	Unwanted and interfering radar returns from waves, trees, mountains, buildings, and other objects.
Conical Scan	Special aircraft tracking technique in which antenna traces a small, circular motion that makes it possible to compare sequential returns and obtain highly accurate angular information about a target.
Continuous Wave Radar	A radar system that relies on a steady beam of energy instead of brief pulses. Often used as an electronic sentry that gives warning when a plane or ship interrupts the beam.
Counter-counter-measures	Systems or steps, such as changing a radar set's operating frequency, for defeating an enemy's countermeasures. Also called ECCM, for electronic counter-countermeasures, or RCCM, for radio counter-countermeasures.
Counter-measures	Systems such as jammers for fooling or defeating an enemy's radar and electronics. Also called ECM, for electronic countermeasures; or RCM, for radio countermeasures.
Crystal	Microwave-sensitive semiconductor, usually silicon or galena, that formed the heart of the radar detector.
C-Scope	Radar screen showing bearing and elevation relative to the center of a scanned region.
DEW Line	Distant Early Warning Line. Network of radars for guarding northern approaches to U.S. from bomber attack.
Dipole	Antenna with two elements, each $\frac{1}{4}$ wavelength long, through which energy radiates.
DMS-1000	Early Rad Lab experimental 10-cm ASV radar.
Doppler Shift	Change in frequency of radar signals due to motion of observer or target.
Duplexer	A T-R box.
Eagle	3-cm blind bombing radar invented by Luis Alvarez. Achieved superior resolution through an extremely long antenna.
GCA	Ground Controlled Approach.
GCI	Ground Control of Interception.
Ground Controlled Approach	Radar-based system for controlling aircraft landing invented by Luis Alvarez and Lawrence Johnston.
Ground Control of Interception	Radar-based technique developed by British for directing fighter aircraft to intercept enemy bombers.
Gunlaying	In radar, a system for transmitting bearing and range information to anti-aircraft guns or artillery and controlling fire.

H₂K	1.25-centimeter blind-bombing and navigation radar.
H₂S	10-cm blind-bombing and navigation radar.
H₂X	3-cm blind-bombing and navigation radar. Also called Mickey.
Identification Friend or Foe	Radar-like interrogator/transponder system for distinguishing between friendly and enemy planes or ships.
IFF	Identification, Friend or Foe.
Jamming	Intentional electronic interference with radio or radar equipment.
J-Scope	Circular radar screen showing range only.
Klystron	Electron tube considered best bet for a microwave radar transmitter before the cavity magnetron arrived on the scene. Widely used in World War II as a local oscillator. After the war, the power klystron supplanted magnetron transmitters.
Loran	LOng-RAnge Navigation. Pulsed radio signal system for navigation of ships and planes.
MEW	Microwave Early Warning.
Microwave Early Warning	Large phased array radar with unique dipole arrangement, used for fighter control and early warning. Another Alvarez brainstorm.
Mixer	Stage of a radar receiver where the detected signal voltage and a voltage from a local oscillator are combined to produce an intermediate frequency signal that can be amplified for display on a cathode ray tube. In World War II microwave radars, a semiconductor crystal.
M-9	Bell Labs analog computer, or predictor, for calculating anti-aircraft gun firing coordinates and shell settings.
Monopulse	A technique for deriving high angular tracking accuracy through a single radar pulse.
Moving Target Indication	The use of Doppler-shifted returns to identify moving targets amidst clutter.
MTI	Moving Target Indication.
Phased Array	An antenna that forms and controls beams by assigning phases to a series of radiating elements.
Plan Position Indicator	Circular cathode ray display in which a sweep, or trace, moves in circles from the screen center, painting in target positions and ranges in direct relationship to the radar operator's position.
Proximity Fuze	Miniature radar in gun shell nose that senses when a target is near and detonates the shell. Used against planes and ground targets. Also called VT Fuze or Pozit.
Pulse	Brief pulse of energy transmitted from radar set. On the order of microseconds in length in World War II systems.
Pulse Compression	A technique in which long pulses are used to increase received signal energy while maintaining the resolution of short pulses.
Radio Direction Finding	Early British code name for radar designed to imply a relatively conventional system for tracking enemy positions by homing in on the source of radio signals.
Range	Distance to a target.

RDF	Radio Direction Finding.
SAGE	Semi-Automatic Ground Environment.
SCR	Signal Corps Radio. Designated an Army radar set. Later replaced by the AN designation.
SCR-584	Rad Lab–developed microwave gunlaying radar. Employed conical scanning and Bell Labs M-9 predictor for highly accurate gun control.
Semi-Automatic Ground Environment	Digital computer–run radar, communications, and fighter control network for protecting the United States from bomber attack.
Side Lobe	Unwanted signals or wave patterns spilling out of an antenna and robbing the main beam of power.
Synthetic Aperture Radar	Radar that uses the motion of a plane or satellite to simulate the aperture of an extremely long antenna, and thereby get much greater resolution. Also called side-looking radar, or SAR.
Texas Towers	Radar-carrying platforms off the Atlantic seaboard that extended U.S. warning coverage eastward.
TR Box	Transmit-Receive box. System permitting the use of a single antenna for transmitting and receiving radar signals. A special tube prevented transmitted pulses from burning out the receiver crystal, then quickly recovered to allow detected signals to get through.
Waveguide	A hollow pipe, often rectangular, used to funnel radar signal to and from antenna.
Whirlwind	Early digital computer prototype for controlling U.S. air warning and fighter interception network.
Window	British name for chaff.

Interviews

No site indicates telephone interview. Correspondence and short conversations are not included.

Henry Abajian: March 22, 1994; May 11, 1995
Ann Alexander: Sept. 7, 1994
Sally Atkinson: April 22, 1994, North Ryde, NSW, Australia
Kenneth Bainbridge: 1994 — date uncertain
Edythe Baker: May 12, 1995
David Barton: Jan. 13, March 16, 1995; March 4, 1996, all in Lexington, MA
Hans Bethe: Nov. 7, 1994, Ithaca, NY
Brebis Bleaney: May 20, 1994
Nicolaas Bloembergen: on or about Dec. 7, 1995
David Bowen: April 21, 22, 1994
Edward Bowen: April 23, 1994
Vesta Bowen: May 18, 1994
Louis Brown: May 22, 1992, Washington, DC
Robert Hanbury Brown: Aug. 25, 1995
Kenneth Carson: March 27, 1996
Brian Cooper: April 18, 1994, Turramurra, NSW, Australia; April 22, 1994
Dale R. Corson: Oct. 4, 1994; Nov. 7–8, 1994, Ithaca, NY; April 17–18, 1995, Ithaca, NY
Owen R. Cote Jr.: July 19, 1995; Aug. 8, 1995, Cambridge, MA
Lee Davenport: March 31, 1994; April 6, 1994; Nov. 30, 1995
James Degnan: Sept. 20, 1994, Tuxedo Park, NY
John H. DeWitt Jr.: Dec. 29, 1992; Oct. 9, 1995
Robert Dicke: April 1, Nov. 11–12, 1992, Princeton, NJ; March 27, 1992; Nov. 29, 1995
Tony Drew: Aug. 28, 1995
Lee A. DuBridge: May 17, 1992, Pasadena, CA
John Dunbar: June 7, 1996
Dan Dustin: April 28, 1995
Rolf Dyce: June 28, 29, 30, 1995
Ron Ekers: April 18, 1994, North Ryde, NSW, Australia
Robert R. Everett: Jan. 20, June 6, 1995, both in Bedford, MA
Harold Ewen: June 17, 1994; July 27, 1994, South Deerfield, MA; Oct. 24, 1995; April 25, 1996
Morris E. Fine: 1994, Cambridge, MA
J. O. Fletcher: June 1, Sept. 2, 1994
Peter Ford: April 5, 1995, Cambridge, MA; April 7, 10, 1995

Jay Forrester: Jan. 30, 1995, Cambridge, MA; April 24, 1996
Ron Fortier: Aug. 23, 1995
Ivan Getting: March 31, Aug. 8, 1994
Alfred Goldberg: Nov. 22, 1994
Richard M. Goldstein: March 19, 1996
James P. Gordon: July 6, 1994
Paul E. Green Jr.: March 7, 15, 1995; April 3, 1996
Norman Greig, M.B.E.: March 12, 1993, Bentley Priory, Stanmore, Britain
Erwin L. Hahn: Feb. 13, 1996
John V. Harrington: Feb. 29, 1996
Paul Hartman: April 17, 1995, Ithaca, NY
Ray Haynes: April 22, 1994, North Ryde, NSW, Australia
Fred J. Heath: Sept. 8, Nov. 11, 1994; April 29, 1995
Albert Hill: Jan. 12, 1993, Needham, MA
D. Brainerd Holmes: April 12, 1995
Nick Holonyak Jr.: Nov. 23, 1994
William T.K. Johnson: April 7, June 16, 1995
Lawrence Johnston: Jan. 11, 25, April 8, Aug. 19, Sept. 23, 1994
Stephen L. Johnston: March 11, 1996
R. V. Jones: March 8, 1993; April 23, 1996
Rick Kasuda: April 13, 1993, Littleton, CO; April 19, 1993; March 12, April 18, 1996
Frank Kerr: Nov. 30, 1995
Norm Krim: May 20, 1994, Waltham, MA
David Lampe: n.d.
Raymond Landry: April 14, 1995, Westford, MA
Robert M. Lerner: March 1, 1995
Henry Loomis: April 15, 1993; on or about May 2, 1994
Daniel T. Lyons: April 15, 1993, Pasadena, CA; April 20, 1993
David W. McCall: Jan. 24, 1996
Max Michelson: March 13, 26, 1996
Bernard Mills: April 21, 1994, Roseville, NSW, Australia; April 22, 1994
Harry Minnett: April 18, 1994, North Ryde, NSW, Australia
Jim Neuman: April 13, 1993, Littleton, CO; March 13, 1996
Charles Newton: May 17, 1992, Pasadena, CA
Robert Novick: on or about Oct. 26, 1995
Dennis Oldel: April 14, 1995, Westford, MA
Kenneth Olsen: March 15, 1996
Rubina Otrupcek: April 20, 1994, Parkes, NSW, Australia
William Papian: Jan. 30, 1995
Gordon Pettengill: Jan. 21, 22, Feb. 1, 1993; April 5, 1995; April 17, 1996 — all in Cambridge, MA; April 21, 1993; April 23, 1995
Roger J. Phillips: April 16, 1993; March 25, 1996
J. R. Pierce: Aug. 4, 1992; Jan. 18, 1993
Glen F. Pippert: May 31, 1995
Jeffrey J. Plaut: April 15, 1993, Pasadena, CA; April 21, 1993
Ernest Pollard: Aug. 29, 30, 1994
Barry Posen: July 10, 1995
R. V. Pound: June 22, July 3, 1992; Nov. 10, 16, 1992 — all in Cambridge, MA; Nov. 17, 1992; June 29, 1994; Feb. 6, 1996
Edmund Preston: Aug. 23, 1995
Marcus Price: April 19, 1994, Parkes, NSW, Australia
Robert Price: April 13, 1995; April 21, 1995, Lexington, MA
Beth Purcell: Dec. 10, 1995, Cambridge, MA; April 25, 1996
Edward Purcell: March 26, Nov. 4, 1992, both in Cambridge, MA; March 30, Oct. 5, 1992; May 11, 1995
Jacqueline Loomis Quillen: May 18, 1996

Norman Ramsey: March 30, 1992, Cambridge, MA; July 2, 1992; Oct. 12, 1993, Cambridge, MA; Aug. 16, 1994; Dec. 10, 1995, Cambridge, MA; Feb. 24, 1996

Dennie Reburn: Aug. 29, 1995

Nicholas Renzetti: March 19, 1996

Denis Robinson: July 30, 1992, Woods Hole, MA; Oct. 7, 1993; Feb. 19, 20, 1994

Harald D. Robinson: Sept. 12, 1995

Peter M. Rolph: May 30, 1996

Gerald Schaber: April 23, 1993

Frederick Seitz: May 16, 1995; March 25, 1996

William M. Siebert: March 1, 1995

Mal Sinclair: April 22, 1994, Dover Heights, NSW, Australia

William L. Sjogren: April 15, 1993, Pasadena, CA; April 19, 1995

Bruce Slee: April 22, 1994, Dover Heights, NSW, Australia

Louis Smullin: Sept. 8, 1994, Cambridge, MA

Sean Solomon: May 22, 1992, Washington, DC; April 21, 1993

Elizabeth MacEvoy Sparks: June 10, 1995; March 12, 1996

Morgan Sparks: June 12, 1995; March 12, 1996

Gordon Stanley: Oct. 26, 1992

Malcom W.P. Strandberg: July 21, 1994; Oct. 16, 18, Nov. 14, 1995

Henry Torrey: Nov. 12, 1992, New Brunswick, NJ; Oct. 11, 1993; Oct. 25, 1994

Charles H. Townes: Aug. 4, 1992; July 15, 16, 1994, Berkeley, CA; July 19, 1995; Sept. 14, 1995, Berkeley, CA; Feb. 13, 1996

Richard M. Trim: May 26, 1994

Donald Turcotte: March 13, 1996

George Valley Jr.: May 8, 1992; May 10, 1992, Concord, MA; Aug. 12, 1993; Oct. 29, 1993; Dec. 19, 1994, Concord, MA; Dec. 22, 1994; Sept. 6, 1995, Concord, MA; Feb. 20, 1996

Robert von Mehren: Oct. 27, 1994

Herbert Weiss: Feb. 9, 1995, Lexington, MA; April 14, 1995, Westford MA; Aug. 23, 1995; Dec. 28, 1995, Lexington, MA

Joan White: Sept. 22, 1994

John White: Sept. 22, 1994

Jerome Wiesner: May 22, 1992, Cambridge, MA

John A. Wood: April 13, 1995

Keith Wood: June 10, 1996

Robert B. "Chris" Wren: March 12, 1993, Uxbridge, Britain; Dec. 20, 1993

Vincent R. Zann: March 29, 1996

Herbert J. Zeiger: March 20, 1995, Sept. 19, 20, 1995

Bibliography

Ajluni, Cheryl. 1994. "Low-cost wideband spread-spectrum device promises to revolutionize radar proximity sensors." *Electronic Design*, July 25, 35, 38.

Allen, W. C. 1978. *Anzio: Edge of disaster.* Talisman/Parrish Bowles, (U.S. edition, New York: Dutton).

Allison, David Kite. 1981. *New eye for the Navy: The origins of radar at the Naval Research Laboratory.* NRL Report 8466. Washington, D.C.: Naval Research Laboratory.

Alvarez, Luis W. 1976. "A physicist examines the Kennedy assassination film." *Americal Journal of Physics*, 44:813; reprinted in Trower, ed., *Discovering Alvarez*, 210–24.

———. 1980. "Alfred Lee Loomis." *Biographical Memoirs*, National Academy of Sciences, vol. 51: 309–41.

———. 1983. "Alfred Lee Loomis — last great amateur of science." *Physics Today*, January, 25–34.

———. 1987. *Alvarez: Adventures of a physicist.* New York: Basic Books (paperback).

Ambrose, Stephen E. 1994. *D-Day. June 6, 1944: The climactic battle of World War II.* New York: Simon & Schuster.

Angier, Natalie. 1985. "Did comets kill the dinosaurs?" *Time*, May 6, 72–83.

anon. 1936. " 'Feelers' for Ships." *Wireless World*, 38 (June 26): 623–24.

anon. 1943. "The first public account of radar." *Bureau of Naval Personnel Information Bulletin*, June, 10–12.

anon. 1945. "Fateful night." *New Yorker*, Aug. 18, 15–16.

anon. 1946a. "Alfred Lee Loomis: Amateur of the sciences." *Fortune*, March.

anon. 1946b. "The multi-cavity magnetron: A high-frequency power generator for radar," *Bell System Technical Journal*, June, 219–23.

anon. 1948. "The transistor." *Bell Laboratories Record*, vol. XXVI, no. 8 (August): 321–24.

anon. 1962. "Air Force may abandon 2 radar towers in Atlantic." *New York Times*, Dec. 27, 7.

Appleton, E. V., and M. A. F. Barnett. 1925. "On some direct evidence for downward atmospheric reflection of electric rays." *Proceedings of the Royal Society* A, vol. 109: 621–41.

Astrahan, Morton M., and John F. Jacobs. 1983. "History of the design of the SAGE computer — the AN/FSQ-7." In *Annals of the History of Computing Special Issue: SAGE*, vol. 5, no. 4 (Oct.): 340–49.

Atlas, David A. 1990. *Radar in meteorology: Battan memorial and 40th anniversary radar meteorology conference.* Boston: American Meteorological Society.

Austin, Pauline M., and Spiros G. Geotis. 1990. "Weather radar at MIT." In Atlas, *Radar in meteorology*, 22–31.

Baldwin, Ralph. 1980. *The deadly fuze: Secret weapon of World War II.* San Rafael, Calif.: Presidio.

Bardeen, John, and Walter H. Brattain. 1948. "The transistor, a semi-conductor triode." *The Physical Review*, vol. 74, no. 2 (2nd series) (July 15): 230–31.

Barton, David K. 1959. "*Sputnik II* as observed by C-band radar." *IRE 1959: Convention record*, part 5, 67–73.

———. 1975. "Real-world radar technology." IEEE International Radar Conference., 1–22.

———. 1984. "A half century of radar." *IEEE Transaction on Microwave Theory and Techniques*, vol. MTT-32, no. 9 (September): 1161–70.

———. 1991. "History of monopulse radar in the U.S." Unclassified Memo No. DKB-91–37. Also in Blumtritt, Petzold, and Aspray, eds., *Tracking the history of radar*, 61–91.

Basalla, George. 1988. *Evolution of technology*. Cambridge, Eng., and New York: Cambridge University Press (paperback).

Batt, R. G. P. 1991. "Why ten centimetres?" In Rolph, ed., 1991. *Fifty years of the cavity magnetron*.

Baxter, James Phinney, III. 1947. *Scientists against time*. Boston: Little, Brown.

Beam, Alex. 1988. "A building with soul." *Boston Globe*, June 29. Reprinted in *RLE Currents*, vol. 4, no. 2 (Spring 1991): 10–11.

Behrens, Charles W. 1978. "The development of the microwave oven." *Appliance Manufacturer*. November.

Bekker, Cajus (pen name of Hans Dieter Berenbrok). 1974. *Hitler's naval war*. Trans. by Frank Ziegler. London: Macdonald, 304–38.

Benington, Herbert D. 1983. "Production of large computer programs." In *Annals of the History of Computing Special Issue: SAGE*, vol. 5, no. 4 (Oct.) 350–61.

Berkner, Lloyd V. 1946. "Naval airborne radar." *Proceedings of the I.R.E. and Waves and Electrons*, vol. 34 (September): 671–706.

Bernstein, Jeremy. 1984. *Three degrees above zero: Bell Labs in the information age*. New York: Scribner's.

Bertolotti, M. 1983. *Masers and lasers: An historical approach*. Bristol, Eng.: Hilger.

Beser, Jacob. 1988. *Hiroshima and Nagasaki revisited*. Memphis, Tenn.: Global Press.

Beyerchen, Alan. 1994. "On strategic goals as perceptual filters: Interwar responses to the military potential of radar in Germany, the UK and the US." In Blumtritt, Petzold, and Aspray eds., *Tracking the history of radar*, 267–83.

Birkenhead, Frederick, second earl of. 1961. *The Prof in two worlds: The official life of professor F. A. Lindemann, viscount Cherwell*. London: Collins.

Blackett, P. M. S. 1960. *Tizard and the science of war*. Tizard Memorial Lecture delivered on Feb. 11, 1960, to Institute for Strategic Studies.

———. 1962. *Studies of war*. New York: Hill and Wang.

Blackwell, James. 1991. "Georgia punch: 24th Mech puts the squeeze on Iraq." *Army Times*, Dec. 2, 16.

Bleaney, B. 1984. "Electron paramagnetic resonance: The early days." *Physics Bulletin*, vol. 35: 466–70.

Bloch, Felix. 1964. "The principle of nuclear induction." In *Nobel Prize Lectures in Physics 1942–1962*. New York: Elsevier, 203–18.

———, W. W. Hansen, and Martin Packard. 1946. "Nuclear induction." *The Physical Review* 69: p. 127.

Blumtritt, Oskar, Hartmut Petzold, and William Aspray, eds. 1994. *Tracking the history of radar*. Piscataway, N.J.: Institute of Electrical and Electronics Engineers.

Bolton, J. G. 1982. "Radio astronomy at Dover Heights." *Proceedings of the Astronomical Society of Australia*, vol. 4, No. 4: 349–58.

Bond, Peter. 1992. "Radar explorers of the solar system." *New Scientist* (Dec. 5): 37–40.

Boot, H. A. H. 1946 (estimated). "Development of the multi-resonator magnetron in the University of Birmingham (1939–1945)," HAHB, folder A.1, NAEST, 61.

Bowden, Lord. 1985. "The story of IFF (identification friend or foe)." *IEE Proceedings*, vol. 132, pt. A, no. 6 (October): 435–37.

Bowen, E. G. 1981. "The pre-history of the Parkes 64-m telescope." *Proceedings of the Astronomical Society of Australia*, vol. 4, no. 2: 267–73.

———. 1984. "The origins of radio astronomy in Australia." In Sullivan, ed., *The early years of radio astronomy*, 84–111.

———. 1987. *Radar days*. Bristol, Eng.: Hilger.

Bracewell, R. N. "Early work on imaging theory in radio astronomy." In Sullivan, ed., *The early years of radio astronomy*, 167–90.

Brakenridge, G. Robert, et al. 1994. "Radar remote sensing aids study of the great flood of 1993." *EOS*, vol. 75, no. 45 (Nov. 8): 521, 526–27.

Brand, Stewart. 1994. *How buildings learn*. New York: Viking.

Braun, Ernest. 1992. "Selected topics from the history of semiconductor physics and its applications." In Hoddeson et al., eds., *Out of the crystal maze*, 443–88.

Breuer, William B. 1988. *The secret war with Germany*. Novato, Calif.: Presidio.

Brittain, James E. 1985. "The magnetron and the beginnings of the microwave age." *Physics Today*, July, 60–67.

———. 1990. "The evolution of electrical and electronics engineering and the Proceedings of the IRE: 1938–1962." *Proceedings of the IEEE*, vol. 78, no. 1: 5–30.

Brooks, John. 1975. *Telephone*. New York: Harper & Row.

Brown, Anthony Cave. 1977. *Bodyguard of lies*. New York: Harper & Row.

Brown, Louis. n.d. Unpublished draft manuscript on history of radar.

———. 1992a. "America enters the radar age." Paper presented at the Munich workshop on radar history, Dec. 14–16, Deutsches Museum.

———. 1992b. "The state of research on the history of American radar." Paper presented at the Munich workshop on radar history, Dec. 14–16, Deutsches Museum.

———. 1993. "The proximity fuze." *IEEE AES Systems Magazine*, July, 3–10.

———. 1994. "Significant effects of radar on the Second World War." In Blumtritt, Petzold, and Aspray, eds., *Tracking the history of radar*, 121–35.

Brown, Robert Hanbury. 1984. "Paraboloids, galaxies and stars: Memories of Jodrell Bank." In Sullivan, ed., *The early years of radio astronomy*, 213–36.

———. 1991. *Boffin*. Bristol, Eng.: Hilger, IOP Publishing.

———, H. C. Minnett, and F. W. G. White. 1992. "Edward George Bowen." *Biographical Memoirs of the Royal Society*, vol. 38, 43–65. Reprint.

Bruce-Briggs, B. 1988. *The shield of faith: A chronicle of strategic defense from zeppelins to Star Wars*. New York: Simon & Schuster.

Bryant, John H., "UK-USA-Canadian networking in microwave electron tubes and radar, 1939–1945. "In Rolph, ed., 1991. *Fifty years of the cavity magnetron*, 71–77.

Buderi, Robert. 1988. "Raking it in at Patent U." *The Scientist*, vol. 2, no. 23: 4, 27.

———. 1993. "Magellan's last mission." *New Scientist*, May 22, 28–33.

———. 1994a. "The darkest days." *New Scientist* supplement, Feb. 19, 31.

———. 1994b. "The V-1 menace." *New Scientist*. June 4, 28–32.

Buell, Thomas B. 1980. *Master of sea power*. Boston: Little, Brown.

Burcham, W. E. 1985. "The development of centimetre AI." In *IEE Proceedings*, vol. 132, pt. A, no. 6 (October): 385–93.

Burchard, John. 1948. *QED: M.I.T. in World War II*. New York: Technology Press.

Burns, Russell, ed. 1988. *Radar development to 1945*. London: Peter Peregrinus.

———. 1995. "Deception, technology and the D-day invasion." *Engineering Science and Education Journal*, vol. 4, no. 2 (April): 81–88.

Bush, Vannevar. 1945. *Science—the endless frontier*. Washington, D.C.: U.S. Government Printing Office.

———. 1968. *Modern arms and free men*. Cambridge: MIT Press.

———. 1970. *Pieces of the action*. New York: Morrow.

Butrica, Andrew J. 1996. "To see the unseen: A history of planetary radar astronomy." Unpublished manuscript.

Callick, E. B. 1990. *Metres to microwaves*. London: Peter Peregrinus.

———. 1991."Early microwave receivers." In Rolph, ed., *Fifty years of the cavity magnetron*, 18–22.

Canfer, B. J., et al. 1990. *History of Bentley Priory*. Royal Air Force special edition to commemorate the 50th anniversary of the Battle of Britain.

Cantafio, Leopold J. 1990. "Space-based radar systems and technology." In Skolnik, ed., *Radar handbook*.

Cantelon, Philip L. 1995. "The origins of microwave telephony—waves of change." *Technology and Culture*, vol. 36, no. 3 (July): 560–82.

Carroll, John M. 1970. *The story of the LASER*. New York: Dutton.

Carter, R. H. A. 1988. "A personal reminiscence: RDF and IFF." In Burns, ed., *Radar development to 1945*, 150–61.

Chalfant, Betsy. 1992. "50 years ago . . . ," column in *Reflections*, newsletter of the Historical Electronics Museum, Inc., vol. 2, issue 1 (February): 3.

Christiansen, W. N. 1984. "The first decade of solar radio astronomy in Australia." In Sullivan, ed., *The early years of radio astronomy*, 112–31.

Churchill, Winston S. 1941. *Blood, sweat, and tears*. New York: G. P. Putnam's Sons.

————. 1948. *The gathering storm*. Boston: Houghton Mifflin (paperback).

————. 1949. *Their finest hour*. Boston: Houghton Mifflin (paperback).

————. 1950a. *The grand alliance*. Boston: Houghton Mifflin (paperback).

————. 1950b. *The hinge of fate*. Boston: Houghton Mifflin (paperback).

————. 1951. *Closing the ring*. Boston: Houghton Mifflin (paperback).

————. 1953. *Triumph and tragedy*. Boston: Houghton Mifflin. (paperback).

Clark, Ronald W. 1962. *The rise of the boffins*. London: Phoenix House Ltd.

————. 1965. *Tizard*. London: Methuen.

Clark, Trevor. 1980. "How Diana touched the moon." *IEEE Spectrum*, May, 44–48.

Cockburn, Sir Robert. n.d. "A survey of British countermeasures during World War II.

————. 1985. "The radio war." *IEE Proceedings*, vol. 132, pt. A, no. 6, (October): 423–34.

Cockcroft, J. D. 1985. "Memories of radar research," *IEE Proceedings*, vol. 132, pt. A, no. 6 (October): 327–39.

Cohn, D'Vera. 1993. "A technology-eye view." *Washington Post*, Nov. 5: B1.

Cole, Hugh M. 1965. *The Ardennes: Battle of the Bulge*. Vol. 3, pt. 7 of *United States Army in World War II: The European theater of operations*. Washington: Office of the Chief of Military History Department of the Army.

Collier, Basil. 1957. *The defence of the United Kingdom*. London: Her Majesty's Stationery Office.

Colton, Roger B. 1945. "Radar in the United States Army." *Proceedings of the IRE*, vol. 33, 740–53.

Cook, Charles E., and W. M. Siebert. 1988. "The early history of pulse compression radar." *IEEE Transactions on Aerospace and Electronic Systems*, vol. 24, no. 6 (November): 825–37.

Cornett, Lloyd H., Jr. 1973. *Aerospace defense command statistical data book radar (U)*. Vol. III (April). Colorado Springs: Aerospace Defense Command History Office.

Craven, Wesley Frank, and James Lea Cate, eds. 1953. *The Pacific: Matterhorn to Nagasaki June 1944 to August 1945*. Vol. 5 of *The Army Air Forces in World War II*. University of Chicago Press.

Cutrona, L. J. 1990. "Synthetic aperture radar." In Skolnik, ed., *Radar handbook*.

Devereux, Tony. 1994. "Strategic aspects of radar at sea." In Blumtritt, Petzold, and Aspray, eds., *Tracking the history of radar*, 157–70

DeWitt, John H., Jr., and E. K. Stodola. 1949. "Detection of radio signals reflected from the moon." *Proceedings of the I.R.E.*, March, 229–42.

Dicke, Robert H. n.d. "The measurement of thermal radiation at microwave frequencies." *Radiation Laboratory Report 787*.

————. 1946. "The measurement of thermal radiation at microwave frequencies." *The Review of Scientific Instruments* 17, no. 7 (July): 268–75.

————, and Robert Beringer. 1946. "Microwave radiation from the sun and moon," *Astrophysical Journal*, 103, no. 3 (May): 375–76.

Dicke, Robert, et al. 1946. "Atmospheric absorption measurements with a microwave radiometer." *The Physical Review* 70: 340–48.

Donaldson, Ralph J. 1990. "Foundations of severe storm detection by radar." In Atlas, *Radar in meteorology*, 115–21.

DuBridge, L. A., and L. N. Ridenour. 1945. "Expanded horizons." *Technology Review*, vol. XLVIII, no. 1 (November).

Dunnigan, James F., and Austin Bay. 1992. *From shield to storm.* New York: Morrow.

Dyson, Freeman. 1981. *Disturbing the universe.* New York: Harper Colophon.

————. 1985. *Weapons and hope.* New York: Harper Colophon.

Edge, David O., and Michael J. Mulkay. 1976. *Astronomy transformed: The emergence of radio astronomy in Britain.* New York: Wiley.

Emory, John M. G. 1986. *The source book of World War II aircraft.* Poole, Eng.: Blandford Press.

Englund, Carl R., Arthur B. Crawford, and William W. Mumford. 1933. "Some results of a study of ultra-short-wave transmission phenomena." *Proceedings of the Institute of Radio Engineers*, vol. 21, no. 2: 464–93.

Evans, D. L., et al. 1993. "The shuttle imaging Radar-C and X-SAR mission." *EOS*, vol. 74, no. 13 (March 30): 145, 157–58.

————. 1994. "Mission employs synthetic aperture radar to study global environment." *EOS*, Sept. 6, 410, 415, 420.

Evans, John V. 1969. "Radar studies of planetary surfaces." In *Annual Review of Astronomy and Astrophysics*, vol. 7: 201–48.

————, and Tor Hagfors, eds. 1968. *Radar astronomy.* New York: McGraw-Hill.

Everett, Robert R. 1983. "SAGE—A data-processing system for air defense." In *Annals of the History of Computing Special Issue: SAGE*, vol. 5, no. 4 (Oct.): 330–39.

Ewen, H. I. 1953. "Radio waves from interstellar hydrogen." *Scientific American*, vol. 189 (December): 42–46.

————, and E. M. Purcell. 1951. "Observation of a line in the galactic radio spectrum: Radiation from galactic hydrogen at 1,420 Mc./sec." *Nature*, vol. 168 (Sept. 1): 356.

Fagen, M. D., ed. 1978. *A history of engineering and science in the Bell System: National service in war and peace (1925–1975).* New York: Bell Telephone Laboratories.

Fahnestock, Mark, et al. 1993. "Greenland ice sheet surface properties and ice dynamics from ERS-1 SAR imagery." *Science*, vol. 262 (Dec. 3): 1530–34.

Farago, Ladislas. 1962. *The tenth fleet.* New York: Obolensky.

Farnett, Edward C., and George H. Stevens. 1990. "Pulse compression radar." In Skolnik, ed., *Radar handbook.*

Fennessy, Sir Edward. 1985. "Introductory paper" for "The History of Radar Development to 1945," seminar sponsored by the Institution of Electrical Engineers, London, June 10.

Fisher, David E. 1988. *A race on the edge of time.* New York: McGraw-Hill.

Fisk, J. B., H. D. Hagstrum, and P. L. Harman. 1946. "The magnetron as a generator of centimeter waves." *Bell System Technical Journal*, vol. XXV, no. 2 (April).

Flanagan, Ruth. 1995. "The pocket radar revolution." *New Scientist*, Aug. 12, 22–26.

Fleischer, Dorothy A. 1991. "The MIT Radiation Laboratory: RLE's microwave heritage." *RLE Currents*, vol. 4, no. 2 (Spring 1991): 1–11.

Fletcher, J. O. 1990. "Early developments of weather radar during World War II." In Atlas, *Radar in meteorology*, 3–6.

Forman, Paul. 1995. " 'Swords into ploughshares': Breaking new ground with radar hardware and technique in physical research after World War II." *Reviews of Modern Physics*, vol. 67, no. 2 (April): 397–455.

Forrester, Jay W. 1992. "From the ranch to system dynamics: An autobiography." In Arthur G. Bedeian, ed., *Management laureates: A collection of autobiographical essays*. vol. 1, Greenwich, Conn.: JAI Press. Manuscript copy supplied by author.

Fowler, Charles, John Entzminger, and James Corum. 1990. "Assessment of ultra-wideband (UWB) technology." *IEEE AES Systems Magazine*, November, 45–48.

Fowler, Glenn. 1990. "Edward L. Bowles, 92, engineer who helped U.S. develop radar." *New York Times*, Sept. 7, p. D19.

Freeman, Eva, ed. 1995. *MIT Lincoln Laboratory: Technology in the national interest.* Lexington, Mass.: Lincoln Laboratory.

Freeman, Joan. 1991. *A passion for physics: The story of a woman physicist.* Bristol, Eng.: Hilger.

Freeman, Roger A. 1986. *The mighty Eighth: Units, men and machines: A history of the US 8th Air Force.* London: Jane's (Rev. ed.).

———, with Alan Crouchman and Vic Maslen. 1990. *The mighty Eighth war diary.* London: Arms and armour (Rev. ed.).

Friedman, Norman. 1991. *Desert victory.* Annapolis, Md.: Naval Institute Press.

Garfinkel, Simson. 1991. "Building 20: The procreative eyesore." *Technology Review,* November/December, 9–13.

Garwin, Richard L. 1987. "Examining the Kennedy assassination evidence." In Trower, ed. *Discovering Alvarez,* 203–24.

Gerstein, Mark. 1994. "Purcell's role in the discovery of nuclear magnetic resonance: Contingency versus inevitability." *American Journal of Physics* 62, no. 7 (July): 596–601.

Getting, Ivan A. 1989. *All in a lifetime: Science in the defense of democracy.* New York: Vantage.

Giessler, Captain Helmuth. 1961. "The breakthrough of the 'Scharnhorst'—some radio-technical details." *IRE Transactions on Military Electronics,* January, 2–7.

Gilbert, Martin. 1989. *The Second World War: A complete history.* New York: Holt.

Goddard, D. E., and D. K. Milne, eds. 1994. *Parkes: Thirty years of radio astronomy.* East Melbourne, Austria: CSIRO Publications.

Goldstein, Jack S. 1992. *A different sort of time: The life of Jerrold R. Zacharias scientist, engineer, educator.* Cambridge: MIT Press.

Goldstein, Richard M., et al. 1993. "Satellite radar interferometry for monitoring ice sheet motion: Application to an Antarctic ice stream." *Science,* vol. 262, (Dec. 3): 1525–30.

Goodchild, Peter. 1981. *J. Robert Oppenheimer: Shatterer of worlds.* Boston: Houghton Mifflin.

Gordon, J. P. 1955. "Hyperfine structure in the inversion spectrum of $N^{14}H_3$ by a new high-resolution microwave spectrometer." *The Physical Review,* vol. 99, no. 4: 1253–63.

———, H. J. Zeiger, and C. H. Townes. 1955a. "Molecular microwave oscillator and new hyperfine structure in the microwave spectrum of NH_3." *The Physical Review,* vol. 95: 282–84.

———. 1955b. "The maser—new type of microwave amplifier, frequency standard, and spectrometer." *The Physical Review,* vol. 99, no. 4: pp.1264–74.

Gordon, Michael R., and Bernard E. Trainor. 1995. *The general's war.* Boston: Little, Brown.

Gorter, C. J. 1967. "Bad luck in attempts to make scientific discoveries." *Physics Today* 20 (January): 76–81.

Gould, Jack. 1946a. "Contact with moon achieved by radar in test by the Army." *New York Times,* Jan. 25, 1.

———. 1946b. "Radar code to the planets envisioned for the future." *New York Times,* Jan. 26, 1.

———. 1946c. "Moon is late for demonstration of how it is reached by radar." *New York Times,* Jan. 28, 21.

Green, Paul E., Jr., and Gordon H. Pettengill. 1960. "Exploring the solar system by radar." *Sky and Telescope,* July, 9–14.

Guerlac, Henry E. 1950. "The radio background of radar." *Journal of the Franklin Institute,* vol. 250, no. 4 (October): 285–308.

———. 1987. *Radar in World War II.* Tomash Publishers, American Institute of Physics.

Hagfors, Tor, and John V. Evans. 1968. "Radar studies of the moon." In Evans and Hagfors, eds., *Radar astronomy,* 219–73.

Halperin, Morton H. 1961. "The Gaither Committee and the policy process." *World Politics,* vol. XIII, no. 3 (April): 360–84.

Harrington, John V. 1983. "Radar data transmission." In *Annals of the History of Computing Special Issue: SAGE,* vol. 5, no. 4 (October): 370–74.

Harrison, Tom. 1976. *Living through the Blitz.* London: Collins.

Hartcup, Guy, and T. E. Allibone. 1984. *Cockcroft and the atom.* Bristol, Eng.: Hilger.

Hastings, Max. 1979. *Bomber Command.* New York: Dial/James Wade.

———. 1984. *Overlord.* New York: Simon & Schuster (Touchstone paperback).

Hazen, Harold Locke. 1976. *Memoirs: an informal story of my life and work.* Cambridge, Mass.: Massachusetts Institute of Technology Archives.

Henriksen, Paul W. 1987. "Solid state physics research at Purdue." *OSIRIS*, 2nd series, 3: 237–60.

Herring, Conyers. 1992. "Recollections from the early years of solid-state physics." *Physics Today*, April, 26–33.

Hey, J. S. 1946. "Solar radiations in the 4–6 metre radio wave-length band." *Nature*, vol. 157 (Jan. 12).

———. 1973. *The evolution of radio astronomy*. New York: Science History Publications.

Hezlet, Vice Admiral Sir Arthur. 1975. *Electronics and sea power*. New York: Stein and Day.

Hinsley, F. H. 1994. "Introduction: The influence of Ultra in the Second World War." In Hinsley and Stripp, eds., *Codebreakers*, 1–13.

———, and Alan Stripp, eds. 1994. *Codebreakers: The inside story of Bletchley Park*. Oxford: Oxford University Press (paperback).

Hoddeson, Lillian Hartmann, 1977. "The roots of solid-state research at Bell Labs." *Physics Today*, March, vol. 30, no. 3, 23–30.

———. 1981. "The discovery of the point-contact transistor." *Historical Studies in the Physical Sciences*, vol. 12, no. 1: 41–76.

———, et al., eds. 1992. *Out of the crystal maze*. New York: Oxford University Press.

Holloway, David. 1994. *Stalin and the bomb*. New Haven: Yale University Press.

Holonyak, Nick, Jr. 1992. "John Bardeen and the point-contact transistor." *Physics Today*, April, 36–43.

———. 1993. "John Bardeen." *Memorial Tributes*, National Academy of Engineering of the United States of America. vol. 6: 3–11.

Home, R. W. 1988. Science on service, 1939–1945. In Home, ed., *Australian science in the making*, 220–51.

———, ed. 1990. *Australian science in the making*. Cambridge, Eng.: Cambridge University Press (paperback).

Hudson, R. Scott, and Steven J. Ostro. 1994. "Shape of asteroid 4769 Castalia (1989 PB) from inversion of radar images." *Science*, vol. 263. (Feb. 18): 940–42.

Hunt, Franklin L. 1943. "New buildings of Bell Telephone Laboratories at Murray Hill, N.J." *Journal of Applied Physics*, vol. 14 (June): 249–57.

Hurley, Charles. 1988. "Luis Alvarez, noted scientist, dies in Berkeley." Lawrence Berkeley Laboratory press release, Sept. 1.

IEEE Center for the History of Electrical Engineering. 1993. *Rad Lab: Oral histories documenting World War II activities at the MIT Radiation Laboratory*. Piscataway, N.J.: Institute of Electrical and Electronics Engineers.

Ivanov, Alex. 1990. "Radar guidance of missiles." In Skolnik, ed., *Radar handbook*.

Jacobs, John F. 1983. "SAGE overview." In *Annals of the History of Computing Special Issue*: SAGE, vol. 5, no. 4 (Oct.): 323–29.

———. 1986. *The SAGE Air Defense System: A personal history*. Bedford, Mass.: The MITRE Corp.

Jacobsen, H. A., and J. Rohwer, eds. 1965. *Decisive battles of World War II: The German view*. Trans. André Deutsch Ltd. New York: G. P. Putnam's Sons. First American edition.

James, Jamie. 1995. "Shuttle radar maps ancient Angkor." *Science*, vol. 267 (Feb. 17): 965.

Johnston, Lawrence. 1987. "The war years." In Trower, ed. *Discovering Alvarez*, 55–71.

Johnston, Richard J. H. 1961. "Colonel cleared in fall of tower," *New York Times*, Aug. 25, 23.

Jones, H. 1960. "Herbert Wakefield Banks Skinner." *Biographical Memoirs of the Fellows of the Royal Society*, vol. 6: 250–68.

Jones, R. V. 1957. "The Right Hon. Viscount Cherwell, P.C., Ch. H., F.R.S." Obituary in *Nature*, vol. 180, no. 4586 (Sept. 21): 579–81.

———. 1961a. "Scientists at war — Lindemann v. Tizard." Part I. London *Times*. Sept. 6, 13.

———. 1961b. Scientists at war — Lindemann v. Tizard. Part II. London *Times*, Sept. 7, 15.

———. 1961c. Scientists at war — Lindemann v. Tizard. Part III. London *Times*. Sept. 8, 9.

———. 1966. "Winston Leonard Spencer Churchill." *Biographical memoirs of the Fellows of the Royal Society*, vol. 12: 35–105.

———. 1978. *The Wizard War: British scientific intelligence 1939–1945*. (American edition of *Most Secret War*.) New York: Coward, McCann & Geoghegan.

————. 1990. *Reflections on intelligence*. London: Mandarin Paperbacks.

Kahn, David. 1967. *The Codebreakers*. New York: Macmillan.

Kaplan, Fred. 1983. *The Wizards of Armageddon*. New York: Simon & Schuster (Touchstone paperback).

Katz, Isadore, and Patrick J. Harney, 1990. "Radar meteorology at Radiation Laboratory, MIT, 1941 to 1947." In atlas, *Radar in meteorology*, 16–21.

Kellerman, K., and B. Sheets, eds. 1983. *Serendipitous discoveries in radio astronomy*. Green Bank, W.V.: National Radio Astronomy Observatory/Associated Universities.

Kennett, Lee. 1982. *A history of strategic bombing*. New York: Scribner's.

Kent, Richard J., Jr. 1980. *Safe, separated, and soaring*. Washington, D.C.: U.S. Department of Transportation.

Kern, Ulrich. 1994. "Review concerning the development of German radar technology up to 1945." In Blumtritt, Petzold, and Aspray, eds., *Tracking the history of radar*, 171–83.

Kerr, Donald E., ed. 1951. *Propagation of short radio waves*. Vol. 13, Radiation Laboratory Series. New York: McGraw-Hill.

Kerr, Frank J. 1952. "On the possibility of obtaining radar echoes from the sun and planets." *Proceedings of the IRE*, vol. 40 (June): 660–66.

————. 1984. "Early days in radio and radar astronomy in Australia." In Sullivan, ed., *The early years of radio astronomy*, 132–45.

Kerr, Richard A. 1993. "Upgrade of storm warnings paying off." *Science*, vol. 262 (Oct. 15): 331–33.

Kevles, Daniel J. 1975. "Scientists, the military, and the control of postwar defense research: The case of the Research Board for National Security, 1944–46." *Technology and Culture* 16 (January): 20–47.

————. 1979. *The Physicists: The history of a scientific community in modern America*. New York: Vintage.

Killian, James R., Jr. 1977. *Sputnik, scientists, and Eisenhower*. Cambridge, Mass.: MIT Press.

Kinsey, Gordon. 1981. *Orfordness — secret site*. Lavenham, Eng.: Terence Dalton.

————. 1983. *Bawdsey — birth of the beam*. Lavenham, Eng.: Terence Dalton.

Kraft, L. G., et al. 1958. "Summary of satellite tracking activities." Lincoln Laboratory Group report No. 312–18 (July 25).

Kraus, John D. 1966. *Radio astronomy*. New York: McGraw-Hill.

Kümmritz, Herbert. 1988. "German radar development up to 1945." In Burns, ed., *Radar development to 1945*, 209–26.

————. 1994. "On the development of radar technologies in Germany up to 1945. In Blumtritt, Petzold, and Aspray, eds., *Tracking the history of radar*, 25–46.

Leslie, Stuart W. 1993. *The Cold War and American science: The military-industrial-academic complex at MIT and Stanford*. New York: Columbia University Press.

Liebowitz, Ruth P. 1985. *Chronology: from the Cambridge Field Station to the Air Force Geophysics Laboratory 1945–1985*. Bedford, Mass.: Air Force Geophysics Laboratory.

Lindgren, Nilo. 1966. "Ten years of the DEW Line." *IEEE Spectrum*, December, 57–62.

Liner, Christopher L., and Jeffrey L. Liner. 1995. "Ground-penetrating radar: A near-face experience from Washington County, Arkansas." *The Leading Edge*. vol. 14, no. 1 (January): 17–21.

Llewellyn, F. B. 1986. "Greetings from the Institute of Radio Engineers, address delivered at the Radiolocation convention March 26, 1946." *Journal of the IEE* 93, pt. 1 (September): 382–84.

Longmate, Norman. 1981. *The Doodlebugs: The story of the flying bombs*. London: Hutchinson.

Lovell, Sir Bernard. 1964. "Joseph Lade Pawsey. *Biographical Memoirs of the Fellows of the Royal Society*, Vol. 10 (November): 229–43.

————. 1983. "Impact of World War II on radio astronomy." In Kellerman and Sheets, eds., *Serendipitous discoveries in radio astronomy*, 89–104.

————. 1984. "The origins and early history of Jodrell Bank." In Sullivan, ed., *The early years of radio astronomy*, 193–212.

————. 1985. "Historical note on H₂S." In *IEE Proceedings*, vol. 132, pt. A, no. 6 (October): 401–22.

————. 1990. *Astronomer by chance.* New York: Basic Books.

————. 1991. *Echoes of war.* Bristol, Eng.: Hilger.

Lubkin, Gloria B. 1992. "John Bardeen." *Physics Today*, April, 23–25.

Mabon, Prescott C. 1975. *Mission communications: The story of Bell Laboratories.* Murray Hill, N.J.: Bell Telephone Laboratories.

MacDonald, Charles B. 1985. *A time for trumpets.* New York: Morrow.

Macdonald, Stuart, and Ernest Braun. 1977. "The transistor and attitude to change." *American Journal of Physics* 45: 1061–65.

Marconi, Guglielmo Senatore. 1922. "Radio telegraphy." *Proceedings of the IRE*, vol. 10: 215–38.

Marks, K. M., D. C. McAdoo, and W. H. F. Smith. 1993. "Mapping the southwest Indian ridge with Geosat." *EOS*, vol. 74, no. 8 (Feb. 23): 81, 86.

Maron, Irving, George Luchak, and William Blitzstein. 1961. "Radar observation of Venus." *Science*, vol. 134 (Nov. 3): 1419–21.

Mason, Francis K. 1990. *Battle over Britain.* Bourne End, Eng.: Aston Publications.

Mason, John. 1989. "The unveiling of Venus." *New Scientist*, May 6, 42–47.

McElheny, Victor K. 1973. "Sir Robert Watson-Watt dies; developed first radar system." *New York Times*, Dec. 7, 44.

McMahon, A. Michal. 1984. *The making of a profession: A century of electrical engineering in America.* New York: IEEE Press.

McNally, I. L. n.d. "Radar reflections." Apparently unpublished paper sent to author.

Michel, B. 1938. *Recherches sur les magnétrons: magnétrons S.F.R. pour ondes ultra-courtes* ("Research on the magnetrons: S.F.R. ultra short waves magnetrons). *Bulletin de la Société Française Radio-Electrique*, no. 2, Deuxième Trimestre: 30–53.

Miller, Russell. 1976. "Secret weapon," London *Times Sunday Magazine*, Sept. 7: 8–14.

Millman, S., ed. 1983. *A history of engineering and science in the Bell System: Physical sciences (1925–1980).* Short Hills, N.J.: AT&T Bell Laboratories.

Mills, B. Y. 1984. "Radio sources and the log N–log S controversy" In Sullivan, ed., *The Early Years of radio astronomy*, 146–65.

MITRE Corp. 1979. *MITRE: The First Twenty Years.* Bedford, Mass.: MITRE Corp.

Morison, Samuel Eliot. 1984. *The Battle of the Atlantic.* Vol. I of *History of United States naval operations in World War II.* Boston: Little, Brown.

————. 1956. *The Atlantic battle won.* Vol. X of *History of United States naval operations in World War II.* Boston: Little, Brown.

————. 1963. *The two-ocean war.* Boston: Little, Brown.

Morrison, Philip, and Phyllis Morrison. 1987. *The ring of truth: How we know what we know.* New York: Random House.

Morse, Philip M. 1977. *In at the beginnings: A physicist's life.* Cambridge, Mass.: MIT Press.

Muhleman, D. O., D. B. Holdbridge, and N. Block. 1962. "The astronomical unit determined by radar reflections from Venus." *The Astronomical Journal*, vol. 67 (May): 191–203.

Muller, C. A., and J. H. Oort. 1951. "Observation of a line in the galactic radio spectrum: The interstellar hydrogen line at 1,420 Mc./sec., and an estimate of galactic rotation." *Nature*, vol. 168. (Sept. 1): 357–58.

Muller, Richard A. 1988. "Luis Alvarez: 1911–1988." *California Monthly*, November, 5.

Murray, Bruce. 1989. *Journey into space.* New York: Norton.

Nakajima, S. 1988. "The history of Japanese radar development to 1945." In Burns, ed., *Radar development to 1945*, 243–58.

Nebeker, Frederik. 1993. *Sparks of genius: Portraits of electrical engineering excellence.* New York: IEEE Press.

Newton, Charles, Therma E. Patterson, and Nancy Joy Perkins. 1991. *Five years at the Radiation Laboratory.* Cambridge, Mass.: Massachusetts Institute of Technology. Reprint of 1946 original, published by the Massachusetts Institute of Technology, with additions of previously classified material.

Nissen, Jack, with A. W. Cockerill. 1987. *Winning the radar war.* New York: St. Martin's.

Norberg, Arthur L., and Robert W. Seidel. 1994. "The contexts for the development of radar: A comparison of efforts in the United States and the United Kingdom in the 1930s." In Blumtritt, Petzold, and Aspray, eds., *Tracking the history of radar*, 199–216.

Office of Scientific Research and Development. 1946. "Military airborne radar systems: Vol. 2 of summary technical report of Division 14," NDRC.

Olson, C. Marcus. 1988. "The pure stuff." *American Heritage of Invention & Technology*, vol. 4 (Spring–Summer): 158–63.

Orgel, N. 1985. "History of fighter direction." *IEE Proceedings*, vol. 132, pt. A, no. 6 (October): 411–22.

Page, Robert Morris. 1962a. *The origin of radar*. New York: Doubleday Anchor (paperback).

———. 1962b. "The early history of radar." *Proceedings of the IRE*, May, 1232–36.

Paige, David A. 1994. "Chance for snowballs in hell." *Nature*, vol. 3369, (May 19): 182.

Pake, George. 1992. "Consultant to industry, adviser to government." *Physics Today*, April, 56–62.

Pawsey, J. L. 1951. Cable to *Nature* vol. 168 (Sept. 1): 358.

———. 1953. "Radio astronomy in Australia." *Journal of the Royal Astronomical Society of Canada*, vol. XLVII, no. 4 (July–August): 137–52. Reprint.

———, and R. N. Bracewell. 1955. *Radio astronomy*. Oxford, Eng.: Oxford at the Clarendon Press.

Peale, S. J., and T. Gold. 1965. "Rotation of the planet Mercury." *Nature*, vol. 206, no. 4990 (June 19): 1240–41.

Pearson, G. L., and Walter H. Brattain. 1973. "History of semiconductor research." *Proceedings of the IRE*, vol. 43. no. 12 (December): 1794–1806.

Pearson, Jamie Parker, ed. 1992. *Digital at work*. Burlington, Mass.: Digital Press.

Peltzer, Gilles, and Paul Rosen. 1995. "Surface displacement of the 17 May 1993 Eureka Valley, California, earthquake observed by SAR inteferometry." *Science*, vol. 268 (June 2): 1333–36.

Pendick, Daniel. 1993. "Volcano watchers draft in radar." *New Scientist*, (Oct. 16): 20.

Pettengill, Gordon H. 1960. "Measurements of lunar reflectivity using the Millstone radar." *Proceedings of the IRE*, vol. 48: 933–34.

———. 1966. "Radar astronomy." *International Science and Technology*, October, 72–82.

———. 1968. "Radar studies of the planets." In Evans and Hagfors, eds., *Radar astronomy*, 275–321.

———, and D. E. Dustin. 1956. "A comparison of selected ICBM early-warning radar configurations." Lincoln Laboratory technical memo no. 127 (Aug. 13).

———, et al. 1962. "A radar investigation of Venus." *The Astronomical Journal*. vol. 67 (May): 181–90.

———, and R. B. Dyce. 1965. "A radar determination of the rotation of the planet Mercury." *Nature*, vol. 206, no. 4990 (June 19): 1240.

———, and Irwin I. Shapiro. 1965. "Radar astronomy." *Annual Review of Astronomy and Astrophysics* 3: 377–410.

———, and T. W. Thompson. 1968. "A radar study of the lunar crater Tycho at 3.8-cm and 70-cm wavelengths." *Icarus* 8: 457–71.

———, Donald B. Campbell, and Harold Masursky. 1980. "The surface of Venus." *Scientific American* 243: 54–64.

Pfann, W. G., and J. H. Scaff. 1949. "Microstructures of silicon ingots." *Metals Transactions*, 185 (June): 389–92.

Phillips, Don. 1994. "New radar not installed before crash." *Washington Post*, July 7, A3.

Piddington, J. H. 1961. *Radio astronomy*. New York: Harper & Brothers.

———, and H. C. Minnett. 1949. "Microwave thermal radiation from the moon." *Australian Journal of Scientific Research*. Series A — *Physical Sciences*, vol. 2, no. 1: 63–77.

Pines, David. 1992. "An extraordinary man: Reflections on John Bardeen." *Physics Today*, April, 64–70.

Pollard, Ernest C. 1982. *Radiation: One story of the M.I.T. Radiation Laboratory*, Durham, N.C.: Woodburn.

Pound, R. V. n.d. Unpublished autobiography.

Prange, Gordon W. 1982. *At dawn we slept*. Harmondsworth, Eng.: Penguin.

Price, Alfred. 1973. *Aircraft versus submarine*. Annapolis, Md.: Naval Institute Press.

———. 1984. *The history of US electronic warfare*. vol. I. Arlington, VA: The Association of Old Crows.

———. 1987. *Instruments of darkness*. Los Altos, CA: Peninsula Publishing. Reprint of 1967 edition.

———. 1989. *The history of US electronic warfare*, vol. II. Arlington, VA: The Association of Old Crows.

———. 1991. *Patrol aircraft vs. submarine*. Shrewsbury, Eng.: Airlife.

Price, Don K. 1965. *The scientific estate*. New York: Oxford University Press (paperback).

Price, Robert. 1959. "The Venus Radar Experiment." A paper presented to the 9th General Assembly of the Advisory Group for Aeronautical Research and Development of the North Atlantic Treaty Organization (NATO), Aachen, Germany. Sept. 21.

———, et al. 1959. "Radar echoes from Venus." *Science*, vol. 129, no. 3351 (March 20): 751–53.

Pringle, J. W. S. 1985. "The work of TRE in the invasion of Europe." *IEE Proceedings*, vol. 132, pt. A, no. 6 (October): 340–58.

Pugh, Emerson W. 1984. *Memories that shaped an industry: Decisions leading to IBM System 360*. Cambridge: MIT Press.

———. 1995. *Building IBM*. Cambridge, Mass.: MIT Press.

Purcell, E. M. 1948. "Nuclear magnetism in relation to problems of the liquid and solid states." *Science*, vol. 107 (April 30): 433–40.

———. 1964. "Research in nuclear magnetism." In *Nobel Lectures in Physics 1942 – 1962*. New York: Elsevier, 219–31.

———. 1986. "Radar and physics." *The Morris Loeb Lectures in Biology*, Cabot Science Library, Harvard University, March 18. Videotaped lecture.

———, H. C. Torrey, and R. V. Pound. 1946. "Resonance absorption by nuclear magnetic moments in a solid." *The Physical Review* 69 (Jan. 1 and 15): 37–38.

Radford, William H. 1962. "M.I.T. Lincoln Laboratory: Its origin and first decade." *Technology Review*, January, 15–25.

Randall, John T. 1946a. "The cavity magnetron." *Physical Society of London, Proceedings*, vol. 58, pt. 3: 247–52.

———. 1946b. "Radar and the magnetron." *Journal of the Royal Society of Arts*, April 12, 303–12.

Ratcliffe, J. A. 1975. "Robert Alexander Watson-Watt." *Biographical Memoirs of the Fellows of the Royal Society*, vol. 21, 548–68.

Rawlinson, J. D. S. 1985. "Development of radar for the Royal Navy 1935–44." *IEE Proceedings*, vol. 132, pt. A, no. 6 (October): 441–44.

Redmond, Kent C., and Thomas M. Smith. 1980. *Project Whirlwind: The history of a pioneer computer*. Bedford, Mass.: Digital Press.

Reeves, A. H., and J. E. N. Hooper. "Oboe: History and development." In *IEE Proceedings*, vol. 132, pt. A, no. 6 (October): 394–98.

Rhodes, Richard. 1986. *The making of the atomic bomb*. New York: Simon & Schuster.

Ridenour, Louis. 1947. *Radar system engineering*. New York: McGraw-Hill.

Rigden, John S. 1985. "The birth of the magnetic-resonance method." In *Observation, experiment, and hypothesis in modern physical science*. Peter Achinstein and Owen Hannaway, eds., Cambridge: MIT Press, 205–37.

———. 1986. "Quantum states and precession: The two discoveries of NMR." *Reviews of Modern Physics* 58, no. 2 (April): 433–48.

——— 1987. *Rabi: Scientist and citizen*. New York: Basic Books.

Robertson, Peter. 1992. *Beyond southern skies*. Cambridge, Eng.: Cambridge University Press.

Robinson, Denis M. 1983. "British microwave radar 1939–41." *Proceedings of the American Philosophical Society*, vol. 127, no. 1: 26–31.

———. 1988. *A mini opus*. Unpublished manuscript.

Rochester, Stuart I. 1976. *Takeoff at mid-century*. Washington, D.C.: U.S. Department of Transportation.

Rohwer, Jürgen. 1965. "The U-boat war against the German supply lines." In H. A. Jacobsen and J. Rohwer, eds., *Decisive battles of World War II: The German view*. Trans. André Deutsch Ltd. New York: G. P. Putnam's Sons, 259–312. First American edition.

Rolph P. M., ed. 1991. *Fifty years of the cavity magnetron*. Proceedings of a one-day symposium, February 21, 1990, School of Physics and Space Research, University of Birmingham.

Roskill, Captain S. W. 1954. *The war at sea 1939–1945*. Vol. 1. London: Her Majesty's Stationery Office.

———. 1956. Vol. II.

———. 1960. Vol. III, Part I.

———. 1961. Vol. III, Part II.

Rowe, A. P. 1948. *One story of radar*. Cambridge, Eng.: Cambridge University Press.

Ryle, Sir Martin. 1985. "D-13: Some personal memories of 24th–28th May 1944." In *IEE Proceedings*, vol. 132, pt. A, no. 6 (October): 438–44.

Sanders, Frederick H. 1947. "Radar development in Canada." *Proceedings of the I.R.E.*, vol. 35: 195–200.

Saunders, R. S., et al. 1992. "Magellan mission summary." *Journal of Geophysical Research*, vol. 97, no. E8 (Aug. 25): 13,067–90.

Scaff, Jack H. 1970. The role of metallurgy in the technology of electronic materials. *Metallurgical transactions*, vol. 1 (March): 561–73.

———, and R. S. Ohl. 1947. "Development of silicon crystal rectifiers for microwave radar receivers." *Bell System Technical Journal*, vol. XXVI, no. 1 (January): 1–30.

———, H. C. Theuerer, and E. E. Schumacher. 1949. "P-type and N-type silicon and the formation of the photovoltaic barrier in silicon ingots." *Metals transactions*, vol. 185 (June): 383–88.

Schaffel, Kenneth. 1990. *The emerging shield*. Washington: Office of Air Force History.

Schoenfeld, Max. 1995. *Stalking the U-boat*. Smithsonian.

Schrieffer, J. Robert. 1992. "John Bardeen and the theory of superconductivity." *Physics Today*, April, 46–53.

Scott, Otto J. 1974. *The creative ordeal: The story of Raytheon*. New York: Atheneum.

Seitz, Frederick. 1995. "Research on silicon and germanium in World War II." *Physics Today*, January, 22–27.

———. 1996. Unpublished manuscript. "The tangled prelude to the age of silicon electronics." Draft of Sept. 12.

Shapiro, Irwin I. 1968. "Spin and orbital motions of planets." In Evans, and Hagfors, eds., *Radar astronomy*, 143–85.

Shearman, E. D. R. 1991. "The post-war story of the cavity magnetron and related microwave sources." In Rolph, ed., *Fifty years of the cavity magnetron*, 78–83.

———, and D. V. Land. n.d. "The beginnings of centimetric radar in the United Kingdom." Typescript.

Shklovsky, I. S. 1949. "Monochromatic radio emission from the galaxy and the possibility of its observation." In Sullivan, ed., *Classics in radio astronomy*, 318–24.

Shockley, William. 1963. "Transistor history, applied research and science teaching." Invited lecture, Japanese Institution of Electrical Engineers, 75th anniversary meeting, Tokyo, Japan, October.

———. 1974. "The invention of the transistor—'An example of creative-failure methodology.'" National Bureau of Standards Special Publication 388. Proceedings of conference on the public need and the role of the inventor, June 11–14, 1973. Issued May, 47–89.

———. 1976. "The path to the conception of the junction transistor." *IEEE transactions on electron devices*, vol. ED-23, no. 7 (July): 597–620.

———, and G. L. Pearson. 1948. "Modulation of conductance of thin films of semiconductors by surface charges." *The Physical Review*, vol. 74, no. 2 (second series) (July 15): pp. 232–33.

Skolnik, Merrill I. 1980. *Introduction to radar systems*, 2nd ed. New York: McGraw-Hill.

————. 1985. "Fifty years of radar." *Proceedings of the IEEE*, vol. 73, no. 2, (February): 182–196.

————, ed. 1990. *Radar handbook*, 2nd ed. New York: McGraw-Hill.

————. 1991. "Radar's environmental role." *IEEE Potentials*, April, 13–16.

Smith, J. B. 1985. "The new H2Ss." *IEE Proceedings*, vol. 132, pt. A, no. 6 (October): 404–10.

Smith, R. A., et al. 1985. "ASV: the detection of surface vessels by airborne radar." *IEE Proceedings*, vol. 132, pt. A, no. 6 (October): 359–84.

Smits, F. M., ed. 1985. *A history of engineering and science in the Bell System: Electronics technology (1925–1975)*. Indianapolis: AT&T Bell Laboratories.

Snow, C. P. 1961. *Science and government*. Cambridge, Mass.: Harvard University Press.

———— 1962. *Appendix to Science and government*, Cambridge, Mass.: Harvard University Press.

Southworth, George C. 1945. "Microwave radiation from the sun." *Journal of the Franklin Institute*, April, 285–97.

Stevenson, M. Lynn. 1987. "The development of the hydrogen bubble chamber." In Trower, ed., *Discovering Alvarez*, 105–107.

Stimson, Henry L., and McGeorge Bundy. 1947. *On active service in war and peace*. New York: Harper & Brothers.

Stodola, E. K. 1988. "Some examples of post World War II radar in the USA." In Burns, ed., *Radar development to 1945*, 478–92.

Stover, Dawn. 1995. "Radar on a chip." *Popular Science*, March, 107–10, 116.

Strebeigh, Fred. 1990. "How England hung the 'curtain' that held Hitler at bay." *The Smithsonian*, July, 120–29.

Suits, C. G., George R. Harrison, and Louis Jordon, eds. 1948. *Applied physics: Electronics, optics, metallurgy*. Boston: Little, Brown.

Sullivan, W. T., III, ed. 1982. *Classics in radio astronomy*. Dordrecht, Holland: Reidel.

————. 1984. *The early years of radio astronomy*. Cambridge, Eng.: Cambridge University Press.

————. 1990. "Early years of Australian radio astronomy." In Home, ed., *Australian science in the Making*, 308–44.

————. n.d. "The discovery of the Twenty-One Centimeter line." Chapter in *A history of radio astronomy*. Unfinished manuscript.

Süsskind, Charles. 1988. "Who invented radar?" In Burns, ed., *Radar development to 1945*, 506–12.

————. 1994. "Radar as a study in simultaneous invention." In Blumtritt, Petzold, and Aspray, eds., *Tracking the history of radar*, 237–45.

Swingle, Donald M. 1990. "Weather radar in the United States Army's Fort Monmouth Laboratories." In Atlas, *Radar in meteorology*, 7–15.

Swords, S. S. 1986. n.d. "The beginnings of radar." Paper unidentified.

————. *Technical history of the beginnings of RADAR*. London: Peter Peregrinus Ltd.

———— 1994. "The significance of radio wave propagation studies in the evolution of radar." In Blumtritt, Petzold, and Aspray, eds., *Tracking the history of radar*, 185–97.

Taylor, John W. R., ed. 1979. *Combat aircraft of the world*. New York: Paragon.

Tedder, Lord. 1947. *Air power in war*. London: Hodder and Stoughton.

Thompson, George Raynor, et al. 1957. *The Signal Corps: The test*. Washington: Office of the Chief of Military History United States Army.

Thompson, George Raynor, and Dixie R. Harris. 1966. *The Signal Corps: The outcome*. Washington: Office of the Chief of Military History United States Army.

Thomson J. H., et al. 1961. "A new determination of the solar parallax by means of radar echoes from Venus." *Nature*, no. 4775 (May 6): 519–20.

————. 1963. "Planetary radar." *Quarterly Journal of the Royal Astronomical Society* 4: 347–375.

Toomay, J. C. 1989. *Radar principles for the non-specialist*, 2nd ed. New York: Van Nostrand Reinhold.

Torrey, Henry C., and Charles A. Whitmer. 1948. *Crystal rectifiers.* Vol. 15 of Radiation Laboratory Series. New York: McGraw-Hill.

Townes, Charles H. 1968. "Quantum electronics, and surprise in development of technology." *Science,* vol. 159, no. 3816 (Feb. 16): p. 699–703.

———— 1978. "The laser's roots: Townes recalls the early days." *Laser Focus.* August, 52–58.

————. 1983. "Science, technology, and invention: Their progress and interactions." *Proceedings of the National Academy of Sciences,* vol. 80 (December): 7679–83.

————. 1984. "Ideas and stumbling blocks in quantum electronics." *IEE Journal of Quantum Electronics,* vol. QE-20, no. 6 (June): 547–50.

————. 1992. "Reflections on my life as a physicist." *Bulletin of the Center for Theology and the Natural Sciences,* vol. 12. no. 3 (summer).

————. 1994. "A life in physics: Bell Telephone Laboratories and World War II, Columbia University and the laser, MIT and government service, California and research in astrophysics." Interviews conducted by Suzanne B. Riess in 1991–1992. Regional Oral History Office, The Bancroft Library, University of California, Berkeley.

Trim, Richard M. 1988. "The development of IFF in the period up to 1945." In Burns, ed., *Radar development to 1945,* 436–57.

————. 1994. "Secondary surveillance radar — past, present and future." In Blumtritt, Petzold, and Aspray, eds., *Tracking the history of radar,* 93–120.

Trower, W. Peter, ed., 1987. *Discovering Alvarez: Selected works of Luis W. Alvarez, with commentary by his students and colleagues.* Chicago: University of Chicago Press.

Tucker, D. G. 1978. "Electrical communication." In Trevor Williams, ed., A *history of technology,* vol. 7, pt. II: 1220–67.

Tuska, C. D. 1944. "Historical notes on the determination of distance by timed radio waves." *Journal of the Franklin Institute,* vol. 237, no. 1: 1–20.

Tuve, Merle A. 1974. "Early days of pulse radio at the Carnegie Institution." *Journal of Atmospheric and Terrestrial Physics,* vol. 36: 2079–83.

Valley, George E., Jr. 1985. "How the SAGE development began." *Annals of the History of Computing,* vol. 7, no. 3 (July): 196–226.

van de Hulst, H. C. 1945a. "Origin of the radio waves." Cornell University translation, including various notations added three years later and a two-page summary, "Radiogolven uit het Wereddruim," or "Radio waves from space summary."

————. 1945b. "Origin of the radio waves from space." 1973 translation by W. T. Sullivan III. In Sullivan, ed., *Classics in radio astronomy,* 302–16.

van der Vat, Dan. 1988. *The Atlantic campaign.* New York: Harper & Row (paperback). First U.S. edition.

van Keuren, David K. 1994. "The military context of early American radar, 1930–1940." In Blumtritt, Petzold and Aspray, eds., *Tracking the history of radar,* 137–56.

Van Vleck, J. H. 1942. "The atmospheric absorption of microwaves." *Radiation Laboratory Report 43–2,* April 27.

————. 1947. "The absorption of microwaves by uncondensed water vapor." *The Physical Review,* vol. 71, no. 7 (April 1): 425–33.

————, E. M. Purcell, and Herbert Goldstein. 1951. "Atmospheric Attenuation." In Kerr, ed. *Propagation of Short Radio Waves,* 641–92.

Varian, Russell H., and Sigurd F. Varian. 1939. "A high frequency oscillator and amplifier." *Journal of Applied Physics,* vol. 10 (May): 321–27.

Ward, William W. 1992. *The AN/FPS-17 coded-pulse radar.* Lincoln Lab technical report released by Air Force on Sept. 3.

Warnock, A. Timothy. n.d. *The battle against the U-boat in the American theater.* Washington, D.C.: Center for Air Force history.

Warren, Clifford A. 1975. "In defense of the nation." *Bell Labs Record,* vol. 53, no. 1 (January): 96–107.

Watson-Watt, Robert A. 1945. "Radar in war and in peace." *Nature,* September 15, 319–24.

————. 1946. "The evolution of radiolocation." *Journal of the IEE,* vol. 93: 374–82.

————. 1957a. *Three steps to victory.* London: Odhams.

————. 1957b. "Battle scars of military electronics — the *Scharnhorst* break-through." *IRE Transactions on Military Electronics*, March, 19–25.

————. 1959. *The pulse of radar: The autobiography of Sir Robert Watson-Watt.* New York: Dial. American edition of *Three Steps to Victory.*

Weinberg, Steven. 1977. *The first three minutes.* New York: Bantam.

Weiner, Charles. 1973. "How the transistor emerged." *IEEE Spectrum* 10, no. 1: 24–35.

Whitnah, Donald R. 1966. *Safer skyways.* Ames: Iowa State University Press.

Wieser, C. Robert. 1983. "The Cape Cod system." In *Annals of the History of Computing Special Issue: SAGE*, vol. 5, no. 4 (Oct.): 362–69.

Wild, J. P. 1972. "The beginnings of radio astronomy in Australia." In *Records of the Australian Academy of Science*, vol. 2, no. 3, Canberra. Offprint.

Wildes, Karl L., and Nilo A. Lindgren. 1985. *A century of electrical engineering and computer science at MIT, 1882–1982.* Cambridge: MIT Press.

Wilford, John Noble. 1992. "On the trail from the sky: Roads point to a lost city." *New York Times*, Feb. 5, A1, A14.

Wilkins, A. F. 1934. "Measurement of the angle of incidence at the ground of downcoming short waves from the ionosphere." *Journal of the IEE*, vol. 74, 582–88.

————. 1981. "A Glimmer" and "The Islanders." In Kinsey, *Orfordness — Secret Site*: 130–54.

Wilkins, M. H. F. 1987. "John Turton Randall." *Biographical Memoirs of the Royal Society of London*, vol. 33: 493–535.

Wilkinson, David T., and P. J. E. Peebles. 1983. "Discovery of the 3 K radiation." In Kellerman and Sheets, *Serendipitous discoveries in radio astronomy*, 175–84.

Wilkinson, Roger I. 1946a. *Short survey of Japanese radar — I. Electrical engineering*, vol. 65: 370–77.

————. 1946b. *Short survey of Japanese radar — II. Electrical engineering*, vol. 65: 455–63.

Williams, Trevor I., ed. 1978. *A history of technology*, vol. 7, pt. II. Oxford: Clarendon Press.

Willshaw, W. E. 1991. "GEC's wartime contribution." In Rolph, ed., *Fifty years of the cavity magnetron.*

Wilson, Robert W. 1983. "Discovery of the cosmic microwave background." In Kellerman and Sheets, *Serendipitous discoveries in radio astronomy*, 185–95.

Wingham, D. J., C. G. Rapley, and J. G. Morley. 1993. "Improved resolution ice sheet mapping with satellite radar altimeters." *EOS*, vol. 74, no. 10 (March 9): 113, 116.

Winterbotham, F. W. 1974. *The Ultra secret.* New York: Harper & Row.

Wolfgram, Dorothea. 1980. "ELB." *Washington University Magazine*, Winter: 34–39.

Wood, Derek, and Derek Dempster. 1990. *The narrow margin: The Battle of Britain and the rise of air power, 1930–1940.* Washington: Smithsonian Institution Press.

Wu, Jin. 1995. "Pairing of radar instruments on satellites could provide optimal mapping of sea surface winds." *EOS*, Jan. 3: 3.

Young, Carolynn, ed. 1990. *The Magellan Venus Explorer's guide.* Pasadena, Calif.: NASA, Jet Propulsion Laboratory.

Zachary, G. Pascal. 1991a. "America's first engineer — the career of Vannevar Bush." Part I. *Upside*, April, 92–101.

————. 1991b. "America's first engineer — the career of Vannevar Bush." Part II. *Upside*, June, 94–103.

————. 1992. "Vannevar Bush backs the bomb." *The Bulletin of the Atomic Scientists.* Dec.: 24–31.

Ziegler, Charles A. 1988. "Waiting for Joe-1: Decisions leading to the detection of Russia's first atomic bomb test." *Social Studies of Science*, vol. 18, no. 2 (May): 197–229.

————, and David Jacobson. 1995. *Spying without spies: Origins of America's secret nuclear surveillance system.* New York: Praeger.

Zook, Nicholas. 1953. "New radar device 'sees' approaching storms." *Worcester Telegram-Gazette*, Nov. 15, n.p.

Zorpette, Glenn. 1995. "Radio astronomy: new windows on the universe." *IEEE Spectrum* (February): 18–25.

Zuckerman, Lord. 1975. "Scientific advice during and since World War II. *Proceedings of the Royal Society*, vol. 342: 467–80.

Acknowledgments

Vic McElheny introduced me to the Rad Lab, remembered my book idea, and prodded me when the time was right. Through long walks in all seasons, he offered invaluable advice and the great gift of perspective.

Without the Alfred P. Sloan Foundation, this book would not have been written. Warm thanks to Art Singer and all those associated with the program. This means more bows to Vic — and also to Dick Rhodes, who got the ball rolling with a key introduction and provided timely words of encouragement.

In the course of four years of writing and reporting, many people have shared information or helped me find it. Personal thanks to all those who gave interviews — too many to mention by name. More than a few went above and beyond the call. Atop the list is Louis Brown. Louis freely shared his extensive bibliography, passed along insights and news clippings, and acted as a sounding board on matters big and small; may his own project be successful. Dale Corson both entertained and enlightened on a variety of fronts — and became a friend. Dave Barton was always available to check radar details. Charles Townes and Jay Forrester reviewed more than they had to, with grace and patience. George Valley Jr. was always candid and ready to get things right — as was an equally approachable Gordon Pettengill. Fred Seitz, despite our never having met, willingly shared critical insights about semiconductors and the transistor. Lee Davenport brought home the realities of buzz bombs, the SCR-584, and much else. Herb Weiss gave a personal tour of Millstone Hill and an introduction to missile-watching. Doc Ewen put me in touch with the cosmos. Largely through the miracles of e-mail, Lawrence Johnston walked me through the work of the Alvarez gang. Rich Muller provided vintage science support. John Carey offered much-appreciated advice and encouragement with the manuscript.

Many of the radar pioneers dug out old records and racked their brains to answer seemingly endless questions in my struggle to hash out events fifty years past. Among them: Hank Abajian, Robert Dicke, the late Lee DuBridge, Ivan Getting, Bernard Lovell, R. V. Pound, Edward Purcell, Norman Ramsey, the late Denis Robinson (a job continued by his son, Hal), Henry Torrey.

Scholars and scientists, beyond those already named, often offered advice that helped clear an outsider's fuzziness. William Aspray and colleagues at the IEEE Center passed on oral histories before they were bound and printed. Russell Burns guided me through the vagaries of the Public Record Office in Britain. Brad E. Gernand at the Library of Congress, with the permission of Dr. Edmund A. Bowles, opened the Edward L. Bowles papers before they were fully catalogued and sorted. Andrew J. Butrica, Owen Cote Jr., Paul Forman, Lillian Hoddeson, David Jacobson, Woody Sullivan, Richard M. Trim, David van Keuren, and Charles A. Ziegler warmly shared their extensive knowledge. Judith Reppy and Shibley Telhami invited me to speak at Cornell. Ted Saad shared his address list of Rad Lab veterans. Bill Ward got me started at Lincoln Lab, and Roger Sudbury largely took over from there. Jan Alvarez put me onto Peter Trower, who passed on pieces of Luis Alvarez's original manuscript.

Everywhere I turned, an archivist, librarian, curator, or amateur historian also seemed ready to help: Warrant Officer "Chris" Wren provided a personal tour of the underground Operations Room where Winston Churchill watched the Battle of Britain, while former Observer Corps member Norman Greig did the same for Bentley Priory. Thanks also to Marjorie Ciarlante, Donna Cunningham, Helen Engle, Norm Krim, Mary Murphy, Jim Owens, Robin Rider, George Young, Helen Samuels and everyone at MIT's archives, Warren Seamons and Michael Yeates at the MIT Museum, and Lenore Symens.

Several researchers provided critical assistance: Andy Feland, Elizabeth McNeil, Tom Pak, and, most notably, Alex de Ravel. Christine Jorgensen did the early transcribing.

Still others defy easy classification: Kathy Norris gave the insider's view of Tuxedo Park. Bruce Wirt was a host *par excellence*. In Britain, Alun Anderson and Elaine Dzierzak offered generous hospitality. Oskar Blumtritt welcomed me to the Deutsches Museum. In Australia, Elaine Pacey and Sally Atkinson provided immense help with arrangements and material — and Sally and Elwyn Donald made a stranger feel at home. Harry Minnett and Bruce Slee took me to the site of their sea cliff interferometer and relived the early days of radio astronomy.

Dick Thompson has long been a great friend and advisor. I swore I'd dedicate my first book to him — but took so long writing it that in the meantime I had a family that usurped his place. With his permission, I thank him here.

My agent, Rafe Sagalyn, greatly eased things from start to finish. Alice Mayhew proved a wonderful editor, inspiring confidence and offering much-needed direction and focus throughout a long process. Thanks also to Simon & Schuster's Roger Labrie for his insights and support.

Finally, although this book began before they came into the world, Kacey and Robbie have proven unlimited sources of strength. My wife, Nancy Walser, provided valuable advice about early versions of the manuscript and kept me steady with her unwavering support, love, and conviction that all would be well in the end.

Index